Exergy Tables

About the Authors

Dr. Lingyan Deng is currently a postdoctoral associate at the Massachusetts Institute of Technology Energy Initiative (MITEI). She received her Master of Science degree in Energy Studies from Ajou University, South Korea, and her PhD degree in Chemical Engineering from McMaster University, Canada, under the guidance of Prof. Thomas A. Adams II. After completing her PhD degree, she moved to the National Energy Technology Laboratory, United States Department of Energy in Pittsburgh working as a Senior Engineer. Later, she decided to transfer to academia and joined MITEI to work as a postdoctoral associate. She has a broad research interest, ranging from lab-scale experimental, to system-level design, modeling, simulation, and optimization, to country-level decarbonization strategies study. She has a passion working on the energy transition to tackle global warming issues worldwide. She proposed and received a $150 K Seed Fund from MIT Future Energy Systems Center to developing a software package for carbon capture and storage targeting for use in thousands of chemical and energy processes. Besides research, she is also active in organizing events at MIT to provide a pleasant communication environment for people from different disciplines.

Prof. Thomas A. Adams II is Professor of Energy and Process Engineering at the Norwegian University of Science and Technology (NTNU). He received dual bachelor's degrees in computer science and chemical engineering from Michigan State University, completed his PhD in chemical and biomolecular engineering at the University of Pennsylvania, and did his postdoctoral studies in chemical engineering at MIT. He is a licensed professional engineer and was a professor from 2010 to 2022 at McMaster University in Canada. Professor Adams' research focuses on the design, modelling, simulation, optimization, and analysis of sustainable energy systems. Application areas include sustainability, biofuels, alternative fuels, synthetic fuels, fuel cells, energy storage, hydrogen economy technologies, integrated community energy, and many other kinds of energy systems.

Professor Adams' honors for education and research include the 2021 David Himmelblau Award for Innovations in Computer-Based Chemical Engineering Education from the American Institute of Chemical Engineers (for the publication of his book *Learn Aspen Plus in 24 Hours*, also available from McGraw Hill), the 2018 *Canadian Journal of Chemical Engineering Lectureship Award* (given to the top early career chemical engineering researcher in Canada), induction into *Industrial & Engineering Chemistry Research*'s 2018 Class of Influential Researchers, the 2017 Emerging Leader of Chemical Engineering Award from the Canadian Society for Chemical Engineering, the 2015 President's Award for Excellence in Graduate Supervision from McMaster University, and the 2014 Ontario Early Researcher Award. He is currently an Associate Editor of the *Canadian Journal of Chemical Engineering, Frontiers in Energy Research*, and *Chemical Product and Process Modelling*. He is also a representative of Canada to the International Standards Organization in the Life Cycle Assessment subcommittee and is the Convenor for the development of ISO Technical Specification 14076 on eco-technoeconomic analyses. He appears on the educational television docu-series *Colossal Machines* on cable channels and streaming worldwide.

Prof. Truls Gundersen is a professor in the Department of Energy and Process Engineering at NTNU. He received a master's degree in physics and a PhD in chemical engineering, both from NTNU. He has 12 years' industrial experience from Norsk Hydro, Corporate Research Centre (1981–1993). He was elected member of the Norwegian Academy of Technological Sciences in 1991. He has published nearly 150 journal articles and is the author of three chapters in two books. At NTNU, he teaches engineering thermodynamics (2nd year) and process integration (4th year). On behalf of NTNU, he has been coordinating two 8-year Centers for Environment-Friendly Energy Research, one on carbon capture and storage (2009–2016) and one on energy efficiency in industry (2017–2024). His primary expertise is in the field of process systems engineering with areas such as pinch analysis, exergy analysis and design of low temperature processes. In addition, he is the recipient of the Best Lecturer Award from the Energy and Environment Study Programme at NTNU.

Exergy Tables

A Comprehensive Set of Exergy Values to Streamline Energy Efficiency Analysis

Lingyan Deng

Thomas A. Adams II

Truls Gundersen

New York   Chicago   San Francisco
Athens   London   Madrid
Mexico City   Milan   New Delhi
Singapore   Sydney   Toronto

Exergy Tables: A Comprehensive Set of Exergy Values to Streamline Energy Efficiency Analysis

1 2 3 4 5 6 7 8 9   LKV   28 27 26 25 24 23

Library of Congress Control Number: 2023938685

ISBN 978-1-264-71572-5
MHID 1-264-71572-2

Sponsoring Editor Robin Najar	**Proofreader** Rup Narayan
Editorial Supervisor Janet Walden	**Indexer** Edwin Durbin
Project Manager Tasneem Kauser, KnowledgeWorks Global Ltd.	**Production Supervisor** Lynn M. Messina
Acquisitions Coordinator Olivia Higgins	**Composition** KnowledgeWorks Global Ltd.
Copy Editor Girish Sharma	**Illustration** KnowledgeWorks Global Ltd.
	Art Director, Cover Jeff Weeks

Contents

Preface

Exergy: An Exciting Frontier in Process Engineering

The thermodynamic concept of *exergy* is a fascinating topic. Using thermodynamic principles, exergy is a metric that measures both the quality and quantity of an energy source. This metric is extremely useful in chemical, mechanical, process, and many other kinds of industrial engineering fields. A practitioner can use exergy as part of a systems analysis to compute useful metrics for an energy system, such as its exergy efficiency, exergy destruction, and exergy flows.

This information is useful to understand where bottlenecks in a process might be, why a process might not be operating as efficiently as possible, and what areas to target for improvement. It is useful also for big picture decision-making, such as selecting the best processes, fuels, or unit operations for the task at hand, or whether to green-light a process in the first place. Because exergy reflects energy quality, it is a proxy for economic value too; when making early stage technology decisions, it can be very difficult to estimate cost, and exergy makes a great stand-in. It even can have an important role in public policy decision-making. In fact, one of the leading trends right now is to get people to "Think Exergy, not Energy"—a mantra encouraged by Science Europe and others. Thinking about energy alone can lead to drastically wrong policy decisions that can negatively impact energy sustainability. We show a few interesting examples in this book.

Because of its potential, interest in exergy is rapidly rising in academia. Figure 1 shows the number of scientific research publications in the Compendex database that contain the word "Exergy." These publications include papers published in peer-reviewed research journals, peer-reviewed conference proceedings, or other kinds of works presented at scientific conferences. You can see the explosive growth in exergy interest over the past 20 years within the academic community. These publications range from detailed thermodynamic discussions to simple applications of exergy within ordinary engineering analyses. But now thanks to *Exergy Tables*, you no longer need a PhD to use it!

Without This Book, Exergy Analyses Can Be Challenging in Practice

Despite the rapid rise in academic interest and its usefulness in engineering analysis, some practitioners are reluctant to add exergy into their toolboxes. However, if someone simply hands them the exergy values of the mass and energy streams of interest, it is relatively easy for a practitioner to conduct an exergy analysis since the associated equations and procedures are quite simple and straightforward.

The problem is that the theory and thermodynamics behind computing exergy for a chemical, fuel, or other energy sources is complex. Most of the existing literature and books on the topic focus on this theory, but do not focus much on tangible ways for the practicing engineer to use it. Quite frankly, we have found a lot of books on the topic to be quite confusing and impenetrable. Let's face it, even the most seasoned engineers find thermodynamics a little scary. The few undergraduate engineering programs that teach exergy have only started this recently, so most engineers in the workforce have little to no formal training about exergy. At the end of the day, most practitioners that want to do an exergy analysis face a high barrier just to get started, and so they are missing out on this amazing and useful bit of knowledge.

We Did the Hard Work So You Don't Have to

That's where *Exergy Tables* comes in. We have created a useful tool for the practicing engineer that makes exergy analysis *much* easier. Chap. 1 explains exergy in simple, plain language, so you can get started quickly. Chap. 2 provides an easy-to-follow discussion of the theory behind it and how the numbers are calculated, in case you want to compute some of your own. Chap. 3 shows examples for how you can use *Exergy Tables* in your own analysis. However, it's the rest of the book that is most useful—seven chapters of lookup tables and reference figures that contain pre-calculated exergy values for thousands of chemicals, fuels, mixtures, and process streams commonly found in the industry.

It can be time-consuming and complicated to calculate these values yourself, but once you have them, they are quite easy to use. For example, the book contains thermo-mechanical exergy values (exergy associated with heat, cold, pressure, and vacuum) for many common substances and working fluids. These are great for quickly determining thermo-mechanical exergy values for systems without reaction, like steam cycles, Brayton cycles, refrigeration cycles, heat pumps, heat exchanger networks, compressed air energy storage, cold storage, liquefied natural gas, and compression systems. When reaction is involved, or concentration differences are impor-

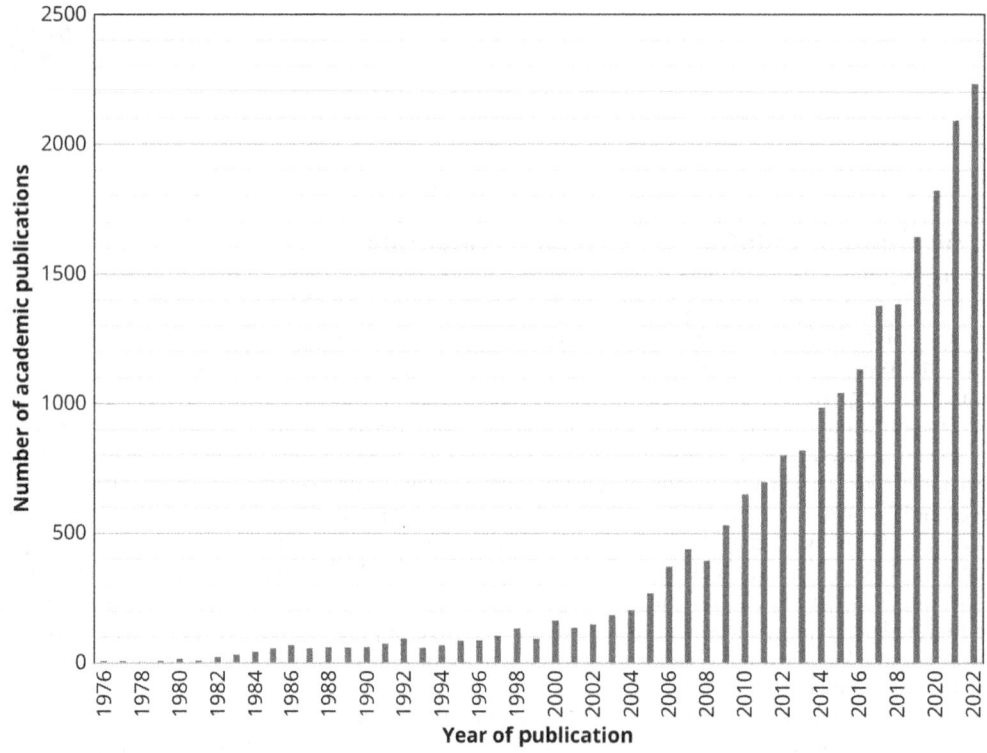

Figure 1 Publications in academic journals and conference proceedings/presentations relating to exergy that have been indexed in the Compendex database.

tant, you can find chemical exergy values for thousands of organic and inorganic pure chemicals. We also provide exergies of common chemical process streams and process intermediates for the chemical, energy, and steel industries, like syngases, reforming products, off-gases, waste streams, and hydrogen fuel products. You can also quickly find the chemical exergy of hundreds of different kinds of natural and shale gases, gasoline, diesel, bio-fuel blends, coals, wood, and biomasses from around the world.

To compute the exergies for each one of these by yourself, you would need information like enthalpy, entropy, and Gibbs free energy. These can be challenging to obtain in practice with commonly available tools, especially when it comes to an entire process with many streams. But this is not enough; you must also define the *reference conditions*, which is a detailed description of the atmosphere, hydrosphere (the oceans), and the lithosphere (the Earth's crust) that forms the thermodynamic reference point critical for computing exergy known as the "dead state." This is what makes it tough: the devil is in the details! To fully compute exergy yourself, you also need to know things like the average amount of bromine ions in the oceans, the argon concentration of the air, or the mole fraction of nickel oxide in the Earth's crust. For many engineers, the time required to learn the theory, gather the data, and perform the calculations is just too big a barrier.

But that's what makes *Exergy Tables* so powerful. We've done all the hard work, so you don't have to! We've defined the reference conditions, collected the necessary data, applied the theory, and compiled the results into easy-to-use tables and figures, so you can get right to the good stuff—applying exergy analysis to gain important insights into your process, investigate options, and make meaningful improvements.

Consistency Across Data Sets

What's more, all the numbers are consistent throughout the book because they all use the same reference conditions—reference conditions which were determined using the latest and greatest theory available in the literature. Although you may find small tables of exergy values in other books, you cannot just cherry-pick numbers from two different books (or sometimes even two different tables in the same book!) because they usually use different reference conditions. This makes it tough to use precalculated values because you should not use two exergy values computed using different references in the same analysis—it could give incorrect meanings and insights. That's why *Exergy Tables* is so useful, its vast collection of exergy values is referentially consistent, so you can quickly pick and choose numbers from one table to another throughout the book and use them all together to get powerful insights into your process.

Innovative New Property Diagrams

Moreover, *Exergy Tables* introduces the innovative "Pressure-Enthalpy-Exergy diagram" for many common chemicals. These are analogous to "PH" diagrams commonly used in the field to graphically show state information like pressure, temperature, enthalpy, entropy, and the phase envelope, but they also have thermo-mechanical exergy lines as well. They are fantastic tools to help you understand thermo-mechanical exergy and are packed with useful information. Flip to the front of Chap. 7 to see some examples—you can't find this anywhere else!

Get Started!

With this book, you'll be able to wield the power of exergy analysis with ease. We hope you find it helpful, insightful, and useful.

For more information, updates, and to make suggestions for streams and chemicals to include in the second edition, visit us at https://PSEcommunity.org/exergy.

CHAPTER 1

Introduction

1.1 Exergy in Plain Language

Exergy is an important thermodynamic principle useful in the design, analysis, and operation of mechanical, chemical, and energy systems. Exergy measures not just the *amount* of energy that an energy source has, but also, the *quality* of that energy. In other words, some forms of energy are more valuable or useful than others.

For example, consider an Olympic-sized swimming pool of warm water at about 28°C. Because this is above ambient temperature (25°C), the warm water contains a lot of stored thermal energy—about 30 GJ. Compare this to 30 GJ of electricity. Which of the two is more useful? The 30 GJ of electricity can do all sorts of useful things: power a typical Canadian's electricity needs for 6 months, drive a 2020 Chevrolet Bolt about 47,000 km, or even heat up an Olympic-sized swimming pool by almost 3°C! With the warm water, well, I can pretty much just swim in it. From a thermodynamic perspective, both the warm water and the electricity have the exact same amount of *energy* (30 GJ), but the electricity has far more *exergy* because it is so much more useful.

We will explore the thermodynamic definition of exergy and its various forms in the next chapter. In simple terms, the exergy of something is a measure of how much useful *work* it can be used to produce. Clearly, the electricity in the example can do a lot more work than an equivalent quantity of energy in the form of warm water. Exergy is measured in terms of energy, with units like joules, calories, or kilowatt-hours. For energy sources like fuels or heat (above ambient temperatures), the exergy of the energy source is always less than or equal to its energy content. In the pool example, the thermo-mechanical exergy content* of 30 GJ of warm water at 28°C and 1 atm (about 2390 tonnes) is only about 0.15 GJ (using the environmental reference temperature of 25°C that we use throughout the book). This means that one could potentially construct a 100% perfectly efficient engine that extracts the 30 GJ of heat from the swimming pool and converts that thermal energy into 0.15 GJ of work while bringing the temperature of the pool back down to 25°C again. On the other hand, electricity is essentially pure exergy, meaning that 30 GJ of electricity can do about 30 GJ of work. So the basic principle here is:

$$0 \leq \text{Exergy} \leq \text{Energy} \tag{1.1}$$

What is also interesting is that "cold" materials (things below ambient temperature, like chilled water) can also be an exergy source, even though they do not store heat as thermal energy (rather, they lack it!). This is because these materials can still be used in a machine that can produce work. The difference is that in this case, in order to produce work using the cold material, you can create an engine that takes heat *from* the environment and rejects it *to* the cold material, producing useful work in the process. In this case, the exergy content is still less than the energy removed from the environment and transferred to the cold material.

For example, consider liquid natural gas (LNG). LNG is shipped on oceangoing vessels at very low temperatures (roughly −170 to −160°C). Once it reaches the shore, it is typically sent to an LNG "regasification" terminal where it is heated up to about ambient temperature (vaporizing it), so that it can go into a normal natural gas pipeline. The heat source is often what is freely available in the environment (air or seawater), and in fact, it is possible to make a power cycle that makes mechanical or electrical power while doing so. As can be found in the book (see Table 7.8 or Fig. 7.11), the thermo-mechanical exergy of liquid methane (also called "cryo-compressed") at −162°C and 1 atm (a good approximation of LNG arriving at an LNG terminal) is about 1083 kJ/kg. This means that using air at 25°C as an energy source, the maximum theoretical work that such a power cycle could produce is about 1083 kJ/kg of LNG. In reality of course, the actual power produced would be lower due to inefficiencies.

Therefore, cold things are stores of exergy too, even if they are not stores of heat. In fact, almost any material that is at a different temperature, pressure, or composition from the environment can theoretically be used to do useful work in some way, no matter how small or complicated. That means the usefulness of virtually every substance and state can be measured by determining its exergy. This book will help you understand the basics in a simple way that is useful for the practicing engineer. More importantly, we provide many precalculated values for your quick reference and use, to save you the hassle of calculating things yourself!

---

*As discussed later, the thermo-mechanical exergy is the portion of a chemical's exergy associated with temperature and pressure. You can check this yourself by using the data in Table 6.1, where you can find that the thermo-mechanical exergy of liquid water at 1 atm and 28°C is 0.0627 kJ/kg.

1.2   Relationship to the Laws of Thermodynamics

1.2.1   First Law

One of the consequences of the First Law of Thermodynamics is that energy is neither created nor destroyed. However, it can be stored, transported, or converted from one form to another.* Energy is counted and measured based on the *quantity* of energy, and the energy content of a material or energy stream is the sum total of all of the different kinds of energies it contains (relative to some reference).

Most engineers understand the concept of the energy efficiency of a system, with one common definition being:

$$\text{System energy efficiency} = \frac{\text{Total energy content of useful outputs}}{\text{Total energy content of inputs}} \tag{1.2}$$

The quality or nature of the energies in either the denominator or numerator are not taken into account, only the quantity. For this reason, the system energy efficiency is often referred to as the *first law efficiency*.

Consider the Olympic pool example again. Suppose we filled the pool with water and left it there for a long time, so that it is now at ambient temperature (25°C), but we want to find out how much natural gas it will take to heat the pool to 28°C using a natural gas heater. We already know from the previous section that we need to deliver 30 GJ of heat to the water to raise it to 28°C.

Figure 1.1 shows a sketch of the system. The natural gas coming into the system has energy and exergy, but ambient air has neither in meaningful quantities. Similarly, the flue gas coming out has some energy content since it is warm, but it is not useful because we are venting it directly to the air, and so we would not count it in the numerator. Supposing the pool heater is 86% efficient on a higher heating value (HHV) basis, we can calculate the amount of natural gas we need using the system energy efficiency equation:

$$0.86 = \frac{30 \text{ GJ}}{\text{HHV of natural gas}} \tag{1.3}$$

Solving this equation shows that 34.9 GJ_{HHV} of natural gas is required (assuming an approximate energy content of about 55 MJ_{HHV}/kg, this is about 634 kg of gas). Of that 34.9 GJ, 30 GJ goes to heating, and the remaining 4.9 GJ leaves with the flue gas, vented to the air. Because of the first law of thermodynamics, that 4.9 GJ is not truly lost—it exists in the form of thermal energy in the exhaust. After venting to the atmosphere, the atmosphere got 4.9 GJ hotter, but since it is so vast, the actual temperature increase of the Earth's air is negligible.

1.2.2   Second Law

The Second Law of Thermodynamics relates energy conversion to entropy. Exergy is a measure that includes both energy and entropy together, so that it is a measure of energy *quality*. Similarly, the exergy is the sum of all of the kinds of exergy it contains, each reflective of its quality. One useful way to compute the system exergy efficiency is similar to energy efficiency[†]:

$$\text{System exergy efficiency} = \frac{\text{Total exergy content of the useful outputs}}{\text{Total exergy content of the inputs}} \tag{1.4}$$

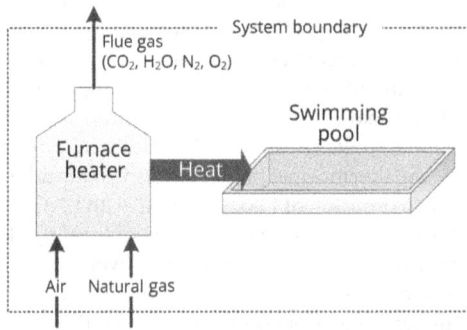

FIGURE 1.1   A sketch of the system boundary diagram for pool heating example.

---

*Technically, energy can be converted into matter (e.g., Einstein's equation $E = mc^2$), such as in a particle accelerator, but this book focuses on common industrial scenarios.

[†]There is no single definition for exergy efficiency. Metrics like exergy efficiency are tools that the engineer can use to analyze their system, with different forms and definitions best suited to different tasks and situations. If you want to learn more, read Marmolejo-Correa D, Gundersen T. A comparison of exergy efficiency definitions with focus on low-temperature processes. *Energy* 44:477–489 (2012).

Likewise, the system exergy efficiency is sometimes called the *second law efficiency*. For the pool heating example, we know from Sec. 1.1 that the exergy content of the pool water at 28°C is 0.15 GJ. From Table 8.2, the chemical exergy* of conventional Canadian natural gas from the Alberta province is about 805.07 kJ/mol, which is about 48.5 MJ/kg (based on the molar gas composition provided in the table). Since we know we need about 634 kg of natural gas from our first law analysis, its exergy content is therefore about 30.8 GJ. Thus the *second law efficiency* is less than 0.5%:

$$\text{System exergy efficiency} = \frac{0.15 \text{ GJ}}{30.8 \text{ GJ}} = 0.0049 \tag{1.5}$$

The *exergy loss* is simply the difference between the exergy of the system before an action (or for a system with flows, the exergy of the inlet flows) minus the exergy after the action (or for a system with flows, the exergy of the outlet flows). For our case, this is:

$$\text{Exergy loss} = \text{Initial exergy or exergy input} - \text{Final exergy or exergy output}$$

$$\text{Exergy loss} = 30.8 \text{ GJ} - 0.15 \text{ GJ} = 30.65 \text{ GJ} \tag{1.6}$$

Note that the exhaust stream (the flue gas) has some exergetic value because it may still be warm. However, in our example, we simply vent this stream to the atmosphere at ambient conditions. Therefore, we chose not to include it in the exergy output term because we are not collecting it as a product, so the waste heat is not a part of our useful products.

It is important to understand that one of the consequences of the second law is that, for any process, the entropy change of the universe as a whole must be greater than or equal to zero. For the pool example, the entropy of the pool (a tiny part of the universe) increases by heating it up at constant pressure. To do this, we combust natural gas, but that also increases the entropy of the universe outside the pool because we release the chemical exergy stored in the chemical bonds of the natural gas molecules (mostly methane). So, the entropy of the universe as a whole increases *permanently*. Since exergy reflects both the energy and entropy of the system components, it means that the exergy lost is truly lost; it cannot be recovered. In the pool heating example, the universe has lost 30.65 GJ of exergy, forever!

1.3   Kinds of Exergy

Figure 1.2 shows a taxonomy of how we categorize exergy for this book. First, we group exergy into two kinds, exergy of matter on the Earth, and exergy of energy on the Earth. The "on the Earth" part is important, because we use the Earth as our reference state (specifically, its atmosphere, hydrosphere, and lithosphere), and so our numbers are only valid on the Earth. For exergy of matter, we break it down into these kinds:

- Mechanical exergy, such as exergy associated with the potential energy stored as height above sea level, or the kinetic energy of motion. Others not shown in the figure might include the exergy associated with stored potential energy of springs, or in the stress or strain of material.

- Thermo-mechanical exergy, which is the exergy associated with temperature differences from the environment (being hot or being cold) and the stored potential energy of pressure (compression or vacuum).

- Chemical exergy, considering the energy stored in atom-atom bonds within a molecule, or, in the form of concentration differences from the environment.

- Electro-chemical exergy, such as electromagnetic charge.

- Atomic exergy, such as the energy stored within the nucleus of an atom.

Of these forms, only the two most relevant for most mechanical, industrial, and chemical engineering applications are included in the book: **thermo-mechanical** and **chemical exergy**. As will be discussed later, the thermo-mechanical exergy of a substance is computed from its enthalpy and entropy, which you can usually compute from its temperature, phase, and pressure. We do not decompose thermo-mechanical exergy further into temperature-based and pressure-based exergies. Rather, the tables and figures in the book report the total thermo-mechanical exergy of substances based on the temperature, pressure, and phase combination. For engineering situations without chemical reactions, very often the thermo-mechanical exergy is the only exergy of relevance.

Also discussed later, the chemical exergy of a substance contains the energy associated with its chemical bonds. Chemicals with high-energy bonds (especially fuels like natural gas, diesel, or wood) will similarly have a high chemical exergy. However, even nonfuels like water have chemical exergy, because chemical exergy also

---

*As discussed later, the chemical exergy is the exergy associated with the chemical potential of a substance and its concentration differential against the environment. For combusting fuels, this is usually the kind of exergy engineers are most concerned with.

Exergy of Matter on Earth

Mechanical		Thermo-Mechanical		Chemical		Electro-chemical	Atomic
Gravitation-al (height above sea level)	Kinetic (related to motion)	Thermal (heat of matter)	Pressure (Physical Potential)	Chemical Potential (bond energy)	Compositional (concentration difference from environment)	Electro-static & Electro-dynamic	Atomic (related to nuclear)
Potential Exergy	Kinetic Exergy	Thermal Exergy	Pressure Exergy	Bond Exergy	Concentration Exergy		
Pot. Exergy e^p	Kin. Ex. e^k	Thermo-Mechanical Exergy e^{tm}		Chemical Exergy e^{ch}			

Not Used In This Book

Thermo-Mechanical Exergy e^{tm}

The portions relevant to most industrial engineering analyses.

$$e^{tm} = (h_1 - h_0) - T_0(s_1 - s_0)$$
when
$$p_1 \geq p_0$$

or

$$e^{tm} = (2h_b - h_0 - h_1) - T_0(s_1 - s_0)$$

when
$$p_1 \leq p_0$$

h_1, s_1 computed at stream conditions.

h_0, s_0 computed at environmental T_0, p_0

h_b computed at s_1, p_0

Chemical Exergy e^{ch}

The portions of exergy associated with fuels, reactions, and concentration.

For pure chemicals

$$e_i^{ch} = \Delta g_i^f + \sum_j n_j e_j^{ch}$$

i is the chemical of interest; j is the corresponding elemental reference species in the air, ocean, or land, with n_j moles in one mole of chemical i

For gas mixtures

$$e_{mix}^{ch} = \sum_{i=1}^{N_c} x_i e_i^{ch} - RT_0 \sum_{i=1}^{N_c} x_i \ln x_i$$

For liquid mixtures and complex solids

$$e^{ch} = f(composition, heating\ value)$$

Not Used In This Book

Conventional Exergy e^{conv}

The portions of exergy relevant to most chemical and industrial engineering analyses. The sum of the above.

Exergy of Energy on Earth

Heat		Work	Electricity	Radiation
Heat Exergy E^{th}		Mechanical Power Exergy E^{mp}	Electric Power Exergy E^{ep}	
When $T_H > T_0$	When $T_C < T_0$			
$\frac{E^{th}}{Q_H} = \left(1 - \frac{T_0}{T_H}\right)$	$\frac{E^{th}}{Q_C} = \left(\frac{T_0}{T_C} - 1\right)$	$E^{mp} = W^m$	$E^{ep} = W^e$	*Not Used In This Book*
Q_H is the duty of the hot exergy source of interest transferred at T_H to some arbitrary cold sink	Q_C is the duty of the heat transferred from some arbitrary hot supply at some higher temperature to the cold exergy source of interest at T_C	Work is pure exergy	Work is pure exergy	

FIGURE 1.2 A taxonomy of exergy as used in the book. The equations and terms are explained throughout the book.

includes the exergy associated with its concentration difference from the environment. In other words, any concentration difference from the environment can in theory be exploited to produce some work, even though it might be small. In this book the combination of chemical bond-related exergy and compositional exergy is collectively referred to as chemical exergy, and tables will list a single chemical exergy number for each chemical.

The engineer can then add the thermo-mechanical exergy to the chemical exergy in order to compute the sum, which we call the **conventional exergy**. We use this term because this is what so many engineers mean when they refer to the "exergy" of a substance in their analyses. We note that other textbooks may use other terms, but the taxonomy explains our meaning.

The other kinds of exergy considered in the book are exergies of energy in abstract forms. Work and electricity, as mentioned previously, are pure exergy. This simply means that the exergy value of work or electricity is its energy value. Radiative forms of energy (photons) are not considered because they are special and complex. However, something very useful to consider is the exergy of heat. This will be explored more later, but essentially, the temperature of a heat stream or cold sink affects its exergy significantly. The farther away from the environmental temperature, the more exergy it has.

1.4 Exergy as a Proxy for Value

One of the most useful aspects of exergy is that it can be used as a proxy or placeholder for *value* in your analysis. For example, consider an engineer who is designing an energy system, a chemical plant, or a machine that either uses or creates energy or energy products. During the design process, engineers have to make decisions that will affect the performance of their system, the kinds and amounts of products it will make, and the kinds and amount of energy it will consume. These decisions might be the selection of operating temperatures and pressures in various pieces of equipment or at various locations, flow rates of various materials, or the particular size and shape of mechanical parts. Very often, the decisions are heavily influenced by cost, profit, or other economic factors that all relate to value in some way. At the design stage, it can be very difficult to estimate what these costs are at that point in time, what they will be when you build it, and what they will be after 20 years of operation.

In these situations, exergy is an extremely useful proxy for value, because thermodynamically, it includes both the quantity and quality of the energy content. Thanks to this book, it is usually considerably easier to estimate the exergy content of a material or energy stream than its cost. Therefore, you can use exergy as a key metric in place of value to aid in the design process.

1.4.1 Exergy Predicts Value for Industrial Utilities

A great example can be found in the pricing of heating and cooling utilities commonly used in chemical plants. The classic textbook by Seider et al.[1] lists typical average prices for utilities like steam for heating or refrigeration at various temperatures. These numbers are estimates based on heuristics and experience. In Fig. 1.3,

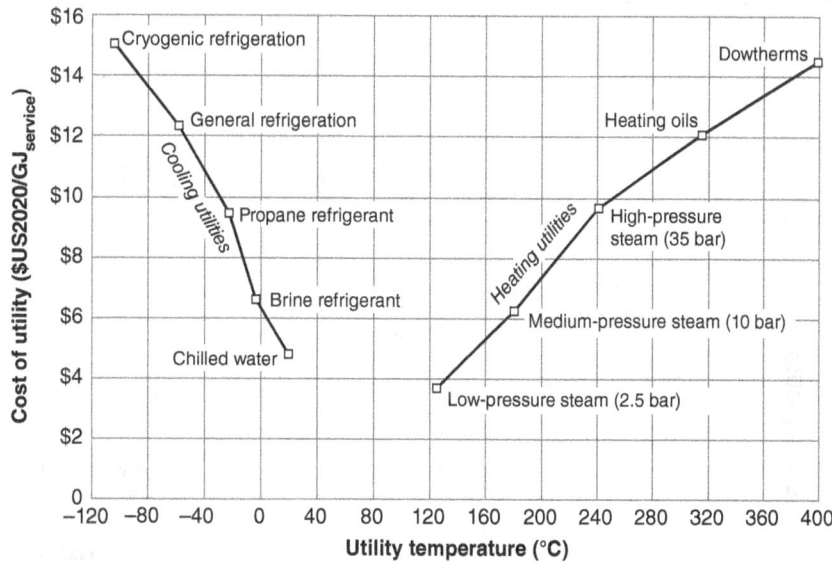

Figure 1.3 Utilities commonly available in chemical plants, based on heuristics and experience. Derived from Seider et al.[1] and updated to 2020 economics. Numbers are approximate. Heating Oils and Dowtherms are assumed to be 25 and 50% more expensive than high-pressure steam, respectively. It is important to understand that the utility is given per GJ of either heating or cooling *service*, not the material itself. For example, the price for medium-pressure steam is not the price per GJ of the actual steam, it is the price per GJ of *heat delivered* to something colder by condensing that steam. Similarly, the price of propane refrigerant is the price per GJ of *heat removed* from something hotter by transferring it to propane within a closed industrial refrigeration loop, not the price of buying propane itself.

we converted the numbers in Seider et al., which are for the year 2006, to modern-day cost estimates in 2020 US dollars per GJ of service. This plot shows the price of the utility per GJ of service (which is its *value*) as a function of temperature. You can clearly see a trend here, that the farther the temperature gets from "ambient" conditions, the higher its price. In this case, "ambient" is not really the temperature of the air around the chemical plant. Chemical plants usually have a lot of waste heat floating around the plant at rather warm temperatures. Taking a look at the prices from Fig. 1.3, if we extrapolate the heating utility and cooling utility lines, they seem to both hit the zero axis at about 78°C (351 K). For all practical purposes, this is the approximate "ambient" or "reservoir" temperature of a typical chemical plant. In other words, you should be able to get heat for about free at 78°C, and it costs money if you want to make streams either cooler or warmer than that.

Using an ambient/reservoir temperature of 78°C, we can compute the approximate exergy value of each of these utilities using the methods discussed in Chap. 2. These values are shown in Fig. 1.4. Interestingly, they follow a very similar shape to Fig. 1.3, with the steeper slope of the cooling utility line reflected in both plots. Taking this one step further, the same information in Figs. 1.3 and 1.4 are plotted again in Fig. 1.5, showing the price of

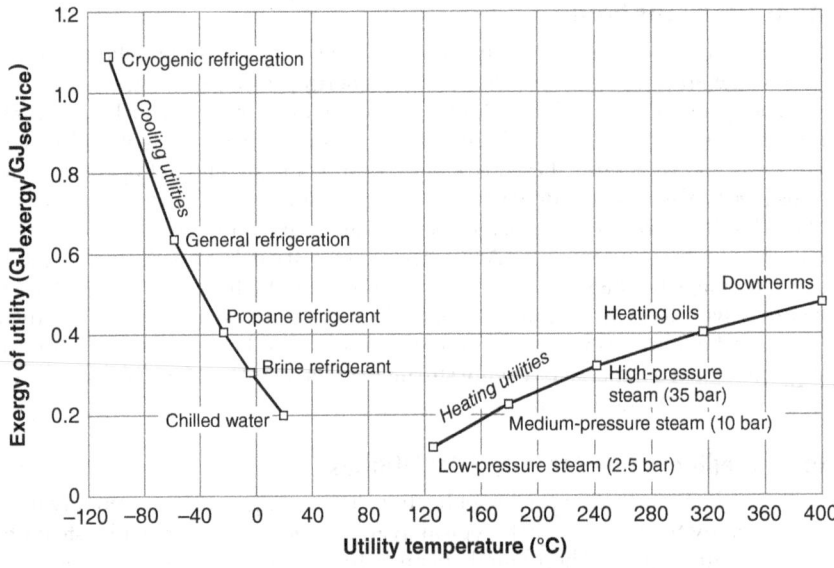

Figure 1.4 Approximate exergy of the heat/cooling *energy service provided* by the utilities corresponding to Fig. 1.3. Ambient/reference temperature used was 78°C. Utilities were assumed to be constant temperature.

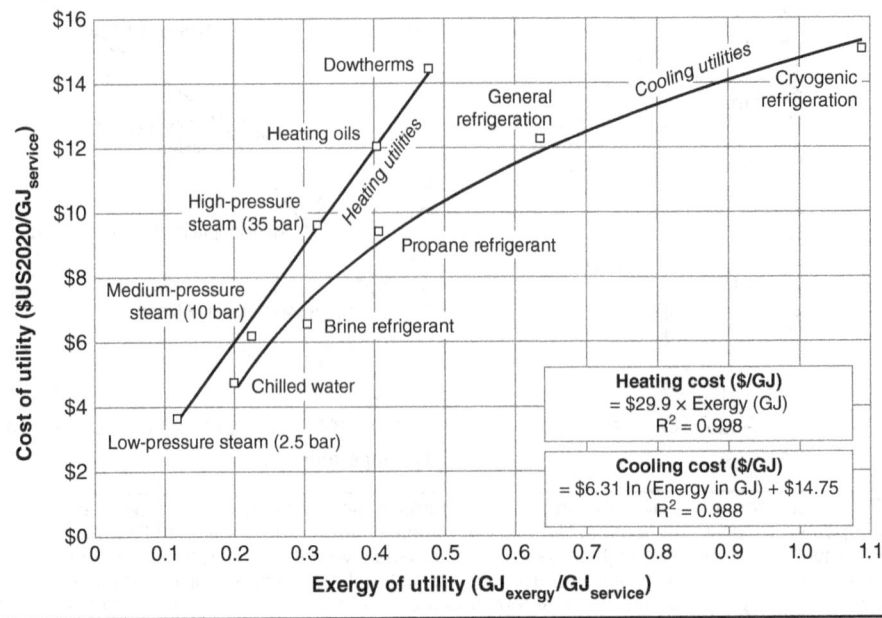

Figure 1.5 Exergy and price parity plot of the data in Figs. 1.3 and 1.4. Correlations between value and exergy content are shown.

the utility as a function of its exergy content. Remarkably, these are heavily correlated, and in fact, the relationship between exergy and heating utility cost is essentially constant at $30 per GJ of exergy! Thus, exergy is an extremely good proxy for value, with $R^2 = 0.99$.

The careful reader might wonder how cryogenic refrigeration can have an exergy-to-energy value above 1.0 if the exergy of a stream is always less than or equal its energy. This is not a mistake and it does not violate the laws of thermodynamics. Rather, it has to do with *which* energy we are making the comparison of exergy and energy. This will be explained in Chap. 2.

1.5 Kinds of Questions Answered by Exergy Analysis

1.5.1 Does the Process Use Exergy Efficiently?

With an exergy analysis, you can compute an exergy efficiency. Most chemical and industrial processes involve some sort of energy conversion process, in which energy is changed from one form to another, be it heat, mechanical work, electricity, potential energy, or chemical bonds. Understanding the exergy efficiency of your process can help assess whether the right design choices have been made from a big-picture perspective.

The exergy efficiency of a process can be very different from its energy efficiency. A modern-at-home natural-gas furnace is a great example. It has a very high energy efficiency—almost 96% on a lower heating value (LHV) basis! Most of the chemical potential of the natural gas is recovered in the form of heat by heating up fresh air that is delivered to a room via an air duct. However, it has an extremely low exergy efficiency (roughly 0.6%[2]), because the exergy of the product (warm air) is orders of magnitude lower than the chemical exergy of the natural gas used as fuel.

1.5.2 Is There a Better Way to Use Energy Resources from a Societal Perspective?

The high-energy-efficiency, low-exergy-efficiency outcome tells us a lot about the nature of the at-home furnace. As the technology developed, designers continually improved the design by making it more and more efficient, slowly approaching 100%. However, these improvements do little to improve the extremely low exergy efficiency of gas furnaces, because fundamentally they all use a very potent exergy source (natural gas) to make a very weak exergy product (warm air).

From a big picture perspective, this tells us that we may want to look at home heating systems of a completely different nature. Integrated community energy systems are one example, in which waste heat from one building (perhaps the low-grade waste heat generated by industrial-sized refrigerators for an ice rink) is delivered to another nearby building and used to heat air or water. Combined heat and power systems are another, in which the lower-grade waste heat left over from a nearby combustion-based power plant is used to provide building heat needs. A third approach might be waste-to-energy districting heating, in which landfill-bound trash or municipal compost off-gases are combusted to produce heat. In all three cases, an energy source with lower exergy (and lower value) such as waste heat or trash are used, improving the exergy efficiency of building heat services by fundamentally changing the design paradigm. Subsequently, all three approaches have the potential to reduce overall lifetime heating service costs and/or overall greenhouse gas emissions, precisely because the exergy efficiency is higher. Interest in these three modern heating strategies has exploded worldwide, with increasingly rapid research, development, and adoption.

1.5.3 What Are the "Energy Bottlenecks" in a Chemical or Mechanical Process?

With an exergy analysis, you can compute the exergy destruction for each individual step or piece of a larger process. Steps with large exergy destruction compared to the others are prime candidates for improvement. Designers can focus on these individual steps, which are likely to lead to meaningful gains at the systems level. Similarly, time spent improving steps with very little exergy destruction may not yield much fruit.

1.5.4 How Can I Optimize for Cost during Early Stages of Process Design?

As shown in Sec. 1.4, the costs of common industrial utilities are related directly to exergy content. This same principle can be extended to the plant in general. The process of designing an industrial or energy process has many stages. One key stage is process synthesis—the act of choosing which unit operations should exist and how they should be connected. A second key stage is process optimization—the act of selecting the optimal design parameters for each unit operation, such as temperatures, pressures, flow rates, catalyst choices, utilities, energy consumption, and sizes and shapes of equipment. In either case, the overall objective very often is to design a plant with the most favorable economics, such as the highest return on investment, the highest profitability, or the lowest costs, depending on the scope.

However, during the early stages of design, rigorous costs are often very difficult to estimate. Although many methods exist for estimating costs at various levels of fidelity or at various stages of the design, they have high degrees of uncertainty and sometimes cannot be used at all depending on the circumstances. This is further complicated when one considers the ever-changing and sometimes unpredictable nature of markets over a potentially long lifetime of the process.

Since exergy is correlated with cost, one can use exergy metrics as a proxy for cost during the early stages of design. For example, one might design a chemical process that maximizes the exergy of the products or maximizes exergy efficiency. This can often be done more quickly than cost analyses, especially with the help of the tables in this book. The approach can be particularly useful for design candidate screening, in which many possible design choices are considered, with the top candidates selected for more detailed examination based on their exergy properties. Any resulting design candidates can then be examined more closely and modified based on more tangible metrics such as cost or profitability.

1.5.5   What Products Should My Process Produce?

Although normally process designers address questions like "what is the best process to produce X from feed Y?", they may also have to answer questions like "what are the best products X to make from feed Y?". The latter case is particularly common for designers working in waste-to-energy applications, where the waste energy source is known (such as biomass, municipal solid waste, spent truck tires, steel refinery off-gases, etc.) but the best way to use them is not. Because exergy can be used as a proxy for value, exergy analyses can be used to help answer this question. For example, choosing products (and the processes that make them) that result in high exergy efficiencies can often best capture the value available in that waste resource. The best process candidates can then be further examined more rigorously to better assess profitability or suitability.

1.6   How to Use This Book

This book is designed to make it as easy as possible for engineers to use exergy analysis tools. Computing the exergy values of chemicals and streams themselves can be quite complex. It takes a lot of time to learn the theory and methodology. It can also be quite challenging to compute the thermo-mechanical and chemical exergy of each process stream given the complexity associated with reference conditions and equations-of-state. However, once the exergy values are known, using them in analyses is quite straightforward and readily understandable by most engineers.

So, we did the hard work for you! This book contains tabulated values of various forms of exergy for a great many chemicals, fuels, and process streams that are common to industrial plants and energy systems. These values were all calculated according to the same reference conditions (the same ambient temperature, pressure, relative humidity, ocean salinity, atmosphere composition, hydrosphere composition, and lithosphere composition). This is important because when doing an exergy analysis, the exergy for all of the streams *must* be computed according to the *same reference conditions*. If exergy values are computed inconsistently within an analysis, the results could be meaningless. However, with this book, you can look up whatever values you need in the exergy tables and use them all together in your analysis in a consistent and meaningful fashion.

Chapter 2 contains the thermodynamic theories of exergy, their derivations, and other important conceptual information. This can help the engineer to have a deeper understanding of exergy and make important connections to more familiar concepts, like energy and entropy.

Chapter 3 shows helpful examples on how to use the results in exergy analyses of different types and for different applications. This includes strategies for identifying and resolving exergy bottlenecks, exergy visualization tools such as Sankey diagrams, and using exergy for thermodynamic targeting and determining theoretical best cases.

1.6.1   Reading the Tables

The exergy of a stream or substance is the sum total of its various forms of exergies, as discussed in Sec. 1.3. The exergy tables in Chaps. 4 through 10 list either the thermo-mechanical exergy e^{tm} or the chemical exergy e^{ch} of a substance, but other forms of exergy are not listed because they are not usually relevant or sufficiently large to warrant consideration in most chemical, industrial, and energy systems. Moreover, the chemical exergy of a fuel or high-energy chemical is often sufficient to approximate the total exergy, especially for solid or liquid fuels at ambient conditions. Similarly, the thermo-mechanical exergy of common working fluids like water, steam, or air are the most relevant since they contain very little chemical exergy.

The tables often list other relevant information for each substance. For example, Table 7.1 shows the chemical exergy of various hydrocarbons. The enthalpy of formation, Gibbs free energy of formation, entropy at the reference state, and specific heat capacity at the reference state are also provided.

For some common chemicals, the thermo-mechanical exergy is presented graphically on Pressure-Enthalpy-Exergy diagrams. For example, Fig. 5.1 shows the Pressure-Enthalpy-Exergy diagram for air. The x-axis is the enthalpy (on a per mass basis) of the substance and the y-axis is its pressure. Black lines are lines of constant temperature, red lines are constant entropy, and blue lines are constant exergy. The phase envelope is shown as the thick black bell-shaped curve, and remember that points to the right of this envelope are vapor phase, points to the left are liquid, points inside the envelope are mixed vapor-liquid, and points above it are supercritical. The black circle is the environmental conditions (25°C and 1.01325 bar). To find the thermo-mechanical exergy, locate the point on the plot corresponding to your substance and then read off the exergy value. Usually, engineers know pressure, temperature, and phase, so find the correct black temperature line, and follow it until it intersects the known pressure on the y-axis (and account for vapor/liquid quality if in the two-phase region). For example, in Fig. 5.1, find the point where the –20°C line reaches 10 bar. Thus, air at –20°C and 10 bar is a gas, and it has a thermo-mechanical exergy of 200 kJ/kg.

All exergy computations in the book use the same set of standard reference environmental conditions, which is detailed at length in Chap. 4. For thermo-mechanical exergy, the only important reference conditions are the temperature (25°C) and pressure (1 atm = 1.01325 bar). However, the environmental reference state also contains assumed values of the concentrations of a long list of reference chemicals in the atmosphere (air), hydrosphere (oceans), and lithosphere (crust). This is very important for computing chemical exergies. The specific assumptions about the environmental reference state, like relative humidity and ocean salinity, play an important role in determining the chemical exergy values. The conditions are chosen using the best research available and are discussed at length in Chap. 4.

The tables which list chemical exergy assume its temperature and pressure are at the reference environmental conditions (25°C, 101.325 kPa). Table 7.1 is one example, where the chemical exergies e^{ch} of selected hydrocarbons are listed. It is important to remember that e^{ch} therefore does not include any thermo-mechanical exergy associated with any pressure or temperature deviations from the environmental conditions. For some situations, you may need to look up both e^{ch} and e^{tm} separately and add them together to get the total conventional exergy e^{conv}. This is typically important for high-energy gases which may also be available at extreme temperatures or pressures in a process.

For example, consider hot (240°C), high pressure (20 bar) methane that is about to be combusted in a gas combustion turbine. In this case, both the chemical exergy and thermo-mechanical exergy might be important. First, find methane (CH_4) in Table 7.1 and see that its chemical exergy e^{ch} is 831.66 kJ/mol (which is equivalently about 51,914 kJ/kg). Then look at Fig. 7.11, which shows the thermo-mechanical exergy e^{tm} of methane graphically at various temperatures and pressures. Find the point where the temperature is 240°C and the pressure 20 bar—at that state, the e^{tm} is 600 kJ/kg. Adding the two together gives the total conventional exergy e^{conv} at 52,514 kJ/kg. Clearly, the thermo-mechanical exergy component contributes a relatively small amount in this situation.

Many tables that list complex solid fuels list the proximate and/or ultimate analysis compositions that were used for the calculation of its exergy. The proximate analysis determines the moisture, ash, volatile matter, and fixed carbon content of the fuel. The ultimate analysis determines the amount of atomic hydrogen, carbon, nitrogen, sulfur, and oxygen. For example, Table 10.6 lists the atomic compositions of woods, straws, and grasses of various types. These are heterogeneous structures that will vary from sample to sample and harvest to harvest, but the compositions provided are good characterizations of the compound as a whole.

1.6.2 Heating Values and Chemical Exergy

For some fuels that are primarily used for combustion, such as gasoline and diesel, the LHV and HHV are listed. The LHV is computed by combusting the fuel in air at standard conditions so that the combustion products (typically CO_2 and H_2O plus leftover air) are very hot, and then cooling the combustion products down to 150°C. The LHV is the amount of thermal energy recovered by this cooling down to 150°C. This is a reasonable approximation of the maximum amount of useful energy that could be recovered in a classic combustion furnace, because in such furnaces one often does not want to condense the water in the gases for practical and safety reasons (150°C is sufficiently high above the dew point to avoid this).

The HHV is simply the LHV plus the heat of vaporization of any water contained in the combustion products. This recognizes that there is additional heat that could be captured if one were to cool this stream further by condensing all the steam into water. Thus, the HHV is a good approximation of the total useful chemical energy of the fuel when combusted. The LHV and HHV are legacy metrics influenced by the technological capabilities of the time, but they are still commonly used because they are defined in standard ways. Note that modern, high-efficiency, condensing furnace technology available in many personal residences can safely condense the water in the combustion exhaust to extract more useful energy, and so the LHV metric starts to lose its meaning in these applications.

Although the HHV is commonly used as the approximate energy content of fuels, it still does not quite include all forms of useful energy. This is because the HHV does not capture the energy associated with the specific heat capacity of the combustion products between the reference state (25°C) and 150°C.

The chemical exergy of a fuel is generally lower than its HHV, very roughly up to 18% lower. Consider for example the approximate HHVs of methane (55.5 MJ/kg), Canadian natural gas from Alberta (55 MJ/kg), hydrogen (142 MJ/kg), and ethanol (29.7 MJ/kg). Their corresponding chemical exergies (see the tables in Chaps. 7 and 8) are all lower: methane (51.9 MJ/kg), Canadian natural gas from Alberta (48.5 MJ/kg), hydrogen (117 MJ/kg), and ethanol (29.4 MJ/kg). Thermodynamically, this makes sense since Exergy ≤ Energy.

1.6.3 Interpolating from the Tables

To find the thermo-mechanical exergy at temperatures or pressures not listed in the tables in Chaps. 5 or 6, we recommend using an interpolation method. For example, looking at Table 6.5, we can estimate the thermo-mechanical exergy of water at 2 bar and 853°C, by linearly interpolating using the data from the table at 2 bar and 850°C (1493.1 kJ/kg) and 855°C (1501.8 kJ/kg) as follows:

$$e^{tm} \approx 1493.1 \frac{kJ}{kg} + \left(\frac{853°C - 850°C}{855°C - 850°C} \right) \left(1501.8 \frac{kJ}{kg} - 1493.1 \frac{kJ}{kg} \right)$$

(1.7)

$$e^{tm} \approx 1498.320 \frac{kJ}{kg}$$

This turns out to be extremely close to the real value at this temperature (1498.321 kJ/kg), accurate to six significant digits. Since data in the tables are reported only to five digits, linear interpolation provides a numerically accurate answer.

To get more accurate estimates when interpolating, more rigorous interpolation methods may be used such as higher-order interpolating polynomials or cubic splines. For the tables in this book, we have found that the difference between linear and cubic interpolation is trivially small when interpolating between temperatures (to the seventh significant digit, which is more precise than the data precision used in the tables), so linear interpolation will be sufficiently accurate.

When interpolating between pressures, however, cubic interpolation does provide additional accuracy compared to linear interpolation. Moreover, properties that relate to pressure scale more naturally with the natural log of pressure, not absolute pressure. Therefore, you should interpolate using the natural log of pressures rather than the absolute value. Consider for example that you want to find the thermo-mechanical exergy of water at 217°C and 17.5 bar. In Table 6.7, the thermo-mechanical exergy of water at 217°C and 17 bar pressure is 905.45 kJ/kg, and at 217°C and 18 bar, it is 910.28 kJ/kg. If you were to use ordinary linear interpolation to estimate e^{tm} at 17.5 bar, you could do the following:

$$e^{tm} \approx 905.45 \frac{kJ}{kg} + \left(\frac{17.5 \text{ bar} - 17 \text{ bar}}{18 \text{ bar} - 17 \text{ bar}} \right) \left(910.28 \frac{kJ}{kg} - 905.45 \frac{kJ}{kg} \right)$$

(1.8)

$$e^{tm} \approx 907.87 \frac{kJ}{kg}$$

However, the actual value is $e^{tm} = 907.93 \frac{kJ}{kg}$ so this is a little bit off. It is better instead to interpolate in the logspace for pressure, like this:

$$e^{tm} \approx 905.45 \frac{kJ}{kg} + \left(\frac{\ln 17.5 - \ln 17}{\ln 18 - \ln 17} \right) \left(910.28 \frac{kJ}{kg} - 905.45 \frac{kJ}{kg} \right)$$

(1.9)

$$e^{tm} \approx 907.90 \frac{kJ}{kg}$$

This is closer to the true value. However, we can do better with a third-order polynomial (a cubic interpolation method), which considers also two additional points of data in Table 6.7 at 217°C, one at 16 bar (900.06 kJ/kg) and another at 19 bar (914.60 kJ/kg). Considering more data points allows you to capture some of the nonlinearity. The goal is to find the coefficients $a_0 \ldots a_3$ of a cubic polynomial like this:

$$e^{tm} \approx a_0 + a_1 \ln P + a_2 (\ln P)^2 + a_3 (\ln P)^3$$

(1.10)

such that if you plug in P at 16, 17, 18, or 19 bar, you get exactly the four known data points in the table. Then you can plug any other P in the range of 16 to 19 bar and get a really good estimate at other pressures not found in the table. You can express this in equation form as follows:

$$900.06 = a_0 + a_1 (\ln 16) + a_2 (\ln 16)^2 + a_3 (\ln 16)^3$$
$$905.45 = a_0 + a_1 (\ln 17) + a_2 (\ln 17)^2 + a_3 (\ln 17)^3$$
$$910.28 = a_0 + a_1 (\ln 18) + a_2 (\ln 18)^2 + a_3 (\ln 18)^3$$
$$914.60 = a_0 + a_1 (\ln 19) + a_2 (\ln 19)^2 + a_3 (\ln 19)^3$$

(1.11)

Or, you can write it in matrix form like this:

$$\begin{bmatrix} 900.06 \\ 905.45 \\ 910.28 \\ 914.60 \end{bmatrix} = \begin{bmatrix} 1 & (\ln 16) & (\ln 16)^2 & (\ln 16)^3 \\ 1 & (\ln 17) & (\ln 17)^2 & (\ln 17)^3 \\ 1 & (\ln 18) & (\ln 18)^2 & (\ln 18)^3 \\ 1 & (\ln 19) & (\ln 19)^2 & (\ln 19)^3 \end{bmatrix} \begin{bmatrix} a_0 \\ a_1 \\ a_2 \\ a_3 \end{bmatrix} \tag{1.12}$$

Then, you can either solve the system of equations or the linear algebra (matrix) form to find the unknown $a_0 \ldots a_3$ using whatever means you like. For example, try the backslash operator in Matlab, `linalg.solve` in NumPy/SciPy, or use the solver or `MINVERSE` function in Excel. However you do it, the solution looks like this:

$$e^{tm} \approx 883.54 - 256.44 \ln P + 158.61 (\ln P)^2 - 23.071 (\ln P)^3, \text{ valid for 16 bar} \leq P \leq 19 \text{ bar for water at } 217°C \tag{1.13}$$

Plugging in 17.5 bar for P ($\ln 17.5 = 2.8622$), we compute the best estimate of $e^{tm} \approx 907.93 \frac{kJ}{kg}$. This is the same value as the "true" value within the precision of the tables.

What if you want to interpolate in both pressure and temperature? Let's say you want to estimate the thermomechanical exergy at 216.5°C and 17.5 bar. We can take a similar approach but use the four points closest to it: 217°C and 17 bar, 217°C and 18 bar, which we used in the previous example, and 216°C and 17 bar, 216°C and 18 bar, which are in the rows above it in Table 6.7. We can use a similar approach, but we have to change our equation a little bit to factor in both temperature and pressure. I suggest something like this:

$$e^{tm} \approx a_0 + a_1 \ln P + a_2 T + a_3 T \ln P \tag{1.14}$$

This form lets us include linear interpolation in both temperature and pressure, but also some nonlinear interaction between the two. We can find the unknown coefficients by solving this set of equations, one for each of the four known points:

$$904.37 = a_0 + a_1 (\ln 17) + a_2 (216) + a_3 (216)(\ln 17)$$
$$909.17 = a_0 + a_1 (\ln 18) + a_2 (216) + a_3 (216)(\ln 18)$$
$$905.45 = a_0 + a_1 (\ln 17) + a_2 (217) + a_3 (217)(\ln 17)$$
$$910.28 = a_0 + a_1 (\ln 18) + a_2 (217) + a_3 (217)(\ln 18)$$
$$\tag{1.15}$$

Solving this, we get:

$$e^{tm} \approx 754.37 - 29.392 \ln P - 0.40703\, T + 0.52486\, T \ln P,$$
$$\text{valid for water } 216°C \leq T \leq 217°C \text{ and 17 bar} \leq P \leq 18 \text{ bar} \tag{1.16}$$

Plugging in 216.5°C and 17.5 bar results in $e^{tm} \approx 907.36 \frac{kJ}{kg}$ which is very close to the actual at $907.38 \frac{kJ}{kg}$.

References

1. Seider WD, Seader JH, Lewin DR, Widago S. *Product and Process Design Principles: Synthesis, Analysis, and Evaluation*. 3rd ed. Wiley: New York (2009). Tables 4.7 and 23.1.
2. Rosen MA, Le MN, Dincer I. Efficiency analysis of a cogeneration and district energy system. *Applied Thermal Engineering* 25:147–159 (2005).

CHAPTER 2

Exergy from Thermodynamics

2.1 Introduction

As discussed in detail in Chap. 1, *energy* only covers quantity, while *exergy* covers both the quantity and quality of energy. This means that energy is related to the first law of thermodynamics (the conservation of energy), while exergy is also related to the second law of thermodynamics (the destruction of energy quality). Since exergy is a measure of quality (either work or the ability to produce work), this concept helps us to handle different energy forms on a common basis.

In daily language (both in industry and society), it is common to refer to *energy consumption*. This term has its limitations, since according to the first law of thermodynamics, energy can be stored, transferred, and converted, but it is always *conserved*, thus not consumed. What is consumed is energy quality or exergy. As a result, exergy consumption would be a much better term. In fact, it has been proposed in the European Union to use exergy consumption as a new footprint for energy use in the process industries.

For those with a basic background in thermodynamics, exergy is also very closely related to *entropy*. It is useful to show the equations for entropy and exergy balances in order to show how these state functions are related. Figure 2.1 shows a simple form of an open/flowing system in general. The system can be anything, such as one small part of a piece of equipment, or an industrial mega-process. Mass and heat enter and exit the system. The system can operate dynamically, or be at steady state. The system can also do work on the outside world or have work done on it.

An entropy balance (using the second law of thermodynamics) for open/flowing systems can be formulated as follows:

Entropy accumulation = Entropy in – Entropy exiting + Entropy generated

$$\frac{dS_{cv}}{dt} = \left[\sum_i \frac{Q_i}{T_i} + \sum_i m_i \cdot s_i \right] - \left[\sum_e \frac{Q_e}{T_e} + \sum_e m_e \cdot s_e \right] + S_{gen} \tag{2.1}$$

Note that work has no entropy, so the work streams do not appear in the entropy balance. The corresponding exergy balance for open/flowing systems can be formulated as follows:

Exergy accumulation = Exergy in – Exergy exiting – Exergy destroyed

$$\frac{dE_{cv}^{conv}}{dt} = \left[E_i^{ep} + E_i^{mp} + \sum_i E_i^{th} + \sum_i m_i \cdot e_i^{conv} \right] - \left[E_e^{ep} + E_e^{mp} + \sum_e E_e^{th} + \sum_e m_e \cdot e_e^{conv} \right] - [T_0 \cdot S_{gen}] \tag{2.2}$$

In these equations, S is total entropy (kJ/K), t is time (in s), Q is heat (kW), T is temperature (K), m is mass flow (kg/s), s is specific entropy (kJ/kg-K), E is exergy (E_{cv} is in kJ, the rest are kJ/s), and e is specific exergy (kJ/kg). The subscripts used have the following meaning: cv is the control volume, i is an identifier for incoming streams, e is an identifier for exiting streams, *gen* is generation (or production), and 0 is the reference condition for exergy. Superscript *conv* is conventional exergy, *ep* is electric power, *mp* is mechanical power, and *th* is thermal energy (heat).

By comparing the entropy and exergy balance equations, you can see the parallels between the individual terms. Each flow term relates to a stream in Fig. 1.2. For each of the heat flows Q in or out of the system, its corresponding entropy flow is Q/T, and its corresponding exergy is what we call the thermal exergy or E^{th}. Similarly, the entropy flow terms for mass streams ($m \cdot s$) correspond to conventional exergy flow terms ($m \cdot e^{conv}$). There are two additional exergy terms, the exergy of electric power E^{ep} and the exergy of mechanical work flows E^{mp}, which correspond directly to the electrical work and mechanical work flows W^e and W^m. We will discuss later how to determine these exergy terms (E^{th}, e^{conv}, E^{ep}, and E^{mp}).

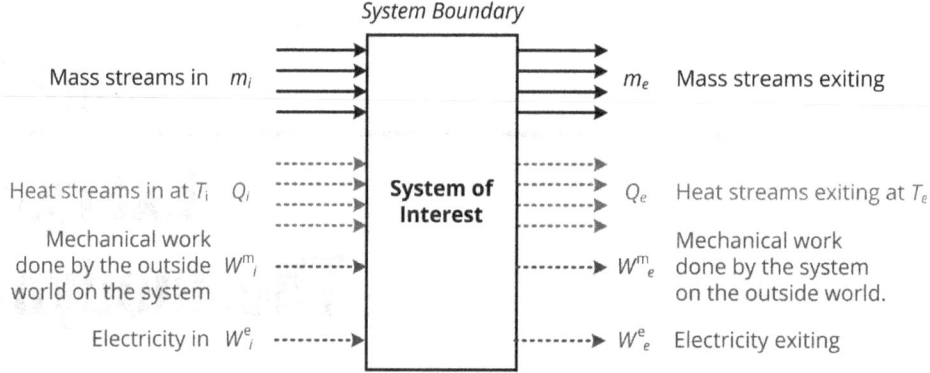

FIGURE 2.1   A general mass system with mass and energy flows.

Note the last term in these equations which shows that exergy destruction ($T_0 \cdot S_{gen}$) is directly related to entropy generation (S_{gen}). Due to irreversibilities (thermodynamic losses) in processes, balance equations for entropy and exergy can only be formulated by adding the entropy generation term to the entropy balance and subtracting the exergy destruction term from the exergy balance. These irreversibilities will be discussed in Sec. 2.5.

This chapter focuses on the thermodynamic treatment of the term *exergy*. We will briefly present the history of the exergy concept and discuss different forms and classifications of exergy, before providing the formal expressions for different exergy types. The concept of exergy is typically used for two purposes when analyzing industrial processes with focus on energy: performing exergy analyses to identify scope for improving the energy system and calculating exergy efficiency as an important key performance indicator (KPI) for the process. The last topic in this chapter is a discussion about the advantages and limitations of the exergy concept and its use to assess the efficiency of industrial processes.

2.2   The Concept of Exergy

Different terms and symbols have been used in different time periods and different parts of the world for the concept that these days is most commonly referred to as exergy. The concept was developed as early as 1873 by J. Willard Gibbs who used the term *Gibbs free energy*. The term *exergy* was coined by Zoran Rant in 1956 using the Greek language as basis. The prefix for "from" or "out" in Greek is $\varepsilon\zeta$ (*ex*) and $\varepsilon\rho\gamma o\nu$ (*ergon*) means work in Greek. Combining these two parts gives the term *exergy*, which is understood as the work that can be obtained from a given system.

Availability (or available energy) has also been used for the concept of exergy, primarily in the United States. Different symbols have also been used to denote exergy. In addition to E with different subscripts and super-scripts used in this book, the letter B has also been frequently used in the literature for exergy. Unfortunately, some authors have suggested that energy can be decomposed into exergy, which is the part above ambient temperature T_0 that is able to produce work, and anergy, which is the remaining part between T_0 and 0 Kelvin that is unable to produce work. Here, we use the word unfortunately since energy below ambient temperature indeed can be used to produce work, for example by acting as a heat sink for heat that is transferred from the environment. A more detailed discussion about this is provided in Sec. 2.4.1.

The thermodynamic definition of exergy as proposed by Szargut[1] is:

Exergy is the amount of work obtainable when some matter is brought to a state of thermodynamic equilibrium with the common components of its surrounding nature by means of reversible processes, involving interaction only with the above mentioned components of nature.

One important word here is *reversible*, which is an idealized case without thermodynamic losses, thus resulting in *maximum* theoretical work produced. Another important term here is *surrounding nature*. This is easy regarding temperature and pressure but more complicated when it comes to chemical composition, as will be discussed in Sec. 2.3 about exergy forms and Sec. 2.4.2 about chemical exergy. The definition of exergy provided here will be illustrated and exemplified when dealing with the two important exergy forms in this book; thermo-mechanical and chemical exergies.

2.3   Exergy Forms and Classifications

Energy exists in a number of different forms and qualities. As explained, exergy is a measure of energy quality. The different exergy forms were discussed in Sec. 1.3 and illustrated in Fig. 1.2 that also contained equations for some of these exergy forms. A distinction is made between exergy related to *energy* on the Earth and exergy related to *matter* on the Earth. The Earth is our reference and can be broken down into the atmosphere, the

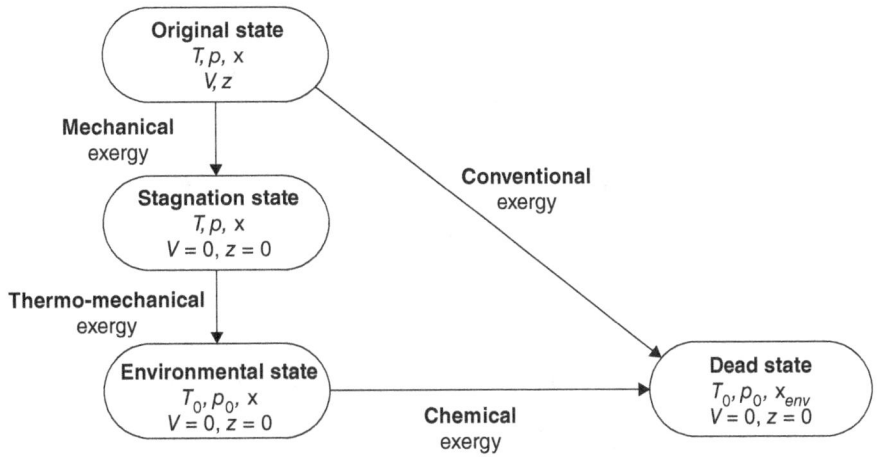

FIGURE 2.2 Important states for the concept of exergy.

hydrosphere (the oceans), and the lithosphere (the Earth's crust). More details about the different exergy forms and their mathematical expressions are provided in Sec. 2.4 and Fig. 1.2.

The energy forms that are most commonly encountered in the process industries are related to parameters such as temperature (T), pressure (p), chemical composition ($x_i \in \mathbf{x}$, $i = 1$ to N_C, where x_i is the mole fraction of component i, N_C is the total number of components, and $\mathbf{x}$ is a vector of mole fractions), linear velocity (V in m/s), and height above sea level (z in m). *Mechanical* exergy, which is the exergy of kinetic energy (related to motion) and potential energy (related to height), is included here for completeness, but will be omitted later since the changes in kinetic and potential energies are neglectable in most industrial applications.

The very general definition of exergy provided in Sec. 2.2 can be broken down into different exergy forms by carefully selecting a set of thermodynamic states. Figure 2.2 shows the *original* state of the system for which we would like to assess the exergy, the *stagnation* state where velocity and height relative to the Earth are both zero, the *environmental* state where temperature and pressure are equal to the reference environment ($T_0 = 298.15$ K and $p_0 = 1$ atm $= 1.01325$ bar), and the *dead* state where composition (or chemical potential) of the substance (single or multi-component) is in equilibrium with the reference environment. Notice in Fig. 2.2 the path from the original state to the stagnation state and further to the environmental state and the dead state.

In this book, we neglect mechanical exergy for all of our data and examples. Therefore, *conventional* exergy in this book is the sum of thermo-mechanical exergy and chemical exergy. As shown in the taxonomy in Fig. 1.2, there are other forms of exergy that are not included in the conventional exergy, such as atomic exergy associated with the nucleus of an atom, and exergy associated with electrostatic charge. They are encountered only rarely in special situations so they are not considered a part of "conventional" exergy and not included in the book.

2.4 Formal Expressions for Exergy

The main objective of this section is to develop expressions for the different types of exergy based on the fundamentals of thermodynamics. First, exergy forms that are related to energy not linked to matter will be discussed. Then exergy forms that are related to matter, primarily thermo-mechanical and chemical exergies, will be discussed.

2.4.1 Exergy of Energy on Earth

Three forms of energy will be discussed here: electricity, work, and heat. Radiation also belongs to this category but will not be included, since it is uncommon.

Electricity is the most versatile energy form, since it can be used to generate other forms of energy such as heat and work. This is why electricity is regarded as pure exergy; i.e., the amount of exergy is equal to the amount of energy. Strictly speaking, to generate work from electricity, we need to let electricity drive a motor, and there are minor losses related to this conversion of energy. For permanent magnet motors, efficiencies as high as 98% can be achieved depending on the load. Nevertheless, electricity here and elsewhere in the literature is regarded as 100% exergy. *Work* is of course by definition 100% exergy, since exergy is a measure of the ability to produce work.

Exergy of *heat* is a much more involved topic. Heat has temperature as a quality parameter, and the further away from ambient conditions (both above and below T_0), the higher quality it has. A heat engine can be used to convert heat into work, and the best performance is obtained for a reversible heat engine, also referred to as a Carnot cycle. Consider Fig. 2.3 where heat engines operate between a hot and a cold thermal reservoir, where "hot" and "cold" are relative properties not reflecting absolute temperature. The reservoirs are characterized by having very large capacities to provide/accept heat transfer to/from the heat engine without changing their temperatures.

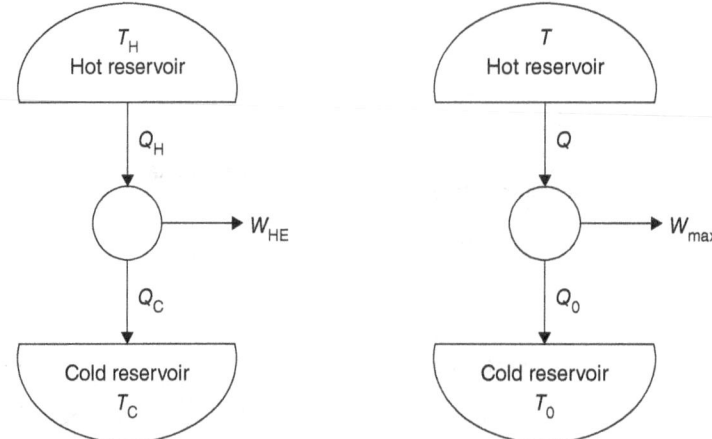

Figure 2.3 Heat engines producing work from heat: real heat engine (left) and Carnot engine (right) that is used to find the exergy of heat.

Consider first the arrangement to the left in Fig. 2.3, where a real (not ideal) heat engine operates between a warmer reservoir with temperature T_H and a colder reservoir with temperature T_C. Heat Q_H is transferred from the hot reservoir to the heat engine (HE) that produces work W_{HE}, while sending some heat Q_C to the cold reservoir. The commonly used KPI for heat engines is the so-called *thermal efficiency*:

$$\eta_{th} = \frac{W_{HE}}{Q_H} = \frac{Q_H - Q_C}{Q_H} = 1 - \frac{Q_C}{Q_H} \tag{2.3}$$

In this expression for thermal efficiency, heat losses to the environment (first law losses) are neglected. If the heat engine is reversible, it operates as a Carnot engine, and the corresponding thermal efficiency becomes (subscript C for Carnot):

$$\eta_C = 1 - \left\{\frac{Q_C}{Q_H}\right\}_{\substack{\text{internally}\\\text{reversible}}} = 1 - \frac{T_C}{T_H} \tag{2.4}$$

This efficiency is also referred to as the Carnot factor. The second law of thermodynamics as formulated in words by Kelvin and Planck states:

> It is impossible for any system to operate in a thermodynamic cycle and deliver a net amount of energy by work to its surroundings while receiving energy and heat transfer from a single thermal reservoir.

This means that thermodynamic losses from the heat engine to a cold reservoir are inevitable in order to be able to produce work. The limit when this loss term (Q_C) approaches zero is the absolute zero where the temperature (T_C) approaches zero Kelvin. Even the ideal (reversible) Carnot engine will have such losses, since operating at zero Kelvin is infeasible. In most cases, the heat sink for a heat engine is the environment, e.g., a river, a lake, or the surrounding air.

The heat engine model can also be used to develop an expression for the exergy of heat Q with a temperature T. The definition of exergy in Sec. 2.2 indicates that the temperature of heat must become equal to the temperature of the reference environment, T_0. When it comes to heat, thermal equilibrium simply means equal temperatures. In addition, this temperature change must be the result of reversible processes. Consider the heat engine to the right in Fig. 2.3 that is assumed to operate reversibly, i.e., as a Carnot cycle. If the cold reservoir is the environment with temperature T_0, then the exergy of an amount of heat Q with a temperature T can be found as the maximum work that can be obtained from this Carnot-type heat engine:

$$E^{th} = W_{max} = Q \cdot \left(1 - \frac{T_0}{T}\right) \tag{2.5}$$

This equation for the exergy of heat is only valid for temperatures above the reference temperature of the environment ($T \geq T_0$). For temperatures below T_0, the environment will act as a hot reservoir, while heat at a lower temperature ($T \leq T_0$) acts as a heat sink. Reorganizing the equations accordingly results in the following expression for exergy of heat below environmental temperature:

$$E^{th} = Q \cdot \left(\frac{T_0}{T} - 1\right) \quad \text{for } T \leq T_0 \tag{2.6}$$

This expression is the reason why we do not accept the often used relation: Energy = Exergy + Anergy. The split between exergy and anergy is at T_0. In the thermodynamic community, the interpretation of the part of the definition of exergy that relates to "equilibrium with the environment" is that the system should change to conform with the environment. Since, however, the environment is a large thermal reservoir, it could transfer heat to the heat sink at lower temperature and then the heat sink would increase its temperature until it becomes equal to the environment. The temperature of the environment remains unchanged because it is such a large reservoir. This means that "equilibrium with the environment" is achieved while at the same time producing work from the heat that is transferred from the environment to the heat engine.

The careful reader may also have observed that the expression for exergy of heat below T_0 increases exponentially when the temperature is reduced. For example, at half of the environmental temperature ($0.5 \cdot 298.15$ K = 149.075 K) the exergy of heat becomes equal to its energy value. This is in contrast to the situation above the environmental temperature T_0, where the exergy value is always less than the corresponding energy value of heat even at extremely (and unrealistically) high temperatures. Below T_0, when the temperature approaches zero Kelvin, the exergy approaches infinity, which is in line with the second law of thermodynamics. It would require infinite amounts of work to maintain a heat sink at temperatures close to zero Kelvin.

Alternatively, we could have focused on the heat that comes from the environment. This is illustrated in Fig. 2.4. In this case, the exergy expression can be formulated as:

$$E^{\text{th}} = Q_0 \cdot \left(1 - \frac{T}{T_0} \right) \tag{2.7}$$

Using this formulation, the exergy of heat that comes from the environment (Q_0) has an exergy value that approaches Q_0 when the temperature of the heat sink (T) approaches zero Kelvin. Figure 2.5 shows how the ratio between exergy and energy (heat) changes with temperature above and below environmental temperature T_0. This figure also shows the change in exergy with temperature when we focus on the heat supplied by the environment rather than the heat delivered to the heat sink. For this latter case, the ratio between exergy (work produced) and energy (heat input) for the reversible heat engine approaches 1.0. This means that work is produced from heat on a 1:1 basis, where the heat engine only interacts with one thermal reservoir (the environment). As stated above, this is in conflict with the Kelvin-Planck formulation of the second law of thermodynamics, and therefore indicates the thermodynamic limit of what can be done.

2.4.2 Exergy of Matter on Earth

Since this book is meant to be a toolbox for the practicing engineer, when we use the word *matter*, we refer to solids, liquids, gases, and supercritical fluids encountered in the process industries. We will often refer to process *streams* as the matter being subject to exergy evaluations. These streams are typically fluids consisting of either single or multiple chemical components. Solids will also be handled, however, with less emphasis in this chapter.

As indicated in Sec. 2.3 and in Fig. 1.2, mechanical exergy is the exergy of energy forms such as kinetic energy (related to velocity) and gravitational or potential energy (related to height above sea level). These exergy forms are not part of the tables in this book, since these energy forms (and their change) are of little importance in the process industries. If needed, however, these exergies can be easily calculated as follows:

- Kinetic exergy: $E^k = \frac{1}{2} \cdot m \cdot V^2$ assuming $V = 0$ is the stagnation state (see Fig. 2.2)
- Potential exergy: $E^p = m \cdot g \cdot z$ assuming $z = 0$ is the stagnation state (see Fig. 2.2)

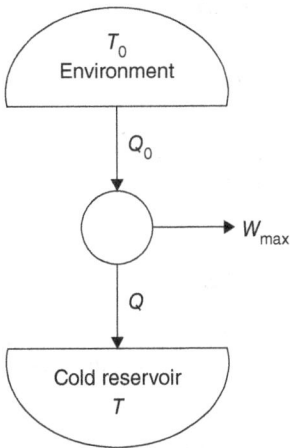

Figure 2.4 Heat engine illustrating the exergy of heat at temperatures below the environment.

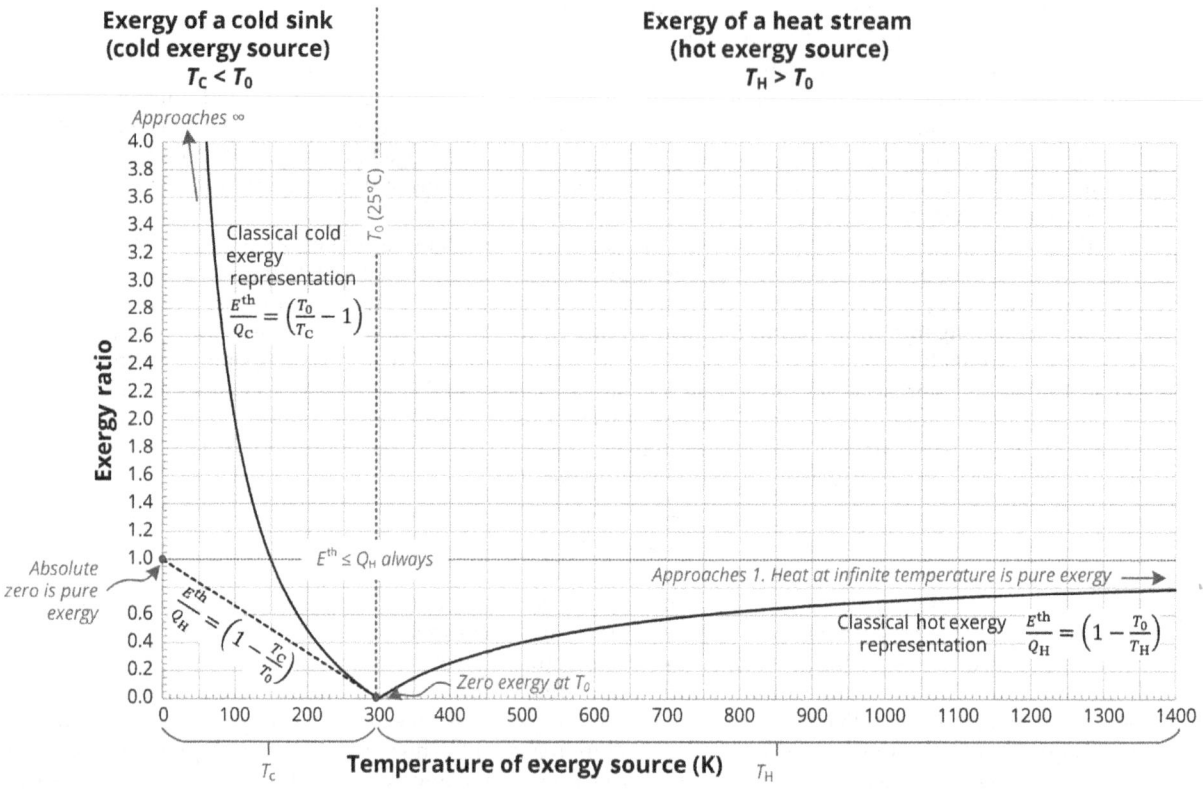

Figure 2.5 Ratio between exergy and heat as a function of temperature. The case when measuring the exergy of heat supplied by the environment is also plotted as the dashed line.

Since the expressions for kinetic and potential exergies are identical to their energy counterparts, it should be obvious that these two energy forms represent pure (i.e., 100%) exergy.

2.4.2.1 Thermo-Mechanical Exergy

With reference to Fig. 2.2, *thermo-mechanical* exergy is the maximum potential for work production between the stagnation state, which is equal to the original state when we ignore mechanical exergy, and the environmental state. Therefore, thermo-mechanical exergy is related only to the temperature, pressure, and phase (φ) of our system. An expression for thermo-mechanical exergy can be developed by combining the mass balance and the first and the second laws of thermodynamics while making proper assumptions.

The mass balance for an open, flowing system in the most general case (Fig. 2.1) can be formulated as follows:

$$\frac{dm_{cv}}{dt} = \sum_i m_i - \sum_e m_e \tag{2.8}$$

The first law of thermodynamics (energy balance) for an open, flowing system in the most general case can be formulated as follows (notice that *En* here is energy in kJ, not exergy):

$$\frac{dEn_{cv}}{dt} = \left[W_i^e + W_i^m + \sum_i Q_i + \sum_i m_i \cdot \left(h_i + \frac{V_i^2}{2} + gz_i \right) \right] - \left[W_e^e + W_e^m + \sum_e Q_e + \sum_e m_e \cdot \left(h_e + \frac{V_e^2}{2} + gz_e \right) \right] \tag{2.9}$$

Here, *h* is specific enthalpy (kJ/kg), while the two other terms in the parenthesis represent specific kinetic and potential energies. Similarly, the second law of thermodynamics (entropy balance) was introduced in Sec. 2.1 but is repeated here for convenience:

$$\frac{dS_{cv}}{dt} = \left[\sum_i \frac{Q_i}{T_i} + \sum_i m_i \cdot s_i \right] - \left[\sum_e \frac{Q_e}{T_e} + \sum_e m_e \cdot s_e \right] + S_{gen} \tag{2.10}$$

The following simplifying assumptions will be made in order to establish an expression for thermo-mechanical exergy. This does not mean that the resulting exergy will be an approximation; the values obtained are correct for the systems we describe:

- Steady state, which means that the three accumulation terms on the left-hand side of the equations will be zero.

- Changes in kinetic and potential energies are neglected.

- Only one incoming (*i*) and one exiting (*e*) mass stream.

- Only one heat flow enters the system at constant boundary temperature T_b and no heat flows exit the system. This is the heat added to the control volume, called Q_i.

- The sum total of all workflows (in minus out) is expressed as one workflow leaving the system. This is the work done on the environment by the control volume, called W_e^m, which is positive for any system that produces a net amount of work.

With the assumption of steady state and only one incoming and one exiting mass stream, the notation for mass flows can be simplified to $m_i = m_e = m$. The balance equations for energy and entropy can then be simplified to the following:

$$0 = Q_i - W_e^m + m \cdot (h_i - h_e)$$

$$0 = \frac{Q_i}{T_b} + m \cdot (s_i - s_e) + S_{\text{gen}}$$

(2.11)

When combining these two equations, we get the following expression for work:

$$W_e^m = m \cdot (h_i - h_e) - T_b \cdot m \cdot (s_i - s_e) - T_b \cdot S_{\text{gen}}$$

(2.12)

The final simplified system is shown in Fig. 2.6. The exergy of a stream in its original state (p_1, T_1, φ_1) is obtained by using reversible processes to achieve equilibrium with the environment, which means that the stream reaches its new state properties (p_0, T_0, φ_0). The equation developed for work above can now be used to find the exergy of the stream in its original state, while noticing that $h_1 = h(p_1, T_1, \varphi_1)$ and $h_0 = h(p_0, T_0, \varphi_0)$. Likewise, we have for the entropy that $s_1 = s(p_1, T_1, \varphi_1)$ and $s_0 = s(p_0, T_0, \varphi_0)$. Also notice that the boundary temperature T_b in this case is the environmental temperature T_0. Since exergy means maximum work produced (or minimum work required), the entropy production (S_{gen}) is zero reflecting the reversible nature of the processes involved in bringing the stream to equilibrium with the environment. The thermo-mechanical exergy (e^{tm}) of the stream in its original state 1 is then:

$$e^{\text{tm}} = \frac{W_{\text{max}}}{m} = (h_1 - h_0) - T_0 \cdot (s_1 - s_0)$$

(2.13)

What happens when the boundary temperature is below T_0? In Sec. 2.4.1, we argued that heat with a temperature $T < T_0$ can be used to produce work if heat is transferred from the environment to this heat sink. Consider the regasification of liquified natural gas (LNG), where large amounts of cold energy (or exergy) is released in order to produce pipeline natural gas at elevated pressures. This large heat sink can of course be used to produce work by having an organic Rankine cycle (ORC) or any other available heat engine operating between the environment and the LNG regasifier.

A similar discussion is related to a system in partial vacuum ($p < p_0$). This case is discussed by Kotas in his textbook on exergy analysis,[2] where an isothermal compressor is used to bring a system in partial vacuum to environmental pressure p_0. This, of course, requires consuming work instead of producing work, which would indicate a *negative* exergy value. However, negative exergy values are not meaningful, since exergy is the maximum work you *could* do. You can always do the trivial case, which is to do nothing with it at all. Doing no work is larger than negative work, so exergy can never be smaller than zero. Since the environment represents a thermal reservoir with very large capacity, our approach is to let air flow from the environment to the system in vacuum through a reversible turbine producing work and thereby indicating a *positive* exergy value. The phrase in the definition of exergy provided in Sec. 2.2 about bringing the system to "equilibrium with its surrounding nature" (here referred to as the environment) should not be understood in such a way that the system in vacuum

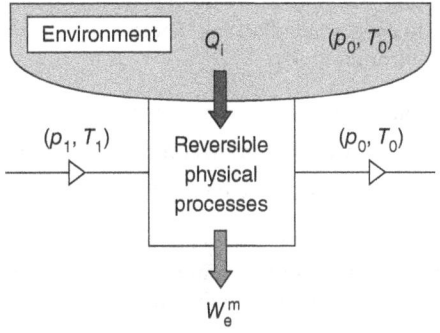

Figure 2.6 Illustration for the development of an expression for thermo-mechanical exergy.

has to be brought to environmental pressure p_0 by pressurizing it. Driving forces related to pressure, temperature, and composition between the system and the environment (with a very large size and capacity) can always be used to produce work while reaching equilibrium conditions.

With the understanding that thermo-mechanical exergy is always positive or zero, the relation between enthalpy and entropy must be handled in the following way:

$$\text{If } p_1 \geq p_0 \text{ then } e^{tm} = (h_1 - h_0) - T_0(s_1 - s_0)$$

$$\text{If } p_1 \leq p_0 \text{ then } e^{tm} = (2h_b - h_1 - h_0) - T_0(s_1 - s_0)$$

(2.14)

where h_b is the enthalpy of an intermediate point at (s_1, p_0) such that $h_b = h(s_1, p_0)$. Note that these exergy equations are general for any T, p, and phase of the fluid (liquid, gas, two-phase mixture, or supercritical). The first equation is the more traditional form of thermo-mechanical exergy and applies in most real cases. If used naively, it can result in (incorrect) underestimates and even negative exergy values in some conditions—for gases, these are usually vacuum conditions ($p < p_0$); for liquids, this usually occurs only in a very tight range of vacuum conditions and very close to T_0. Both equations are derived from a more general way of computing exergy, using a minimum theoretical work required to move away from the dead state approach.

2.4.2.2 Chemical Exergy

With reference to Fig. 2.2, *chemical* exergy is the maximum potential for work production between the environmental state and the dead state. In the dead state, the chemical components of the system are in equilibrium with the natural components of the environment. A formal definition of chemical exergy is provided by Kotas[2]:

> Chemical exergy is equal to the maximum amount of work obtainable when the substance under consideration is brought from the environmental state to the dead state by processes involving heat transfer and exchange of substances only with the environment.

Since the word *maximum* is used in this definition, it is understood that the processes involved in bringing the system to equilibrium with the environment (i.e., the dead state) are reversible. The mentioned equilibrium means that the chemical potentials are equal, which in some cases with gases means equal partial pressures.

The so-called environment must be properly described in detail (see Chap. 4 for an extensive discussion). For thermo-mechanical exergy, this was easy, just specifying $T_0 = 298.15$ K and $p_0 = 1$ atm $= 1.01325$ bar. For chemical exergy, we need to identify *reference substances* for each chemical component in the system. One of the requirements to be nominated as a reference substance is that it is in equilibrium with the rest of the environment. As an example, CO_2 could be a reference substance for carbon (C), even though current global warming is an effect of increased concentration of CO_2 in the atmosphere. CO, on the other hand, could not be a reference substance for carbon, since CO is not stable in the atmosphere. Another requirement for reference substances is that they should have the lowest chemical potential among the different candidate environmental substances. The last requirement is that the concentration of a potential reference substance must be known with adequate precision.

Reference substances are placed in one of the three spheres (atmosphere, hydrosphere, lithosphere) as a part of their reference state. Each element on the periodic table has exactly one corresponding reference substance. For example, the composition of the atmospheric reference state is provided in Table 4.1, where nine components are listed with their mole fraction, partial pressure, and molar chemical exergy. Two important assumptions are made when creating Table 4.1: The humidity of air at standard reference conditions is 70%, and the assumed concentration of CO_2 in the atmosphere is 338 parts per million (ppm). It is important to be aware of the sensitivity of the values in the tables of this book to the definition of the atmospheric reference state. Ertesvåg[3] discussed the sensitivity of chemical exergy for atmospheric gases and gaseous fuels to variations in ambient conditions, such as ambient temperature (from −30 to 45°C), atmospheric pressure (from 0.6 to 1.1 bar) and relative humidity (from 10 to 100%).

The path from the environmental state to the dead state is not straightforward and depends on the composition of the substance subject to exergy calculations. There are three different cases requiring different actions to reach equilibrium with the environment:

1. The component is a component of the atmosphere.

2. The component can be easily converted to components of the atmosphere by simple chemical reactions.

3. The component cannot be converted to components of the atmosphere.

For case 1, the component must be subject to compression or expansion until equilibrium with the same component in the atmosphere (e.g., the same partial pressure). For case 2, we use CH_4 as an example. This component can easily be converted to components in the atmosphere by the following (combustion) reaction:

$$CH_4 + 2O_2 \rightleftarrows CO_2 + 2H_2O$$

(2.15)

Similarly, ammonia (NH_3) can be converted to components of the atmosphere by the following reaction:

$$4NH_3 + 3O_2 \rightleftarrows 2N_2 + 6H_2O \tag{2.16}$$

The way Kotas[2] handles case 2 (here with CH_4 and NH_3 as examples) is an abstract reversible device referred to as the van 't Hoff equilibrium box. The way this device works is that the substance in question (here CH_4 or NH_3) is brought into the box at environmental conditions with respect to pressure and temperature (p_0 and T_0). Oxygen (referred to as a coreactant) on the other hand is compressed from its partial pressure in air (around 20.8 kPa) to p_0 (101.325 kPa). The energy involved in the conversion of a nonatmospheric substance to components of the atmosphere is referred to as enthalpy of *devaluation*. For case 3, we use chlorine as an example [Cl appears in standard form Cl_2 or in components such as HCl (hydrochloric acid)]. This substance cannot be converted to components of the atmosphere, but it is part of the hydrosphere and dealt with as such.

For substances in case 2, after the devaluation reaction, the reaction products are treated in the same way as components in case 1, i.e., with compression or expansion to partial pressures equal to the same components in the atmosphere. The above was an attempt to describe in words how a component in the environmental state can be changed to a situation of equilibrium with the environment. A more formal procedure to calculate the chemical exergy of various substances is as follows:

a) A list is made of all *elements* of interest (from the periodic table) for this book. For each of these elements, a representative elemental form must be selected. For nitrogen (N), the obvious choice is N_2 that exists in the *atmosphere*. For hydrogen (H), we cannot use H_2, since this component does not persist in nature normally. Instead, water vapor (H_2O) is selected as a natural component of the atmosphere. Some elements have their representative elemental form in other environments, such as the *hydrosphere* (e.g., Na uses the ion Na^+ as the representative elemental form) or the *lithosphere* (e.g., Ca uses $CaCO_3$ as the representative elemental form).

b) For each element (e.g., hydrogen H), we now have a list of chemicals that consist of an elemental form, which is how it appears in nature (e.g., H_2) and an associated reference substance (e.g., H_2O).

c) Since atoms consist of elements and molecules consist of atoms and elements, the energy levels of elements, atoms, and molecules are related. The same is of course the case with the exergy of elements, atoms, and molecules. A decision is therefore needed to set the scale of these exergy values. The choice made is to define the chemical bond (potential) part of the exergy of the reference substance in the *atmosphere* or *lithosphere* to be zero. The chemical exergy (e_i^{ch}) of the reference substance (in kJ/mol) then only contains concentration exergy, which can then be obtained from the following equation (that can be developed from basic thermodynamic equations):

$$e_i^{ch} = -R \cdot T_0 \cdot \ln(x_i) \tag{2.17}$$

where i is the reference substance, R is the universal gas constant (8.314 kJ/kmol·K), and x_i is the mole fraction of reference substance i in the atmosphere or lithosphere. For the hydrosphere, all of the reference species are ions, and e^{ch} is defined as exactly zero for each (including both the concentration and chemical bond component).

d) The exergy of all relevant elemental substances are found by solving a set of linear equations expressing the chemical exergy of reference substances.

$$e_i^{ch} = \Delta g_i^f + \sum_j n_j \cdot e_j^{ch} \tag{2.18}$$

Here, the chemical exergies of the reference substances i (e_i^{ch}) are known from the previous step. The Gibbs free energies of the reference substances (Δg_i^f), and the number of moles j of the elements in the reference substances n_j are also known.

e) Now that e^{ch} is known for all reference substances and all elemental forms, the chemical exergy of all other pure chemicals can be calculated directly using the same equation as listed in (d).

f) For gas mixtures, the chemical exergy consists of a weighted linear combination of the chemical exergy of the individual components of the mixture and a second term reflecting the exergy loss by diluting a pure component into a mixture:

$$e_{mix}^{ch} = \sum_i x_i \cdot e_i^{ch} + R \cdot T_0 \cdot \sum_i x_i \cdot \ln(x_i) \tag{2.19}$$

g) Liquid fuel mixtures (such as diesel blends) and solid fuels (such as coal, wood, etc.) are often complex or even heterogenous mixtures of chemical compounds that can be difficult or impossible to identify and classify exactly. As a result, chemical exergy cannot be determined directly from Gibbs free energy and must be estimated. This can be done by creating an empirical model predicting exergy as a function of the atomic composition or the heating value of the fuel by using data from pure chemicals. See Chaps. 9 and 10 for more details on these models.

2.4.3   Decomposition of Exergy Forms

Figure 1.2 indicates that exergy can be classified in a number of ways; some of these decompositions are of a fundamental nature, others are more for convenience or practical reasons. Examples of fundamental decompositions of exergy are listed next.

2.4.3.1   *Matter vs. Energy*

Exergy forms related to matter (tangible) and energy (abstract) on Earth are completely independent and can be decomposed. Examples of the latter are electricity, work, and heat as described in Sec. 2.4.1. These exergy forms are also independent and can be decomposed.

2.4.3.2   *Breaking Down Exergy of Matter*

Exergy related to matter on Earth can be decomposed into five forms: mechanical, thermo-mechanical (referred to as physical in some literature), chemical, electrochemical, and atomic. Mechanical exergy can be further decomposed into gravitational and kinetic exergy that are completely independent. One is related to velocity, the other to position in a gravitational field.

2.4.3.3   *Thermo-Mechanical Exergy vs. Chemical Exergy*

The only fundamental decomposition of exergy forms in this book is the one between thermo-mechanical exergy and chemical exergy, and the tables in Chaps. 4 through 10 distinguish between them in each instance.

2.4.3.4   *Thermal Exergy vs. Pressure Exergy*

Engineers sometimes like to decompose thermo-mechanical exergy further into thermal exergy and pressure exergy as indicated in Fig. 1.2. The former is a function of temperature, while the latter is a function of pressure. This decomposition is not reflected in the tables of this book, because it does not reflect any fundamental phenomena. Thermo-mechanical exergy is the maximum work that can be obtained when the system is changed from its original state (p_1, T_1, and φ_1) to the environmental state (p_0, T_0, and φ_0); see Fig. 2.7. A temperature-based exergy (e^T) could be envisaged by keeping the pressure constant (at p_1) while changing temperature from T_1 to T_0. Then the pressure-based exergy (e^P) would be obtained by keeping the temperature constant at T_0 while changing pressure from p_1 to p_0. However, one could also have used to opposite route; changing pressure first and then change temperature. The thermo-mechanical exergy would of course be the same, but the two components (temperature-based and pressure-based) will change. In fact, any path between (p_1, T_1, and φ_1) and (p_0, T_0, and φ_0) could be used. This clearly indicates that this decomposition has no fundamental meaning. The exception is for simplified cases, such as the ideal gas model, where this decomposition results in separate expressions for the temperature-based part and the pressure-based part.

2.4.3.5   *Bond Exergy vs. Concentration Exergy*

Chemical exergy may also be decomposed into bond exergy and concentration exergy as indicated in Fig. 1.2. A simplified way to explain this is that the former is related to chemical reactions and chemical potential, while the latter is related to separation and mixing causing changes in composition. This breakdown is important in the definition of the environmental reference chemical exergies as noted previously. However, it is complicated by

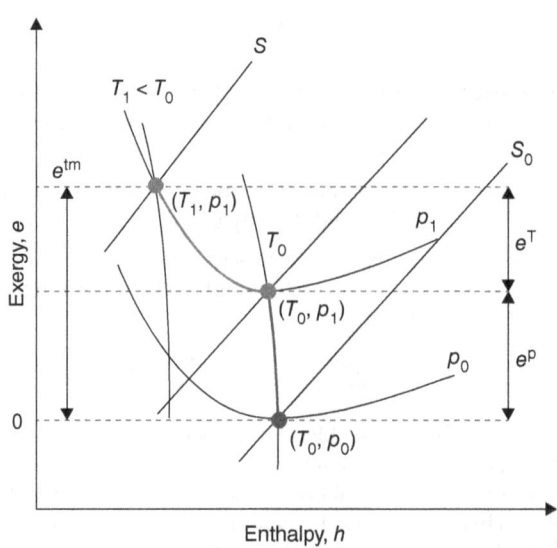

Figure 2.7   A practical but not fundamental decomposition of thermo-mechanical exergy.

the fact that the same chemical can exist in different phases at 25°C and 1 atm on the Earth (e.g., water), and the corresponding Gibbs free energies of formation at the reference temperature and pressure will be different for the different phases. That means that the phase of a chemical affects exergy aspects of both chemical potential and concentration. Because of this complication, it is very difficult to break down exergy into bond exergy and concentration exergy in the general sense and so they are not listed as such in the tables.

However, their relative contribution to the total chemical exergy is evident. For example, consider octane (a common component of gasoline) in Table 7.1. The chemical exergy of octane in the liquid state is about 5415 kJ/mol, while in the gas state it is only slightly higher at 5423 kJ/mol. Clearly, the chemical bond portion of the exergy has much more of an impact than its phase. However, as shown in Chap. 6, the chemical exergy of water vapor (about 9.49 kJ/mol) is an order of magnitude higher than liquid water (about 0.89 kJ/mol). Here, the phase has a far greater impact than the chemical bond exergy, largely because water has almost no chemical potential at all.

2.5 Exergy Efficiency

Exergy numbers are used for two main purposes in the process industries, both related to analyzing the use of energy in such processes: To establish exergy balances that can be used for example to draw Grassmann diagrams, and to calculate exergy efficiencies for individual pieces of equipment, for processes or plants and even industrial sites or parks. This section will focus on exergy efficiency, where a number of different versions have been proposed in the literature.[4] The two main categories of exergy efficiencies are the *input-output* efficiency and the *consumed-produced* efficiency.

Considering first the input-output exergy efficiency, which as the name indicates performs book-keeping of exergy flows crossing the system boundary (dashed line in Fig. 2.8). Exergy enters and leaves the system in the form of material and energy flows. While energy is conserved (as discussed in Sec. 2.1), exergy is destroyed due to irreversibilities in the system. This exergy destruction is referred to as internal losses. In addition, there are commonly waste streams containing some exergy that are referred to as external losses. One example is exhaust gas from combustion (e.g., from a fired heater), where the temperature is regarded to be too low to make it worthwhile to recover the energy and exergy of the effluent stream.

Based on Fig. 2.8, a typical input-output exergy efficiency would be the following:

$$\eta_{\text{in-out}} = \frac{\sum \text{Exergy out}}{\sum \text{Exergy in}} = 1 - \frac{\sum \text{Exergy destruction} + \sum \text{Exergy losses}}{\sum \text{Exergy in}} \tag{2.20}$$

This efficiency is in line with the so-called second law efficiency defined in Sec. 1.2.2. What should be noticed from this exergy efficiency is that even when recovering all exergy losses crossing the system boundary, there will still be internal losses due to irreversibilities that make the efficiency often considerably less than 100%, depending of course on the process.

Consider next the consumed-produced exergy efficiency, focusing more on the purpose of the system (equipment or total process). The term *useful output* is often used in these efficiencies. Whatever efficiency is used, it is important to ask the question about what the purpose of the investigation is. Book-keeping has limited value when looking for opportunities to improve the system. Consider for example the case where a hydrocarbon stream passes a piece of equipment such as a heat exchanger or a compressor. Since chemical exergy is orders of magnitude larger than thermo-mechanical exergy, the chemical exergy will flow through the equipment without changing its value, but it will dilute the effect of thermo-mechanical exergy (Chap. 3 illustrates some examples of this). Input-output as well as consumed-produced exergy efficiencies will approach 100% in such cases, no matter how inefficient the equipment is. This is why Brodyansky et al.[5] introduced the term *transit exergy* and used it to propose the following exergy efficiency:

$$\eta_{\text{transit}} = \frac{\sum \text{Exergy out} - \text{Transit Exergy}}{\sum \text{Exergy in} - \text{Transit Exergy}} \tag{2.21}$$

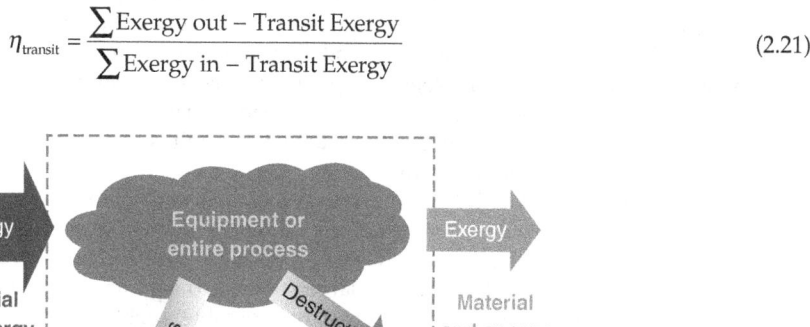

Figure 2.8 Exergy flows entering and leaving the system to be analyzed.

A somewhat different approach was taken by Marmolejo-Correa and Gundersen,[6] who carefully identified exergy sources and exergy sinks and thereby defined the exergy transfer effectiveness (ETE) with focus (as the name indicates) on the transfer of exergy in the system. This will, for example, completely avoid the problem of exergy that is in transit through the system. The ETE is defined as follows:

$$ETE = \frac{\sum Exergy\ sinks}{\sum Exergy\ sources} \tag{2.22}$$

The exergy sources are defined as process streams with a decrease in exergy, while exergy sinks are defined as process streams with an increase in exergy. Perhaps counterintuitive, a compressor (even though it consumes power) is an exergy source. This can be explained by the fact that the compressor will increase the exergy of the stream being compressed. Likewise, an expander (that produces power) is an exergy sink. This can be explained by the fact that the expander will decrease the exergy of the stream being expanded.

Among all the proposed exergy efficiencies in the literature, the one that comes closest to the ETE is the $\eta_{transit}$ proposed by Brodyansky et al.[5] that was mentioned above. This has been demonstrated in case studies related to the liquefaction of natural gas.[4]

2.6   Advantages and Limitations with the Exergy Concept

The concept of exergy allows us to treat different energy forms on a common basis. This is valuable, since different energy forms have different qualities, and comparing exergy is much better than comparing or adding different energy forms according to their energy values in the form of J, kJ, etc. The analogue would be to compare or add apples and bananas. While a number of energy forms such as kinetic energy, potential energy, work, electricity, etc. have exergy values close to their energy values (almost pure exergy), heat is the energy form that stands out with temperature as its quality parameter. The quality (or exergy) of heat depends strongly on its temperature as indicated in Sec. 2.4.1.

The major problem or disadvantage with the exergy concept is that it reflects the maximum work that can be obtained by reversible processes. Of course, such processes do not exist, and most real processes are quite far away from the performance of an ideal, reversible process. Using the Carnot model, as indicated in Sec. 2.4.1, will result in a gross overestimation of the quality (or true value) of heat. Most processes (heat engines) that convert heat to work have thermal efficiencies that are considerably less than the Carnot factor. As a result, the exergy of heat is a very optimistic estimation of its ability to produce work.

Consider an example with combustion of a fuel. A typical so-called adiabatic flame temperature (depending on the fuel and amount of excess air) is 2055°C or 2328.15 K. Assuming $T_0 = 298.15$ K, the Carnot efficiency would be:

$$\eta_C = 1 - \frac{T_0}{T} = 1 - \frac{298.15}{2323.15} = 87.2\% \tag{2.23}$$

The thermal efficiency of some common heat engines in the process industries can be listed as follows: gas turbine (35–40%), steam turbine (36–42%), and combined cycle (50–60%). All these efficiencies are far below the Carnot efficiency, which is overly optimistic. When calculating chemical exergy, we also rely on the assumption of reversibility. Chapter 3 contains more examples, which explore this in more detail.

As indicated in Fig. 2.8, we have two main classes of exergy losses. Internal losses are caused by irreversibilities, while external losses are related to waste streams and subject to economic evaluations whether or not to recover these exergy losses. For the internal losses, one problem in using exergy analysis to identify thermodynamic inefficiencies (exergy destructions) is that industrial processes are heavily integrated. This means that exergy losses may appear in other places (units or equipment) than where they actually were caused. This problem has been addressed by Morosuk and Tsatsaronis[7] who decomposed exergy destruction into unavoidable/avoidable and endogenous/exogenous parts. *Endogenous* exergy destruction in one piece of equipment or unit is obtained when all other units in the system are ideal. This part is then exergy losses caused by imperfections in the unit itself. The *exogenous* exergy destruction in a unit is caused by irreversibilities in the remaining units and the structure of the overall system.

With reference to Fig. 2.8, external exergy losses caused by not recovering the exergy content of waste streams from the process are obviously *avoidable*; it all narrows down to economy whether or not to avoid these exergy losses. The internal exergy losses, referred to as exergy destructions, are caused by irreversibilities due to heat transfer with temperature difference larger than zero, spontaneous chemical reactions, pressure reduction in valves, mixing, etc. Some of these losses are avoidable, such as replacing a valve with a turbine that produces power while reducing the pressure of the stream. However, some exergy losses are *unavoidable*, but often it is the economics that defines the border between avoidable and unavoidable.

Reducing temperature-driving forces in heat exchangers reduces exergy losses, but investment cost due to the need for larger heat transfer area will increase. Operating chemical reactions at lower reaction rates will reduce exergy losses at the expense of a larger reactor. In distillation, where the purpose is to separate chemical components, one should of course avoid mixing. For binary distillation, it is generally optimal to choose a feed tray to the column such that the composition of the feed closely matches the composition of the mixture on the tray, specifically to avoid this extra mixing. For cases with three or more chemicals, this may not be possible using classic binary columns. Instead, it is often better to design systems that use prefractionation approaches (such as dividing wall columns). Prefractionation approaches partially separate a feed stream into "sloppy" cuts in a first column or column section and then feed those sloppy cuts to different locations of a second column or section to reduce the mixing effect and thereby reduce exergy losses.

References

1. Szargut J. International progress in second law analysis. *Energy* 5:709–718 (1980).
2. Kotas TJ. *The Exergy Method of Thermal Plant Analysis*. 3rd ed. Exergon Publishing Company UK Ltd.: London (2012).
3. Ertesvåg IS. Sensitivity of chemical exergy for atmospheric gases and gaseous fuels to variations in ambient conditions. *Energy Conversion and Management* 48:1983–1985 (2007).
4. Marmolejo-Correa D, Gundersen T. A comparison of exergy efficiency definitions with focus on low temperature processes. *Energy* 44:477–489 (2012).
5. Brodyansky VM, Sorin MV, Le Goff P. *The Efficiency of Industrial Processes: Exergy Analysis and Optimization*. Elsevier: Amsterdam (1994).
6. Marmolejo-Correa D, Gundersen T. A new efficiency parameter for exergy analysis in low temperature processes. *International Journal of Exergy* 17(2):135–170 (2015).
7. Morosuk T, Tsatsaronis G. Advanced exergetic evaluation of refrigeration machines using different working fluids. *Energy* 34(12):2248–2258 (2009).

CHAPTER 3

Using Exergy Data in Analyses

One of the most useful ways to understand a process is through exergy analysis. This can make very clear where exergy is being created or lost, where inefficiencies or bottlenecks might occur, or what should be targeted for improvement. How the analysis is conducted is up to the engineer; the methods, metrics, and visualizations should be tailored for each specific situation. In this chapter, we will share some examples that illustrate some useful analysis techniques.

3.1 Exergy Flows

When considering an energy system, such as a chemical plant, thermodynamic cycle, or even a global oil transportation network, it can be incredibly useful to visualize how exergy flows through a system. A very useful place to start is the creation of an exergy flow diagram, which is a flow chart that depicts how exergy flows between different parts of a process. While the diagram itself is often very illuminating, the information used to create the diagram is the same information used to compute all sorts of energy metrics, like exergy efficiency or exergy loss, which is very useful on its own.

Let's first consider a simple example of a factory that needs to produce medium pressure steam at 10 bar and 300°C, shown in Fig. 3.1. The main components are a boiler, two blowers, a heat exchanger, and a pump. In the boiler, fuel (conventional US pipeline natural gas, 93.9 mol% methane, 3.2% propane, 0.7% ethane, 0.4% butane, 0.8% nitrogen, and 1% carbon dioxide, see Table 8.1) and air (both at 25°C and 1 atm = 1.01325 bar) are both blown into the combustion section of the boiler. The fuel combusts, producing heat and exhaust gas. A pump is used to pump liquid water (also at 25°C and 1 atm = 1.01325 bar) up to 10.2 bar and inject it into the heat exchanger. The heat exchanger is inside the boiler, so that heat transfers from the combustion section to the water, creating steam. After a little pressure drop in the heat exchanger, the steam leaves at 300°C and 10 bar. The combustion exhaust leaves at 150°C and 1.05 bar and contains unspent air as well as CO_2 and H_2O produced from natural gas combustion. The corresponding stream conditions are shown in Table 3.1, along with the chemical, thermo-mechanical, and conventional exergy values that correspond with each stream.

3.1.1 Computing the Exergy Flows

First, gather enough information about each stream to determine its exergy. For this example, we used Aspen Plus, a computer simulation package.[*] Aspen Plus uses computer models of process units like pumps, compressors, reactors, and heat exchangers to compute mass and energy balances. Many other similar software tools exist, like ProMax, Pro/II, Unisim, Chemcad, Hysys, and more. Measured plant data is even better if you can get it. The key is to identify enough information so that you can estimate its exergy. For thermo-mechanical exergy, the temperature and pressure are usually the easiest place to start. For chemical exergy, the composition of the stream is critical. The conventional exergy is simply their sum: $E^{conv} = E^{tm} + E^{ch}$.

Let's start with air in this example. The thermo-mechanical and chemical exergy of air at atmospheric conditions (stream 3) are both zero. Air at a slightly higher pressure (1.1 bar) and temperature (34.8°C) has a small $e^{tm} = 7.2$ kJ/kg, which was interpolated from Table 5.2 using the methods shown at the end of Chap. 1. Multiply that by the flow rate of 167.8 kg/min and you get a total exergy flow of $E^{tm} = 1.21$ MJ/min (note we use capital letters for flows, and lower case for values on per-mass or per-mole basis in this book). Alternatively, if you have access to a process simulator or a thermodynamics package, the simulator may provide you with the enthalpy and entropy of streams 3 and 4 directly, which in turn would allow you to compute the thermo-mechanical exergy yourself.

---

[*]Interested readers should see *Learn Aspen Plus in 24 Hours*, 2nd edition, by Thomas A. Adams II, published by McGraw-Hill Education. If you would like to download the Aspen Plus files used in this chapter, you can get it at https://PSEcommunity.org/LAPSE:2023.0004.

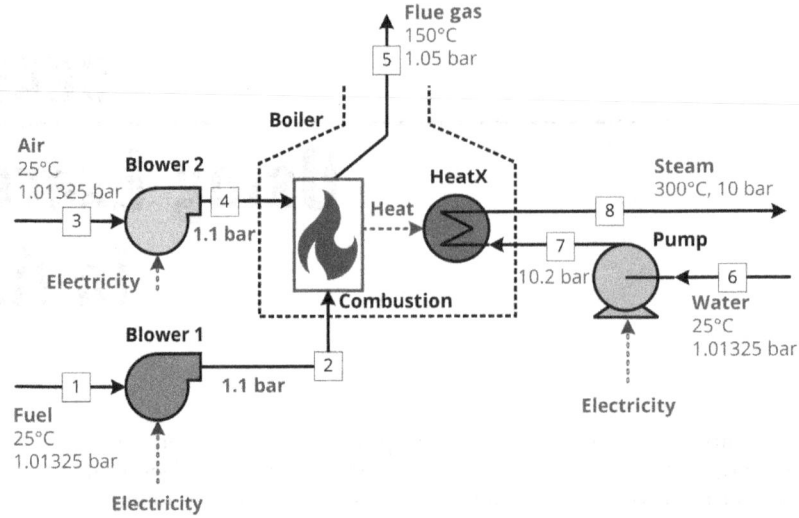

FIGURE 3.1 A natural gas combustion process to produce medium pressure steam.

#	T (°C)	P (bar)	Composition	F (kg/min)	E^{tm} (MJ/min)	E^{ch} (MJ/min)	E^{conv} (MJ/min)
Mass Flows							
1	25.0	1.013	US pipeline natural gas	2.0	—	99.18	99.18
2	32.7	1.10	US natural gas (higher P)	2.0	0.026	99.18	99.20
3	25.0	1.013	Air	167.8	—	—	—
4	34.8	1.10	Air (higher pressure)	167.8	1.21	—	1.21
5	150	1.05	74.7 wt% N_2, 18.2% O_2, 3.2% CO_2, 3.8% H_2O	169.8	7.30	1.52	8.82
6	25.0	1.013	Liquid water	24.7	—	1.23	1.23
7	25.6	10.2	Compressed water	24.7	0.023	1.23	1.25
8	300	10.0	Steam	24.7	23.0	12.98	35.97

Stream	Energy (MJ/min)	Exergy E^{th} (MJ/min)	Exergy E^W (MJ/min)
Energy Flows			
Blower 1 electricity	0.033	—	0.033
Blower 2 electricity	1.68	—	1.68
Pump electricity	0.090	—	0.090
Heat transfer at 350°C in boiler	75.08	39.16	—

TABLE 3.1 Process stream conditions for the steam generation example using natural gas

Liquid water at atmospheric conditions (stream 6) has no thermo-mechanical exergy, but the chemical exergy is 49.83 kJ/kg from Chap. 6, giving a total flow of E^{ch} = 1.23 MJ/min. This chemical exergy is the same when water is compressed, but in steam form its chemical exergy is higher at 9.4822 kJ/mol (see Table 4.1), which becomes E^{ch} = 13.0 MJ/min. The thermo-mechanical exergies of streams 7 ($e^{tm} \approx$ 0.94 kJ/kg) and 8 (e^{tm} = 931.99 kJ/kg) can be determined by interpolation from Tables 6.2 and 6.6, respectively. Note that the thermo-mechanical exergy of steam is larger than its chemical exergy, which makes sense since steam is not a fuel.

The fuel (stream 1) is simple because e^{tm} is zero at atmospheric temperature and pressure, and e^{ch} is easily found in Table 8.1 (854.51 kJ/mol). However, e^{tm} for stream 2 is less trivial. A quick check of Table 8.1 shows that natural gas is mostly methane, but also contains CO_2, N_2, ethane, and a small amount of higher hydrocarbons. We have a few options here. One is to read the e^{tm} from the charts for all those chemicals found in Chaps. 5 and 7, and take the weighted average. Another is to use a computer program to determine the enthalpy and entropy of stream 2 and compute e^{tm} directly. A shortcut method might be just to use the thermo-mechanical exergy figure for methane alone, since natural gas is mostly methane anyway, and its thermo-mechanical exergy is likely to be

Chemical	Mass Fraction	e^{tm} (kJ/kg) Estimated from Chaps. 5, 6, and 7	Mole Fraction x_i	e^{ch} for the Pure Chemical (kJ/mol) (Table 4.1)	$x_i e_i^{ch}$	$x_i \ln x_i$
N_2	0.747	25	0.757	0.670	0.5071	−0.2106
O_2	0.182	22	0.161	3.925	0.6335	−0.2944
CO_2	0.032	21	0.021	19.817	0.4103	−0.0803
H_2O	0.038	517	0.061	9.482	0.5748	−0.1699
			e_{mix}^{ch} (kJ/mol)	0.254		
	e_{mix}^{tm} (kJ/kg)	43	e_{mix}^{ch} (kJ/kg)	8.93		
			$e_{mix}^{conv} = 51.9$ kJ/kg			

TABLE 3.2 Intermediate calculation data for determining e^{tm} and e^{ch} for stream 5

relatively small and have little impact on the analysis. A careful reading of Fig. 7.11 (pure methane) gives $e^{tm} \approx 13$ kJ/kg, which is close enough to what you would get using a very rigorous method (12 kJ/kg), especially because it contributes so little to the overall conventional exergy. Our analysis will use the shortcut number.

For the flue gas (stream 5), the thermo-mechanical exergy can be computed the same way it was for stream 2. However, both kinds of exergies are trickier here. For this mixture, the thermo-mechanical exergies of its individual components are quite different, and the mixture is too diverse to assume a single chemical as model. A weighted sum of the component thermo-mechanical exergies is necessary, but they can be found in the tables and figures in the book. The chemical exergy for the flue gas is the trickier part. All four chemical exergies are easily found in the tables, but the challenge is that chemical exergy includes mixing, so the chemical exergy of the mixture is not a simple mass-weighted or mole-weighted average. Instead, the chemical exergy computation of a gas mixture needs to use the equation for gas mixtures discussed in the beginning of Chap. 8. A summary of this calculation is found in Table 3.2, with a final conventional exergy of $e_{mix}^{conv} = 51.9$ kJ/kg.

Finally, the exergy of the heat can be computed relatively easily using the formulas discussed in Sec. 2.2. In this example, we assume that all of the heat transfer occurs at an average temperature of 350°C to keep things simple. That gives $E^{th}/Q_H = 0.522$, which when multiplied by the heat duty Q_H of 75.08 MJ/min gives $E^{th} = 39.16$ MJ/min.

Note that the exergy flow of electricity is equal to its energy flow, since electricity is pure exergy.

3.1.2 Plotting the Exergy Flows

You can plot the conventional exergy flows in an exergy flow diagram. One example is Fig. 3.2, in which the width of each line corresponds to the magnitude of the conventional exergy. It becomes immediately clear that the fuel is the primary source of exergy, and that steam is the primary exergy product. The flue gas has a non-trivial amount of exergy. With the current design, this exergy is not used for anything useful since it is vented directly to the atmosphere. The other streams (electricity, water, and air) contribute very little. In fact, the lines are

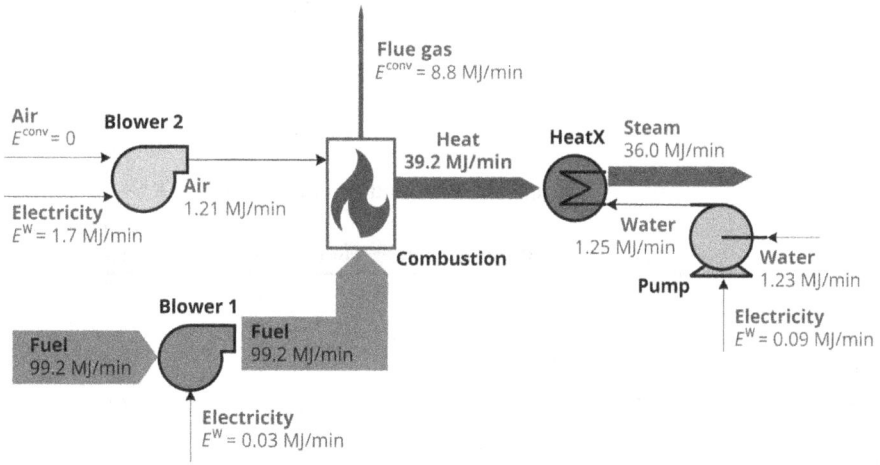

FIGURE 3.2 Exergy flows of material and energy streams for the example problem. The line width is proportional to the magnitude of the flow.

thicker than they should be, otherwise they would be too thin to see at all. This tells us that the shortcut approximations for thermo-mechanical exergies of air, water, and natural gas are adequate because they are so small in comparison to the rest of the flows such that very rigorous computations were not necessary for those quantities.

3.2   Exergy Bottlenecks and Unit Operation Analysis

The flow diagram shows us the big picture, but it is useful to consider exergy on a unit-by-unit basis. In Table 3.3, the exergy balances around each of the five unit operations are shown. The exergy loss is simply the exergy in minus the exergy out. This is a great check: the total exergy loss should always be zero or positive. If it is not, you have made a mistake by either leaving something out, or there is some problem with the way you have managed your model or estimations. For example, maybe there are kinds of exergies that you did not include in the metric but should have, even ones not covered in this book (such as exergy stored in high elevations—something important to hydroelectric power plants and pumped hydroelectric energy storage).

The first section in Table 3.3 shows the total exergy, which is the conventional exergy of material plus the exergy of heat or electricity flowing into or out of a block. It also has the total exergy efficiency of the unit, which for this example is defined simply as the total exergy out divided by the total exergy in. The total exergy destruction, expressed as a percentage, is 100% minus this number. The second table section shows the same information, but it does not include the chemical exergy of any of these components, just thermo-mechanical exergy, heat exergy, or electrical exergy. We called the corresponding metrics the "thermal" exergy efficiency and thermal exergy destruction, but you might call it something else. As the engineer, it is up to you to create the metrics you want in your exergy analysis, and name and define them accordingly, as there is no standard way of defining these terms.

Let's look at blower 1. The total exergy efficiency is essentially 100% for this unit, which at first glance would make it seem like a perfectly efficient blower, with almost no losses at all. But the chemical exergy component vastly outweighs the other kinds of exergies. Because there is no chemical reaction, it is not really meaningful to consider it when examining the exergy efficiency of this unit operation. Instead, the thermal exergy efficiency is much more useful. For blower 1, the value is 78.7%, meaning that 21.3% of the thermo-mechanical exergy is being lost in the process.

Blower 2 is similar to blower 1 in terms of thermal exergy efficiency (72%), which is equal to its total exergy efficiency. This is because there is no chemical exergy at all of the fluid inside it (air). The units are fundamentally very similar, and yet there is a vast difference in total efficiencies between blower 1 and blower 2. This is another clue to the analyzing engineer that the total exergy efficiency metric is not the right metric to use for this kind of equipment or situation. The same goes for the pump. In all three cases, it is not appropriate to include the chemical exergy in the analysis because there is no chemical reaction or phase change going on.

Now, consider the combustion step. This is a step in which chemical exergy of the fuel is converted into thermal exergy of the products (heat and the high temperature of the flue gas). This means that total exergy is the more appropriate metric here. The exergy destruction is quite high at 52.2%, which we will discuss later. Look at the thermo-mechanical exergy lost for this unit—it is negative. This is not a mistake since the math is correct, it is

Total Exergy ($E^{conv} + E^{th} + E^{W}$)					
	In (MJ/min)	Out (MJ/min)	Lost (MJ/min)	Total Exergy Efficiency	Total Exergy Destruction
Blower 1	99.21	99.20	0.007	100.0%	0.007%
Blower 2	1.68	1.21	0.469	72.0%	28.0%
Pump	1.32	1.25	0.067	94.9%	5.1%
Combustion	100.4	48.0	52.4	47.8%	52.2%
Heat Exchanger	40.4	36.0	4.4	89.0%	11.0%
Thermo-Mechanical, Heat, and Work Exergy ($E^{tm} + E^{th} + E^{W}$)					
	In (MJ/min)	Out (MJ/min)	Lost (MJ/min)	Thermal Exergy Efficiency	Thermal Exergy Destruction
Blower 1	0.033	0.026	0.007	78.7%	21.3%
Blower 2	1.68	1.21	0.469	72.0%	28.0%
Pump	0.090	0.023	0.067	25.7%	74.3%
Combustion	1.2	46.5	−45.2	3765%	−3665%
Heat exchanger	39.2	23.0	16.2	58.7%	41.3%

TABLE 3.3   Exergy losses and efficiencies for each unit in the example problem. See the text for explanation, as some of the numbers (shown in italics) are not quite meaningful in this context.

just not a very meaningful metric for this circumstance. To interpret this term properly, do not conclude that 45.2 MJ/min of exergy are created from nothing. Rather, this indicates that there is another source of exergy that you did not include in this metric but should have (chemical exergy in this case).

The heat exchanger is more interesting because both kinds of exergies are relevant here. Not including the chemical exergy may make sense at first because water is not undergoing a chemical reaction. However, water is in fact undergoing a phase change, and the phase change results in different chemical exergies, since the state of matter influences the chemical exergy (see Chaps. 4 and 6). Therefore, this particular heat exchanger actually does impact the chemical exergy, and so the total efficiency metric is probably more important, although both versions of efficiency can provide useful information to the engineer.

Now that we understand what metrics to look at, we can look at each unit and try to see if there are any good places for improvement. The pump has the most exergy destruction at about 74% (using the more appropriate metric). This unit has the worst efficiency, but is it the bottleneck? In this case, no, because the absolute value of the exergy lost is so trivial compared to the rest (you can tell from the thinness of the lines in Fig. 3.2). The combustion step also has very high exergy destruction at about 52%, and the loss is quite large too. In fact, almost all of the exergy losses originate from this step, and so this is clearly the bottleneck.

What can be done to reduce the exergy losses in the combustion step? Well, not a lot actually. One thing we could do is reduce the air feed to the combustion step. Currently, it is in very large excess, and results in an average combustion temperature before heat transfer of about 550°C (not shown in the above data, but this can be determined from the Aspen Plus simulation). It could be possible to reduce the air supply somewhat so it does not have as much excess air but yet still completely combust the fuel. This would result in a hotter combustion zone, and a higher average temperature of heat transfer to the heat exchanger. That exergy flow would get higher (though the actual heat duty would stay the same), and then the total exergy efficiency of this step would go up. However, that would not actually be that useful for the system as a whole, since you can design the system such that you produce the same amount and quality of steam using the same amount of fuel, whether your average combustion temperature was 550°C or else something higher. There is not actually a lot you can do about this unit to improve its exergy efficiency—its losses are fundamental to the nature of combustion itself. Improvements might better be made from bigger-picture system changes, which we will discuss in Sec. 3.3.

The analysis for the heat exchanger is similar. Even though it has about 11% exergy destruction, at our current level of design, all of the energy is used for its intended purpose—our example assumes no heat losses to the atmosphere (e.g., leaks in the insulation). There is little you can do to change the design of the unit itself to improve this under these design parameters too; it is a passive metal structure with no moving parts. Rather, the exergy destruction originates from the fundamental nature of heat transfer.

The machinery (pump and blowers) have high levels of exergy destruction, however. This is in large part due to the inefficiencies in the equipment itself. The models used to represent the equipment factor this into the simulation, and so the exergy destruction reflects the assumed mechanical or isentropic inefficiencies used in the model. This also means that these unit operations have room for improvement, as more efficient machinery could be chosen, resulting in lower exergy destruction. As the engineer, you could decide whether this is worth the effort. In our case, you could consider finding a better blower design to make blower 2 more efficient, but the net benefits to the process would be small. Blower 1 and the pump have such trivial contributions to exergy losses that it is probably best not to spend much time on them.

3.3 Exergy and Process Design

Let's take another look at the big picture. Figure 3.3 shows a Sankey diagram of the process (also called a Grassmann diagram). This is a lot like an exergy flow diagram, but it is usually stacked, and has more conceptual groupings. In this case, electricity has all been grouped together. All of the exergy losses, including exergy lost through the flue, are combined into one category as well, and shown in Fig. 3.3. In a figure like this, where the lost category is included as a final stage, the total value of the exergy flows leaving the first stage should equal the total value of the exergy flows into the final stage. Immediately, it becomes clear that the fuel exergy contributes the most to the input, but that most of the exergy is ultimately lost.

Let's look at the total system exergy efficiency using the information we found in the previous section and displayed nicely in Fig. 3.3. In the spirit of Chap. 1, this metric for the case study is defined as

$$\text{Total system exergy efficiency} = \frac{\text{Exergy of useful products}}{\text{Exergy of inputs}} \qquad (3.1)$$

In this case, the only useful product is the steam. Just like in the pool heater example in Chap. 1, the hot flue gas is just lost to the atmosphere, so we should not count it as a useful product. However, it is not entirely clear what portion of the exergy of that steam we should be using in our numerator. If the steam is going to be used as a heating utility, and used only to deliver heat, then maybe you might only consider e^{tm} to be relevant. However, because it is going through phase change, we still want the e^{ch} component, since the phase change is considered in chemical exergy. That means we might as well use e^{conv} for the steam product in the numerator. As for the denominator, we should be including e^{ch} for liquid water as well. Even though its chemical exergy is small, it

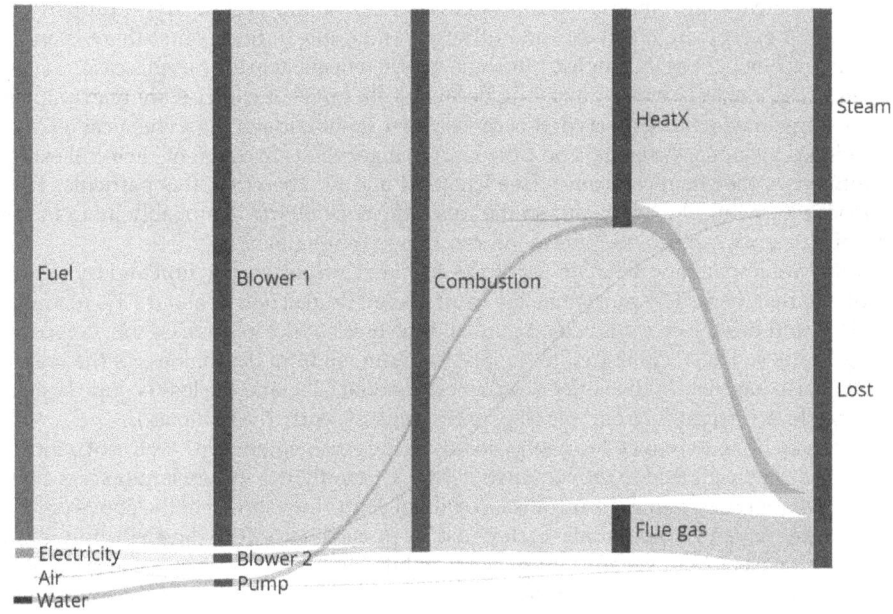

FIGURE 3.3 A Sankey diagram of conventional exergy flows for the example process, grouped by flow type.

needs to be considered because we are considering phase change and e^{ch} in the steam product too. That gives us a final definition of:

$$\text{Total system exergy efficiency} = \frac{E_{\text{steam}}^{\text{conv}}}{E_{\text{fuel}}^{\text{conv}} + E_{\text{water}}^{\text{conv}} + E_{\text{elec}}^{\text{W}}} \tag{3.2}$$

and a value of 35.2%.

Since in the previous section we already determined that there was not much we could do to each individual piece of equipment to improve the exergy efficiency, maybe there are some bigger picture changes we could be making. Is the design itself poor? Or is the idea of making 300°C steam from natural gas itself the problem? Maybe there are wastes that we could be capturing?

Let's start with the wastes. The only waste stream that we could do something with is the flue gas, which has 8.82 MJ/min of exergy that is currently sent to the atmosphere and lost. This in theory could be used for something. In fact, it is a nice upper bound: exergy tells you the maximum work that could be produced using it (against the environment). If we were to have this magic, perfect engine, our total system exergy efficiency would rise to 43.8%. That kind of improvement is worth pursuing. Of course, the real system would not have a perfect engine, but even if we could only get half of that value, it could be worth it.

One option might be just to recycle as much of that heat as possible by constructing an additional heat exchanger that preheats the water fed to the boiler. Although flue gases are traditionally kept quite hot to avoid condensing water (and the condensed acid gases that go with it), modern furnaces are now able to handle exhausts at lower temperatures. What if we were able to do that in our system?

Figure 3.4 shows a possible design change that would use an economizer, with corresponding stream conditions in Table 3.4. In this particular design, a countercurrent heat exchanger could preheat the compressed liquid water to 140°C while bringing the flue gas down to 63°C. This works well because the compressed liquid at 140°C remains a liquid and the flue gas is hot enough to still be all gas phase, since the relative humidity is low thanks to plenty of excess air. The new stream conditions are shown in Table 3.4. For the same fuel rate and air, this new design produces about 20% more steam. The additional pump electricity required is trivial. The new total system exergy efficiency has risen to 43.4%.

Now the system is considerably better. Comparing stream 5 with 5B in Table 3.4, you can see that this new design used over 90% of the available thermo-mechanical exergy. A condensing heat exchanger that captured the chemical exergy remaining in the flue gas (by condensing water in the gas phase) could go further, but not much. If we recalculate the new maximum theoretical system exergy efficiency that could be achieved by capturing the remaining exergy from the flue gas at 63°C, it is 44.5%. Since our actual efficiency is very close at 43.4%, there is not much room for improvement at this point.

So, if this is about as good as the design can get, but we are only 43.4% efficient, how do we make further changes? These require bigger picture decisions, discussed in Sec. 3.4.

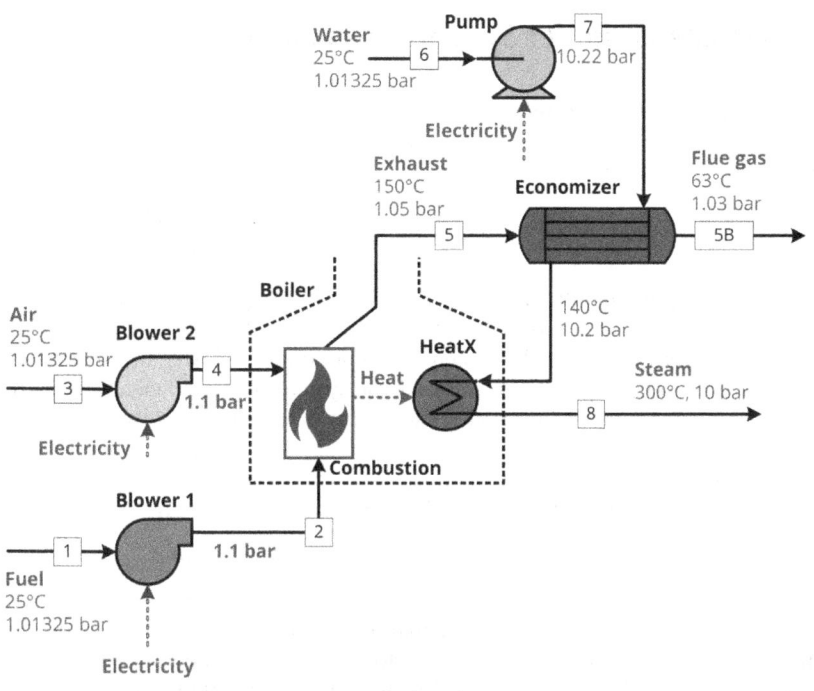

FIGURE 3.4   A proposed design modification that uses an economizer.

\multicolumn{8}{c}{**Mass Flows**}							
#	T (°C)	P (bar)	Composition	F (kg/min)	E^{tm} (MJ/min)	E^{ch} (MJ/min)	E^{conv} (MJ/min)
1	25	1.013	US pipeline natural gas	2.0	—	99.18	99.18
2	32.7	1.1	US natural gas (higher P)	2.0	0.026	99.18	99.20
3	25	1.013	Air	167.8	—	—	—
4	34.8	1.1	Air (higher pressure)	167.8	1.21	—	1.21
5	150	1.05	74.7 wt% N_2, 18.2% O_2, 3.21% CO_2, 3.85% H_2O	169.8	7.30	1.52	8.82
5B	63	1.03	74.7 wt% N_2, 18.2% O_2, 3.21% CO_2, 3.85% H_2O	169.8	0.66	1.52	2.17
6	25	1.013	Liquid water	29.8	—	1.48	1.48
7	25.6	10.22	Compressed water	29.8	0.028	1.48	1.51
8	300	10	Steam	29.8	27.8	15.67	43.43

\multicolumn{4}{c}{**Energy Flows**}			
Stream	Energy (MJ/min)	Exergy E^{th} (MJ/min)	Exergy E^W (MJ/min)
Blower 1 electricity	0.033	—	0.033
Blower 2 electricity	1.68	—	1.68
Pump electricity	0.109	—	0.109
Heat transfer at 350°C in boiler	75.08	39.16	—

TABLE 3.4   Process stream conditions for the revised design that uses an economizer

3.4   Exergy Matching for Better Systems Engineering

3.4.1   Refinery Off-Gas Example

Let's look at the system energy efficiency (not exergy efficiency) of our improved design. In this case, I chose to define it as follows:

$$\text{System energy efficiency (LHV basis)} = \frac{H_8 - H_6}{\text{LHV fuel} + \text{Electricity}} \tag{3.3}$$

The numerator attempts to quantify the amount of "useful energy" delivered to the product by considering the enthalpy difference between the product steam and the room temperature water from which it started. The denominator is the energy input to the system. If we assume the lower heating value (LHV) of conventional natural gas is about 47 MJ_{LHV}/kg, we can compute the system energy efficiency using the data in Table 3.4 (and taking the enthalpy flows from a process simulator or from Fig. 6.1) as follows:

$$\text{System energy efficiency (LHV basis)} = \frac{90.7 \ \frac{MJ}{min}}{94 \ \frac{MJ}{min} + 1.8 \ \frac{MJ}{min}} = 94.7\%_{LHV} \tag{3.4}$$

This is very high; 94.7% of all the energy we purchased (natural gas and electricity) ended up being used to raise steam. Most people would see this and believe that because this is a very energy efficient way to produce steam from natural gas, it must also be a *good* system. However, exergy experts (like you!) understand that because exergy efficiency is only 43.4%, there are some bigger picture issues at play. We can see from this result that there is an exergy "mismatch" between the high-exergy energy sources (electricity and natural gas) and the lower-exergy energy products (medium pressure steam). It means that we are not really exploiting the value of that natural gas to its fullest extent.

Since natural gas is a nonrenewable fossil-fuel resource, there are many negative sustainability consequences (resource depletion, carbon dioxide emissions, etc.) associated with its use. From a societal perspective, if you are going to use natural gas, perhaps we should be using natural gas to produce something with a much higher exergy product than medium pressure steam. Or similarly, maybe we can try to use a lower-exergy, lower-value resource than natural gas to produce steam.

Steel refineries often produce a variety of waste gases that have a calorific value, but not at the potency of natural gas. One of the best uses of these waste gases is for heat production for this reason. For example, blast furnace gas (BFG), a waste gas produced in a steel refinery blast furnace, has high-energy chemicals like H_2 and CO, but also a lot of low-energy CO_2 and N_2. As shown in Table 8.1, BFG has a chemical exergy of only 76.96 kJ/mol, but you can still use it instead of natural gas in the same process as in Fig. 3.4 since it will still get hot enough to make 300°C steam. Basic oxygen furnace gas (BOFG) has a similar makeup, but with more CO and a more potent chemical exergy of 131.82 kJ/mol shown in Table 8.1.

We resimulated the process using both BFG and BOFG. Both versions achieve the same temperatures and pressures of all streams as shown in Fig. 3.4 with the only differences being the flow rates of the streams and the composition of the exhaust gas. BFG is pretty weak, so the process consumes 1.2 kg of BFG per kg of 300°C steam produced. BOFG requires less at 0.73 kg of BOFG per kg of stream produced. Compare this though to the just 0.07 kg of potent natural gas required per kg of steam produced and the relative weakness of the waste gases becomes evident.

Nevertheless, both systems work well using the waste gases, with an exergy efficiency of 44.1% for the BFG process and 44.7% for the BOFG process. Both are higher than the natural gas process (43.4%) without any changes to the system design. Even better, the BOFG process exergy efficiency is higher than the theoretical maximum efficiency we could achieve through additional waste heat capture for the natural gas process in the previous subsection. This tells us from a big picture perspective that this low-quality, lower-exergy waste gas is a more suitable fuel than natural gas for the purpose of creating 300°C steam, because the quality of the resource and product (as measured with exergy) are more closely matched.

3.4.2   Pool Heating Example Revisited—Integrated Community Energy Systems

The value of exergy matching can be more dramatic in some situations. Consider again the heated pool example from Chap. 1, where we found that it took 30 GJ of energy to heat up an Olympic-sized swimming pool from 25 to 28°C. The exergy content of that amount of 28°C water (2390 tonnes) was only 0.15 GJ. In that chapter, we found that using natural gas to heat it resulted in an exergy efficiency less than 0.5%. Clearly, there is a major exergy mismatch between potent natural gas and lukewarm pool water. Is there a better way to heat the pool? Suppose our pool is in a city sitting next to a large grocery store. Grocery stores have large refrigerators and freezers that produce a lot of waste heat and reject it to the atmosphere. It is not very high-quality heat either, perhaps only available at about 32 to 35°C. However, it is warm enough to heat up the pool.

There is a new process concept called integrated community energy (ICE) systems, in which buildings in a community should have their energy demands and products integrated so they can share energy and reduce primary energy combustion. Suppose we could get a steady stream of 30 GJ of 32°C (305.15 K) waste heat to the neighboring pool to heat it through a heat transfer loop. The exergy of this waste heat would be:

$$E^{th} = 30 \text{ GJ}\left(1 - \frac{298.15 \text{ K}}{305.15 \text{ K}}\right) = 0.69 \text{ GJ} \tag{3.5}$$

Then the exergy efficiency of the system as a whole becomes 0.15 GJ (for the warm pool water) divided by the 0.69 GJ of the waste heat, giving an exergy efficiency of 21.8%. This is 2 orders of magnitude more exergy-efficient than using natural gas, because using low-quality waste heat is a much better exergy match than using natural gas. It is a fantastic way to use low-quality waste energy. This is one of the main drivers of ICE as a way to make communities more sustainable and demonstrates how better exergy matching can lead to better societal outcomes.

3.5 Exergy as Thermodynamic Targets for System Design

3.5.1 Direct Air Capture Example

Carbon dioxide capture systems are becoming increasingly important as a means for combating greenhouse gas emissions and climate change. We highlight only two: post-combustion capture from power plants and direct air capture (DAC). In post-combustion capture, a solvent is used to capture CO_2 from the exhaust stream of a power plant just before it is emitted to the atmosphere. In DAC, CO_2 is "captured" from the atmosphere (anywhere) and purified. In both cases, the captured CO_2 (both roughly at about atmospheric pressure and temperature) has to then be sequestered in some fashion. One common strategy is to pump it to supercritical conditions (about 120 bar and roughly 30°C) and then inject it underground into some kind of natural reservoir.

Exergy is the *maximum* amount of work something can produce if it is brought into equilibrium conditions with the environment. As explained in Chap. 2, this is calculated using a reversibility assumption. That means we can also think of that same exergy value as the reverse—it is the *minimum* amount of work required to do the reverse process, such as change a material's temperature or pressure from the equilibrium state to a new state, or to isolate that chemical from the environment. From Table 4.1, the chemical exergy of CO_2 is 19.817 kJ/mol (450 kJ/kg). This is the maximum theoretical work that could be produced by creating a machine that exploits the concentration differential between pure CO_2 and CO_2 in the atmosphere (we use 337 ppm, see Chap. 4 for explanation). It is also the *minimum* theoretical work required to go the other direction, removing CO_2 out of the air and creating pure CO_2 at the same temperature and pressure (25°C and 1 atm). This is exactly what DAC is, and so this is a thermodynamic lower bound for that process.

DAC systems are in the demonstration stage. Public data are scarce, but we have some basic information we can use to evaluate the technology in light of this thermodynamic minimum work requirement. The three DAC demonstration plants that exist are projected* to require roughly 400 kWh of electricity per tonne of CO_2 removed from the atmosphere, isolated, purified, and available at roughly atmospheric temperature and pressure (equal to about 1440 kJ/kg).[1] They also require heat at various temperature ranges, with one example being the Climeworks facility in Switzerland which if commercialized at scale is projected to require 5.76 GJ of heat at about 120°C per tonne of CO_2 captured[2] (5760 kJ/kg).

To see how this stacks up against theoretical minimum work required of 450 kJ/kg, we can estimate the required amount of work that would be required to operate this facility. One way to provide for the 5760 kJ/kg of heat at 120°C is through electric heaters, which are essentially 100% efficient and so would require about 5760 kJ of electric work to provide the necessary heat. However, as discussed in Sec. 3.4.1, efficiency can be very misleading. Like the natural gas boiler example, direct electric heaters have a poor exergy mismatch between high-exergy electricity and low-exergy product (heat at 120°C in this case).

From the equations in Chap. 2, heat at 120°C (393.15 K) has an exergy per heat of only about:

$$\frac{E^{th}}{Q_H} = \left(1 - \frac{298.15 \text{ K}}{393.15 \text{ K}}\right) = 0.242 \tag{3.6}$$

For a 100% energy-efficient electric heater, the electric work exactly equals the heat delivered ($W^e = Q_H$), but the exergy efficiency of the heater is only 24.2%.

A better design might consider an air-source heat pump, which extracts heat from the atmosphere and delivers the heat at a higher temperature to something else, while consuming electricity. A reasonable coefficient of performance (COP) for heat pumps is 3, meaning that for every 1 unit of electricity spent, 2 units of heat are removed from the air, and 3 units of heat are delivered at the higher temperature:

$$COP = \frac{\text{Units of heat delivered at a higher temperature}}{\text{Units of electricity consumed to do it}} = \frac{Q_H}{W^e} \tag{3.7}$$

---

*These are projections for commercial scale versions, after years of unspecified improvements.

It makes sense for the DAC system to use a heat pump in practice, since it must be powered entirely from renewable (wind, solar, hydroelectric, etc.) or carbonless (nuclear power) energy to make sense as a CO_2 mitigation device. If we could use an air-sourced heat pump with COP of 3, it would therefore require 0.333 kJ of electricity (which also has $E^e = 0.333$ kJ of exergy) to produce 1 kJ of heat (which at this temperature has $E^{th} = 0.242$ kJ of exergy). The air used for the heat pump is potentially another source of exergy to the system, but it would have $E^{air} = 0$ kJ of exergy if the source air was at 25°C. This means that the exergy efficiency of the heat pump would be:

$$\text{Heat pump exergy efficiency} = \frac{E^{th}}{E^e + E^{air}} = \frac{0.242 \text{ kJ}}{0.333 \text{ kJ} + 0 \text{ kJ}} = 0.725$$

Thus an air-sourced heat pump with COP = 3 is about 72.5% exergy efficient, which is far superior to the all-electric heater. If we use this heat pump for the DAC system heat needed, it would require about 1920 kJ of electricity per kg_{CO_2} to meet the system's heat needs (this is the 5760 kJ/kg of heat needed divided by the COP of 3). Add that to the 1440 kJ/kg electricity consumption for the machinery, and you get the total work requirement for DAC at 3360 kJ/kg. This is about 7.5 times the thermodynamic minimum (450 kJ/kg) and remember this is for projected future capabilities. The exergy efficiency is the chemical exergy of the captured CO_2 product divided by the exergy of the inputs (which is just the total electric work):

$$\text{Total system exergy efficiency} = \frac{450 \text{ kJ/kg}}{3360 \text{ kJ/kg}} = 13.4\% \tag{3.8}$$

So even when using a heat pump, there is a lot of room for improvement in DAC.

3.5.2 Post-Combustion CO₂ Capture Example

Let's consider the post-combustion capture process as an alternative, shown in Fig. 3.5. Normally, power plant combustion exhaust (which for this example contains about 78 mol% N_2, 18% unspent O_2, 1.3% H_2O, and 2.6% CO_2 which originated from the carbon in the fuel) is vented to the atmosphere at just above atmospheric pressure. In the post-combustion capture process, a liquid solvent (diglycolamine, or DGA) is fed to the top of an absorber column, while the combustion exhaust is fed to the bottom of the column. The column is filled with packing materials that allow for lots of gas-liquid interactions. As the solvent falls down through the column by gravity, it selectively absorbs CO_2 from the rising exhaust (dissolving it into the liquid) but not N_2, which bubbles up through and vents to the atmosphere. The solvent containing dissolved CO_2 falls out the bottom. It is then sent to a distillation column (also called a stripper) which heats up the solvent, releasing the dissolved CO_2. The result is a high-purity CO_2 stream at roughly atmospheric pressure. The lean solvent, now released of most of its CO_2, is recycled back to the absorber and used again.

We can again use exergy to estimate a minimum work requirement for this process. Power plant exhaust is not quite at the same composition of air (which has zero chemical exergy by definition), but it is quite close. In fact, using the data in Table 4.1 and the equation for gas mixtures in Chap. 8, this exhaust chemical exergy can be calculated at only 8.8 kJ/kg, which is a tiny fraction of that of pure CO_2 (450 kJ/kg). Therefore, the chemical exergy of CO_2 is once again a good estimate for a minimum work requirement for this process, like the DAC example in Sec. 3.5.1.

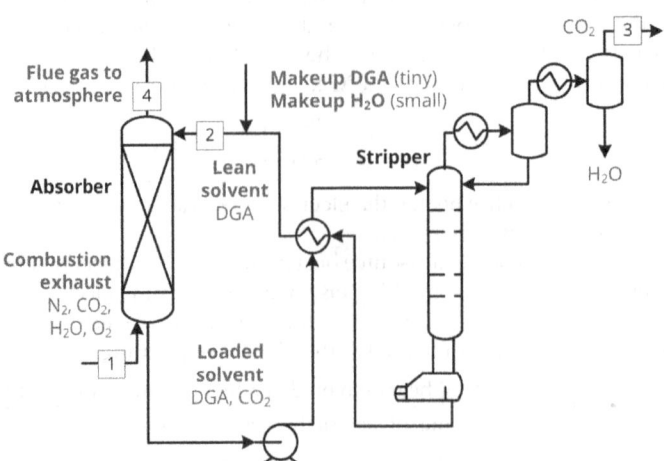

FIGURE 3.5 A process for post-combustion capture of power plant exhaust using diglycolamine (DGA).

Mass Flows				
#	T (°C)	P (bar)	F (kg/min)	Composition (mass %)
1 (Exhaust to absorber)	16.2	1.06	620	75% N_2, 20% O_2, 3.9% CO_2, 0.1% H_2O
2 (Lean solvent feed)	37.9	1.01	231	Diglycolamine (DGA) and water
3 (Captured CO_2)	35.0	1.01	23.8	97.6% CO_2, rest water
4 (Final exhaust to atmosphere)	38	1.01	611	76.4% N_2, 20.3% O_2, 0.1% CO_2, 3.1% H_2O

Energy Flows		
Energy Stream	Energy (MJ/min)	Electricity Consumed (MJ/min)
DGA pump electricity	0.001	0.001
Column cooling (77°C, rejected to environment)	74.1	0.0
Column heating (110°C, provided by heat pump with COP = 3)	149.9	50.0
	Total	50.0
	Work required	2196 (kJ/kg_{CO_2} captured)

Table 3.5   Selected conditions for the post-combustion capture process

We simulated the process in Fig. 3.5 in Aspen Plus* to generate the necessary data. A summary of key streams and utilities can be found in Table 3.5; 95.3% of the CO_2 in the feed (stream 1) is captured (stream 3). The electric work for the pump is very small compared to the distillation column utility requirement. We can use a similar approach to our analysis of the DAC system to consider the work required for the column heating and cooling utilities. The cooling system (for the distillation column condenser) takes heat from the process at 77°C and rejects it to the environment. Ideally, we should be producing work using this heat against the atmosphere, such as using an organic Rankine cycle, but this is usually not done because of cost. Instead, this is simply value lost to the environment. In fact, we need to spend a little electricity to operate fans and blowers for industrial cooling towers, but this amount is small so we can neglect it for our analysis.

The column heating requirement at 110°C is similar in nature to the DAC case. Normally, this process would use thermal methods (likely combustion of the same fuel used in the power plant), but since we are trying to look at work requirements, we can use a heat pump with COP of 3 just like the DAC case for comparison purposes. The 149.9 MJ/min of heat required for the distillation column reboiler would then require 50 MJ/min of electricity. The total work required is 2196 kJ/kg_{CO_2} captured, still more than the thermodynamic minimum of 450 kJ/kg_{CO_2} but much lower than DAC which required 3360 kJ/kg_{CO_2}. The exergy efficiency of the post-combustion capture system is then approximately 20.5%; this is still not so great, but it is better than DAC. Although post-combustion capture is the most mature of carbon capture technologies, there is a huge body of research that is trying to beat it, precisely because the exergy efficiency is so low. If you want to find out more, see Ref. 4 for a summary of various technologies that have been proposed to try to beat post-combustion capture.

3.5.3 CO_2 Compression and Sequestration Example

Whatever CO_2 capture technology used, you still have to sequester the CO_2 underground somehow. One popular option is to compress it to about 120 bar and 30°C, and then store that in an underground reservoir. In the two CO_2 capture process examples discussed, both systems capture CO_2 at roughly atmospheric pressure and temperature. This means we can again use exergy to set a minimum work requirement for compressing it to sequestration conditions. The thermo-mechanical exergy of CO_2 at 120 bar and 30°C can be read directly from Fig. 5.3 to be about $e^{tm} = 220$ kJ/kg. This is the minimum work required to take pure CO_2 from 25°C and 1 atm to sequestration conditions.

Figure 3.6 shows a process that we simulated in Aspen Plus* using the captured CO_2 from the post-combustion process of Sec 3.5.2 (stream 3). The captured CO_2 has a little bit of water in it, so the process has to remove the water. The process consists of a sequence of four stages, each consisting of a compressor, a cooler, and a flash drum in which some liquid water gets condensed out of the gas. At the end of the compressor sequence, the CO_2 is now supercritical at about 80 bar, and a pump can be used to raise its pressure to the final 120 bar (stream 5). The final product is 99.9 wt% CO_2. The total mechanical work required by all the compressors is about 387 kJ/kg_{CO_2} captured, and the pump requires 24 kJ/kg_{CO_2}, for a total of 411 kJ/kg_{CO_2}, less than double

---

*You can download our simulation files from https://PSEcommunity.org/LAPSE:2023.0004. Also, see Ref. 3 for lots of examples of CO_2 capture system simulations.

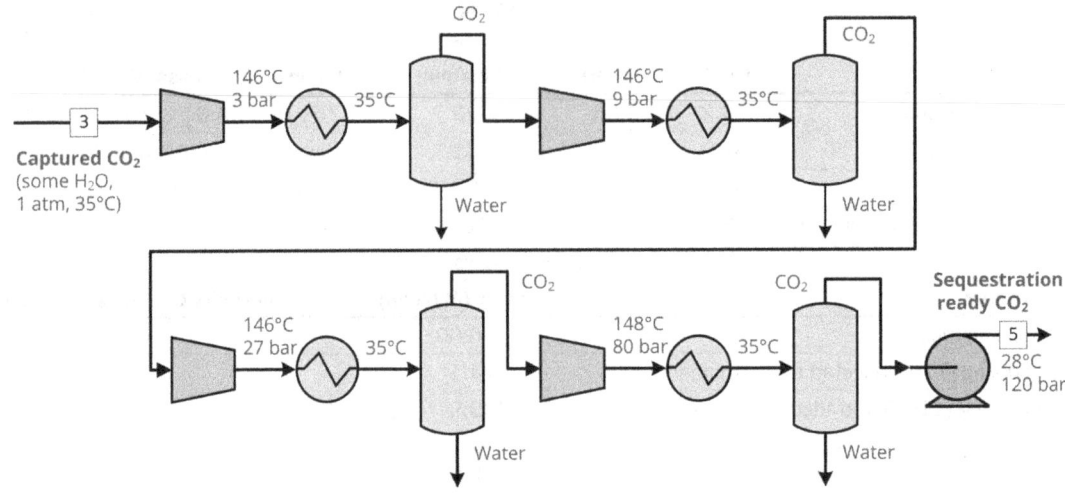

FIGURE 3.6   A process for compressing captured CO_2 from a power plant to sequestration conditions, which contains some unwanted water that must be removed.

the thermodynamic minimum. For the same reasons as in the DAC, we assume for analysis purposes that we can cool the system with minimal or no work since the heat is rejected above ambient. This puts the exergy efficiency at:

$$\text{Total system exergy efficiency} = \frac{220 \text{ kJ/kg}}{411 \text{ kJ/kg}} = 53.5\% \tag{3.9}$$

which is not so bad. The exergy losses are almost entirely due to the energy lost in all the cooling stages, combined with inefficiencies of each compressor.

The 53.5% exergy efficiency could improve further with more efficient compressors (the simulation model assumes an isentropic compressor efficiency of only 72%). In addition, it might be possible to remove the water upfront using triethylene glycol or another desiccant so water does not enter the compressors. This would reduce the work required for compression, hopefully more than offsetting the extra work of the desiccant system. With upfront water removal, it might also be possible to allow the compressors to run hotter (perhaps requiring a more advanced material or cooling technology) by reducing or eliminating the inter-stage cooling. The heat generated from these compressors could be partially recovered with an organic Rankine cycle or some other heat engine, producing work to offset some of the compressor work while also providing cooling. You might want to next compute the exergy flows and losses to help you identify the best place to start.

One interesting thing to point out is that energy efficiency cannot really be used here as a metric for any of the examples in Sec. 3.5 because they are not energy conversion processes. The final product in each case has no useful energy in the traditional sense (the captured CO_2 has neither high temperature nor any kind of heating value). However, exergy efficiency is immediately useful in helping us interpret how close to thermodynamic perfection our design is. This is one of the major advantages of exergy analysis.

References

1. McQueen N, Gomes KV, McCormick C, Blumanthal K, Pisciotta M, Wilcox J. A review of direct air capture (DAC): scaling up commercial technologies and innovating for the future. *Progress in Energy* 3(3):032001 (2021).
2. Beuttler C, Charles L, Wurzbacher J. The role of direct capture mitigation of anthropogenic greenhouse gas emissions. *Frontiers in Climate* 1:10 (2019).
3. Adams TA II, Salkuyeh Khojestah Y, Nease J. Processes and simulations for solvent-based CO_2 capture and syngas cleanup. In: *Reactor and Process Design in Sustainable Energy Technology*. Editor Shi F. Elsevier: Amsterdam (2014).
4. Adams TA II, Hoseinzade L, Madabhushi PM, Okeke IJ. Comparison of CO_2 capture approaches for fossil-based power generation: review and meta-study. *Processes* 5:44 (2017).

CHAPTER 4

Environmental Reference State

Exergy is computed relative to the environment. The environmental reference state used for all calculations in this book (except where otherwise noted) is the Earth at sea level at 1 atm (1.01325 bar = 101.325 kPa) and 25°C. For thermo-mechanical exergy, only the temperature and pressure are relevant. For chemical exergy, the environmental reference state is more complex. Chemical exergy is computed by chemical concentration differences against equilibrium, and for that, we break up the environmental reference state into three portions of the Earth: the atmosphere near sea level, the hydrosphere (the oceans), and the lithosphere (containing the Earth's crust). This is necessary because when released to the environment, different chemicals will equilibrate to these different areas given infinite time. For example, nitrogen gas released to the Earth will equilibrate into the atmosphere, a normal solid like iron will equilibrate into the ground, and elements which form ions in water, like chlorine (as in ordinary salt), will equilibrate into the oceans.

The practicing engineer using this book will not usually need to care much about the exact numbers used to define the reference state, since all the numbers in the book use them consistently except where otherwise mentioned in special cases. However, you might need it if you will be computing exergy values for chemicals not already contained in the book.

The next sections explain the reference state and why it was chosen, broken into the atmosphere, hydrosphere, and lithosphere. The chemical exergies of the reference chemicals are provided as well.

4.1 The Atmosphere

For volatile substances, or anything that combusts completely into atmospheric gases, we use the Earth's air as the environmental reference state (such as all chemicals containing only C, H, and O). Table 4.1 shows the concentration of various chemicals in the Earth's air used as the reference state, which was chosen to be consistent with popular exergy textbooks.[1]

Considering that the species in atmospheric air follow ideal gas behavior, their standard chemical exergy can be evaluated by[2]:

$$e_i^{ch} = RT_0 \ln\left(\frac{1}{x_i}\right) \tag{4.1}$$

where e^{ch} represents chemical exergy, i is the reference species, and x is the mole fraction of i in the atmosphere. All terms are evaluated at the reference temperature T_0 at 298.15 K and reference pressure P_0 at 1.01325 bar (1 atm). The partial pressure of water in the air, P_{H_2O}, is obtained from the relative humidity (φ) definition, which is:

$$\varphi = \frac{P_{H_2O}}{P_{H_2O}^{vap}} \tag{4.2}$$

where $P_{H_2O}^{vap}$ is the vapor pressure of water. We computed this using Antoine's expression[3]:

$$\log_{10} P_{H_2O}^{vap} = A - \frac{B}{C+T} \tag{4.3}$$

For temperatures between 1 and 100°C, the parameters A, B, and C equal 8.07131, 1730.63, and 233.426, respectively. $P_{H_2O}^{vap}$ is in the unit of mmHg (1 kPa = 7.501 mmHg), and T is in the unit of °C. The result at 25°C is $P_{H_2O}^{vap} = 3.15793$ kPa.

Then chemical exergy of the element could be back-calculated using Eq. (4.4)[4]:

$$e_i^{ch} = \Delta g_i^f + \sum_j n_j e_j^{ch} \tag{4.4}$$

Chemical Element *j*		Reference Species *i*				
Elements	e_j^{ch} (kJ/mol)	Reference Species Formula *i*	Mole Percentage x_i	Partial Pressure (kPa)	Δg_i^f (kJ/mol)	e_i^{ch} (kJ/mol)
H_2 (g)	236.11	H_2O (g)	2.1816	2.2106	−228.59	9.4822
N_2 (g)	0.66965	N_2	76.328	77.339	—	0.66965
O_2 (g)	3.9246	O_2	20.532	20.804	—	3.9246
C (s)	410.27	CO_2	0.0337	0.034194	−394.38	19.817
He (g)	30.313	He	4.89×10^{-4}	0.000496	—	30.313
Ne (g)	27.138	Ne	1.76×10^{-3}	0.001784	—	27.138
Ar (g)	11.643	Ar	0.9126	0.92474	—	11.643
Kr (g)	34.303	Kr	9.7818×10^{-5}	9.9114×10^{-5}	—	34.303
Xe (g)	40.272	Xe	8.8×10^{-6}	8.92×10^{-6}	—	40.272

TABLE 4.1 The composition of Earth's air used as the atmospheric reference state throughout this work.[1] The chemical formula, concentration, standard gibbs free energy of formation, and chemical exergy of each reference species are provided. The chemical exergy for the corresponding element associated with each reference species is also provided. Note that the chemical exergies are for pure components (for example, the chemical exergy of pure water vapor is 9.482 kJ/mol of H_2O). Abbreviations: s, solid; g, gas

where i is the reference species, j is the element species of interest associated with the reference species i (e.g., for H_2O, the corresponding element species are H_2 and O_2), and n is the number of moles of element species j in one mole of species i (e.g., for H_2O, $n_{H_2} = 1$ and $n_{O_2} = 0.5$).

Note that two important assumptions were made in creating this table. First, the humidity of air at standard reference conditions is 70%. Humidity routinely increases and decreases throughout the day in real conditions. However, the value of 70% is an approximate middle ground. Second, the reference state assumes a CO_2 concentration in the atmosphere of 337 ppm. CO_2 concentrations follow seasonal cycles, but with a steady overall increase over the past century, with the most recent average concentration (August 2022) being about 414 ppm.[5] This begs the question as to what the most appropriate value for CO_2 concentration in the atmosphere is to use for the standard. Whatever we pick today will become out of date in the future, and we cannot necessarily even assume that the trend will continue to rise forever because of the massive global effort in battling greenhouse gas emissions. Nevertheless, whatever we pick, it is important that all calculations use the same value. Therefore, we used 337 ppm because so much of the literature has used it as well, despite the not having seen atmospheric CO_2 concentrations this low since the early 1980s. The authors remain hopeful that humanity will achieve CO_2 reductions so we can return to this level again soon.

We note that other chemicals are present in the atmosphere. For example, CH_4 is present to about 1900 ppb, N_2O is present to about 335 ppb, and SF_6 to 11 ppt.[6-8] However, these are at relatively trace levels, and they are unstable in the atmosphere and will break down over time, and thus are not included as a part of the reference state.

Table 4.2 shows the impact of different humidity assumptions on the chemical exergies of the selected chemicals in the atmosphere by performing a sensitivity analysis on the assumed relative humidity. This shows that the relative humidity assumption most directly impacts the chemical exergy of water, with minor impacts on N_2, and trivial impacts on the remaining chemicals. Note that this table serves to illustrate the impacts of assumptions on the definition of the Earth's atmosphere. Because the numbers in this table are computed using different relative humidities than the rest of the book (except the 70% relative humidity numbers), engineers generally should not use the numbers from Table 4.2 in their analyses.

4.2   The Hydrosphere

The hydrosphere (the ocean) forms the liquid phase portion of the environmental reference state, relevant for elements like chlorine and sodium. We have chosen to use the work of Rivero and Garfias[2] for the definition of the hydrosphere and lithosphere reference states. This assumes that the salinity of seawater is 35% and the pH level is 8.1. For thermodynamic purists, it is important to note that there has historically been some debate as to which sphere certain elements should be placed (e.g., should the reference state for calcium be in the hydrosphere or lithosphere?), but we believe the model of Rivero and Garfias to be the most up to date for the purposes of this book. It is important to note that the reference state of water is based on the gas form, not liquid, and therefore water does not appear in the equilibrium reference state for the hydrosphere.

Rivero and Garfias[2] argue that the elements Ni, Cu, Zn, Ag, Cd, and Hg should be referenced from the lithosphere instead of hydrosphere as Szargut[1] previously proposed. This is because the reference states Szargut

Relative Humidity (%)	H_2O (g) (kJ/kmol)	N_2 (kJ/kmol)	O_2 (kJ/kmol)	CO_2 (kJ/kmol)	He (kJ/kmol)	Ne (kJ/kmol)	Ar (kJ/kmol)	Kr (kJ/kmol)	Xe (kJ/kmol)
0	—	615.0	3870.0	19762.2	30258.3	27082.9	11587.9	34248.0	40217.2
1	20014.1	615.7	3870.7	19762.9	30259.1	27083.7	11588.7	34248.8	40218.0
2	18295.8	616.5	3871.5	19763.7	30259.9	27084.5	11589.5	34249.6	40218.8
3	17290.7	617.3	3872.3	19764.5	30260.6	27085.3	11590.3	34250.4	40219.6
4	16577.5	618.1	3873.1	19765.2	30261.4	27086.0	11591.0	34251.1	40220.3
5	16024.3	618.8	3873.8	19766.0	30262.2	27086.8	11591.8	34251.9	40221.1
6	15572.4	619.6	3874.6	19766.8	30263.0	27087.6	11592.6	34252.7	40221.9
7	15190.2	620.4	3875.4	19767.6	30263.7	27088.4	11593.3	34253.5	40222.7
8	14859.2	621.2	3876.1	19768.3	30264.5	27089.1	11594.1	34254.2	40223.4
9	14567.2	621.9	3876.9	19769.1	30265.3	27089.9	11594.9	34255.0	40224.2
10	14306.1	622.7	3877.7	19769.9	30266.1	27090.7	11595.7	34255.8	40225.0
11	14069.8	623.5	3878.5	19770.7	30266.8	27091.5	11596.4	34256.6	40225.8
12	13854.1	624.3	3879.2	19771.4	30267.6	27092.2	11597.2	34257.3	40226.5
13	13655.7	625.0	3880.0	19772.2	30268.4	27093.0	11598.0	34258.1	40227.3
14	13472.0	625.8	3880.8	19773.0	30269.2	27093.8	11598.8	34258.9	40228.1
15	13300.9	626.6	3881.6	19773.8	30269.9	27094.6	11599.5	34259.7	40228.9
16	13140.9	627.4	3882.4	19774.5	30270.7	27095.3	11600.3	34260.4	40229.6
17	12990.7	628.1	3883.1	19775.3	30271.5	27096.1	11601.1	34261.2	40230.4
18	12849.0	628.9	3883.9	19776.1	30272.3	27096.9	11601.9	34262.0	40231.2
19	12714.9	629.7	3884.7	19776.9	30273.0	27097.7	11602.7	34262.8	40232.0
20	12587.8	630.5	3885.5	19777.7	30273.8	27098.4	11603.4	34263.5	40232.7
21	12466.8	631.2	3886.2	19778.4	30274.6	27099.2	11604.2	34264.3	40233.5
22	12351.5	632.0	3887.0	19779.2	30275.4	27100.0	11605.0	34265.1	40234.3
23	12241.3	632.8	3887.8	19780.0	30276.2	27100.8	11605.8	34265.9	40235.1
24	12135.8	633.6	3888.6	19780.8	30276.9	27101.6	11606.5	34266.7	40235.9
25	12034.6	634.4	3889.3	19781.5	30277.7	27102.3	11607.3	34267.4	40236.6
26	11937.4	635.1	3890.1	19782.3	30278.5	27103.1	11608.1	34268.2	40237.4
27	11843.8	635.9	3890.9	19783.1	30279.3	27103.9	11608.9	34269.0	40238.2
28	11753.7	636.7	3891.7	19783.9	30280.0	27104.7	11609.7	34269.8	40239.0
29	11666.7	637.5	3892.5	19784.7	30280.8	27105.4	11610.4	34270.6	40239.7
30	11582.6	638.3	3893.2	19785.4	30281.6	27106.2	11611.2	34271.3	40240.5
31	11501.4	639.0	3894.0	19786.2	30282.4	27107.0	11612.0	34272.1	40241.3
32	11422.7	639.8	3894.8	19787.0	30283.2	27107.8	11612.8	34272.9	40242.1
33	11346.4	640.6	3895.6	19787.8	30283.9	27108.6	11613.6	34273.7	40242.9
34	11272.4	641.4	3896.4	19788.6	30284.7	27109.3	11614.3	34274.5	40243.6
35	11200.5	642.2	3897.1	19789.3	30285.5	27110.1	11615.1	34275.2	40244.4
36	11130.7	642.9	3897.9	19790.1	30286.3	27110.9	11615.9	34276.0	40245.2
37	11062.8	643.7	3898.7	19790.9	30287.1	27111.7	11616.7	34276.8	40246.0
38	10996.7	644.5	3899.5	19791.7	30287.9	27112.5	11617.5	34277.6	40246.8
39	10932.3	645.3	3900.3	19792.5	30288.6	27113.3	11618.2	34278.4	40247.6
40	10869.5	646.1	3901.1	19793.3	30289.4	27114.0	11619.0	34279.1	40248.3
41	10808.3	646.9	3901.8	19794.0	30290.2	27114.8	11619.8	34279.9	40249.1
42	10748.5	647.6	3902.6	19794.8	30291.0	27115.6	11620.6	34280.7	40249.9
43	10690.2	648.4	3903.4	19795.6	30291.8	27116.4	11621.4	34281.5	40250.7

TABLE 4.2   Chemical exergy e^{ch} of selected chemicals if the environmental reference state assumed relative humidity values are different than those in Table 4.1. This is a sensitivity analysis that is for illustrative purposes only. If you use values other than those for 70% relative humidity, you should not also use values from other tables in the book since they are not at the same reference conditions; rather, you must compute your own values using these new reference conditions.

Relative Humidity (%)	H₂O (g) (kJ/kmol)	N₂ (kJ/kmol)	O₂ (kJ/kmol)	CO₂ (kJ/kmol)	He (kJ/kmol)	Ne (kJ/kmol)	Ar (kJ/kmol)	Kr (kJ/kmol)	Xe (kJ/kmol)
44	10633.2	649.2	3904.2	19796.4	30292.5	27117.2	11622.2	34282.3	40251.5
45	10577.5	650.0	3905.0	19797.2	30293.3	27118.0	11622.9	34283.1	40252.3
46	10523.0	650.8	3905.8	19798.0	30294.1	27118.7	11623.7	34283.8	40253.0
47	10469.7	651.6	3906.5	19798.7	30294.9	27119.5	11624.5	34284.6	40253.8
48	10417.5	652.3	3907.3	19799.5	30295.7	27120.3	11625.3	34285.4	40254.6
49	10366.4	653.1	3908.1	19800.3	30296.5	27121.1	11626.1	34286.2	40255.4
50	10316.3	653.9	3908.9	19801.1	30297.3	27121.9	11626.9	34287.0	40256.2
51	10267.2	654.7	3909.7	19801.9	30298.0	27122.7	11627.7	34287.8	40257.0
52	10219.1	655.5	3910.5	19802.7	30298.8	27123.4	11628.4	34288.6	40257.7
53	10171.9	656.3	3911.2	19803.4	30299.6	27124.2	11629.2	34289.3	40258.5
54	10125.6	657.0	3912.0	19804.2	30300.4	27125.0	11630.0	34290.1	40259.3
55	10080.1	657.8	3912.8	19805.0	30301.2	27125.8	11630.8	34290.9	40260.1
56	10035.4	658.6	3913.6	19805.8	30302.0	27126.6	11631.6	34291.7	40260.9
57	9991.5	659.4	3914.4	19806.6	30302.8	27127.4	11632.4	34292.5	40261.7
58	9948.4	660.2	3915.2	19807.4	30303.5	27128.2	11633.2	34293.3	40262.5
59	9906.0	661.0	3916.0	19808.2	30304.3	27128.9	11633.9	34294.1	40263.2
60	9864.4	661.8	3916.8	19809.0	30305.1	27129.7	11634.7	34294.8	40264.0
61	9823.4	662.6	3917.5	19809.7	30305.9	27130.5	11635.5	34295.6	40264.8
62	9783.1	663.3	3918.3	19810.5	30306.7	27131.3	11636.3	34296.4	40265.6
63	9743.4	664.1	3919.1	19811.3	30307.5	27132.1	11637.1	34297.2	40266.4
64	9704.4	664.9	3919.9	19812.1	30308.3	27132.9	11637.9	34298.0	40267.2
65	9665.9	665.7	3920.7	19812.9	30309.1	27133.7	11638.7	34298.8	40268.0
66	9628.1	666.5	3921.5	19813.7	30309.8	27134.5	11639.5	34299.6	40268.8
67	9590.8	667.3	3922.3	19814.5	30310.6	27135.3	11640.2	34300.4	40269.6
68	9554.1	668.1	3923.1	19815.3	30311.4	27136.0	11641.0	34301.1	40270.3
69	9517.9	668.9	3923.8	19816.0	30312.2	27136.8	11641.8	34301.9	40271.1
70	9482.2	669.7	3924.6	19816.8	30313.0	27137.6	11642.6	34302.7	40271.9
71	9447.1	670.4	3925.4	19817.6	30313.8	27138.4	11643.4	34303.5	40272.7
72	9412.4	671.2	3926.2	19818.4	30314.6	27139.2	11644.2	34304.3	40273.5
73	9378.2	672.0	3927.0	19819.2	30315.4	27140.0	11645.0	34305.1	40274.3
74	9344.5	672.8	3927.8	19820.0	30316.2	27140.8	11645.8	34305.9	40275.1
75	9311.2	673.6	3928.6	19820.8	30317.0	27141.6	11646.6	34306.7	40275.9
76	9278.4	674.4	3929.4	19821.6	30317.7	27142.4	11647.4	34307.5	40276.7
77	9246.0	675.2	3930.2	19822.4	30318.5	27143.2	11648.1	34308.3	40277.5
78	9214.0	676.0	3931.0	19823.2	30319.3	27143.9	11648.9	34309.1	40278.2
79	9182.4	676.8	3931.8	19824.0	30320.1	27144.7	11649.7	34309.8	40279.0
80	9151.2	677.6	3932.5	19824.7	30320.9	27145.5	11650.5	34310.6	40279.8
81	9120.4	678.4	3933.3	19825.5	30321.7	27146.3	11651.3	34311.4	40280.6
82	9090.0	679.1	3934.1	19826.3	30322.5	27147.1	11652.1	34312.2	40281.4
83	9060.0	679.9	3934.9	19827.1	30323.3	27147.9	11652.9	34313.0	40282.2
84	9030.3	680.7	3935.7	19827.9	30324.1	27148.7	11653.7	34313.8	40283.0
85	9000.9	681.5	3936.5	19828.7	30324.9	27149.5	11654.5	34314.6	40283.8
86	8971.9	682.3	3937.3	19829.5	30325.7	27150.3	11655.3	34315.4	40284.6
87	8943.3	683.1	3938.1	19830.3	30326.5	27151.1	11656.1	34316.2	40285.4
88	8914.9	683.9	3938.9	19831.1	30327.3	27151.9	11656.9	34317.0	40286.2

TABLE **4.2** Chemical exergy e^ch of selected chemicals if the environmental reference state assumed relative humidity values are different than those in Table 4.1. This is a sensitivity analysis that is for illustrative purposes only. If you use values other than those for 70% relative humidity, you should not also use values from other tables in the book since they are not at the same reference conditions; rather, you must compute your own values using these new reference conditions. (*Continued*)

Relative Humidity (%)	H_2O (g) (kJ/kmol)	N_2 (kJ/kmol)	O_2 (kJ/kmol)	CO_2 (kJ/kmol)	He (kJ/kmol)	Ne (kJ/kmol)	Ar (kJ/kmol)	Kr (kJ/kmol)	Xe (kJ/kmol)
89	8886.9	684.7	3939.7	19831.9	30328.1	27152.7	11657.7	34317.8	40287.0
90	8859.2	685.5	3940.5	19832.7	30328.8	27153.5	11658.5	34318.6	40287.8
91	8831.8	686.3	3941.3	19833.5	30329.6	27154.3	11659.3	34319.4	40288.6
92	8804.8	687.1	3942.1	19834.3	30330.4	27155.1	11660.0	34320.2	40289.4
93	8778.0	687.9	3942.9	19835.1	30331.2	27155.9	11660.8	34321.0	40290.2
94	8751.4	688.7	3943.7	19835.9	30332.0	27156.7	11661.6	34321.8	40290.9
95	8725.2	689.5	3944.5	19836.7	30332.8	27157.4	11662.4	34322.6	40291.7
96	8699.2	690.3	3945.3	19837.5	30333.6	27158.2	11663.2	34323.3	40292.5
97	8673.6	691.1	3946.1	19838.3	30334.4	27159.0	11664.0	34324.1	40293.3
98	8648.1	691.9	3946.9	19839.1	30335.2	27159.8	11664.8	34324.9	40294.1
99	8623.0	692.7	3947.7	19839.8	30336.0	27160.6	11665.6	34325.7	40294.9
100	8598.1	693.5	3948.4	19840.6	30336.8	27161.4	11666.4	34326.5	40295.7

TABLE 4.2 Chemical exergy e^{ch} of selected chemicals if the environmental reference state assumed relative humidity values are different than those in Table 4.1. This is a sensitivity analysis that is for illustrative purposes only. If you use values other than those for 70% relative humidity, you should not also use values from other tables in the book since they are not at the same reference conditions; rather, you must compute your own values using these new reference conditions. (*Continued*)

proposed for these elements have negative chemical exergy values. Rivero and Garfias' work referenced Morris et al.[9,10] regarding the chemical exergy calculation of elements. Morris et al. [9,10] state that it is necessary to identify a reference species of the element in the hydrosphere in order to calculate its element's chemical exergy. By definition, the reference species (ions) have a chemical exergy of 0 kJ/mol.[9,10] To calculate the standard chemical exergy of elements contained within the reference species dissolved in seawater, we used the equations from Rivero and Garfias.[2] This is consistent both with the earlier works by Morris et al.,[9,10] and later adaptations by Szargut.[4] See Szargut's book for example calculations.[4]

Table 4.3 contains the reference conditions of those elements contained in the hydrosphere. You can find the chemical formula, molality in mol/kg_{H_2O}, the normal standard free energy of formation of the element's reference species, and the chemical exergy of the element itself.

Chemical Element		Reference Species		
Element	e^{ch} (kJ/mol)	Reference Species Formula (aq)	Molality in Oceans (mol/kg_{H_2O})	Δg_i^f (kJ/mol)
As (s)	492.6	$HAsO_4^{2-}$	3.87×10^{-8}	−714.7
B (s)	628.1	$B(OH)_3$	3.42×10^{-4}	−1361.9
Bi (s)	274.8	BiO^+	9.92×10^{-11}	−146.4
Br_2 (l)	101.0	Br^-	8.73×10^{-4}	−103.97
Cl_2 (g)	123.7	Cl^-	5.66×10^{-1}	−131.26
Cs (s)	404.6	Cs^+	2.34×10^{-9}	−282.23
I_2 (s)	175.7	IO_3^-	5.23×10^{-7}	−128
K (s)	366.7	K^+	1.04×10^{-2}	−282.44
Li (s)	392.7	Li^+	2.54×10^{-5}	−294
Mo (s)	731.3	MoO_4^{2-}	1.08×10^{-7}	−836.4
Na (s)	336.7	Na^+	4.74×10^{-1}	−262.048
P (s)	861.3	HPO_4^{2-}	4.86×10^{-7}	−1089.3
Rb (s)	388.7	Rb^+	1.46×10^{-6}	−282.4
S (s)	609.3	SO_4^{2-}	1.24×10^{-2}	−744.63
Se (s)	347.5	SeO_4^{2-}	1.18×10^{-9}	−441.4
W (s)	828.5	WO_4^{2-}	5.64×10^{-10}	−920.5

TABLE 4.3 The makeup of the hydrosphere (oceans) used as the hydrosphere reference state throughout this work.[2] The reference species' chemical formula, concentration (molality), standard Gibbs free energy of formation, and the element's chemical exergy are provided. Note that the chemical exergy of each reference species is 0 by definition. Abbreviations: s, solid; l, liquid; g, gas; aq, aqueous

4.3   The Lithosphere

The reference chemicals for the lithosphere are listed in Table 4.4, again based on Ref. 2. To calculate the element's chemical exergy, we first calculated its reference species chemical exergy (e_i^{ch}) using the following equation[2,9]:

$$e_i^{ch} = -RT_0\ln(x_i) \tag{4.5}$$

where i is the reference species (e.g., $CaCO_3$), and x_i is the mole fraction of the reference species i in the lithosphere. Then, we determined the chemical exergy of the element by solving the set of equations (4.6)[3]:

$$e_i^{ch} = \Delta g_i^f + \sum_j n_j e_j^{ch} \tag{4.6}$$

where j is the element species of interest associated with the reference species i (e.g., for $CaCO_3$ there would be three element species, Ca, C, and O_2), and n is the number of moles of element species j in one mole of species i (e.g., n_{O_2} in $CaCO_3$ is 1.5). The Gibbs free energy of formation is computed at the environmental reference temperature and pressure (298.15 K and 101.325 kPa). Rivero and Garfias[2] explain the calculation sequence in more detail. The basic procedure is to calculate the species in the atmosphere first, then the species in the hydrosphere, and finally the species in the lithosphere.

Chemical Element		Reference Species			
Element (s)	e_j^{ch} (kJ/mol)	Reference Species Formula (s)	Mole Fraction in Crust x_{i0}	Δg_i^f (kJ/mol)	e_i^{ch} (kJ/mol)
Ag	99.3	AgCl	1.00×10^{-9}	−109.8	51.37
Al	795.7	Al_2SiO_5	2.07×10^{-3}	−2441.0	15.32
Au	50.6	Au	1.36×10^{-9}	—	50.61
Ba	775.4	$BaSO_4$	4.20×10^{-6}	−1361.9	30.69
Be	604.3	Be_2SiO_4	2.10×10^{-7}	−2033.3	38.12
Ca	729.1	$CaCO_3$	1.40×10^{-3}	−1129.0	16.29
Cd	298.4	$CdCO_3$	1.22×10^{-8}	−669.4	45.17
Ce	1054.7	CeO_2	1.17×10^{-6}	−1024.8	33.86
Co	313.4	$CoFe_2O_4$	2.85×10^{-7}	−1032.6	37.36
Cr	584.4	$K_2Cr_2O_7$	1.35×10^{-6}	−1882.3	33.50
Cu	132.6	$CuCO_3$	5.89×10^{-6}	−518.9	29.85
Dy	976	$Dy(OH)_3$	4.88×10^{-8}	−1294.3	41.73
Er	972.8	$Er(OH)_3$	4.61×10^{-8}	−1291.0	41.88
Eu	1003.8	$Eu(OH)_3$	2.14×10^{-8}	−1320.1	43.78
F_2	505.8	$CaF_2 \cdot 3Ca_3(PO_4)_2$	2.24×10^{-4}	−12985	20.83
Fe	374.3	Fe_2O_3	6.78×10^{-3}	−742.2	12.38
Ga	515	Ga_2O_3	2.98×10^{-7}	−998.6	37.25
Gd	969	$Gd(OH)_3$	9.21×10^{-8}	−1288.9	40.16
Ge	557.7	GeO_2	9.49×10^{-8}	−521.5	40.09
Hf	1063.1	HfO_2	1.15×10^{-7}	−1027.4	39.61
Hg	107.9	$HgCl_2$	5.42×10^{-10}	−178.7	52.89
Ho	978.7	$Ho(OH)_3$	1.95×10^{-8}	−1294.8	44.01
In	436.9	In_2O_3	2.95×10^{-9}	−830.9	48.69
Ir	247	IrO_2	3.59×10^{-12}	−185.6	65.33
La	994.7	$La(OH)_3$	5.96×10^{-7}	−1319.2	35.53
Lu	945.8	$Lu(OH)_3$	7.86×10^{-9}	−1259.6	46.26
Mg	626.9	$Mg_3Si_4O_{10}(OH)_2$	8.67×10^{-4}	−5543.0	17.48

TABLE 4.4   The makeup of the hydrosphere (oceans) used as the hydrosphere reference state throughout this work. The reference species' chemical formula, mole fraction in the lithosphere, standard Gibbs free energy of formation, the element's chemical exergy, and its reference species' chemical exergy (if a pure chemical) are provided. Abbreviation: s, solid

Chemical Element		Reference Species			
Element (s)	e_j^{ch} (kJ/mol)	Reference Species Formula (s)	Mole Fraction in Crust x_{j0}	Δg_i^f (kJ/mol)	e_i^{ch} (kJ/mol)
Mn	487.7	MnO_2	2.30×10^{-5}	−465.2	26.48
Nb	970.1	Nb_2O_3	1.49×10^{-7}	−1766.4	38.97
Nd	242.6	$Nd(OH)_3$	5.15×10^{-7}	−1294.3	35.89
Ni	368.4	NiO	1.76×10^{-6}	−211.7	32.85
Os	249.2	OsO_4	3.39×10^{-13}	−305.1	71.18
Pb	138.7	$PbCO_3$	1.04×10^{-7}	−625.5	39.86
Pd	963.9	PdO	6.37×10^{-11}	−82.5	58.20
Pr	141.2	$Pr(OH)_3$	1.57×10^{-7}	−1285.1	38.84
Pt	1100.1	PtO_2	1.76×10^{-11}	−83.7	61.39
Pu	824.2	PuO_2	8.40×10^{-20}	−995.1	108.88
Ra	559.6	$RaSO_4$	2.98×10^{-14}	−1364.2	77.21
Re	179.7	Re_2O_7	3.66×10^{-12}	−1067.6	65.28
Rh	318.6	Rh_2O_3	3.29×10^{-12}	−299.8	65.54
Ru	438.2	RuO_2	6.78×10^{-13}	−253.1	69.46
Sb	925.3	Sb_2O_5	1.08×10^{-10}	−829.3	56.89
Sc	855	Sc_2O_3	3.73×10^{-7}	−1819.7	36.69
Si	993	SiO_2	4.07×10^{-1}	−856.7	2.23
Sm	993.7	$Sm(OH)_3$	1.08×10^{-7}	−1314.0	39.77
Sn	551.8	SnO_2	4.61×10^{-7}	−519.6	36.17
Sr	749.8	$SrCO_3$	2.91×10^{-5}	−1140.1	25.89
Ta	974.1	Ta_2O_5	7.45×10^{-9}	−1911.6	46.39
Tb	998.5	$Tb(OH)_3$	1.71×10^{-8}	−1314.2	44.33
Te	329.3	TeO_2	9.48×10^{-12}	−270.3	62.92
Th	1202.7	ThO_2	2.71×10^{-7}	−1169.1	37.48
Ti	907.2	TiO_2	1.63×10^{-4}	−889.5	21.62
Tl	194.9	Tl_2O_4	1.49×10^{-9}	−347.3	50.38
Tm	951.8	$Tm(OH)_3$	7.59×10^{-9}	−1265.5	46.35
U	1196.6	$UO_3 \cdot H_2O$	1.49×10^{-8}	−1395.9	44.68
V	721.3	V_2O_5	1.83×10^{-6}	−1419.6	32.75
Y	965.6	$Y(OH)_3$	1.00×10^{-6}	−1291.4	34.25
Yb	944.3	$Yb(OH)_3$	4.61×10^{-8}	−1262.5	41.88
Zn	344.7	$ZnCO_3$	7.45×10^{-6}	−731.6	29.27
Zr	1083	$ZrSiO_4$	2.44×10^{-5}	−1919.5	26.33

TABLE 4.4 The makeup of the hydrosphere (oceans) used as the hydrosphere reference state throughout this work. The reference species' chemical formula, mole fraction in the lithosphere, standard Gibbs free energy of formation, the element's chemical exergy, and its reference species' chemical exergy (if a pure chemical) are provided. Abbreviation: s, solid (*Continued*)

References

1. Szargut J, Morris DR, Steward FR. *Exergy Analysis of Thermal, Chemical, and Metallurgical Processes.* United States: Hemisphere (1987).
2. Rivero R, Garfias M. Standard chemical exergy of elements updated. *Energy* 31:3310–3326 (2006). doi:10.1016/j.energy.2006.03.020.
3. Saturated vapor pressure. http://ddbonline.ddbst.com/AntoineCalculation/AntoineCalculationCGI.exe.
4. Szargut J. *Exergy Method: Technical and Ecological Applications.* Vol. 18. WIT Press (2005).

5. Global Monitoring Laboratory. Trends in atmospheric carbon dioxide: global monthly mean CO_2. National Oceanic & Atmospheric Administration. Accessed November 2022. https://gml.noaa.gov/ccgg/trends/.

6. Global Monitoring Laboratory. Trends in atmospheric nitrous oxide: global CH_4 monthly means. National Oceanic & Atmospheric Administration. Accessed November 2022. https://gml.noaa.gov/ccgg/trends_ch4/.

7. Global Monitoring Laboratory. Trends in atmospheric nitrous oxide: global N_2O monthly means. National Oceanic & Atmospheric Administration. Accessed November 2022. https://gml.noaa.gov/ccgg/trends_n2o/.

8. Global Monitoring Laboratory. Trends in atmospheric sulfur hexafluoride: global SF_6 monthly means. National Oceanic & Atmospheric Administration. Accessed November 2022. https://gml.noaa.gov/ccgg/trends_sf6/.

9. Morris DR, Szargut J. Standard chemical exergy of some elements and compounds on the planet earth. *Energy* 11:8:733–755 (1986).

10. Morris DR, Steward FR. Exergy analysis of a chemical metallurgical process. *Metallurgical Transactions B* 15(4):645–654 (1984).

CHAPTER **5**
Atmospheric Gases

5.1 Chemical Exergy

The chemical exergy of pure gases naturally present and stable in the atmosphere are shown in Chap. 4, Table 4.1. Because the gases naturally present in the atmosphere are largely inert with very little chemical potential, the chemical exergies are due largely to the concentration difference against the atmosphere. For N_2 and O_2, because they are at a high concentration in the atmosphere, there is little exergy stored in concentration differential as well. For example, the chemical exergy of N_2 (0.66905 kJ/mol or 23.90 kJ/kg) may be dwarfed by its thermo-mechanical exergy (in the 0 to 300 kJ/kg range, see Fig. 5.6) depending on its temperature and pressure.

At the other extreme, Xe has a chemical exergy an order of magnitude higher than N_2 at 40.272 kJ/mol (306.74 kJ/kg), which is larger than the range of thermo-mechanical exergies shown in Fig. 5.8 of about 0 to 100 kJ/kg. This is because the concentration of Xe in the atmosphere is very small (about 88 ppb). One alternative way to interpret this value is that the *minimum* work required to *produce* pure Xenon *from* the atmosphere is 306.74 kJ/kg of Xenon produced, which makes sense because Xe is so dilute compared to N_2.

"Air" as a whole, by definition, has zero chemical exergy because it is itself at the atmospheric reference condition.

Other gases that are present in the air in trace amounts, but not stable (like methane, nitrous oxide, etc.) can be found in other chapters, such as Chap. 7.

Although water vapor is present in the air, see Chap. 6 for water exergy information.

5.2 Thermo-Mechanical Exergy of Common Gases in Air

Figures 5.1 to 5.8 show the thermo-mechanical exergy of "air" and related gases on a pressure-enthalpy diagram. For all such diagrams in the book, the underlying physical properties were computed with the CoolProps thermodynamic package,[1] which in turn uses physical properties from models provided in the literature as noted. Ultimately, the accuracy of the exergy lines are limited to the accuracy of the underlying physical property models. Note that for a few of the pressure-enthalpy diagrams, the exergy lines are nonconvex in the low-temperature, supercritical region (the "upper left"). This is a consequence of the underlying physical property characteristics in relation to T_0.

Because of its importance, Tables 5.1 to 5.6 also show the thermo-mechanical exergy of "air" as a generic component, for additional precision. The graphical form and the tables are two ways of presenting the same information.

Thermo-mechanical exergy diagrams are provided for:

- Air   Fig. 5.1
- Argon   Fig. 5.2
- Carbon Dioxide Fig. 5.3
- Krypton   Fig. 5.4
- Neon   Fig. 5.5
- Nitrogen   Fig. 5.6
- Oxygen   Fig. 5.7
- Xenon   Fig. 5.8

Though an elemental reference for the atmospheric reference state, H_2 is not naturally present. For the thermo-mechanical exergy of H_2, see Chap. 7.

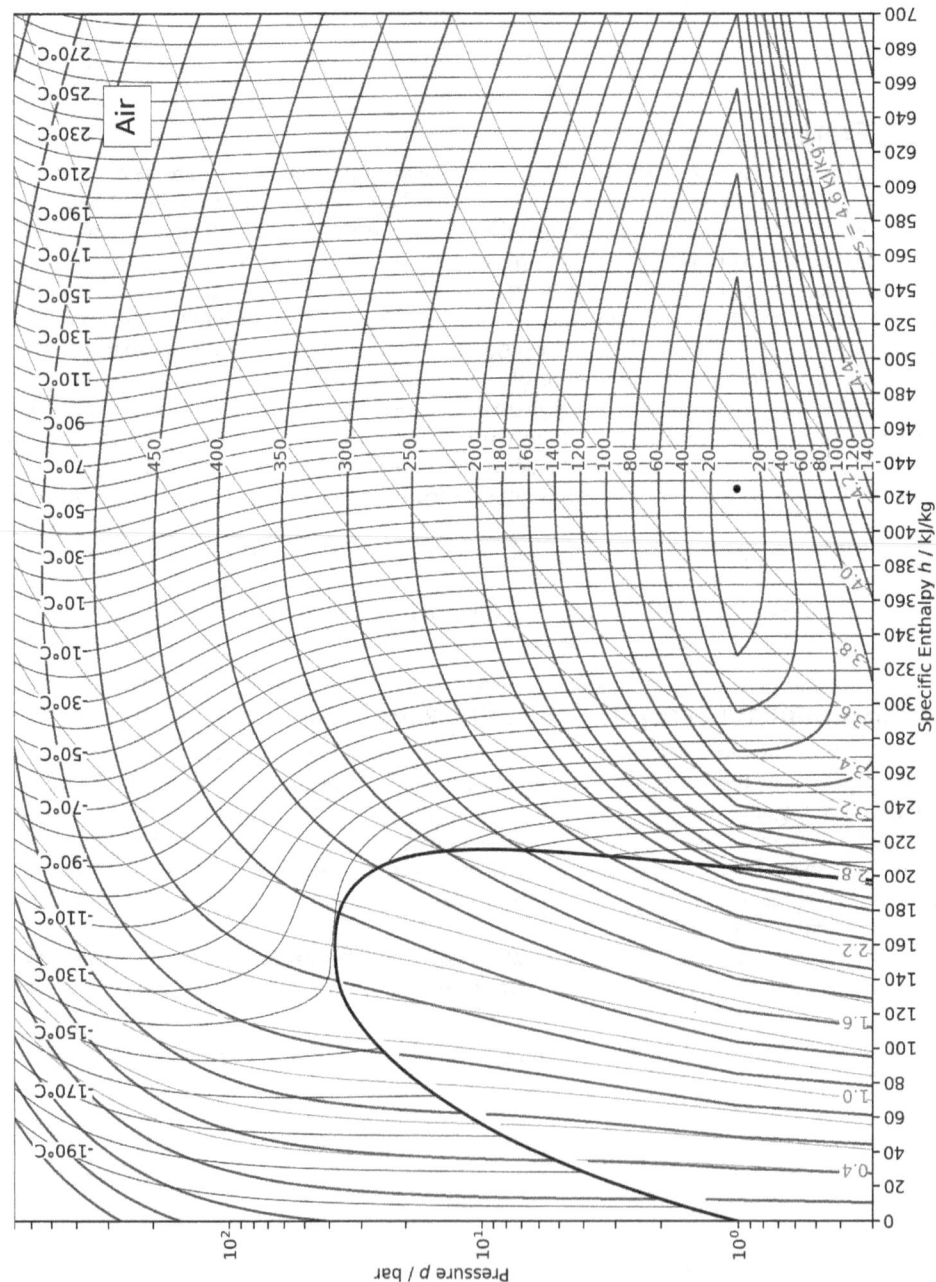

Figure 5.1 e^{tm} for air. Since air has no chemical exergy, this is also the conventional exergy of air. Based on the physical property models of Ref. 2.

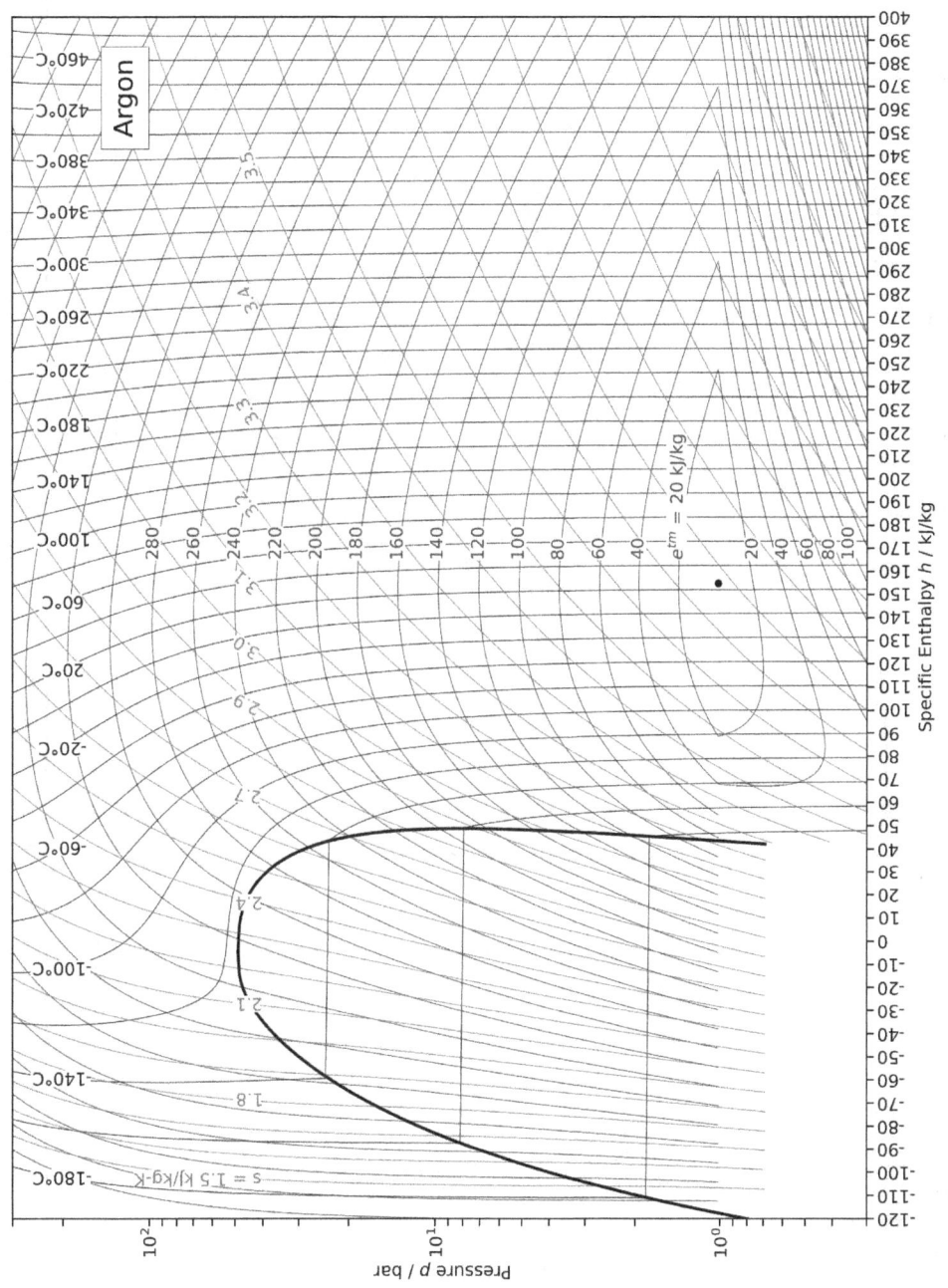

Figure 5.2 e^{tm} for Ar. See Table 4.1 for the corresponding chemical exergy. Based on the physical property models of Ref. 3.

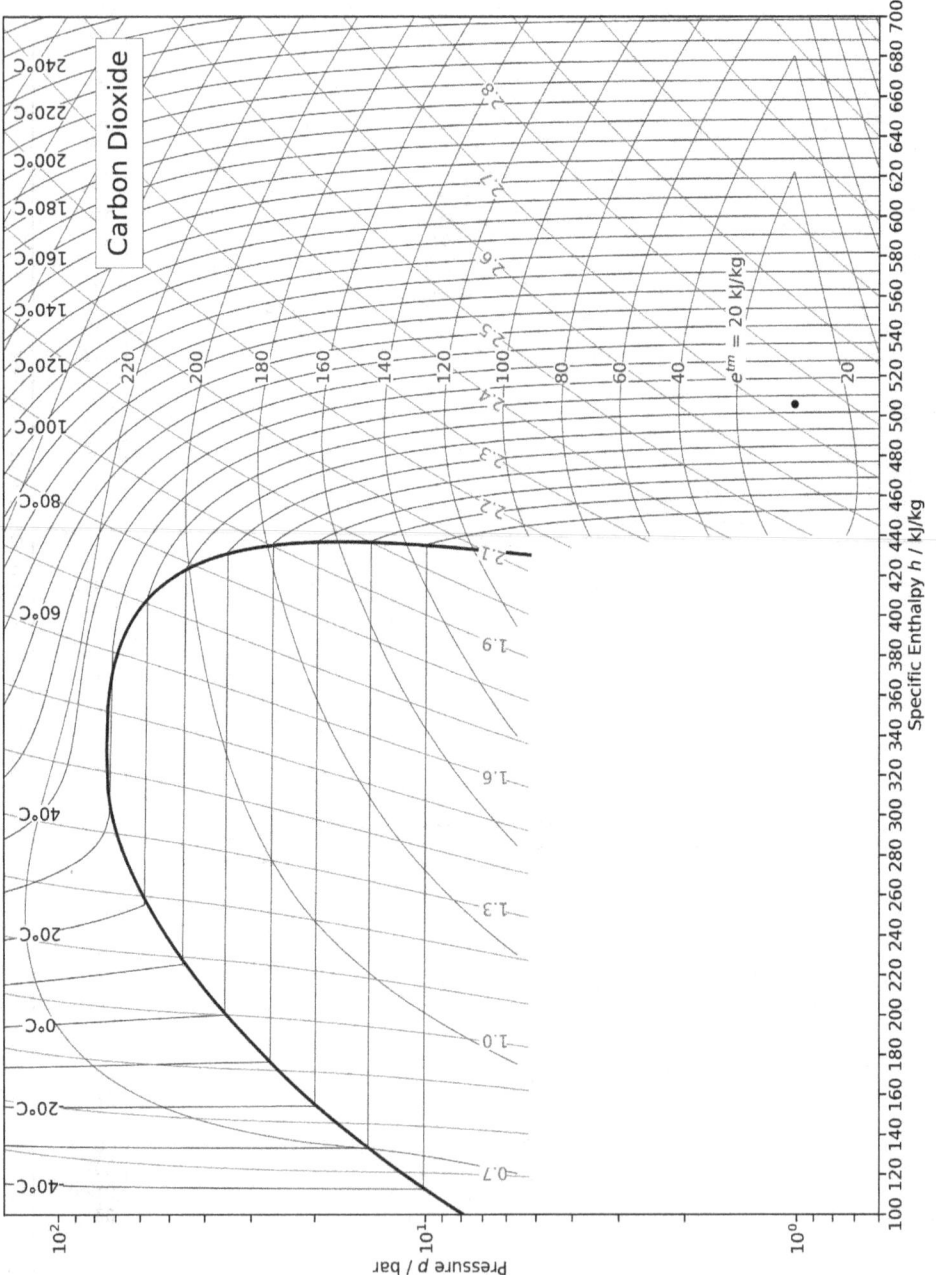

Figure 5.3 e^{tm} for CO_2. See Table 4.1 for the corresponding chemical exergy. Based on the physical property models of Ref. 4.

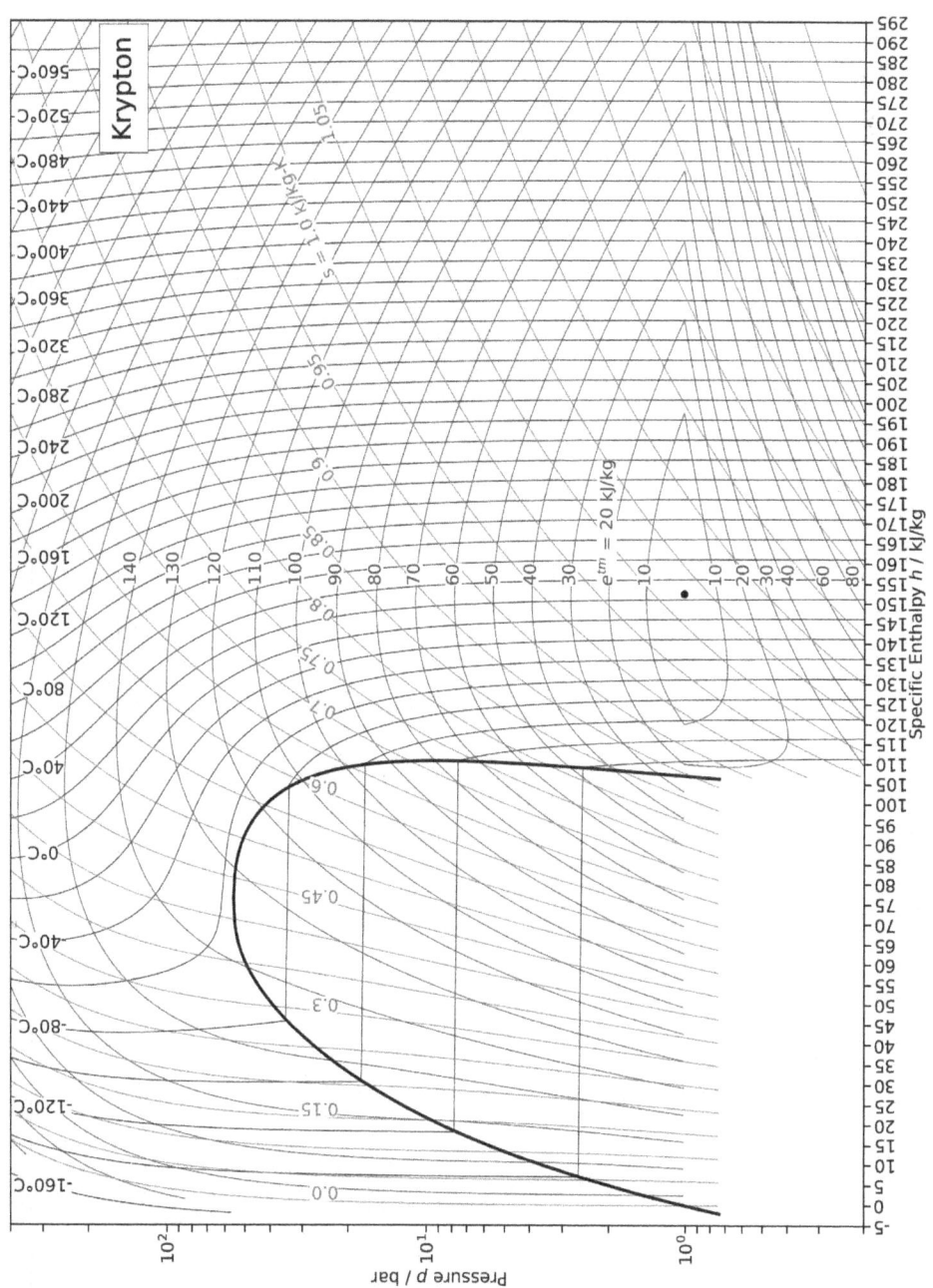

Figure 5.4 e^{tm} for Kr. See Table 4.1 for the corresponding chemical exergy. Based on the physical property models of Ref. 5.

51

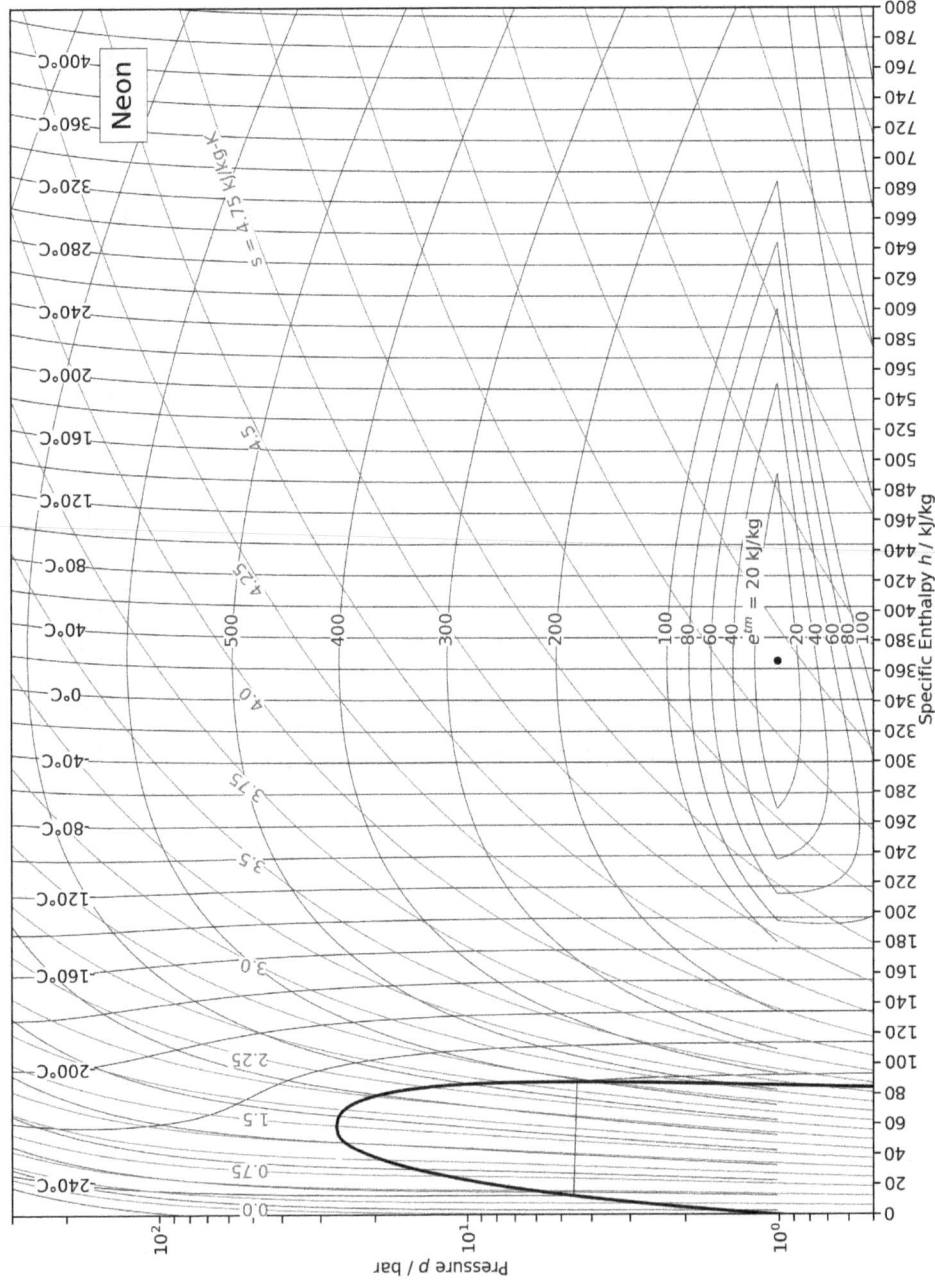

Figure 5.5 e^{tm} for Ne. See Table 4.1 for the corresponding chemical exergy. Based on the physical property models in CoolProps.

52

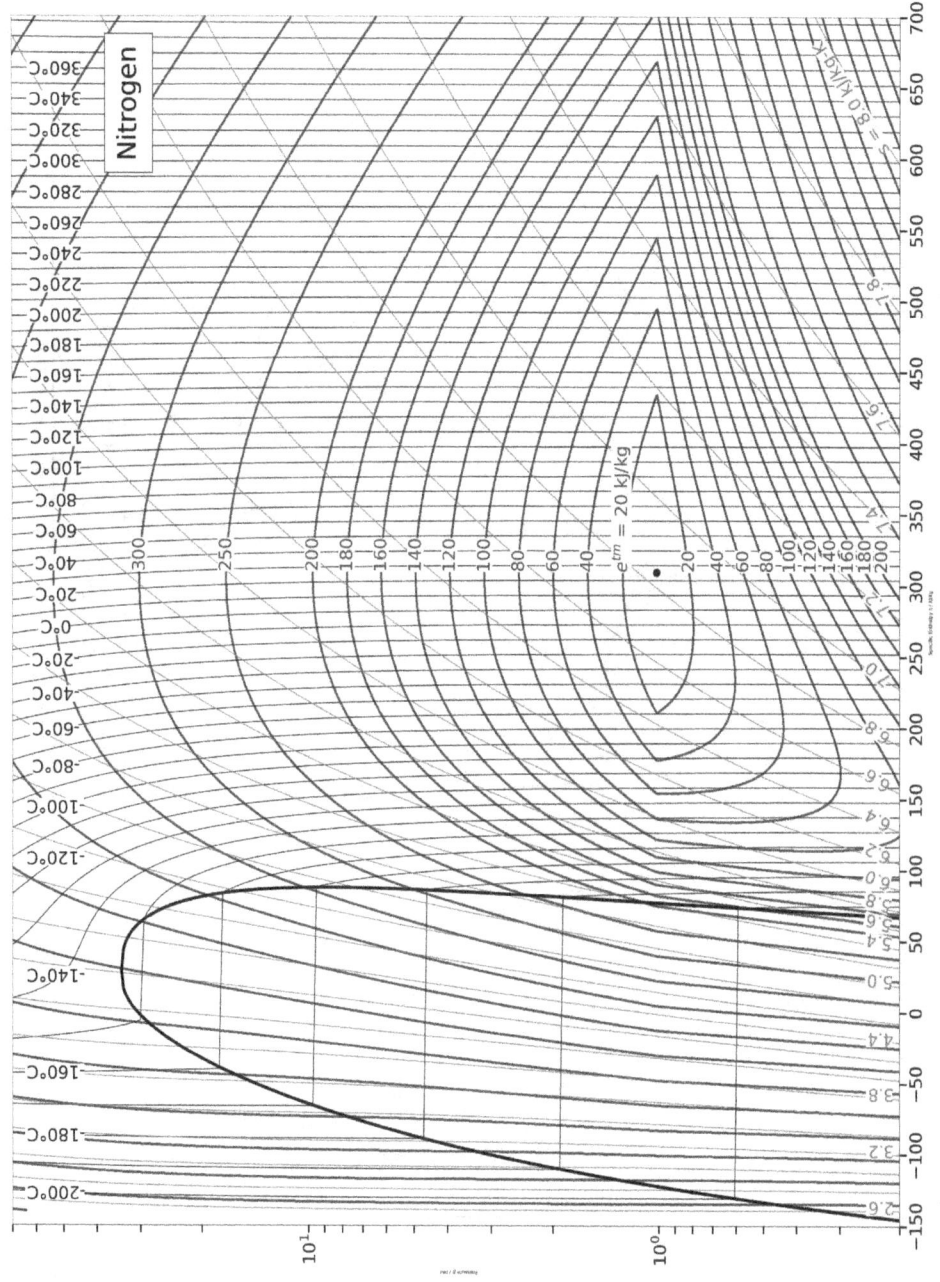

Figure 5.6 e^{tm} for N_2. See Table 4.1 for the corresponding chemical exergy. Based on the physical property models of Ref. 6.

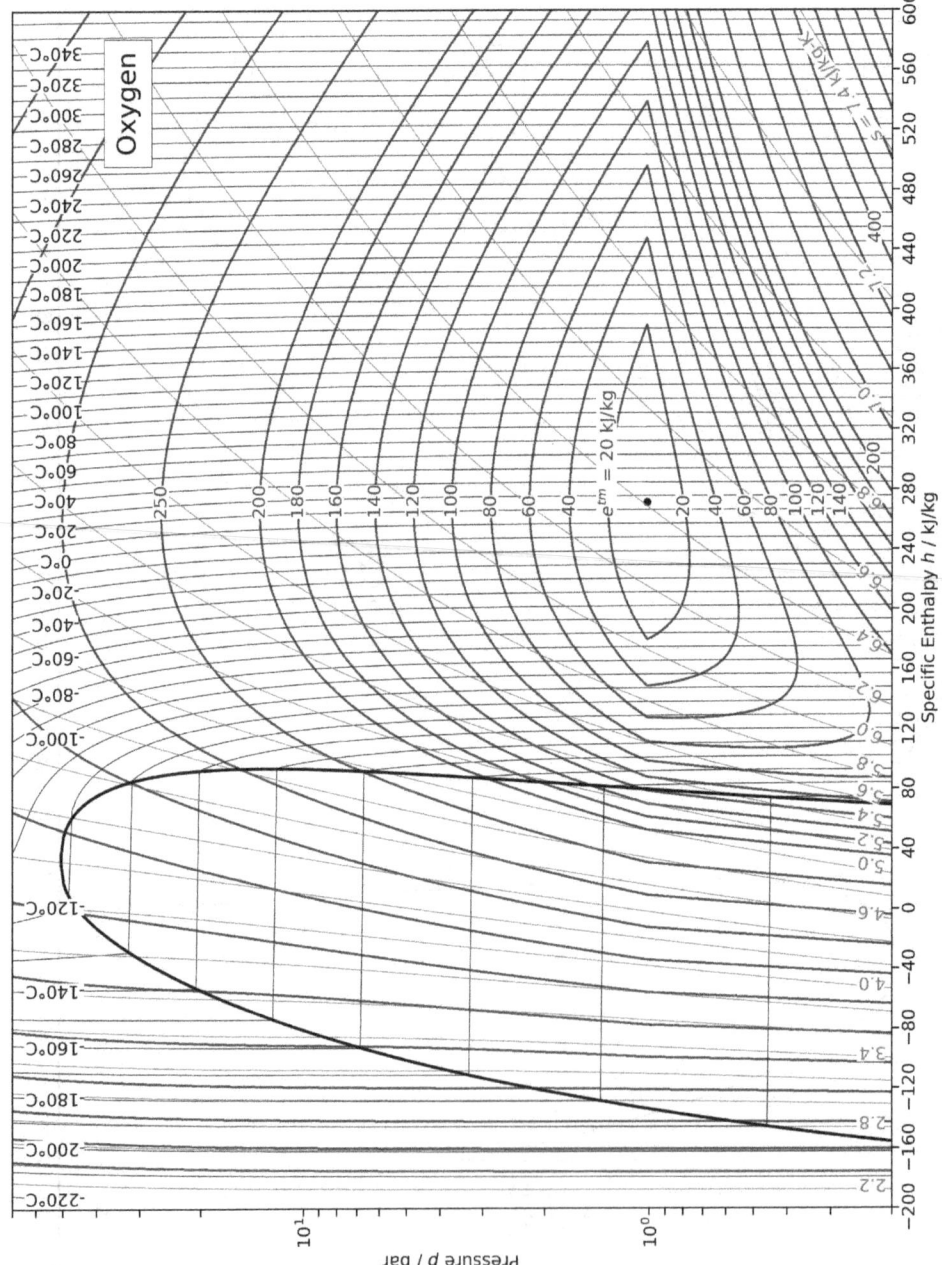

Figure 5.7 e^{tm} for O_2. See Table 4.1 for the corresponding chemical exergy. Based on the physical property models of Refs. 7 and 8.

54

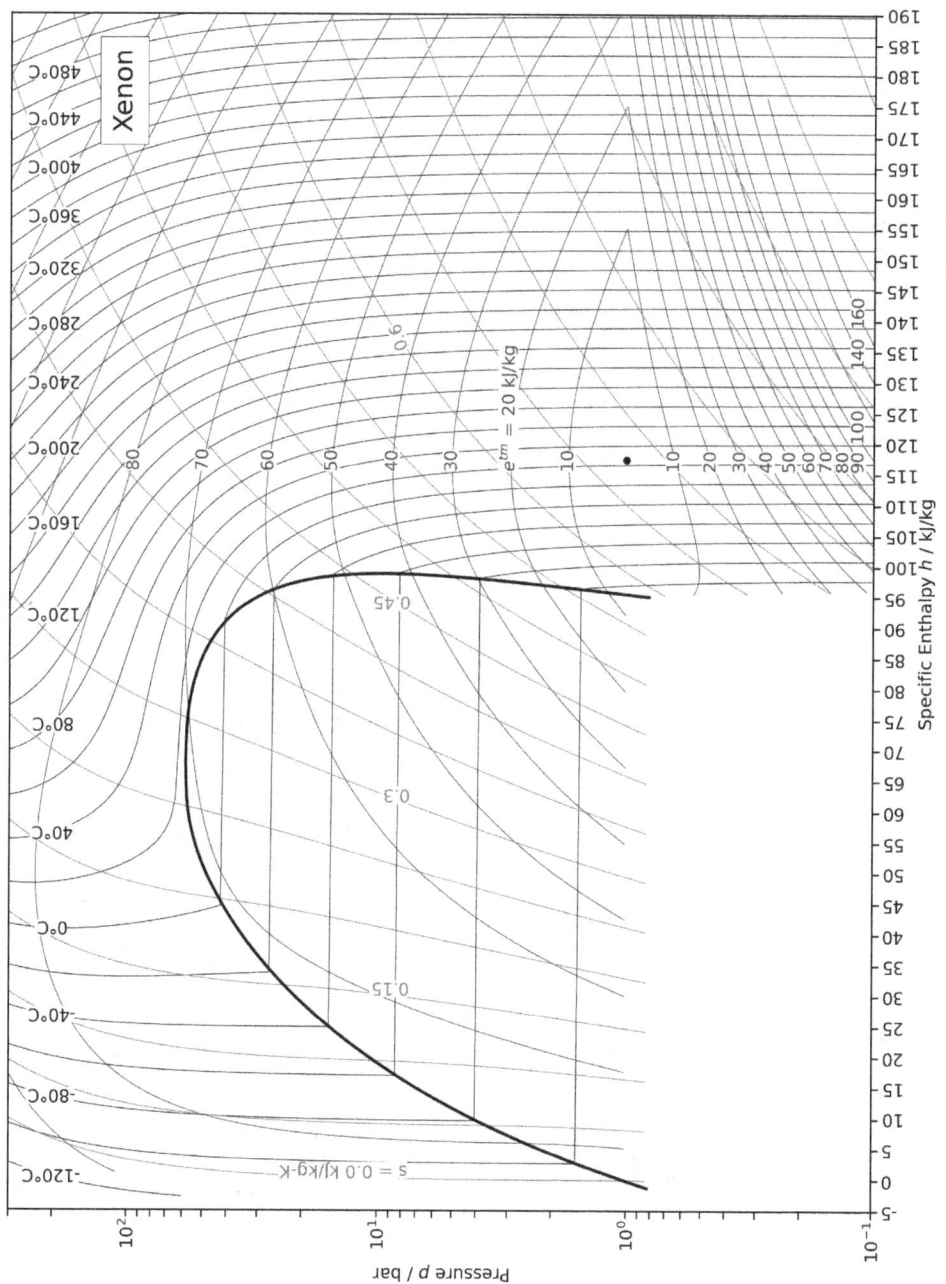

Figure 5.8 e^{tm} for Xe. See Table 4.1 for the corresponding chemical exergy. Based on the physical property models of Ref. 5.

e^tm (kJ/kg)

| T (°C) | \multicolumn{15}{c}{Pressure (bar)} |
	0.00100	0.00164	0.00268	0.00439	0.00720	0.0118	0.0193	0.0316	0.0518	0.0848	0.1390	0.2276	0.3728	0.6105	1.0000
-150	1024.0	838.02	679.90	546.70	435.70	344.36	270.34	211.50	165.91	131.86	107.88	92.645	85.067	84.221	89.371
-140	1129.8	927.10	753.97	607.35	484.41	382.51	299.20	232.23	179.54	139.32	109.96	90.045	78.384	73.955	75.929
-130	1236.1	1017.0	829.17	669.33	534.60	422.24	329.69	254.62	194.85	148.46	113.72	89.139	73.404	65.412	64.243
-120	1342.8	1107.6	905.23	732.37	586.00	463.28	361.57	278.43	211.60	159.05	118.95	89.696	69.895	58.351	54.062
-110	1449.6	1198.6	981.95	796.27	638.41	505.43	394.63	303.47	229.61	170.91	125.45	91.530	67.666	52.579	45.186
-100	1556.4	1289.9	1059.2	860.85	691.64	548.53	428.72	329.59	248.72	183.88	133.07	94.486	66.565	47.941	37.454
-90	1663.2	1381.2	1136.7	925.97	745.56	592.43	463.68	356.63	268.79	197.83	141.68	98.439	66.463	44.307	30.734
-80	1769.7	1472.7	1214.6	991.50	800.05	637.01	499.41	384.50	289.71	212.66	151.17	103.28	67.253	41.569	24.917
-70	1876.0	1564.1	1292.5	1057.4	855.00	682.16	535.80	413.09	311.39	228.27	161.45	108.92	68.845	39.635	19.908
-60	1982.0	1655.4	1370.6	1123.5	910.33	727.80	572.76	442.31	333.75	244.57	172.45	115.28	71.160	38.428	15.630
-50	2087.7	1746.6	1448.7	1189.7	965.96	773.85	610.22	472.09	356.72	261.51	184.10	122.29	74.133	37.880	12.015
-40	2193.1	1837.6	1526.8	1256.1	1021.8	820.25	648.11	502.37	380.22	279.02	196.33	129.90	77.704	37.934	9.0028
-30	2298.0	1928.4	1604.8	1322.6	1077.9	866.94	686.37	533.09	404.21	297.04	209.09	138.05	81.822	38.537	6.5434
-20	2402.6	2019.0	1682.7	1389.1	1134.1	913.89	724.97	564.20	428.63	315.53	222.34	146.70	86.444	39.647	4.5917
-10	2506.8	2109.3	1760.5	1455.7	1190.5	961.03	763.85	595.66	453.45	334.45	236.04	155.80	91.530	41.223	3.1082
0	2610.6	2199.4	1838.2	1522.2	1246.9	1008.4	802.97	627.42	478.63	353.76	250.15	165.33	97.044	43.230	2.0578
10	2713.9	2289.2	1915.7	1588.6	1303.3	1055.8	842.32	659.46	504.13	373.42	264.63	175.25	102.96	45.639	1.4094
20	2816.9	2378.7	1993.1	1655.0	1359.9	1103.4	881.84	691.75	529.92	393.41	279.47	185.53	109.24	48.419	1.1351
30	2919.5	2467.9	2070.3	1721.4	1416.4	1151.1	921.54	724.25	555.97	413.70	294.62	196.15	115.86	51.548	1.2101
40	3021.7	2556.9	2147.3	1787.6	1472.9	1198.8	961.37	756.96	582.27	434.26	310.08	207.08	122.81	55.002	1.6118
50	3123.5	2645.6	2224.2	1853.8	1529.4	1246.6	1001.3	789.84	608.79	455.08	325.81	218.31	130.06	58.761	2.3200
60	3225.0	2734.1	2300.9	1919.9	1586.0	1294.5	1041.4	822.88	635.51	476.13	341.81	229.81	137.59	62.807	3.3164
70	3326.1	2822.2	2377.4	1985.9	1642.4	1342.4	1081.6	856.07	662.42	497.40	358.04	241.56	145.39	67.123	4.5844
80	3426.8	2910.2	2453.8	2051.8	1698.9	1390.3	1121.8	889.39	689.50	518.87	374.50	253.56	153.44	71.695	6.1088
90	3527.2	2997.8	2529.9	2117.6	1755.3	1438.3	1162.1	922.82	716.73	540.54	391.17	265.78	161.72	76.507	7.8759
100	3627.3	3085.3	2605.9	2183.2	1811.7	1486.3	1202.5	956.37	744.11	562.37	408.04	278.22	170.23	81.548	9.8731
110	3727.1	3172.5	2681.8	2248.8	1868.0	1534.3	1243.0	990.02	771.63	584.38	425.10	290.87	178.95	86.806	12.089
120	3826.6	3259.4	2757.5	2314.3	1924.3	1582.3	1283.5	1023.8	799.27	606.53	442.33	303.70	187.87	92.271	14.513
130	3925.8	3346.2	2833.0	2379.8	1980.6	1630.3	1324.0	1057.6	827.04	628.84	459.73	316.72	196.99	97.932	17.135
140	4024.7	3432.8	2908.4	2445.1	2036.8	1678.3	1364.6	1091.5	854.92	651.28	477.29	329.91	206.28	103.78	19.946
150	4123.4	3519.2	2983.7	2510.3	2093.0	1726.3	1405.3	1125.5	882.90	673.85	495.01	343.26	215.76	109.81	22.938
160	4221.8	3605.3	3058.8	2575.5	2149.2	1774.3	1446.0	1159.5	910.98	696.55	512.86	356.78	225.40	116.01	26.104
170	4320.0	3691.4	3133.8	2640.6	2205.3	1822.4	1486.7	1193.7	939.16	719.37	530.85	370.44	235.20	122.38	29.437

TABLE 5.1   Thermo-mechanical exergy of air at vacuum pressures. All are in the gas phase.

T (°C)	e^{tm} (kJ/kg) Pressure (bar)														
	0.00100	0.00164	0.00268	0.00439	0.00720	0.0118	0.0193	0.0316	0.0518	0.0848	0.1390	0.2276	0.3728	0.6105	1.0000
180	4418.0	3777.2	3208.8	2705.6	2261.4	1870.4	1527.4	1227.9	967.43	742.30	548.97	384.25	245.15	128.90	32.929
190	4515.8	3862.9	3283.6	2770.5	2317.4	1918.4	1568.2	1262.1	995.79	765.34	567.22	398.20	255.25	135.58	36.575
200	4613.4	3948.5	3358.3	2835.4	2373.5	1966.5	1609.1	1296.4	1024.2	788.48	585.59	412.28	265.49	142.40	40.369
210	4710.8	4033.9	3432.9	2900.3	2429.5	2014.5	1649.9	1330.8	1052.8	811.72	604.08	426.49	275.87	149.37	44.307
220	4808.0	4119.2	3507.4	2965.1	2485.5	2062.6	1690.9	1365.3	1081.4	835.06	622.68	440.82	286.38	156.47	48.382
230	4905.1	4204.4	3581.8	3029.8	2541.5	2110.7	1731.8	1399.8	1110.0	858.50	641.38	455.27	297.02	163.70	52.591
240	5002.0	4289.5	3656.2	3094.5	2597.5	2158.8	1772.8	1434.3	1138.8	882.03	660.20	469.84	307.78	171.06	56.929
250	5098.8	4374.5	3730.5	3159.2	2653.4	2206.9	1813.8	1468.9	1167.6	905.65	679.11	484.52	318.66	178.55	61.392
260	5195.5	4459.4	3804.8	3223.8	2709.4	2255.0	1854.8	1503.6	1196.5	929.35	698.12	499.32	329.66	186.15	65.977
270	5292.1	4544.2	3879.0	3288.5	2765.4	2303.2	1895.9	1538.3	1225.5	953.14	717.23	514.21	340.77	193.87	70.680
280	5388.6	4629.0	3953.2	3353.1	2821.3	2351.3	1937.1	1573.1	1254.6	977.01	736.44	529.21	351.99	201.71	75.497
290	—	4713.6	4027.3	3417.6	2877.3	2399.5	1978.2	1607.9	1283.7	1001.0	755.73	544.31	363.32	209.65	80.425
300	—	4798.3	4101.4	3482.2	2933.3	2447.7	2019.5	1642.8	1312.9	1025.0	775.12	559.51	374.75	217.71	85.462
310	—	4882.8	4175.4	3546.8	2989.3	2496.0	2060.7	1677.8	1342.1	1049.1	794.59	574.80	386.28	225.86	90.605
320	—	4967.4	4249.5	3611.3	3045.3	2544.3	2102.0	1712.8	1371.4	1073.3	814.15	590.18	397.91	234.12	95.850
330	—	5051.8	4323.5	3675.9	3101.3	2592.6	2143.4	1747.8	1400.8	1097.6	833.79	605.66	409.63	242.48	101.20
340	—	5136.3	4397.5	3740.4	3157.3	2640.9	2184.7	1783.0	1430.3	1121.9	853.52	621.23	421.45	250.93	106.64
350	—	5220.7	4471.5	3805.0	3213.4	2689.3	2226.2	1818.1	1459.8	1146.3	873.32	636.88	433.36	259.48	112.18
360	—	5305.1	4545.4	3869.6	3269.4	2737.7	2267.7	1853.4	1489.4	1170.8	893.21	652.61	445.36	268.12	117.81
370	—	—	4619.4	3934.2	3325.6	2786.1	2309.2	1888.6	1519.0	1195.3	913.17	668.44	457.45	276.85	123.54
380	—	—	4693.4	3998.8	3381.7	2834.6	2350.7	1924.0	1548.7	1220.0	933.21	684.34	469.63	285.67	129.35
390	—	—	4767.4	4063.4	3437.8	2883.1	2392.4	1959.4	1578.5	1244.7	953.33	700.32	481.88	294.58	135.25
400	—	—	4841.4	4128.0	3494.0	2931.7	2434.0	1994.8	1608.3	1269.4	973.52	716.38	494.22	303.57	141.24
410	—	—	4915.4	4192.7	3550.3	2980.3	2475.8	2030.3	1638.2	1294.3	993.78	732.52	506.64	312.64	147.31
420	—	—	4989.4	4257.4	3606.5	3028.9	2517.5	2065.9	1668.2	1319.2	1014.1	748.74	519.14	321.80	153.46
430	—	—	5063.4	4322.1	3662.8	3077.6	2559.3	2101.5	1698.2	1344.1	1034.5	765.02	531.72	331.03	159.69
440	—	—	5137.5	4386.8	3719.1	3126.3	2601.2	2137.2	1728.3	1369.2	1055.0	781.39	544.37	340.35	166.00
450	—	—	5211.6	4451.6	3775.5	3175.1	2643.1	2172.9	1758.4	1394.3	1075.5	797.82	557.10	349.74	172.39
460	—	—	—	4516.4	3831.9	3223.9	2685.1	2208.7	1788.6	1419.4	1096.2	814.33	569.90	359.20	178.85
470	—	—	—	4581.2	3888.3	3272.8	2727.1	2244.5	1818.9	1444.6	1116.8	830.90	582.78	368.74	185.39
480	—	—	—	4646.1	3944.8	3321.7	2769.2	2280.4	1849.2	1469.9	1137.6	847.55	595.72	378.35	192.00
490	—	—	—	4711.0	4001.3	3370.6	2811.3	2316.3	1879.6	1495.3	1158.4	864.26	608.74	388.03	198.68
500	—	—	—	4775.9	4057.9	3419.6	2853.5	2352.4	1910.0	1520.7	1179.3	881.04	621.82	397.78	205.43
520	—	—	—	4905.9	4171.2	3517.8	2938.0	2424.5	1971.0	1571.7	1221.2	914.79	648.18	417.49	219.15
540	—	—	—	5036.1	4284.6	3616.1	3022.7	2496.9	2032.3	1622.9	1263.4	948.79	674.81	437.46	233.12
560	—	—	—	—	4398.2	3714.6	3107.6	2569.5	2093.8	1674.4	1305.8	983.04	701.69	457.68	247.36

Table 5.1 Thermo-mechanical exergy of air at vacuum pressures. All are in the gas phase. (Continued)

	e^tm (kJ/kg)														
	Pressure (bar)														
T (°C)	0.00100	0.00164	0.00268	0.00439	0.00720	0.0118	0.0193	0.0316	0.0518	0.0848	0.1390	0.2276	0.3728	0.6105	1.0000
580	—	—	—	—	4512.0	3813.4	3192.7	2642.3	2155.5	1726.1	1348.4	1017.5	728.81	478.16	261.84
600	—	—	—	—	4625.9	3912.3	3277.9	2715.3	2217.4	1778.0	1391.3	1052.2	756.16	498.87	276.56
620	—	—	—	—	4740.1	4011.3	3363.4	2788.5	2279.6	1830.1	1434.4	1087.2	783.74	519.81	291.51
640	—	—	—	—	4854.4	4110.6	3449.1	2862.0	2341.9	1882.5	1477.7	1122.3	811.55	540.97	306.69
660	—	—	—	—	4968.8	4210.1	3535.0	2935.6	2404.5	1935.0	1521.2	1157.7	839.57	562.36	322.08
680	—	—	—	—	—	4309.7	3621.1	3009.4	2467.2	1987.8	1565.0	1193.3	867.79	583.95	337.69
700	—	—	—	—	—	4409.5	3707.3	3083.4	2530.1	2040.7	1608.9	1229.1	896.22	605.74	353.50
720	—	—	—	—	—	4509.5	3793.8	3157.6	2593.2	2093.8	1653.0	1265.0	924.84	627.73	369.50
740	—	—	—	—	—	4609.7	3880.4	3231.9	2656.5	2147.1	1697.3	1301.2	953.66	649.91	385.70
760	—	—	—	—	—	4710.0	3967.1	3306.5	2720.0	2200.6	1741.8	1337.5	982.65	672.28	402.08
780	—	—	—	—	—	4810.5	4054.1	3381.2	2783.7	2254.3	1786.4	1374.1	1011.8	694.82	418.64
800	—	—	—	—	—	—	4141.2	3456.0	2847.5	2308.1	1831.3	1410.8	1041.2	717.54	435.38
820	—	—	—	—	—	—	4228.5	3531.1	2911.5	2362.1	1876.2	1447.6	1070.7	740.43	452.29
840	—	—	—	—	—	—	4315.9	3606.3	2975.6	2416.3	1921.4	1484.6	1100.4	763.49	469.36
860	—	—	—	—	—	—	4403.5	3681.6	3039.9	2470.6	1966.7	1521.8	1130.2	786.70	486.59
880	—	—	—	—	—	—	4491.2	3757.1	3104.4	2525.1	2012.2	1559.2	1160.2	810.07	503.98
900	—	—	—	—	—	—	4579.1	3832.8	3169.0	2579.7	2057.8	1596.7	1190.4	833.59	521.51
920	—	—	—	—	—	—	4667.1	3908.6	3233.7	2634.5	2103.6	1634.3	1220.7	857.25	539.20
940	—	—	—	—	—	—	—	3984.5	3298.6	2689.4	2149.5	1672.1	1251.1	881.06	557.03
960	—	—	—	—	—	—	—	4060.6	3363.7	2744.5	2195.5	1710.0	1281.7	905.01	574.99
980	—	—	—	—	—	—	—	4136.8	3428.8	2799.7	2241.7	1748.0	1312.4	929.09	593.09
1000	—	—	—	—	—	—	—	4213.1	3494.1	2855.0	2288.0	1786.2	1343.2	953.31	611.32
1100	—	—	—	—	—	—	—	—	3822.4	3133.4	2521.4	1979.0	1499.3	1076.2	704.34
1200	—	—	—	—	—	—	—	—	4153.4	3414.7	2757.7	2174.5	1658.1	1202.0	800.18
1300	—	—	—	—	—	—	—	—	—	3698.3	2996.4	2372.6	1819.5	1330.2	898.50
1400	—	—	—	—	—	—	—	—	—	3984.2	3237.3	2572.9	1983.1	1460.7	999.05
1500	—	—	—	—	—	—	—	—	—	—	3480.1	2775.1	2148.7	1593.1	1101.6
1600	—	—	—	—	—	—	—	—	—	—	3724.8	2979.2	2316.0	1727.4	1205.9
1700	—	—	—	—	—	—	—	—	—	—	—	3184.8	2485.0	1863.2	1311.9
1800	—	—	—	—	—	—	—	—	—	—	—	3392.0	2655.5	2000.6	1419.4
1900	—	—	—	—	—	—	—	—	—	—	—	—	2827.3	2139.3	1528.2
2000	—	—	—	—	—	—	—	—	—	—	—	—	3000.4	2279.3	1638.3

TABLE 5.1 Thermo-mechanical exergy of air at vacuum pressures. All are in the gas phase. (Continued)

T (°C)	e^{tm} (kJ/kg) Pressure (bar)														
	1.01325	2	3	4	5	6	7	8	9	10	11	12	13	14	15
-150	89.59	148.51	183.98	209.40	229.32	245.79	259.88	272.25	283.31	293.36	302.60	311.20	319.26	326.89	334.17
-140	76.06	134.77	170.00	195.17	214.83	231.01	244.81	256.86	267.58	277.26	286.11	294.27	301.86	308.97	315.68
-130	64.30	122.85	157.93	182.93	202.42	218.42	232.03	243.88	254.39	263.86	272.47	280.39	287.72	294.56	300.98
-120	54.04	112.49	147.45	172.34	191.71	207.60	221.08	232.80	243.17	252.50	260.96	268.73	275.91	282.58	288.83
-110	45.09	103.46	138.35	163.15	182.44	198.24	211.63	223.26	233.55	242.78	251.15	258.81	265.89	272.45	278.59
-100	37.28	95.60	130.43	155.18	174.41	190.14	203.47	215.04	225.26	234.42	242.73	250.32	257.32	263.82	269.88
-90	30.48	88.76	123.55	148.26	167.44	183.14	196.42	207.94	218.12	227.23	235.49	243.03	249.99	256.43	262.44
-80	24.59	82.83	117.59	142.27	161.42	177.09	190.34	201.83	211.97	221.05	229.27	236.78	243.70	250.10	256.08
-70	19.50	77.73	112.46	137.12	156.25	171.89	185.11	196.58	206.69	215.75	223.94	231.43	238.32	244.70	250.65
-60	15.15	73.36	108.07	132.71	151.82	167.44	180.65	192.10	202.20	211.23	219.41	226.88	233.75	240.11	246.04
-50	11.45	69.65	104.36	128.98	148.08	163.69	176.88	188.32	198.40	207.42	215.59	223.04	229.89	236.24	242.15
-40	8.37	66.55	101.25	125.86	144.95	160.55	173.74	185.16	195.23	204.25	212.40	219.84	226.69	233.03	238.93
-30	5.83	64.01	98.70	123.31	142.39	157.98	171.16	182.57	192.64	201.65	209.79	217.23	224.06	230.40	236.29
-20	3.80	61.98	96.66	121.26	140.34	155.93	169.10	180.51	190.57	199.57	207.71	215.14	221.97	228.30	234.19
-10	2.24	60.42	95.09	119.69	138.77	154.35	167.52	178.93	188.98	197.98	206.12	213.54	220.37	226.69	232.58
0	1.12	59.29	93.96	118.56	137.63	153.21	166.38	177.78	187.84	196.83	204.97	212.39	219.22	225.54	231.42
10	0.39	58.56	93.24	117.83	136.90	152.48	165.65	177.05	187.10	196.10	204.23	211.65	218.48	224.79	230.68
20	0.04	58.21	92.88	117.48	136.55	152.13	165.29	176.69	186.75	195.74	203.87	211.29	218.12	224.44	230.32
30	0.04	58.21	92.88	117.48	136.55	152.13	165.29	176.75	186.75	195.74	203.87	211.29	218.12	224.44	230.32
40	0.37	58.54	93.21	117.80	136.88	152.45	165.62	177.02	187.08	196.07	204.20	211.62	218.45	224.77	230.65
50	1.00	59.17	93.84	118.44	137.51	153.09	166.26	177.66	187.72	196.71	204.84	212.27	219.09	225.41	231.29
60	1.92	60.09	94.77	119.36	138.44	154.02	167.19	178.59	188.65	197.64	205.77	213.20	220.03	226.35	232.23
70	3.11	61.28	95.96	120.56	139.63	155.22	168.39	179.79	189.85	198.85	206.98	214.41	221.24	227.56	233.44
80	4.56	62.74	97.41	122.01	141.09	156.67	169.84	181.25	191.31	200.31	208.45	215.87	222.70	229.03	234.91
90	6.25	64.43	99.11	123.71	142.79	158.37	171.55	182.95	193.02	202.02	210.15	217.58	224.42	230.74	236.63
100	8.17	66.35	101.03	125.64	144.72	160.30	173.48	184.89	194.95	203.95	212.09	219.52	226.36	232.69	238.57
110	10.31	68.49	103.18	127.78	146.86	162.45	175.63	187.04	197.11	206.11	214.25	221.69	228.52	234.85	240.74
120	12.66	70.84	105.53	130.14	149.22	164.81	177.99	189.40	199.47	208.48	216.62	224.05	230.89	237.22	243.12
130	15.21	73.39	108.08	132.69	151.77	167.37	180.55	191.96	202.03	211.04	219.19	226.62	233.46	239.80	245.69
140	17.94	76.13	110.82	135.43	154.52	170.11	183.30	194.71	204.79	213.79	221.94	229.38	236.22	242.56	248.46
150	20.86	79.05	113.74	138.35	157.44	173.04	186.22	197.65	207.72	216.73	224.88	232.32	239.16	245.50	251.40
160	23.95	82.14	116.83	141.45	160.54	176.14	189.33	200.75	210.83	219.84	227.99	235.43	242.28	248.62	254.52
170	27.21	85.40	120.10	144.71	163.81	179.41	192.60	204.02	214.10	223.11	231.27	238.71	245.56	251.90	257.80

TABLE 5.2 Thermo-mechanical exergy of air, 1.01325 to 15 bar. All are in the gas phase.

| | e^{tm} (kJ/kg) | | | | | | | | | | | | | | |
| | Pressure (bar) | | | | | | | | | | | | | | |
T (°C)	1.01325	2	3	4	5	6	7	8	9	10	11	12	13	14	15
180	30.62	88.82	123.52	148.14	167.23	182.83	196.03	207.45	217.53	226.55	234.70	242.15	249.00	255.34	261.25
190	34.19	92.39	127.09	151.71	170.81	186.41	199.61	211.04	221.12	230.14	238.30	245.74	252.60	258.94	264.85
200	37.91	96.11	130.81	155.44	174.54	190.14	203.34	214.77	224.85	233.87	242.03	249.48	256.34	262.69	268.59
210	41.77	99.97	134.68	159.30	178.41	194.01	207.21	218.64	228.73	237.75	245.92	253.37	260.22	266.57	272.48
220	45.77	103.98	138.68	163.31	182.41	198.02	211.22	222.66	232.75	241.77	249.93	257.39	264.25	270.60	276.51
230	49.91	108.11	142.82	167.45	186.55	202.17	215.37	226.80	236.89	245.92	254.09	261.54	268.40	274.76	280.67
240	54.17	112.37	147.09	171.72	190.82	206.44	219.64	231.08	241.17	250.20	258.37	265.83	272.69	279.04	284.96
250	58.56	116.76	151.48	176.11	195.22	210.84	224.04	235.48	245.57	254.60	262.77	270.23	277.10	283.45	289.37
260	63.07	121.27	155.99	180.62	199.74	215.35	228.56	240.00	250.10	259.13	267.30	274.76	281.63	287.99	293.91
270	67.69	125.90	160.62	185.26	204.37	219.99	233.20	244.64	254.74	263.77	271.95	279.41	286.28	292.64	298.56
280	72.43	130.65	165.37	190.00	209.12	224.74	237.95	249.40	259.49	268.53	276.71	284.17	291.04	297.40	303.32
290	77.29	135.50	170.22	194.86	213.98	229.60	242.81	254.26	264.36	273.40	281.58	289.04	295.91	302.28	308.20
300	82.25	140.46	175.19	199.83	218.95	234.57	247.78	259.23	269.34	278.37	286.55	294.02	300.89	307.26	313.19
310	87.32	145.53	180.26	204.90	224.02	239.65	252.86	264.31	274.42	283.46	291.64	299.11	305.98	312.35	318.28
320	92.49	150.70	185.43	210.07	229.20	244.82	258.04	269.49	279.60	288.64	296.82	304.29	311.17	317.54	323.47
330	97.76	155.98	190.70	215.35	234.47	250.10	263.32	274.77	284.88	293.92	302.11	309.58	316.46	322.83	328.76
340	103.12	161.35	196.07	220.72	239.85	255.48	268.70	280.15	290.26	299.31	307.49	314.97	321.85	328.22	334.15
350	108.59	166.81	201.54	226.19	245.32	260.95	274.17	285.63	295.74	304.78	312.97	320.45	327.33	333.70	339.63
360	114.14	172.37	207.10	231.75	250.88	266.51	279.74	291.19	301.31	310.35	318.54	326.02	332.90	339.27	345.21
370	119.79	178.02	212.75	237.40	256.53	272.17	285.39	296.85	306.97	316.01	324.20	331.68	338.57	344.94	350.88
380	125.53	183.76	218.49	243.14	262.28	277.91	291.14	302.60	312.71	321.76	329.96	337.44	344.32	350.70	356.64
390	131.35	189.58	224.32	248.97	268.11	283.74	296.97	308.43	318.55	327.60	335.79	343.28	350.16	356.54	362.48
400	137.26	195.50	230.23	254.89	274.02	289.66	302.89	314.35	324.47	333.52	341.72	349.20	356.09	362.47	368.41
410	143.26	201.49	236.23	260.89	280.02	295.66	308.89	320.36	330.48	339.53	347.73	355.21	362.10	368.48	374.42
420	149.33	207.57	242.31	266.97	286.10	301.74	314.98	326.44	336.56	345.62	353.81	361.30	368.19	374.57	380.52
430	155.49	213.73	248.47	273.13	292.26	307.91	321.14	332.61	342.73	351.79	359.98	367.47	374.36	380.74	386.69
440	161.73	219.96	254.70	279.37	298.50	314.15	327.38	338.85	348.97	358.03	366.23	373.72	380.61	386.99	392.94
450	168.04	226.28	261.02	285.68	304.82	320.47	333.70	345.17	355.29	364.35	372.55	380.04	386.94	393.32	399.27
460	174.43	232.66	267.41	292.07	311.21	326.86	340.10	351.57	361.69	370.75	378.95	386.44	393.34	399.73	405.67
470	180.89	239.13	273.87	298.54	317.68	333.33	346.57	358.04	368.16	377.23	385.43	392.92	399.82	406.20	412.15
480	187.42	245.66	280.41	305.08	324.22	339.87	353.11	364.58	374.71	383.77	391.97	399.47	406.36	412.75	418.70
490	194.03	252.27	287.02	311.69	330.83	346.48	359.72	371.20	381.32	390.39	398.59	406.09	412.98	419.37	425.33
500	200.71	258.95	293.70	318.37	337.51	353.16	366.40	377.88	388.01	397.07	405.28	412.77	419.67	426.06	432.02
520	214.27	272.51	307.26	331.93	351.08	366.73	379.98	391.45	401.58	410.65	418.86	426.36	433.26	439.65	445.61

TABLE 5.2 Thermo-mechanical exergy of air, 1.01325 to 15 bar. All are in the gas phase. (*Continued*)

	e^{tm} (kJ/kg)														
	Pressure (bar)														
T (°C)	1.01325	2	3	4	5	6	7	8	9	10	11	12	13	14	15
540	228.09	286.34	321.09	345.76	364.91	380.57	393.81	405.29	415.43	424.49	432.70	440.20	447.11	453.50	459.46
560	242.17	300.42	335.18	359.85	379.00	394.66	407.91	419.39	429.52	438.59	446.81	454.31	461.21	467.61	473.57
580	256.50	314.75	349.51	374.19	393.34	409.00	422.25	433.73	443.87	452.94	461.15	468.66	475.56	481.96	487.92
600	271.07	329.33	364.08	388.76	407.92	423.58	436.83	448.31	458.45	467.53	475.74	483.25	490.15	496.55	502.52
620	285.88	344.13	378.89	403.57	422.73	438.39	451.64	463.13	473.27	482.34	490.56	498.07	504.98	511.38	517.34
640	300.90	359.16	393.92	418.60	437.76	453.42	466.68	478.16	488.31	497.38	505.60	513.11	520.02	526.42	532.39
660	316.15	374.40	409.16	433.85	453.01	468.67	481.93	493.42	503.56	512.64	520.86	528.37	535.28	541.69	547.65
680	331.60	389.86	424.62	449.31	468.47	484.13	497.39	508.88	519.03	528.11	536.33	543.84	550.75	557.16	563.13
700	347.26	405.52	440.28	464.97	484.13	499.80	513.06	524.55	534.69	543.78	552.00	559.51	566.43	572.83	578.80
720	363.11	421.37	456.14	480.83	499.99	515.66	528.92	540.41	550.56	559.64	567.87	575.38	582.30	588.71	594.68
740	379.16	437.42	472.19	496.87	516.04	531.71	544.97	556.47	566.62	575.70	583.93	591.44	598.36	604.77	610.74
760	395.39	453.65	488.42	513.11	532.28	547.95	561.21	572.71	582.86	591.94	600.17	607.68	614.60	621.02	626.99
780	411.80	470.06	504.83	529.52	548.69	564.37	577.63	589.13	599.28	608.36	616.59	624.11	631.03	637.44	643.42
800	428.38	486.65	521.42	546.11	565.28	580.96	594.22	605.72	615.87	624.96	633.19	640.71	647.63	654.04	660.02
820	445.14	503.41	538.18	562.87	582.04	597.72	610.98	622.48	632.64	641.73	649.96	657.48	664.40	670.81	676.79
840	462.06	520.33	555.10	579.80	598.97	614.64	627.91	639.41	649.57	658.66	666.89	674.41	681.33	687.75	693.73
860	479.14	537.41	572.18	596.88	616.05	631.73	645.00	656.50	666.65	675.74	683.98	691.50	698.42	704.84	710.82
880	496.37	554.64	589.42	614.12	633.29	648.97	662.24	673.74	683.90	692.99	701.22	708.74	715.67	722.09	728.07
900	513.76	572.03	606.81	631.51	650.68	666.36	679.63	691.13	701.29	710.38	718.62	726.14	733.07	739.49	745.47
920	531.29	589.56	624.34	649.04	668.22	683.90	697.17	708.67	718.83	727.92	736.16	743.68	750.61	757.03	763.01
940	548.97	607.24	642.02	666.72	685.89	701.58	714.85	726.35	736.51	745.61	753.84	761.37	768.30	774.72	780.70
960	566.78	625.06	659.84	684.54	703.71	719.40	732.67	744.17	754.33	763.43	771.67	779.19	786.12	792.54	798.53
980	584.73	643.01	677.79	702.49	721.67	737.35	750.62	762.13	772.29	781.39	789.62	797.15	804.08	810.50	816.49
1000	602.81	661.09	695.87	720.57	739.75	755.43	768.71	780.22	790.38	799.48	807.71	815.24	822.17	828.60	834.58
1100	695.07	753.35	788.14	812.84	832.02	847.71	860.99	872.50	882.67	891.77	900.01	907.54	914.47	920.90	926.89
1200	790.15	848.43	883.22	907.93	927.11	942.80	956.08	967.60	977.77	986.87	995.12	1002.65	1009.59	1016.02	1022.01
1300	887.72	946.01	980.80	1005.51	1024.69	1040.39	1053.67	1065.19	1075.36	1084.47	1092.71	1100.25	1107.19	1113.62	1119.62
1400	987.51	1045.80	1080.59	1105.31	1124.50	1140.19	1153.48	1165.00	1175.17	1184.28	1192.53	1200.07	1207.01	1213.44	1219.44
1500	1089.30	1147.59	1182.38	1207.10	1226.29	1241.98	1255.27	1266.79	1276.97	1286.08	1294.33	1301.87	1308.81	1315.25	1321.25
1600	1192.88	1251.17	1285.97	1310.68	1329.88	1345.58	1358.86	1370.39	1380.56	1389.68	1397.93	1405.47	1412.42	1418.86	1424.86
1700	1298.10	1356.39	1391.19	1415.90	1435.10	1450.80	1464.09	1475.61	1485.79	1494.91	1503.16	1510.70	1517.65	1524.09	1530.09
1800	1404.80	1463.10	1497.90	1522.61	1541.81	1557.51	1570.80	1582.33	1592.51	1601.62	1609.88	1617.42	1624.37	1630.81	1636.81
1900	1512.88	1571.17	1605.97	1630.69	1649.89	1665.59	1678.88	1690.41	1700.59	1709.71	1717.96	1725.51	1732.46	1738.90	1744.90
2000	1622.21	1680.51	1715.31	1740.03	1759.23	1774.93	1788.22	1799.75	1809.93	1819.05	1827.31	1834.85	1841.80	1848.25	1854.25

TABLE 5.2 Thermo-mechanical exergy of air, 1.01325 to 15 bar. All are in the gas phase. *(Continued)*

e^{tm} (kJ/kg)

T (°C)	Pressure (bar)														
	16	17	18	19	20	21	22	23	24	25	26	27	28	29	30
-140	322.04	328.10	333.91	339.50	344.89	350.13	355.22	360.21	365.11	369.94	374.72	379.48	384.25	389.05	393.92
-130	307.03	312.76	318.21	323.42	328.41	333.20	337.82	342.28	346.61	350.81	354.90	358.89	362.79	366.62	370.37
-120	294.70	300.24	305.50	310.50	315.27	319.84	324.22	328.43	332.49	336.42	340.22	343.90	347.47	350.95	354.34
-110	284.35	289.77	294.91	299.78	304.43	308.86	313.10	317.17	321.09	324.86	328.49	332.01	335.42	338.72	341.92
-100	275.56	280.91	285.97	290.76	295.32	299.66	303.82	307.80	311.62	315.30	318.84	322.26	325.56	328.75	331.85
-90	268.07	273.36	278.36	283.10	287.60	291.89	295.99	299.91	303.67	307.28	310.76	314.11	317.35	320.48	323.51
-80	261.67	266.93	271.89	276.59	281.05	285.30	289.35	293.23	296.95	300.52	303.95	307.26	310.46	313.54	316.52
-70	256.21	261.44	266.38	271.05	275.48	279.70	283.73	287.58	291.27	294.81	298.21	301.49	304.65	307.70	310.65
-60	251.58	256.79	261.71	266.36	270.77	274.97	278.97	282.80	286.47	289.99	293.37	296.63	299.77	302.80	305.73
-50	247.69	252.88	257.78	262.42	266.82	271.00	274.99	278.81	282.46	285.96	289.33	292.57	295.69	298.71	301.62
-40	244.45	249.63	254.52	259.15	263.53	267.71	271.69	275.49	279.13	282.63	285.98	289.21	292.33	295.33	298.23
-30	241.80	246.98	251.86	256.48	260.86	265.03	269.00	272.79	276.43	279.91	283.26	286.48	289.59	292.58	295.48
-20	239.70	244.87	249.74	254.36	258.73	262.89	266.86	270.65	274.28	277.76	281.10	284.32	287.41	290.40	293.29
-10	238.08	243.25	248.12	252.73	257.10	261.26	265.22	269.01	272.63	276.11	279.45	282.66	285.76	288.74	291.63
0	236.92	242.09	246.96	251.56	255.93	260.09	264.05	267.83	271.45	274.92	278.26	281.47	284.56	287.55	290.43
10	236.17	241.34	246.21	250.81	255.18	259.33	263.29	267.07	270.69	274.17	277.50	280.71	283.80	286.78	289.66
20	235.82	240.98	245.85	250.45	254.82	258.97	262.93	266.71	270.33	273.80	277.14	280.34	283.43	286.42	289.30
30	235.81	240.98	245.85	250.45	254.82	258.97	262.93	266.71	270.33	273.80	277.13	280.34	283.43	286.42	289.30
40	236.15	241.31	246.18	250.78	255.15	259.30	263.26	267.05	270.67	274.14	277.47	280.68	283.77	286.75	289.64
50	236.79	241.96	246.83	251.43	255.80	259.95	263.91	267.70	271.32	274.79	278.13	281.34	284.43	287.41	290.29
60	237.73	242.90	247.77	252.37	256.74	260.90	264.86	268.64	272.27	275.74	279.08	282.29	285.38	288.36	291.25
70	238.94	244.11	248.98	253.59	257.96	262.12	266.08	269.87	273.49	276.96	280.30	283.51	286.61	289.60	292.48
80	240.42	245.59	250.46	255.07	259.44	263.60	267.56	271.35	274.97	278.45	281.79	285.00	288.10	291.09	293.97
90	242.13	247.30	252.18	256.79	261.16	265.32	269.29	273.08	276.71	280.18	283.53	286.74	289.84	292.83	295.71
100	244.08	249.26	254.13	258.75	263.12	267.28	271.25	275.04	278.67	282.15	285.49	288.71	291.81	294.80	297.69
110	246.25	251.43	256.31	260.92	265.30	269.46	273.43	277.22	280.85	284.33	287.68	290.90	294.00	296.99	299.88
120	248.63	253.81	258.69	263.30	267.68	271.85	275.82	279.61	283.25	286.73	290.08	293.30	296.40	299.40	302.29
130	251.20	256.38	261.27	265.89	270.27	274.43	278.41	282.20	285.84	289.32	292.67	295.90	299.00	302.00	304.89
140	253.97	259.15	264.04	268.66	273.04	277.21	281.19	284.98	288.62	292.11	295.46	298.68	301.79	304.79	307.69
150	256.92	262.10	266.99	271.61	276.00	280.17	284.14	287.94	291.58	295.07	298.43	301.65	304.76	307.76	310.66
160	260.04	265.22	270.11	274.74	279.13	283.30	287.28	291.08	294.72	298.21	301.57	304.80	307.91	310.91	313.81
170	263.33	268.51	273.40	278.03	282.42	286.60	290.58	294.38	298.02	301.52	304.88	308.11	311.22	314.22	317.13
180	266.77	271.96	276.86	281.48	285.88	290.05	294.04	297.84	301.49	304.98	308.34	311.57	314.69	317.70	320.60

TABLE 5.3 Thermo-mechanical exergy of air, 16 to 30 bar. All are in the gas phase.

e^{tm} (kJ/kg)

T (°C)	Pressure (bar)														
	16	17	18	19	20	21	22	23	24	25	26	27	28	29	30
190	270.37	275.57	280.46	285.09	289.48	293.66	297.65	301.46	305.10	308.60	311.96	315.20	318.31	321.32	324.23
200	274.12	279.32	284.21	288.85	293.24	297.42	301.41	305.22	308.87	312.37	315.73	318.97	322.09	325.10	328.01
210	278.01	283.21	288.11	292.74	297.14	301.32	305.31	309.12	312.77	316.28	319.64	322.88	326.00	329.01	331.92
220	282.04	287.24	292.14	296.78	301.18	305.36	309.35	313.17	316.82	320.32	323.69	326.93	330.05	333.07	335.98
230	286.20	291.40	296.31	300.94	305.35	309.53	313.52	317.34	320.99	324.50	327.87	331.11	334.24	337.25	340.17
240	290.50	295.70	300.60	305.24	309.64	313.83	317.83	321.64	325.30	328.81	332.18	335.42	338.55	341.57	344.48
250	294.91	300.11	305.02	309.66	314.07	318.26	322.25	326.07	329.73	333.24	336.61	339.86	342.99	346.00	348.92
260	299.45	304.65	309.56	314.20	318.61	322.80	326.80	330.62	334.28	337.79	341.16	344.41	347.54	350.56	353.48
270	304.10	309.31	314.22	318.86	323.27	327.46	331.46	335.29	338.95	342.46	345.84	349.08	352.22	355.24	358.16
280	308.87	314.08	318.99	323.63	328.04	332.24	336.24	340.07	343.73	347.24	350.62	353.87	357.00	360.03	362.95
290	313.75	318.96	323.87	328.52	332.93	337.13	341.13	344.96	348.62	352.14	355.52	358.77	361.90	364.93	367.85
300	318.73	323.94	328.86	333.51	337.92	342.12	346.13	349.95	353.62	357.14	360.52	363.77	366.91	369.94	372.86
310	323.82	329.04	333.95	338.60	343.02	347.22	351.23	355.06	358.72	362.24	365.62	368.88	372.02	375.05	377.98
320	329.02	334.23	339.15	343.80	348.22	352.42	356.43	360.26	363.93	367.45	370.83	374.09	377.23	380.26	383.19
330	334.31	339.53	344.45	349.10	353.52	357.72	361.73	365.56	369.24	372.76	376.14	379.40	382.54	385.58	388.51
340	339.70	344.92	349.84	354.50	358.92	363.12	367.13	370.97	374.64	378.16	381.55	384.81	387.95	390.99	393.92
350	345.19	350.41	355.33	359.99	364.41	368.61	372.63	376.46	380.14	383.66	387.05	390.31	393.45	396.49	399.42
360	350.77	355.99	360.91	365.57	369.99	374.20	378.21	382.05	385.73	389.25	392.64	395.91	399.05	402.09	405.02
370	356.44	361.66	366.58	371.24	375.67	379.88	383.89	387.73	391.41	394.93	398.33	401.59	404.74	407.78	410.71
380	362.19	367.42	372.34	377.01	381.43	385.64	389.66	393.50	397.18	400.71	404.10	407.36	410.51	413.55	416.49
390	368.04	373.26	378.19	382.86	387.28	391.49	395.51	399.35	403.03	406.56	409.96	413.22	416.37	419.41	422.35
400	373.97	379.20	384.13	388.79	393.22	397.43	401.45	405.29	408.97	412.50	415.90	419.17	422.32	425.36	428.30
410	379.98	385.21	390.14	394.81	399.24	403.45	407.47	411.31	415.00	418.53	421.93	425.20	428.35	431.39	434.33
420	386.08	391.31	396.24	400.91	405.34	409.55	413.57	417.42	421.10	424.64	428.03	431.30	434.46	437.50	440.44
430	392.25	397.48	402.42	407.08	411.52	415.73	419.76	423.60	427.29	430.82	434.22	437.49	440.65	443.69	446.64
440	398.51	403.74	408.67	413.34	417.77	421.99	426.02	429.86	433.55	437.09	440.49	443.76	446.91	449.96	452.91
450	404.84	410.07	415.00	419.68	424.11	428.33	432.35	436.20	439.89	443.43	446.83	450.10	453.26	456.31	459.25
460	411.24	416.48	421.41	426.08	430.52	434.74	438.77	442.62	446.30	449.84	453.24	456.52	459.68	462.73	465.67
470	417.72	422.96	427.89	432.57	437.00	441.22	445.25	449.10	452.79	456.33	459.73	463.01	466.17	469.22	472.17
480	424.27	429.51	434.45	439.12	443.56	447.78	451.81	455.66	459.35	462.89	466.30	469.58	472.74	475.79	478.74
490	430.90	436.13	441.07	445.75	450.19	454.41	458.44	462.29	465.98	469.53	472.93	476.21	479.37	482.42	485.37
500	437.59	442.83	447.77	452.44	456.88	461.11	465.14	468.99	472.68	476.23	479.63	482.91	486.08	489.13	492.08
520	451.18	456.42	461.36	466.04	470.48	474.71	478.74	482.60	486.29	489.84	493.24	496.53	499.69	502.75	505.70
540	465.03	470.28	475.22	479.90	484.34	488.57	492.61	496.47	500.16	503.71	507.12	510.40	513.57	516.63	519.58
560	479.14	484.39	489.33	494.02	498.46	502.69	506.73	510.59	514.29	517.84	521.25	524.53	527.70	530.76	533.72
580	493.50	498.75	503.70	508.38	512.83	517.06	521.10	524.96	528.66	532.21	535.62	538.91	542.08	545.14	548.10

TABLE 5.3 Thermo-mechanical exergy of air, 16 to 30 bar. All are in the gas phase. (Continued)

63

e^{tm} (kJ/kg)

T (°C)	Pressure (bar)														
	16	17	18	19	20	21	22	23	24	25	26	27	28	29	30
600	508.10	513.34	518.29	522.98	527.43	531.66	535.70	539.57	543.27	546.82	550.23	553.52	556.70	559.76	562.72
620	522.92	528.17	533.12	537.81	542.26	546.50	550.54	554.40	558.11	561.66	565.08	568.37	571.54	574.60	577.57
640	537.97	543.22	548.18	552.87	557.32	561.56	565.60	569.46	573.17	576.72	580.14	583.43	586.61	589.67	592.64
660	553.24	558.49	563.45	568.14	572.59	576.83	580.87	584.74	588.45	592.00	595.42	598.72	601.89	604.96	607.93
680	568.72	573.97	578.92	583.62	588.07	592.31	596.36	600.23	603.93	607.49	610.91	614.21	617.39	620.45	623.42
700	584.39	589.65	594.61	599.30	603.75	608.00	612.04	615.91	619.62	623.18	626.61	629.90	633.08	636.15	639.12
720	600.27	605.52	610.48	615.18	619.63	623.88	627.93	631.80	635.51	639.07	642.49	645.79	648.97	652.04	655.01
740	616.33	621.59	626.55	631.25	635.70	639.95	644.00	647.87	651.58	655.14	658.57	661.87	665.05	668.12	671.09
760	632.58	637.84	642.80	647.50	651.96	656.20	660.26	664.13	667.84	671.41	674.83	678.13	681.32	684.39	687.36
780	649.01	654.27	659.23	663.93	668.39	672.64	676.69	680.57	684.28	687.85	691.27	694.57	697.76	700.83	703.81
800	665.62	670.88	675.84	680.54	685.00	689.25	693.30	697.18	700.89	704.46	707.89	711.19	714.37	717.45	720.42
820	682.39	687.65	692.61	697.31	701.78	706.03	710.08	713.96	717.67	721.24	724.67	727.97	731.16	734.24	737.21
840	699.32	704.59	709.55	714.25	718.72	722.97	727.02	730.90	734.62	738.19	741.62	744.92	748.11	751.19	754.16
860	716.42	721.68	726.65	731.35	735.82	740.07	744.12	748.00	751.72	755.29	758.72	762.03	765.21	768.29	771.27
880	733.67	738.93	743.90	748.60	753.07	757.32	761.38	765.26	768.98	772.55	775.98	779.29	782.48	785.56	788.53
900	751.07	756.33	761.30	766.01	770.47	774.73	778.78	782.67	786.39	789.96	793.39	796.70	799.89	802.97	805.95
920	768.61	773.88	778.85	783.56	788.02	792.28	796.34	800.22	803.94	807.51	810.95	814.25	817.44	820.53	823.51
940	786.30	791.57	796.54	801.25	805.72	809.97	814.03	817.91	821.64	825.21	828.64	831.95	835.14	838.23	841.21
960	804.13	809.40	814.37	819.08	823.55	827.80	831.86	835.75	839.47	843.04	846.48	849.79	852.98	856.06	859.05
980	822.09	827.36	832.33	837.04	841.51	845.77	849.83	853.71	857.44	861.01	864.45	867.76	870.95	874.04	877.02
1000	840.19	845.45	850.43	855.14	859.61	863.86	867.93	871.81	875.54	879.11	882.55	885.86	889.05	892.14	895.12
1100	932.50	937.77	942.75	947.46	951.93	956.20	960.26	964.15	967.88	971.46	974.90	978.21	981.41	984.50	987.49
1200	1027.62	1032.90	1037.88	1042.59	1047.07	1051.33	1055.40	1059.29	1063.02	1066.61	1070.05	1073.37	1076.57	1079.66	1082.65
1300	1125.23	1130.51	1135.49	1140.21	1144.69	1148.95	1153.03	1156.92	1160.65	1164.24	1167.68	1171.00	1174.21	1177.30	1180.29
1400	1225.06	1230.34	1235.32	1240.04	1244.52	1248.79	1252.86	1256.76	1260.50	1264.08	1267.53	1270.85	1274.06	1277.15	1280.15
1500	1326.87	1332.15	1337.14	1341.86	1346.34	1350.61	1354.69	1358.59	1362.32	1365.91	1369.36	1372.69	1375.89	1378.99	1381.99
1600	1430.47	1435.76	1440.75	1445.47	1449.96	1454.23	1458.30	1462.20	1465.94	1469.53	1472.98	1476.31	1479.52	1482.62	1485.62
1700	1535.71	1541.00	1545.99	1550.71	1555.20	1559.47	1563.55	1567.45	1571.19	1574.78	1578.24	1581.56	1584.77	1587.87	1590.87
1800	1642.44	1647.72	1652.71	1657.44	1661.93	1666.20	1670.28	1674.18	1677.92	1681.52	1684.97	1688.30	1691.51	1694.61	1697.61
1900	1750.53	1755.81	1760.81	1765.53	1770.02	1774.30	1778.38	1782.28	1786.02	1789.62	1793.07	1796.40	1799.61	1802.72	1805.72
2000	1859.87	1865.16	1870.15	1874.88	1879.37	1883.65	1887.73	1891.63	1895.38	1898.97	1902.43	1905.76	1908.97	1912.08	1915.08

TABLE 5.3 Thermo-mechanical exergy of air, 16 to 30 bar. All are in the gas phase. (Continued)

e^tm (kJ/kg)

T (°C)	Pressure (bar)														
	31	32	33	34	35	36	37	38	39	40	41	42	43	44	45
-140	398.91	404.06	409.46	415.22	421.54	428.76	437.67	450.89	480.66	496.57	502.79	506.70	509.62	512.00	514.03
-130	374.07	377.72	381.32	384.89	388.44	391.97	395.48	399.00	402.52	406.05	409.61	413.20	416.84	420.53	424.29
-120	357.64	360.87	364.03	367.13	370.16	373.14	376.07	378.95	381.79	384.58	387.34	390.07	392.76	395.43	398.07
-110	345.04	348.07	351.02	353.90	356.71	359.46	362.15	364.78	367.36	369.89	372.37	374.81	377.20	379.56	381.87
-100	334.86	337.78	340.61	343.38	346.07	348.69	351.25	353.75	356.20	358.59	360.93	363.22	365.47	367.67	369.83
-90	326.44	329.29	332.06	334.75	337.36	339.91	342.39	344.82	347.18	349.49	351.75	353.96	356.12	358.23	360.31
-80	319.41	322.21	324.93	327.57	330.14	332.64	335.07	337.44	339.76	342.01	344.22	346.37	348.48	350.54	352.56
-70	313.51	316.28	318.97	321.57	324.11	326.57	328.97	331.31	333.59	335.81	337.98	340.10	342.17	344.20	346.18
-60	308.57	311.31	313.97	316.56	319.07	321.51	323.89	326.20	328.46	330.66	332.80	334.90	336.94	338.95	340.90
-50	304.44	307.17	309.82	312.39	314.88	317.31	319.67	321.96	324.20	326.38	328.51	330.59	332.62	334.61	336.55
-40	301.04	303.76	306.39	308.95	311.43	313.85	316.19	318.48	320.71	322.88	324.99	327.06	329.08	331.05	332.98
-30	298.28	300.99	303.61	306.16	308.64	311.04	313.38	315.66	317.88	320.04	322.15	324.21	326.22	328.18	330.10
-20	296.09	298.79	301.41	303.96	306.42	308.82	311.16	313.43	315.64	317.80	319.90	321.95	323.96	325.91	327.83
-10	294.42	297.12	299.73	302.27	304.74	307.13	309.46	311.73	313.94	316.09	318.19	320.24	322.24	324.19	326.10
0	293.22	295.92	298.53	301.07	303.53	305.92	308.25	310.51	312.72	314.87	316.97	319.01	321.01	322.96	324.87
10	292.45	295.15	297.76	300.30	302.76	305.15	307.47	309.74	311.94	314.09	316.19	318.23	320.23	322.18	324.08
20	292.08	294.78	297.39	299.93	302.39	304.78	307.10	309.36	311.57	313.72	315.81	317.85	319.85	321.80	323.70
30	292.08	294.78	297.39	299.92	302.38	304.78	307.10	309.36	311.57	313.72	315.81	317.85	319.85	321.80	323.70
40	292.42	295.12	297.73	300.27	302.73	305.12	307.44	309.71	311.91	314.06	316.15	318.20	320.19	322.14	324.05
50	293.08	295.78	298.39	300.93	303.39	305.78	308.11	310.37	312.58	314.73	316.82	318.87	320.86	322.81	324.72
60	294.03	296.73	299.35	301.89	304.35	306.74	309.07	311.33	313.54	315.69	317.79	319.83	321.83	323.78	325.69
70	295.27	297.97	300.59	303.12	305.59	307.98	310.31	312.58	314.79	316.94	319.04	321.08	323.08	325.04	326.94
80	296.76	299.47	302.08	304.62	307.09	309.49	311.82	314.09	316.29	318.45	320.55	322.60	324.60	326.55	328.46
90	298.51	301.21	303.83	306.37	308.84	311.24	313.57	315.84	318.05	320.21	322.31	324.36	326.36	328.32	330.23
100	300.48	303.19	305.81	308.35	310.82	313.22	315.56	317.83	320.04	322.20	324.30	326.35	328.36	330.31	332.23
110	302.68	305.39	308.01	310.56	313.03	315.43	317.76	320.04	322.25	324.41	326.52	328.57	330.58	332.54	334.45
120	305.09	307.80	310.42	312.97	315.44	317.85	320.18	322.46	324.67	326.83	328.94	331.00	333.01	334.97	336.88
130	307.69	310.40	313.03	315.58	318.06	320.46	322.80	325.08	327.30	329.46	331.57	333.62	335.63	337.60	339.52
140	310.49	313.20	315.83	318.38	320.86	323.27	325.61	327.89	330.11	332.27	334.38	336.44	338.45	340.42	342.34
150	313.47	316.18	318.81	321.37	323.84	326.25	328.60	330.88	333.10	335.27	337.38	339.44	341.46	343.42	345.35
160	316.62	319.33	321.97	324.52	327.00	329.41	331.76	334.04	336.27	338.44	340.55	342.61	344.63	346.60	348.52
170	319.93	322.65	325.29	327.85	330.33	332.74	335.09	337.37	339.60	341.77	343.89	345.95	347.97	349.94	351.87
180	323.41	326.13	328.77	331.33	333.81	336.23	338.58	340.86	343.09	345.26	347.38	349.45	351.47	353.44	355.37

TABLE 5.4 Thermo-mechanical exergy of air, 31 to 45 bar. All are gas phase or supercritical.

e^{tm} (kJ/kg)

Pressure (bar)

T (°C)	31	32	33	34	35	36	37	38	39	40	41	42	43	44	45
190	327.04	329.76	332.40	334.96	337.45	339.87	342.22	344.51	346.74	348.91	351.03	353.10	355.12	357.10	359.03
200	330.82	333.54	336.18	338.75	341.24	343.66	346.01	348.30	350.53	352.71	354.83	356.90	358.92	360.90	362.83
210	334.74	337.46	340.11	342.67	345.16	347.58	349.94	352.23	354.47	356.64	358.77	360.84	362.87	364.84	366.78
220	338.80	341.52	344.17	346.74	349.23	351.65	354.01	356.30	358.54	360.72	362.84	364.92	366.94	368.93	370.86
230	342.99	345.72	348.36	350.93	353.42	355.85	358.21	360.50	362.74	364.92	367.05	369.13	371.16	373.14	375.08
240	347.30	350.04	352.68	355.25	357.75	360.18	362.54	364.83	367.07	369.26	371.39	373.46	375.50	377.48	379.42
250	351.75	354.48	357.13	359.70	362.20	364.63	366.99	369.29	371.53	373.71	375.85	377.93	379.96	381.94	383.89
260	356.31	359.04	361.70	364.27	366.77	369.20	371.56	373.86	376.11	378.29	380.43	382.51	384.54	386.53	388.47
270	360.99	363.72	366.38	368.95	371.45	373.89	376.25	378.55	380.80	382.99	385.12	387.21	389.24	391.23	393.18
280	365.78	368.52	371.17	373.75	376.25	378.69	381.05	383.36	385.60	387.79	389.93	392.02	394.05	396.04	397.99
290	370.68	373.42	376.08	378.66	381.16	383.60	385.97	388.27	390.52	392.71	394.85	396.94	398.98	400.97	402.92
300	375.69	378.44	381.09	383.67	386.18	388.62	390.99	393.30	395.54	397.74	399.88	401.96	404.01	406.00	407.95
310	380.81	383.55	386.21	388.79	391.30	393.74	396.11	398.42	400.67	402.87	405.01	407.10	409.14	411.13	413.09
320	386.03	388.77	391.43	394.02	396.52	398.96	401.34	403.65	405.90	408.10	410.24	412.33	414.37	416.37	418.32
330	391.34	394.09	396.75	399.34	401.85	404.29	406.66	408.98	411.23	413.43	415.57	417.66	419.71	421.71	423.66
340	396.75	399.50	402.17	404.75	407.27	409.71	412.08	414.40	416.65	418.85	421.00	423.09	425.14	427.14	429.09
350	402.26	405.01	407.68	410.27	412.78	415.22	417.60	419.92	422.17	424.37	426.52	428.61	430.66	432.66	434.62
360	407.86	410.61	413.28	415.87	418.38	420.83	423.21	425.53	427.78	429.98	432.13	434.23	436.28	438.28	440.24
370	413.55	416.30	418.97	421.56	424.08	426.53	428.91	431.23	433.48	435.69	437.84	439.94	441.99	443.99	445.95
380	419.33	422.08	424.76	427.35	429.86	432.31	434.69	437.01	439.27	441.48	443.63	445.73	447.78	449.79	451.75
390	425.20	427.95	430.62	433.22	435.73	438.18	440.57	442.89	445.15	447.35	449.51	451.61	453.66	455.67	457.63
400	431.15	433.90	436.58	439.17	441.69	444.14	446.52	448.85	451.11	453.32	455.47	457.57	459.63	461.63	463.60
410	437.18	439.94	442.61	445.21	447.73	450.18	452.56	454.89	457.15	459.36	461.51	463.62	465.67	467.68	469.65
420	443.29	446.05	448.73	451.32	453.85	456.30	458.69	461.01	463.27	465.48	467.64	469.74	471.80	473.81	475.78
430	449.49	452.25	454.92	457.52	460.04	462.50	464.89	467.21	469.48	471.69	473.84	475.95	478.01	480.02	481.99
440	455.76	458.52	461.20	463.79	466.32	468.77	471.16	473.49	475.76	477.97	480.13	482.23	484.29	486.31	488.27
450	462.10	464.87	467.54	470.15	472.67	475.13	477.52	479.85	482.11	484.33	486.49	488.59	490.65	492.67	494.64
460	468.53	471.29	473.97	476.57	479.10	481.56	483.95	486.28	488.55	490.76	492.92	495.03	497.09	499.10	501.08
470	475.02	477.79	480.47	483.07	485.60	488.06	490.45	492.78	495.05	497.26	499.43	501.54	503.60	505.61	507.59
480	481.59	484.36	487.04	489.64	492.17	494.63	497.02	499.35	501.63	503.84	506.00	508.12	510.18	512.20	514.17
490	488.23	491.00	493.68	496.28	498.81	501.27	503.67	506.00	508.27	510.49	512.65	514.77	516.83	518.85	520.82
500	494.94	497.70	500.39	502.99	505.53	507.99	510.38	512.72	514.99	517.21	519.37	521.49	523.55	525.57	527.54
520	508.56	511.33	514.01	516.62	519.15	521.62	524.01	526.35	528.63	530.85	533.01	535.13	537.19	539.21	541.19
540	522.44	525.21	527.90	530.51	533.04	535.51	537.91	540.25	542.52	544.75	546.91	549.03	551.10	553.12	555.10

TABLE 5.4 Thermo-mechanical exergy of air, 31 to 45 bar. All are gas phase or supercritical. (*Continued*)

e^{tm} (kJ/kg)

T (°C)	Pressure (bar)														
	31	32	33	34	35	36	37	38	39	40	41	42	43	44	45
560	536.58	539.35	542.04	544.65	547.19	549.66	552.06	554.40	556.68	558.90	561.07	563.19	565.26	567.28	569.26
580	550.96	553.74	556.43	559.04	561.58	564.05	566.45	568.79	571.07	573.30	575.47	577.59	579.66	581.69	583.67
600	565.58	568.36	571.05	573.67	576.21	578.68	581.08	583.42	585.71	587.93	590.11	592.23	594.30	596.33	598.32
620	580.43	583.21	585.91	588.52	591.06	593.54	595.94	598.29	600.57	602.80	604.97	607.10	609.17	611.20	613.19
640	595.51	598.29	600.98	603.60	606.14	608.62	611.03	613.37	615.66	617.89	620.06	622.19	624.26	626.30	628.28
660	610.80	613.58	616.27	618.89	621.44	623.91	626.32	628.67	630.96	633.19	635.37	637.49	639.57	641.60	643.59
680	626.29	629.07	631.77	634.39	636.94	639.42	641.83	644.18	646.46	648.70	650.88	653.00	655.08	657.12	659.11
700	641.99	644.77	647.47	650.10	652.64	655.12	657.53	659.88	662.17	664.41	666.59	668.72	670.80	672.83	674.82
720	657.88	660.67	663.37	665.99	668.54	671.02	673.44	675.79	678.08	680.31	682.49	684.63	686.71	688.74	690.74
740	673.97	676.76	679.46	682.08	684.63	687.11	689.53	691.88	694.17	696.41	698.59	700.72	702.81	704.84	706.84
760	690.24	693.02	695.73	698.35	700.90	703.39	705.80	708.16	710.45	712.69	714.87	717.00	719.09	721.13	723.12
780	706.68	709.47	712.18	714.80	717.36	719.84	722.26	724.61	726.90	729.14	731.33	733.46	735.55	737.59	739.58
800	723.30	726.09	728.80	731.43	733.98	736.46	738.88	741.24	743.53	745.77	747.96	750.09	752.18	754.22	756.22
820	740.09	742.88	745.59	748.22	750.77	753.26	755.68	758.03	760.33	762.57	764.76	766.89	768.98	771.02	773.02
840	757.04	759.83	762.54	765.17	767.73	770.21	772.63	774.99	777.29	779.53	781.72	783.86	785.95	787.99	789.99
860	774.15	776.95	779.65	782.29	784.84	787.33	789.75	792.11	794.41	796.65	798.84	800.98	803.07	805.11	807.11
880	791.42	794.21	796.92	799.55	802.11	804.60	807.02	809.38	811.68	813.92	816.11	818.25	820.34	822.39	824.39
900	808.83	811.63	814.34	816.97	819.53	822.02	824.44	826.80	829.10	831.35	833.54	835.68	837.77	839.81	841.82
920	826.39	829.19	831.90	834.53	837.09	839.58	842.00	844.37	846.67	848.91	851.10	853.25	855.34	857.39	859.39
940	844.09	846.89	849.60	852.24	854.80	857.29	859.71	862.07	864.38	866.62	868.82	870.96	873.05	875.10	877.10
960	861.93	864.73	867.44	870.08	872.64	875.13	877.56	879.92	882.22	884.47	886.66	888.81	890.90	892.95	894.95
980	879.91	882.70	885.42	888.05	890.62	893.11	895.53	897.90	900.20	902.45	904.64	906.79	908.88	910.93	912.94
1000	898.01	900.81	903.52	906.16	908.72	911.22	913.64	916.01	918.31	920.56	922.76	924.90	927.00	929.05	931.05
1100	990.38	993.18	995.90	998.54	1001.10	1003.60	1006.03	1008.40	1010.71	1012.96	1015.16	1017.30	1019.40	1021.46	1023.47
1200	1085.54	1088.35	1091.07	1093.71	1096.28	1098.78	1101.21	1103.58	1105.90	1108.15	1110.35	1112.50	1114.61	1116.66	1118.67
1300	1183.19	1186.00	1188.72	1191.37	1193.94	1196.44	1198.88	1201.25	1203.56	1205.82	1208.02	1210.18	1212.28	1214.34	1216.35
1400	1283.05	1285.86	1288.58	1291.23	1293.80	1296.31	1298.75	1301.12	1303.44	1305.70	1307.90	1310.06	1312.16	1314.22	1316.24
1500	1384.89	1387.70	1390.43	1393.08	1395.65	1398.16	1400.60	1402.97	1405.29	1407.55	1409.76	1411.92	1414.03	1416.09	1418.11
1600	1488.52	1491.33	1494.06	1496.71	1499.29	1501.80	1504.24	1506.62	1508.93	1511.20	1513.41	1515.57	1517.68	1519.74	1521.76
1700	1593.78	1596.59	1599.32	1601.97	1604.55	1607.06	1609.50	1611.88	1614.20	1616.47	1618.68	1620.84	1622.95	1625.01	1627.04
1800	1700.52	1703.33	1706.07	1708.72	1711.30	1713.81	1716.25	1718.63	1720.95	1723.22	1725.43	1727.59	1729.71	1731.77	1733.79
1900	1808.62	1811.44	1814.17	1816.83	1819.41	1821.92	1824.36	1826.75	1829.07	1831.33	1833.55	1835.71	1837.82	1839.89	1841.91
2000	1917.98	1920.80	1923.54	1926.19	1928.77	1931.28	1933.73	1936.11	1938.44	1940.70	1942.92	1945.08	1947.20	1949.26	1951.29

TABLE 5.4 Thermo-mechanical exergy of air, 31 to 45 bar. All are gas phase or supercritical. (Continued)

T (°C)	e^tm (kJ/kg) Pressure (bar)														
	50	100	150	200	250	300	350	400	450	500	600	700	800	900	1000
-200	776.01	786.92	797.42	807.58	817.45	827.05	836.42	845.59	854.57	863.37	880.52	897.12	913.26	928.99	944.35
-190	722.58	734.28	745.41	756.08	766.36	776.31	785.98	795.40	804.59	813.58	831.02	847.86	864.18	880.05	895.52
-180	676.61	689.40	701.33	712.62	723.40	733.75	743.74	753.43	762.85	772.04	789.80	806.88	823.38	839.38	854.96
-170	635.95	650.32	663.32	675.38	686.74	697.54	707.91	717.90	727.57	736.96	755.06	772.37	789.06	805.20	820.88
-160	598.73	615.59	630.03	643.05	655.10	666.43	677.20	687.52	697.45	707.07	725.50	743.06	759.92	776.20	791.99
-150	562.64	583.99	600.47	614.70	627.56	639.48	650.69	661.35	671.57	681.41	700.18	717.99	735.03	751.44	767.33
-140	521.52	554.47	573.91	589.64	603.44	616.00	627.69	638.71	649.21	659.28	678.40	696.44	713.66	730.20	746.18
-130	444.28	525.94	549.83	567.40	582.25	595.50	607.67	619.06	629.84	640.13	659.58	677.86	695.24	711.90	727.97
-120	410.95	497.81	527.92	547.64	563.61	577.57	590.23	601.98	613.04	623.55	643.32	661.81	679.35	696.12	712.27
-110	392.94	472.18	508.18	530.16	547.28	561.93	575.06	587.15	598.47	609.18	629.24	647.94	665.62	682.50	698.73
-100	380.08	452.20	490.93	514.86	533.04	548.34	561.91	574.31	585.86	596.76	617.09	635.96	653.77	670.74	687.04
-90	370.09	437.56	476.47	501.70	520.74	536.60	550.55	563.23	574.99	586.05	606.61	625.65	643.56	660.62	676.98
-80	362.07	426.60	464.75	490.60	510.24	526.54	540.81	553.72	565.65	576.85	597.62	616.79	634.81	651.94	668.35
-70	355.50	418.12	455.40	481.39	501.39	518.01	532.52	545.62	557.70	569.01	589.95	609.24	627.34	644.52	660.98
-60	350.10	411.44	447.96	473.86	494.04	510.85	525.54	538.78	550.97	562.38	583.45	602.84	621.00	638.24	654.74
-50	345.65	406.12	442.05	467.78	488.00	504.93	519.73	533.07	545.35	556.83	578.01	597.47	615.69	632.97	649.50
-40	342.02	401.88	437.37	462.92	483.12	500.10	514.97	528.38	540.72	552.25	573.51	593.03	611.30	628.61	645.16
-30	339.10	398.53	433.71	459.09	479.25	496.24	511.14	524.59	536.97	548.54	569.86	589.42	607.72	625.06	641.63
-20	336.80	395.94	430.89	456.14	476.25	493.23	508.15	521.62	534.03	545.61	566.97	586.57	604.89	622.25	638.84
-10	335.06	394.00	428.79	453.95	474.01	490.98	505.90	519.38	531.80	543.40	564.78	584.40	602.74	620.11	636.71
0	333.81	392.63	427.32	452.41	472.43	489.38	504.30	517.79	530.21	541.82	563.22	582.85	601.20	618.58	635.18
10	333.01	391.77	426.40	451.45	471.45	488.38	503.30	516.78	529.21	540.82	562.22	581.86	600.22	617.60	634.21
20	332.63	391.36	425.96	450.99	470.98	487.91	502.82	516.30	528.73	540.34	561.75	581.39	599.75	617.14	633.75
30	332.63	391.36	425.96	450.99	470.98	487.91	502.82	516.30	528.73	540.34	561.75	581.39	599.75	617.14	633.74
40	332.98	391.73	426.35	451.40	471.39	488.33	503.25	516.73	529.16	540.77	562.18	581.82	600.18	617.56	634.17
50	333.66	392.44	427.10	452.17	472.19	489.14	504.06	517.55	529.98	541.60	563.01	582.65	601.00	618.38	634.99
60	334.63	393.47	428.18	453.29	473.33	490.30	505.24	518.74	531.17	542.79	564.20	583.84	602.19	619.57	636.17
70	335.89	394.80	429.56	454.72	474.79	491.79	506.74	520.25	532.69	544.31	565.73	585.36	603.72	621.10	637.70
80	337.42	396.40	431.22	456.43	476.55	493.57	508.54	522.07	534.52	546.15	567.57	587.20	605.56	622.93	639.53
90	339.19	398.25	433.15	458.41	478.57	495.62	510.62	524.16	536.63	548.26	569.69	589.33	607.69	625.06	641.66
100	341.20	400.34	435.31	460.63	480.84	497.93	512.96	526.52	539.00	550.64	572.09	591.73	610.09	627.47	644.06
110	343.43	402.66	437.71	463.09	483.35	500.48	515.53	529.12	541.61	553.27	574.73	594.39	612.75	630.12	646.72
120	345.88	405.19	440.32	465.77	486.08	503.25	518.34	531.94	544.46	556.13	577.61	597.27	615.64	633.02	649.62
130	348.52	407.93	443.13	468.65	489.01	506.23	521.35	534.98	547.52	559.21	580.70	600.38	618.76	636.14	652.74

TABLE 5.5 Thermo-mechanical exergy of air, 50 to 1000 bar. All are supercritical.

e^{tm} (kJ/kg)

Pressure (bar)

T (°C)	50	100	150	200	250	300	350	400	450	500	600	700	800	900	1000
140	351.35	410.85	446.13	471.72	492.14	509.40	524.56	538.22	550.78	562.48	584.01	603.70	622.08	639.47	656.08
150	354.37	413.96	449.32	474.98	495.45	512.76	527.96	541.65	554.23	565.95	587.50	607.21	625.60	643.00	659.61
160	357.55	417.24	452.68	478.40	498.94	516.30	531.53	545.25	557.86	569.60	591.18	610.91	629.31	646.72	663.33
170	360.91	420.68	456.21	482.00	502.60	520.00	535.27	549.02	561.66	573.42	595.03	614.78	633.20	650.61	667.23
180	364.42	424.28	459.89	485.75	506.41	523.86	539.17	552.96	565.62	577.40	599.04	618.82	637.25	654.67	671.31
190	368.09	428.04	463.72	489.66	510.37	527.87	543.22	557.04	569.73	581.54	603.22	623.01	641.46	658.90	675.54
200	371.90	431.94	467.70	493.70	514.48	532.03	547.42	561.28	573.99	585.82	607.54	627.36	645.83	663.28	679.93
210	375.86	435.98	471.82	497.89	518.72	536.32	551.76	565.65	578.39	590.25	612.00	631.85	650.34	667.80	684.46
220	379.95	440.16	476.08	502.21	523.10	540.75	556.23	570.15	582.93	594.81	616.60	636.47	654.98	672.46	689.14
230	384.17	444.47	480.46	506.66	527.61	545.31	560.83	574.79	587.59	599.50	621.33	641.23	659.76	677.26	693.95
240	388.52	448.90	484.97	511.24	532.24	549.99	565.55	579.55	592.38	604.31	626.18	646.12	664.67	682.19	698.89
250	393.00	453.46	489.60	515.93	536.99	554.79	570.39	584.42	597.28	609.25	631.16	651.12	669.70	687.23	703.95
260	397.59	458.13	494.34	520.74	541.85	559.70	575.34	589.41	602.31	614.29	636.25	656.25	674.85	692.40	709.13
270	402.30	462.92	499.20	525.66	546.83	564.73	580.41	594.52	607.44	619.45	641.45	661.48	680.11	697.68	714.43
280	407.13	467.82	504.17	530.69	551.91	569.86	585.58	599.72	612.68	624.72	646.76	666.83	685.48	703.07	719.84
290	412.06	472.83	509.25	535.83	557.10	575.09	590.86	605.04	618.03	630.09	652.18	672.28	690.96	708.57	725.35
300	417.10	477.94	514.42	541.06	562.40	580.43	596.24	610.45	623.47	635.57	657.70	677.83	696.54	714.17	730.97
310	422.25	483.16	519.71	546.40	567.79	585.87	601.72	615.97	629.01	641.14	663.31	683.48	702.22	719.87	736.69
320	427.49	488.47	525.08	551.84	573.27	591.40	607.29	621.57	634.65	646.80	669.02	689.23	707.99	725.67	742.51
330	432.83	493.88	530.56	557.37	578.86	597.03	612.96	627.27	640.39	652.56	674.83	695.07	713.86	731.56	748.42
340	438.27	499.39	536.13	562.99	584.53	602.75	618.71	633.07	646.21	658.41	680.72	701.00	719.82	737.55	754.42
350	443.81	504.99	541.79	568.71	590.29	608.55	624.56	638.94	652.12	664.35	686.70	707.01	725.87	743.62	760.51
360	449.44	510.68	547.54	574.51	596.14	614.44	630.49	644.91	658.11	670.37	692.77	713.12	732.00	749.78	766.69
370	455.15	516.46	553.37	580.40	602.08	620.42	636.50	650.96	664.19	676.47	698.92	719.31	738.22	756.02	772.95
380	460.95	522.32	559.29	586.37	608.09	626.48	642.60	657.09	670.35	682.66	705.15	725.57	744.51	762.34	779.30
390	466.84	528.27	565.30	592.42	614.19	632.62	648.78	663.30	676.59	688.93	711.46	731.92	750.89	768.75	785.72
400	472.82	534.31	571.38	598.56	620.37	638.84	655.04	669.59	682.91	695.27	717.85	738.35	757.35	775.23	792.23
410	478.87	540.42	577.55	604.77	626.63	645.14	661.37	675.95	689.30	701.69	724.32	744.85	763.88	781.79	798.81
420	485.01	546.61	583.79	611.06	632.97	651.51	667.78	682.39	695.77	708.19	730.86	751.43	770.49	788.42	805.46
430	491.23	552.88	590.11	617.43	639.38	657.96	674.26	688.91	702.31	714.76	737.47	758.08	777.17	795.12	812.19
440	497.52	559.23	596.51	623.87	645.86	664.48	680.81	695.49	708.93	721.40	744.15	764.80	783.92	801.90	818.99
450	503.89	565.65	602.98	630.39	652.41	671.07	687.44	702.15	715.61	728.11	750.91	771.59	790.74	808.75	825.86
500	536.82	598.83	636.39	664.00	686.22	705.06	721.58	736.44	750.04	762.66	785.67	806.53	825.84	843.97	861.20
520	550.48	612.58	650.22	677.91	700.20	719.11	735.70	750.61	764.26	776.92	800.02	820.95	840.31	858.50	875.77
540	564.40	626.58	664.30	692.07	714.43	733.40	750.05	765.02	778.72	791.43	814.61	835.61	855.03	873.27	890.59

TABLE 5.5 Thermo-mechanical exergy of air, 50 to 1000 bar. All are supercritical. (Continued)

e^{tm} (kJ/kg)

T (°C)	50	100	150	200	250	300	350	400	450	500	600	700	800	900	1000
										Pressure (bar)					
560	578.57	640.84	678.64	706.48	728.91	747.94	764.64	779.66	793.41	806.17	829.43	850.49	869.98	888.27	905.63
580	592.98	655.34	693.21	721.12	743.61	762.70	779.46	794.54	808.33	821.13	844.47	865.60	885.15	903.49	920.89
600	607.63	670.06	708.01	735.99	758.54	777.69	794.50	809.63	823.47	836.31	859.73	880.93	900.53	918.92	936.37
620	622.52	685.02	723.03	751.07	773.69	792.89	809.76	824.93	838.82	851.70	875.19	896.46	916.12	934.56	952.06
640	637.62	700.19	738.27	766.38	789.05	808.30	825.22	840.44	854.37	867.30	890.86	912.19	931.91	950.40	967.94
660	652.93	715.58	753.72	781.88	804.61	823.92	840.89	856.15	870.12	883.09	906.73	928.12	947.89	966.43	984.01
680	668.45	731.16	769.37	797.59	820.37	839.73	856.74	872.05	886.07	899.07	922.78	944.23	964.06	982.65	1000.28
700	684.18	746.95	785.21	813.49	836.32	855.73	872.79	888.14	902.20	915.24	939.02	960.53	980.41	999.05	1016.72
720	700.10	762.93	801.25	829.58	852.46	871.92	889.02	904.41	918.51	931.59	955.43	977.01	996.94	1015.63	1033.34
740	716.20	779.09	817.47	845.85	868.78	888.28	905.43	920.86	935.00	948.11	972.02	993.66	1013.64	1032.37	1050.12
760	732.49	795.44	833.87	862.30	885.28	904.82	922.01	937.48	951.65	964.80	988.78	1010.47	1030.50	1049.28	1067.08
780	748.96	811.96	850.44	878.92	901.94	921.53	938.76	954.27	968.48	981.66	1005.70	1027.44	1047.53	1066.35	1084.19
800	765.60	828.66	867.18	895.71	918.77	938.40	955.67	971.22	985.46	998.67	1022.77	1044.58	1064.71	1083.58	1101.45
820	782.41	845.51	884.09	912.65	935.76	955.43	972.74	988.32	1002.60	1015.85	1040.00	1061.86	1082.04	1100.96	1118.87
840	799.38	862.53	901.15	929.76	952.91	972.62	989.96	1005.58	1019.89	1033.17	1057.39	1079.29	1099.52	1118.48	1136.43
860	816.51	879.71	918.37	947.02	970.21	989.95	1007.33	1022.99	1037.33	1050.64	1074.91	1096.87	1117.15	1136.15	1154.14
880	833.79	897.03	935.74	964.43	987.66	1007.44	1024.85	1040.54	1054.91	1068.25	1092.58	1114.59	1134.91	1153.95	1171.98
900	851.22	914.51	953.25	981.98	1005.25	1025.06	1042.51	1058.23	1072.63	1086.00	1110.38	1132.44	1152.81	1171.89	1189.96
920	868.80	932.13	970.91	999.68	1022.98	1042.83	1060.31	1076.06	1090.49	1103.88	1128.32	1150.43	1170.84	1189.96	1208.06
940	886.52	949.88	988.71	1017.51	1040.84	1060.73	1078.24	1094.02	1108.48	1121.90	1146.39	1168.54	1189.00	1208.16	1226.30
960	904.37	967.78	1006.64	1035.48	1058.84	1078.76	1096.30	1112.11	1126.60	1140.05	1164.58	1186.78	1207.28	1226.48	1244.65
980	922.36	985.80	1024.70	1053.57	1076.97	1096.92	1114.49	1130.32	1144.84	1158.32	1182.90	1205.15	1225.68	1244.92	1263.13
1000	940.48	1003.96	1042.89	1071.79	1095.22	1115.20	1132.80	1148.66	1163.20	1176.71	1201.34	1223.63	1244.21	1263.48	1281.73
1100	1032.91	1096.55	1135.64	1164.69	1188.26	1208.37	1226.11	1242.10	1256.76	1270.38	1295.23	1317.73	1338.49	1357.94	1376.35
1200	1128.13	1191.91	1231.13	1260.31	1284.00	1304.23	1322.08	1338.17	1352.94	1366.67	1391.71	1414.39	1435.33	1454.94	1473.50
1300	1225.83	1289.73	1329.05	1358.34	1382.14	1402.47	1420.41	1436.61	1451.47	1465.28	1490.50	1513.34	1534.43	1554.18	1572.88
1400	1325.72	1389.72	1429.15	1458.53	1482.42	1502.83	1520.86	1537.14	1552.08	1565.97	1591.34	1614.32	1635.55	1655.44	1674.26
1500	1427.60	1491.68	1531.19	1560.65	1584.62	1605.11	1623.21	1639.56	1654.58	1668.53	1694.04	1717.15	1738.50	1758.50	1777.44
1600	1531.26	1595.42	1635.00	1664.53	1688.56	1709.12	1727.29	1743.70	1758.78	1772.79	1798.42	1821.64	1843.10	1863.21	1882.25
1700	1636.54	1700.77	1740.41	1770.00	1794.09	1814.71	1832.93	1849.40	1864.53	1878.60	1904.33	1927.65	1949.21	1969.42	1988.55
1800	1743.31	1807.59	1847.29	1876.93	1901.07	1921.74	1940.01	1956.53	1971.70	1985.82	2011.64	2035.06	2056.71	2077.00	2096.21
1900	1851.43	1915.76	1955.51	1985.20	2009.38	2030.09	2048.41	2064.97	2080.18	2094.35	2120.25	2143.74	2165.47	2185.84	2205.13
2000	1960.81	2025.18	2064.97	2094.70	2118.92	2139.67	2158.02	2174.62	2189.87	2204.07	2230.05	2253.61	2275.41	2295.85	2315.20

TABLE 5.5 Thermo-mechanical exergy of air, 50 to 1000 bar. All are supercritical. (Continued)

T (°C)	e^tm (kJ/kg) — Pressure (bar)														
	1	2	3	4	5	6	7	8	9	10	11	12	13	14	15
−212	843.8	844.0	844.2	844.4	844.6	844.8	845.0	845.2	845.5	845.7	845.9	846.1	846.3	846.5	846.7
−211	836.3	836.5	836.8	837.0	837.2	837.4	837.6	837.8	838.0	838.3	838.5	838.7	838.9	839.1	839.3
−210	829.0	829.3	829.5	829.7	829.9	830.1	830.3	830.6	830.8	831.0	831.2	831.4	831.6	831.8	832.1
−209	821.9	822.1	822.3	822.6	822.8	823.0	823.2	823.4	823.6	823.9	824.1	824.3	824.5	824.7	824.9
−208	814.9	815.1	815.3	815.6	815.8	816.0	816.2	816.4	816.7	816.9	817.1	817.3	817.5	817.7	818.0
−207	808.0	808.3	808.5	808.7	808.9	809.1	809.4	809.6	809.8	810.0	810.2	810.5	810.7	810.9	811.1
−206	801.3	801.5	801.8	802.0	802.2	802.4	802.6	802.9	803.1	803.3	803.5	803.8	804.0	804.2	804.4
−205	794.7	794.9	795.2	795.4	795.6	795.8	796.1	796.3	796.5	796.7	796.9	797.2	797.4	797.6	797.8
−204	788.2	788.5	788.7	788.9	789.1	789.4	789.6	789.8	790.0	790.3	790.5	790.7	790.9	791.2	791.4
−203	781.9	782.1	782.3	782.6	782.8	783.0	783.2	783.5	783.7	783.9	784.1	784.4	784.6	784.8	785.0
−202	775.6	775.9	776.1	776.3	776.5	776.8	777.0	777.2	777.5	777.7	777.9	778.1	778.4	778.6	778.8
−201	769.5	769.7	769.9	770.2	770.4	770.6	770.9	771.1	771.3	771.6	771.8	772.0	772.2	772.5	772.7
−200	763.5	763.7	763.9	764.2	764.4	764.6	764.8	765.1	765.3	765.5	765.8	766.0	766.2	766.5	766.7
−199	757.6	757.8	758.1	758.3	758.5	758.8	759.0	759.2	759.5	759.7	759.9	760.2	760.4	760.6	760.9
−198	751.7	752.0	752.2	752.5	752.7	752.9	753.2	753.4	753.6	753.9	754.1	754.3	754.6	754.8	755.0
−197	746.0	746.2	746.5	746.7	747.0	747.2	747.4	747.7	747.9	748.2	748.4	748.6	748.9	749.1	749.3
−196	740.4	740.6	740.8	741.1	741.3	741.6	741.8	742.1	742.3	742.5	742.8	743.0	743.2	743.5	743.7
−195	734.8	735.1	735.3	735.5	735.8	736.0	736.3	736.5	736.8	737.0	737.2	737.5	737.7	738.0	738.2
−194	—	729.6	729.8	730.1	730.3	730.6	730.8	731.1	731.3	731.5	731.8	732.0	732.3	732.5	732.8
−193	—	724.2	724.4	724.7	724.9	725.2	725.4	725.7	725.9	726.2	726.4	726.7	726.9	727.2	727.4
−192	—	718.9	719.1	719.4	719.6	719.9	720.1	720.4	720.6	720.9	721.1	721.4	721.6	721.9	722.1
−191	—	713.7	713.9	714.2	714.4	714.7	714.9	715.2	715.4	715.7	715.9	716.2	716.4	716.7	716.9
−190	—	708.5	708.8	709.0	709.3	709.5	709.8	710.0	710.3	710.6	710.8	711.1	711.3	711.6	711.8
−189	—	703.4	703.7	703.9	704.2	704.5	704.7	705.0	705.2	705.5	705.7	706.0	706.3	706.5	706.8
−188	—	—	698.7	698.9	699.2	699.5	699.7	700.0	700.2	700.5	700.8	701.0	701.3	701.5	701.8
−187	—	—	693.7	694.0	694.2	694.5	694.8	695.0	695.3	695.6	695.8	696.1	696.4	696.6	696.9
−185	—	—	688.8	689.1	689.4	689.6	689.9	690.2	690.4	690.7	691.0	691.2	691.5	691.8	692.0
−184	—	—	684.0	684.3	684.6	684.8	685.1	685.4	685.6	685.9	686.2	686.5	686.7	687.0	687.3
−183	—	—	679.2	679.5	679.8	680.1	680.3	680.6	680.9	681.2	681.4	681.7	682.0	682.3	682.5
−182	—	—	—	674.8	675.1	675.4	675.6	675.9	676.2	676.5	676.8	677.0	677.3	677.6	677.9
−181	—	—	—	670.1	670.4	670.7	671.0	671.3	671.6	671.9	672.1	672.4	672.7	673.0	673.3

TABLE 5.6 Thermo-mechanical exergy of **liquid** air, 1 to 15 bar. Blank spaces indicate that air is a mixture of liquid and gas at that temperature and pressure.

| | e^tm (kJ/kg) | | | | | | | | | | | | | | |
| | Pressure (bar) | | | | | | | | | | | | | | |
T (°C)	1	2	3	4	5	6	7	8	9	10	11	12	13	14	15
−180	—	—	—	665.5	665.8	666.1	666.4	666.7	667.0	667.3	667.6	667.8	668.1	668.4	668.7
−179	—	—	—	—	661.3	661.6	661.9	662.2	662.4	662.7	663.0	663.3	663.6	663.9	664.2
−178	—	—	—	—	656.8	657.1	657.4	657.7	658.0	658.3	658.6	658.8	659.1	659.4	659.7
−177	—	—	—	—	652.3	652.6	652.9	653.2	653.5	653.8	654.1	654.4	654.7	655.0	655.3
−176	—	—	—	—	—	648.2	648.5	648.8	649.1	649.4	649.7	650.0	650.3	650.6	650.9
−175	—	—	—	—	—	643.8	644.1	644.4	644.7	645.0	645.4	645.7	646.0	646.3	646.6
−174	—	—	—	—	—	—	639.7	640.1	640.4	640.7	641.0	641.3	641.7	642.0	642.3
−173	—	—	—	—	—	—	635.4	635.8	636.1	636.4	636.7	637.1	637.4	637.7	638.0
−172	—	—	—	—	—	—	—	631.5	631.9	632.2	632.5	632.8	633.2	633.5	633.8
−171	—	—	—	—	—	—	—	627.3	627.6	627.9	628.3	628.6	629.0	629.3	629.6
−170	—	—	—	—	—	—	—	—	623.4	623.7	624.1	624.4	624.8	625.1	625.4
−169	—	—	—	—	—	—	—	—	619.2	619.5	619.9	620.2	620.6	620.9	621.3
−168	—	—	—	—	—	—	—	—	—	615.3	615.7	616.1	616.4	616.8	617.1
−167	—	—	—	—	—	—	—	—	—	611.1	611.5	611.9	612.3	612.6	613.0
−166	—	—	—	—	—	—	—	—	—	—	607.3	607.7	608.1	608.5	608.9
−165	—	—	—	—	—	—	—	—	—	—	—	603.6	604.0	604.4	604.8
−164	—	—	—	—	—	—	—	—	—	—	—	599.4	599.8	600.3	600.7
−163	—	—	—	—	—	—	—	—	—	—	—	—	595.7	596.1	596.5
−162	—	—	—	—	—	—	—	—	—	—	—	—	—	592.0	592.4
−161	—	—	—	—	—	—	—	—	—	—	—	—	—	—	588.2

TABLE 5.6 Thermo-mechanical exergy of **liquid** air, 1 to 15 bar. Blank spaces indicate that air is a mixture of liquid and gas at that temperature and pressure. (*Continued*)

T (°C)	1	2	3	4	5	6	7	8	9	10	11	12	13	14	15
-178	—	—	—	—	656.8	657.1	657.4	657.7	658.0	658.3	658.6	658.8	659.1	659.4	659.7
-177	—	—	—	—	652.3	652.6	652.9	653.2	653.5	653.8	654.1	654.4	654.7	655.0	655.3
-176	—	—	—	—	—	648.2	648.5	648.8	649.1	649.4	649.7	650.0	650.3	650.6	650.9
-175	—	—	—	—	—	643.8	644.1	644.4	644.7	645.0	645.4	645.7	646.0	646.3	646.6
-174	—	—	—	—	—	—	639.7	640.1	640.4	640.7	641.0	641.3	641.7	642.0	642.3
-173	—	—	—	—	—	—	635.4	635.8	636.1	636.4	636.7	637.1	637.4	637.7	638.0
-172	—	—	—	—	—	—	—	631.5	631.9	632.2	632.5	632.8	633.2	633.5	633.8
-171	—	—	—	—	—	—	—	627.3	627.6	627.9	628.3	628.6	629.0	629.3	629.6
-170	—	—	—	—	—	—	—	—	623.4	623.7	624.1	624.4	624.8	625.1	625.4
-169	—	—	—	—	—	—	—	—	619.2	619.5	619.9	620.2	620.6	620.9	621.3
-168	—	—	—	—	—	—	—	—	—	615.3	615.7	616.1	616.4	616.8	617.1
-167	—	—	—	—	—	—	—	—	—	611.1	611.5	611.9	612.3	612.6	613.0
-166	—	—	—	—	—	—	—	—	—	—	607.3	607.7	608.1	608.5	608.9
-165	—	—	—	—	—	—	—	—	—	—	—	603.6	604.0	604.4	604.8
-164	—	—	—	—	—	—	—	—	—	—	—	599.4	599.8	600.3	600.7
-163	—	—	—	—	—	—	—	—	—	—	—	—	595.7	596.1	596.5
-162	—	—	—	—	—	—	—	—	—	—	—	—	—	592.0	592.4
-161	—	—	—	—	—	—	—	—	—	—	—	—	—	—	588.2

e^{tm} (kJ/kg)

Pressure (bar)

TABLE 5.6 Thermo-mechanical exergy of **liquid** air, 1 to 15 bar. Blank spaces indicate that air is a mixture of liquid and gas at that temperature and pressure. *(Continued)*

References

1. Hill IH, Wronski J, Quoilin S, Lemort V. Pure and pseudo-pure fluid thermophysical property evaluation and the open-source thermophysical property library CoolProp. *Ind Eng Chem Res* 53:2498–2508 (2014). doi:10.1021/ie4033999.

2. Lemmon EW, Jacobsen RT, Penoncello SG, Friend DG. Thermodynamic properties of air and mixtures of nitrogen, argon, and oxygen from 60 to 2000 K at pressures to 2000 MPa. *J Phys Chem Ref Data* 29(3):331–385 (2000). doi:10.1063/1.1285884.

3. Tegeler CH, Span R, Wagner W. A new equation of state for argon covering the fluid region for temperatures from the melting line to 700 K at pressures up to 1000 MPa. *J Phys Chem Ref Data* 28:779–850 (1999). doi:10.1063/1.556037.

4. Span R, Wagner W. A new equation of state for carbon dioxide covering the fluid region from the triple point temperature to 1100 K at pressures up to 800 MPa. *J Phys Chem Ref Data* 25:1509–1596 (1996). doi:10.1063/1.555991.

5. Lemmon EW, Span R. Short Fundamental equations of state for 20 industrial fluids. *J Chem Eng Data* 51:785–850 (2006). doi:10.1021/je050186n.

6. Span R, Lemmon EW, Jacobsen RT, Wagner W, Yokozeki A. A reference equation of state for the thermodynamic properties of nitrogen for temperatures from 63.151 to 1000 K and pressures to 2200 MPa. *J Phys Chem Ref Data* 29:1361–1433 (2000). doi:10.1063/1.1349047.

7. Stewart RB, Jacobsen RT, Wagner W. Thermodynamic properties of oxygen from the triple point to 300 K with pressures to 80 MPa. *J Phys Chem Ref Data* 20(5):917 (1991). doi:10.1063/1.555897.

8. Schmidt R, Wagner W. A new form of the equation of state for pure substances and its application to oxygen. *Fluid Phase Equilib* 19(3):175–200 (1985) doi:10.1016/0378-3812(85)87016-3.

CHAPTER **6**

Water, Steam, and Heavy Water

6.1 Chemical Exergy

The chemical exergy of water in the gas phase at 25°C and 1 atm is 9.4882 kJ/mol (526.68 kJ/kg) as noted in Table 4.1 in Chap. 4. The chemical exergy of pure liquid water is 0.89223 kJ/mol (49.83 kJ/kg). These are roughly the same orders of magnitude as their corresponding thermo-mechanical exergies for the range explored in Fig. 6.1. It is important to mention that the phase difference between liquid and gaseous water at atmospheric conditions is reflected in the difference in the chemical exergies of the two phases. Therefore, the chemical exergy of water is often important to consider whenever it is involved in phase change.

6.2 Thermo-Mechanical Exergy

Figure 6.1 shows thermo-mechanical exergy of water graphically on a pressure-enthalpy diagram. The underlying physical property models for Fig. 6.1 are based on IAPWS Formulation 1995[1] computed via Ref. 2. Figure 6.2 shows the thermo-mechanical exergy of heavy water (D_2O), using Ref. 3. The thermo-mechanical exergy ranges for D_2O are quite similar to H_2O on a per-kg basis.

The tables (Tables 6.1 through 6.10) in the chapter list the thermo-mechanical exergy of water and steam more precisely. The information in Tables 6.1 to 6.10 and Fig. 6.1 is the same; you can use both interchangeably.

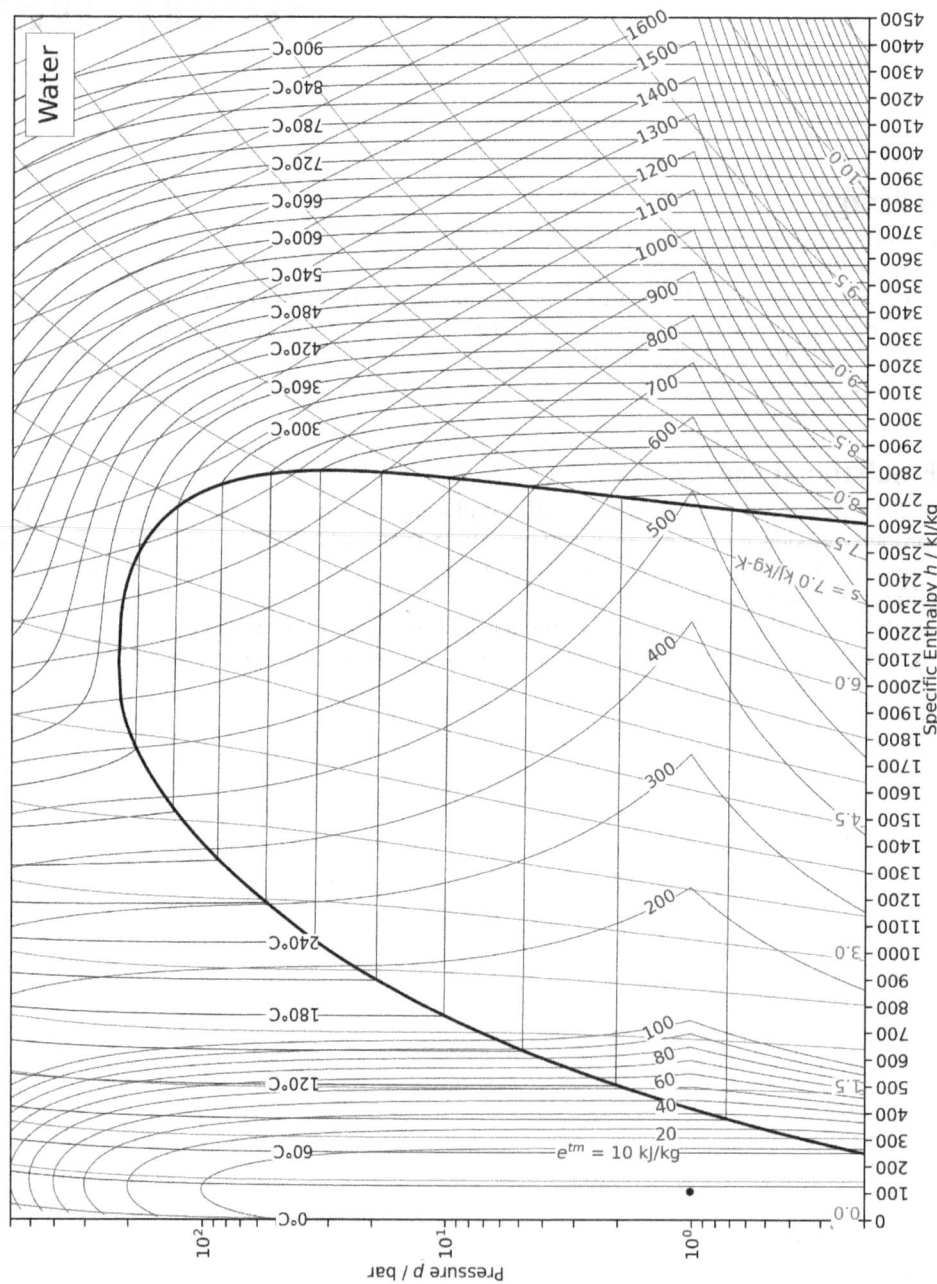

Figure 6.1 Pressure-enthalpy diagram for water, with the thermo-mechanical exergy of water e^{tm} (kJ/kg).

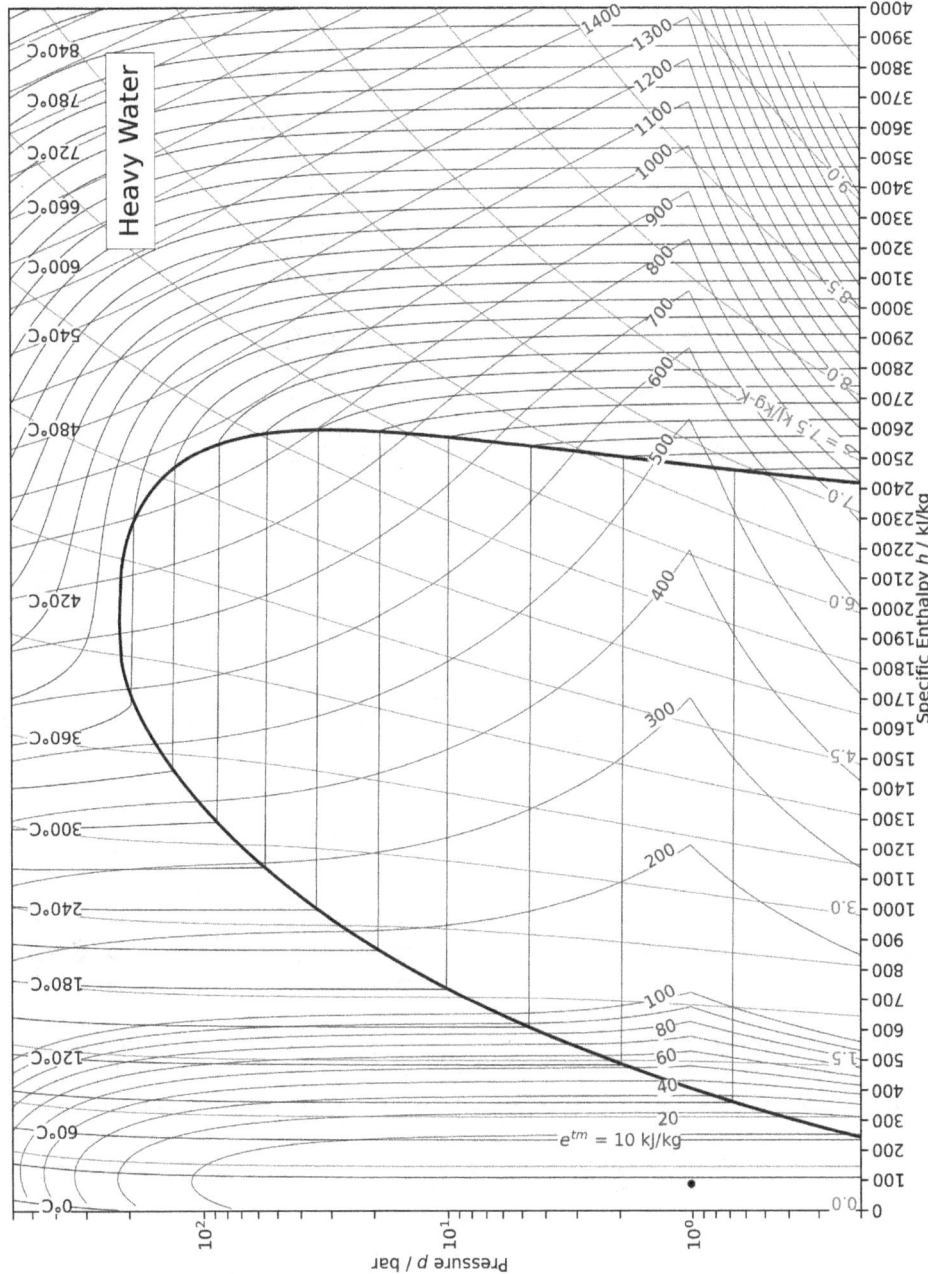

Figure 6.2 Pressure-enthalpy diagram for heavy water (D_2O), with the thermo-mechanical exergy of water e^{tm} (kJ/kg).

e^tm (kJ/kg)

T (°C)	Pressure (bar)														
	1.01325	1.084	1.154	1.225	1.295	1.366	1.436	1.507	1.577	1.648	1.718	1.789	1.859	1.930	2.000
0	—	—	—	4.6866	4.6936	4.7007	4.7077	4.7147	4.7218	4.7288	4.7359	4.7429	4.7499	4.7570	4.7640
1	4.2879	4.2949	4.3019	4.3090	4.3160	4.3231	4.3301	4.3371	4.3442	4.3512	4.3583	4.3653	4.3723	4.3794	4.3864
2	3.9272	3.9343	3.9413	3.9484	3.9554	3.9625	3.9695	3.9765	3.9836	3.9906	3.9977	4.0047	4.0118	4.0188	4.0259
3	3.5835	3.5905	3.5976	3.6046	3.6117	3.6187	3.6257	3.6328	3.6398	3.6469	3.6539	3.6610	3.6680	3.6751	3.6821
4	3.2564	3.2634	3.2705	3.2775	3.2846	3.2916	3.2987	3.3057	3.3128	3.3198	3.3269	3.3339	3.3410	3.3480	3.3550
5	2.9458	2.9529	2.9599	2.9670	2.9740	2.9811	2.9881	2.9952	3.0022	3.0093	3.0163	3.0234	3.0304	3.0375	3.0445
6	2.6516	2.6587	2.6657	2.6728	2.6798	2.6869	2.6939	2.7010	2.7080	2.7151	2.7221	2.7292	2.7362	2.7433	2.7503
7	2.3736	2.3807	2.3878	2.3948	2.4019	2.4089	2.4160	2.4230	2.4301	2.4371	2.4442	2.4512	2.4583	2.4654	2.4724
8	2.1118	2.1188	2.1259	2.1329	2.1400	2.1471	2.1541	2.1612	2.1682	2.1753	2.1823	2.1894	2.1964	2.2035	2.2106
9	1.8658	1.8729	1.8800	1.8870	1.8941	1.9011	1.9082	1.9153	1.9223	1.9294	1.9364	1.9435	1.9505	1.9576	1.9647
10	1.6357	1.6428	1.6499	1.6569	1.6640	1.6710	1.6781	1.6852	1.6922	1.6993	1.7063	1.7134	1.7205	1.7275	1.7346
11	1.4213	1.4284	1.4354	1.4425	1.4496	1.4566	1.4637	1.4707	1.4778	1.4849	1.4919	1.4990	1.5061	1.5131	1.5202
12	1.2225	1.2295	1.2366	1.2437	1.2507	1.2578	1.2648	1.2719	1.2790	1.2860	1.2931	1.3001	1.3072	1.3143	1.3213
13	1.0390	1.0461	1.0532	1.0602	1.0673	1.0744	1.0814	1.0885	1.0955	1.1026	1.1097	1.1167	1.1238	1.1309	1.1379
14	0.8709	0.8780	0.8851	0.8921	0.8992	0.9063	0.9133	0.9204	0.9274	0.9345	0.9416	0.9486	0.9557	0.9628	0.9698
15	0.7180	0.7251	0.7321	0.7392	0.7463	0.7533	0.7604	0.7675	0.7745	0.7816	0.7887	0.7957	0.8028	0.8099	0.8169
16	0.5802	0.5872	0.5943	0.6014	0.6084	0.6155	0.6226	0.6296	0.6367	0.6438	0.6508	0.6579	0.6650	0.6720	0.6791
17	0.4573	0.4644	0.4714	0.4785	0.4856	0.4926	0.4997	0.5068	0.5138	0.5209	0.5280	0.5350	0.5421	0.5492	0.5562
18	0.3493	0.3564	0.3634	0.3705	0.3776	0.3846	0.3917	0.3988	0.4058	0.4129	0.4200	0.4270	0.4341	0.4412	0.4482
19	0.2560	0.2631	0.2701	0.2772	0.2843	0.2913	0.2984	0.3055	0.3125	0.3196	0.3267	0.3337	0.3408	0.3479	0.3550
20	0.1774	0.1844	0.1915	0.1986	0.2056	0.2127	0.2198	0.2268	0.2339	0.2410	0.2480	0.2551	0.2622	0.2692	0.2763
21	0.1132	0.1203	0.1274	0.1344	0.1415	0.1486	0.1557	0.1627	0.1698	0.1769	0.1839	0.1910	0.1981	0.2051	0.2122
22	0.0635	0.0706	0.0777	0.0848	0.0918	0.0989	0.1060	0.1130	0.1201	0.1272	0.1342	0.1413	0.1484	0.1554	0.1625
23	0.0282	0.0352	0.0423	0.0494	0.0565	0.0635	0.0706	0.0777	0.0847	0.0918	0.0989	0.1059	0.1130	0.1201	0.1271
24	0.0070	0.0141	0.0212	0.0282	0.0353	0.0424	0.0494	0.0565	0.0636	0.0706	0.0777	0.0848	0.0919	0.0989	0.1060
25	0.0000	0.0071	0.0141	0.0212	0.0283	0.0353	0.0424	0.0495	0.0566	0.0636	0.0707	0.0778	0.0848	0.0919	0.0990
26	0.0070	0.0141	0.0211	0.0282	0.0353	0.0423	0.0494	0.0565	0.0635	0.0706	0.0777	0.0848	0.0918	0.0989	0.1060
27	0.0279	0.0350	0.0421	0.0491	0.0562	0.0633	0.0703	0.0774	0.0845	0.0915	0.0986	0.1057	0.1127	0.1198	0.1269
28	0.0627	0.0697	0.0768	0.0839	0.0910	0.0980	0.1051	0.1122	0.1192	0.1263	0.1334	0.1404	0.1475	0.1546	0.1616

TABLE 6.1 Thermo-mechanical exergy of liquid water from 1.01325 bar (1 atm) to 2 bar. Dashed lines indicate that water is not a liquid at that temperature and pressure (it is a solid, a mixture of gas and liquid, or a gas).

e^{tm} (kJ/kg)

T (°C)	1.01325	1.084	1.154	1.225	1.295	1.366	1.436	1.507	1.577	1.648	1.718	1.789	1.859	1.930	2.000
									Pressure (bar)						
29	0.1112	0.1182	0.1253	0.1324	0.1394	0.1465	0.1536	0.1607	0.1677	0.1748	0.1819	0.1889	0.1960	0.2031	0.2101
30	0.1733	0.1804	0.1875	0.1945	0.2016	0.2087	0.2157	0.2228	0.2299	0.2369	0.2440	0.2511	0.2581	0.2652	0.2723
31	0.2490	0.2561	0.2632	0.2702	0.2773	0.2844	0.2914	0.2985	0.3056	0.3126	0.3197	0.3268	0.3338	0.3409	0.3480
32	0.3382	0.3453	0.3523	0.3594	0.3665	0.3735	0.3806	0.3877	0.3947	0.4018	0.4089	0.4159	0.4230	0.4301	0.4371
33	0.4407	0.4478	0.4549	0.4619	0.4690	0.4761	0.4831	0.4902	0.4973	0.5044	0.5114	0.5185	0.5256	0.5326	0.5397
34	0.5566	0.5637	0.5707	0.5778	0.5849	0.5919	0.5990	0.6061	0.6131	0.6202	0.6273	0.6343	0.6414	0.6485	0.6555
35	0.6856	0.6927	0.6998	0.7068	0.7139	0.7210	0.7280	0.7351	0.7422	0.7492	0.7563	0.7634	0.7704	0.7775	0.7846
36	0.8278	0.8349	0.8419	0.8490	0.8561	0.8631	0.8702	0.8773	0.8843	0.8914	0.8985	0.9055	0.9126	0.9197	0.9267
37	0.9830	0.9901	0.9971	1.0042	1.0113	1.0183	1.0254	1.0325	1.0395	1.0466	1.0537	1.0607	1.0678	1.0749	1.0819
38	1.1512	1.1582	1.1653	1.1724	1.1794	1.1865	1.1936	1.2006	1.2077	1.2148	1.2218	1.2289	1.2359	1.2430	1.2501
39	1.3322	1.3393	1.3463	1.3534	1.3605	1.3675	1.3746	1.3817	1.3887	1.3958	1.4028	1.4099	1.4170	1.4240	1.4311
40	1.5260	1.5331	1.5402	1.5472	1.5543	1.5614	1.5684	1.5755	1.5825	1.5896	1.5967	1.6037	1.6108	1.6179	1.6249
41	1.7326	1.7396	1.7467	1.7538	1.7608	1.7679	1.7749	1.7820	1.7891	1.7961	1.8032	1.8103	1.8173	1.8244	1.8314
42	1.9518	1.9588	1.9659	1.9729	1.9800	1.9871	1.9941	2.0012	2.0082	2.0153	2.0224	2.0294	2.0365	2.0435	2.0506
43	2.1835	2.1905	2.1976	2.2047	2.2117	2.2188	2.2258	2.2329	2.2400	2.2470	2.2541	2.2611	2.2682	2.2753	2.2823
44	2.4277	2.4347	2.4418	2.4489	2.4559	2.4630	2.4700	2.4771	2.4841	2.4912	2.4983	2.5053	2.5124	2.5194	2.5265
45	2.6843	2.6913	2.6984	2.7055	2.7125	2.7196	2.7266	2.7337	2.7407	2.7478	2.7549	2.7619	2.7690	2.7760	2.7831
46	2.9532	2.9603	2.9673	2.9744	2.9814	2.9885	2.9955	3.0026	3.0097	3.0167	3.0238	3.0308	3.0379	3.0449	3.0520
47	3.2344	3.2414	3.2485	3.2556	3.2626	3.2697	3.2767	3.2838	3.2908	3.2979	3.3049	3.3120	3.3191	3.3261	3.3332
48	3.5277	3.5348	3.5419	3.5489	3.5560	3.5630	3.5701	3.5771	3.5842	3.5912	3.5983	3.6053	3.6124	3.6194	3.6265
49	3.8332	3.8403	3.8473	3.8544	3.8614	3.8685	3.8755	3.8826	3.8896	3.8967	3.9037	3.9108	3.9178	3.9249	3.9319
50	4.1507	4.1577	4.1648	4.1719	4.1789	4.1860	4.1930	4.2001	4.2071	4.2142	4.2212	4.2283	4.2353	4.2424	4.2494
51	4.4801	4.4872	4.4942	4.5013	4.5083	4.5154	4.5224	4.5295	4.5366	4.5436	4.5507	4.5577	4.5648	4.5718	4.5789
52	4.8215	4.8285	4.8356	4.8426	4.8497	4.8567	4.8638	4.8708	4.8779	4.8849	4.8920	4.8990	4.9061	4.9131	4.9202
53	5.1746	5.1817	5.1887	5.1958	5.2028	5.2099	5.2169	5.2240	5.2310	5.2381	5.2451	5.2522	5.2592	5.2663	5.2733
54	5.5396	5.5466	5.5537	5.5607	5.5677	5.5748	5.5818	5.5889	5.5959	5.6030	5.6100	5.6171	5.6241	5.6312	5.6382
55	5.9162	5.9232	5.9302	5.9373	5.9443	5.9514	5.9584	5.9655	5.9725	5.9796	5.9866	5.9936	6.0007	6.0077	6.0148
56	6.3044	6.3114	6.3184	6.3255	6.3325	6.3396	6.3466	6.3537	6.3607	6.3678	6.3748	6.3818	6.3889	6.3959	6.4030
57	6.7041	6.7112	6.7182	6.7252	6.7323	6.7393	6.7464	6.7534	6.7605	6.7675	6.7745	6.7816	6.7886	6.7957	6.8027
58	7.1154	7.1224	7.1294	7.1365	7.1435	7.1506	7.1576	7.1646	7.1717	7.1787	7.1858	7.1928	7.1998	7.2069	7.2139
59	7.5380	7.5451	7.5521	7.5591	7.5662	7.5732	7.5802	7.5873	7.5943	7.6014	7.6084	7.6154	7.6225	7.6295	7.6366
60	7.9720	7.9791	7.9861	7.9931	8.0002	8.0072	8.0142	8.0213	8.0283	8.0354	8.0424	8.0494	8.0565	8.0635	8.0705

TABLE 6.1 Thermo-mechanical exergy of liquid water from 1.01325 bar (1 atm) to 2 bar. Dashed lines indicate that water is not a liquid at that temperature and pressure (it is a solid, a mixture of gas and liquid, or a gas). (*Continued*)

e^{tm} (kJ/kg)

T (°C)	Pressure (bar)														
	1.01325	1.084	1.154	1.225	1.295	1.366	1.436	1.507	1.577	1.648	1.718	1.789	1.859	1.930	2.000
61	8.4173	8.4244	8.4314	8.4384	8.4455	8.4525	8.4595	8.4666	8.4736	8.4806	8.4877	8.4947	8.5017	8.5088	8.5158
62	8.8738	8.8809	8.8879	8.8949	8.9020	8.9090	8.9160	8.9231	8.9301	8.9371	8.9442	8.9512	8.9582	8.9653	8.9723
63	9.3415	9.3486	9.3556	9.3626	9.3696	9.3767	9.3837	9.3907	9.3978	9.4048	9.4118	9.4189	9.4259	9.4329	9.4400
64	9.8203	9.8273	9.8344	9.8414	9.8484	9.8555	9.8625	9.8695	9.8765	9.8836	9.8906	9.8976	9.9047	9.9117	9.9187
65	10.310	10.317	10.324	10.331	10.338	10.345	10.352	10.359	10.366	10.373	10.380	10.387	10.394	10.401	10.409
66	10.811	10.818	10.825	10.832	10.839	10.846	10.853	10.860	10.867	10.874	10.881	10.888	10.895	10.902	10.909
67	11.323	11.330	11.337	11.344	11.351	11.358	11.365	11.372	11.379	11.386	11.393	11.400	11.407	11.414	11.421
68	11.845	11.852	11.859	11.866	11.873	11.880	11.887	11.894	11.901	11.908	11.915	11.922	11.929	11.937	11.944
69	12.379	12.386	12.393	12.400	12.407	12.414	12.421	12.428	12.435	12.442	12.449	12.456	12.463	12.470	12.477
70	12.923	12.930	12.937	12.944	12.951	12.958	12.965	12.972	12.979	12.986	12.993	13.000	13.007	13.014	13.021
71	13.478	13.485	13.492	13.499	13.506	13.513	13.520	13.527	13.534	13.541	13.548	13.555	13.562	13.569	13.576
72	14.043	14.050	14.057	14.064	14.071	14.078	14.085	14.092	14.099	14.106	14.113	14.120	14.127	14.134	14.141
73	14.619	14.626	14.633	14.640	14.647	14.654	14.661	14.668	14.675	14.682	14.689	14.696	14.703	14.710	14.717
74	15.206	15.213	15.220	15.227	15.234	15.241	15.248	15.255	15.262	15.269	15.276	15.283	15.290	15.297	15.304
75	15.803	15.810	15.817	15.824	15.831	15.838	15.845	15.852	15.859	15.866	15.873	15.880	15.887	15.894	15.901
76	16.410	16.417	16.424	16.431	16.438	16.445	16.452	16.459	16.466	16.473	16.480	16.487	16.494	16.501	16.508
77	17.028	17.035	17.042	17.049	17.056	17.063	17.070	17.077	17.084	17.091	17.098	17.105	17.112	17.119	17.126
78	17.656	17.663	17.670	17.677	17.684	17.691	17.698	17.705	17.712	17.719	17.726	17.733	17.740	17.747	17.754
79	18.294	18.301	18.308	18.315	18.322	18.329	18.336	18.343	18.350	18.357	18.364	18.371	18.378	18.385	18.392
80	18.943	18.950	18.957	18.964	18.971	18.978	18.985	18.992	18.999	19.006	19.013	19.020	19.027	19.034	19.041
81	19.601	19.608	19.615	19.622	19.629	19.636	19.643	19.650	19.657	19.664	19.671	19.678	19.685	19.692	19.699
82	20.270	20.277	20.284	20.291	20.298	20.305	20.312	20.319	20.326	20.333	20.340	20.347	20.354	20.361	20.368
83	20.949	20.956	20.963	20.970	20.977	20.984	20.991	20.998	21.005	21.012	21.019	21.026	21.033	21.040	21.047
84	21.638	21.645	21.652	21.659	21.666	21.673	21.680	21.687	21.694	21.701	21.708	21.715	21.722	21.729	21.736
85	22.336	22.343	22.350	22.357	22.364	22.371	22.378	22.385	22.392	22.399	22.406	22.413	22.420	22.427	22.434
86	23.045	23.052	23.059	23.066	23.073	23.080	23.087	23.094	23.101	23.108	23.115	23.122	23.129	23.136	23.143
87	23.764	23.771	23.778	23.785	23.792	23.799	23.806	23.813	23.820	23.827	23.834	23.840	23.847	23.854	23.861
88	24.492	24.499	24.506	24.513	24.520	24.527	24.534	24.541	24.548	24.555	24.562	24.569	24.576	24.583	24.590
89	25.230	25.237	25.244	25.251	25.258	25.265	25.272	25.279	25.286	25.293	25.300	25.307	25.314	25.321	25.328

TABLE 6.1 Thermo-mechanical exergy of liquid water from 1.01325 bar (1 atm) to 2 bar. Dashed lines indicate that water is not a liquid at that temperature and pressure (it is a solid, a mixture of gas and liquid, or a gas). (Continued)

e^{tm} (kJ/kg)

T (°C)	Pressure (bar)														
	1.01325	1.084	1.154	1.225	1.295	1.366	1.436	1.507	1.577	1.648	1.718	1.789	1.859	1.930	2.000
90	25.978	25.985	25.992	25.999	26.006	26.013	26.020	26.027	26.034	26.041	26.048	26.055	26.062	26.069	26.076
91	26.736	26.743	26.749	26.756	26.763	26.770	26.777	26.784	26.791	26.798	26.805	26.812	26.819	26.826	26.833
92	27.503	27.510	27.517	27.524	27.531	27.538	27.545	27.551	27.558	27.565	27.572	27.579	27.586	27.593	27.600
93	28.279	28.286	28.293	28.300	28.307	28.314	28.321	28.328	28.335	28.342	28.349	28.356	28.363	28.370	28.377
94	29.066	29.073	29.080	29.087	29.094	29.101	29.107	29.114	29.121	29.128	29.135	29.142	29.149	29.156	29.163
95	29.862	29.868	29.875	29.882	29.889	29.896	29.903	29.910	29.917	29.924	29.931	29.938	29.945	29.952	29.959
96	30.667	30.674	30.681	30.688	30.695	30.702	30.708	30.715	30.722	30.729	30.736	30.743	30.750	30.757	30.764
97	31.481	31.488	31.495	31.502	31.509	31.516	31.523	31.530	31.537	31.544	31.551	31.558	31.565	31.572	31.579
98	32.305	32.312	32.319	32.326	32.333	32.340	32.347	32.354	32.361	32.368	32.375	32.382	32.389	32.396	32.403
99	33.139	33.146	33.153	33.160	33.167	33.174	33.181	33.187	33.194	33.201	33.208	33.215	33.222	33.229	33.236
100	—	33.988	33.995	34.002	34.009	34.016	34.023	34.030	34.037	34.044	34.051	34.058	34.065	34.072	34.079
101	—	34.840	34.847	34.854	34.861	34.868	34.875	34.882	34.889	34.896	34.903	34.910	34.917	34.924	34.931
102	—	—	35.708	35.715	35.722	35.729	35.736	35.743	35.750	35.757	35.764	35.771	35.778	35.785	35.792
103	—	—	36.579	36.586	36.593	36.600	36.607	36.613	36.620	36.627	36.634	36.641	36.648	36.655	36.662
104	—	—	—	37.465	37.472	37.479	37.486	37.493	37.500	37.507	37.514	37.521	37.528	37.534	37.541
105	—	—	—	38.354	38.361	38.368	38.375	38.382	38.388	38.395	38.402	38.409	38.416	38.423	38.430
106	—	—	—	—	39.258	39.265	39.272	39.279	39.286	39.293	39.300	39.307	39.314	39.321	39.328
107	—	—	—	—	40.165	40.172	40.179	40.186	40.193	40.200	40.207	40.213	40.220	40.227	40.234
108	—	—	—	—	—	41.088	41.095	41.102	41.108	41.115	41.122	41.129	41.136	41.143	41.150
109	—	—	—	—	—	—	42.019	42.026	42.033	42.040	42.047	42.054	42.061	42.068	42.075
110	—	—	—	—	—	—	42.953	42.960	42.967	42.974	42.980	42.987	42.994	43.001	43.008
111	—	—	—	—	—	—	—	43.902	43.909	43.916	43.923	43.930	43.937	43.944	43.951
112	—	—	—	—	—	—	—	—	44.860	44.867	44.874	44.881	44.888	44.895	44.902
113	—	—	—	—	—	—	—	—	—	45.827	45.834	45.841	45.848	45.855	45.862
114	—	—	—	—	—	—	—	—	—	46.796	46.803	46.810	46.817	46.824	46.831
115	—	—	—	—	—	—	—	—	—	—	47.781	47.788	47.795	47.802	47.809
116	—	—	—	—	—	—	—	—	—	—	—	48.774	48.781	48.788	48.795
117	—	—	—	—	—	—	—	—	—	—	—	—	49.776	49.783	49.790
118	—	—	—	—	—	—	—	—	—	—	—	—	—	50.787	50.794
119	—	—	—	—	—	—	—	—	—	—	—	—	—	51.800	51.807
120	—	—	—	—	—	—	—	—	—	—	—	—	—	—	52.828

TABLE 6.1 Thermo-mechanical exergy of liquid water from 1.01325 bar (1 atm) to 2 bar. Dashed lines indicate that water is not a liquid at that temperature and pressure (it is a solid, a mixture of gas and liquid, or a gas). (*Continued*)

| | e^{tm} (kJ/kg) | | | | | | | | | | | | | | |
| | Pressure (bar) | | | | | | | | | | | | | | |
T (°C)	2.5	3	3.5	4	5	6	7	8	9	10	11	12	13	14	15
0	4.8139	4.8638	4.9138	4.9637	5.0635	5.1633	5.2632	5.3630	5.4628	5.5626	5.6624	5.7622	5.8620	5.9618	6.0616
1	4.4364	4.4863	4.5363	4.5862	4.6861	4.7859	4.8858	4.9857	5.0855	5.1854	5.2852	5.3851	5.4849	5.5848	5.6846
2	4.0758	4.1258	4.1757	4.2257	4.3256	4.4255	4.5254	4.6253	4.7252	4.8251	4.9250	5.0249	5.1248	5.2247	5.3245
3	3.7321	3.7821	3.8321	3.8820	3.9820	4.0819	4.1819	4.2818	4.3818	4.4817	4.5816	4.6815	4.7815	4.8814	4.9813
4	3.4050	3.4550	3.5050	3.5550	3.6550	3.7550	3.8550	3.9550	4.0549	4.1549	4.2549	4.3548	4.4548	4.5547	4.6547
5	3.0945	3.1445	3.1946	3.2446	3.3446	3.4446	3.5446	3.6446	3.7446	3.8446	3.9446	4.0446	4.1446	4.2446	4.3446
6	2.8004	2.8504	2.9004	2.9505	3.0505	3.1506	3.2506	3.3506	3.4507	3.5507	3.6507	3.7507	3.8508	3.9508	4.0508
7	2.5225	2.5725	2.6225	2.6726	2.7727	2.8727	2.9728	3.0729	3.1729	3.2730	3.3730	3.4731	3.5731	3.6732	3.7732
8	2.2606	2.3107	2.3607	2.4108	2.5109	2.6110	2.7111	2.8112	2.9112	3.0113	3.1114	3.2115	3.3115	3.4116	3.5117
9	2.0147	2.0648	2.1149	2.1649	2.2650	2.3652	2.4653	2.5654	2.6655	2.7656	2.8657	2.9658	3.0659	3.1660	3.2661
10	1.7847	1.8347	1.8848	1.9349	2.0350	2.1352	2.2353	2.3354	2.4356	2.5357	2.6358	2.7359	2.8360	2.9361	3.0362
11	1.5703	1.6203	1.6704	1.7205	1.8207	1.9208	2.0210	2.1212	2.2213	2.3214	2.4216	2.5217	2.6218	2.7220	2.8221
12	1.3714	1.4215	1.4716	1.5217	1.6219	1.7221	1.8222	1.9224	2.0226	2.1227	2.2229	2.3230	2.4232	2.5233	2.6235
13	1.1880	1.2381	1.2882	1.3383	1.4385	1.5387	1.6389	1.7391	1.8393	1.9395	2.0396	2.1398	2.2400	2.3401	2.4403
14	1.0199	1.0701	1.1202	1.1703	1.2705	1.3707	1.4709	1.5711	1.6713	1.7715	1.8717	1.9718	2.0720	2.1722	2.2724
15	0.8670	0.9172	0.9673	1.0174	1.1176	1.2178	1.3181	1.4183	1.5185	1.6187	1.7189	1.8191	1.9193	2.0194	2.1196
16	0.7292	0.7793	0.8295	0.8796	0.9798	1.0801	1.1803	1.2805	1.3807	1.4809	1.5811	1.6814	1.7816	1.8817	1.9819
17	0.6064	0.6565	0.7066	0.7567	0.8570	0.9572	1.0575	1.1577	1.2579	1.3582	1.4584	1.5586	1.6588	1.7590	1.8592
18	0.4984	0.5485	0.5986	0.6487	0.7490	0.8493	0.9495	1.0497	1.1500	1.2502	1.3504	1.4507	1.5509	1.6511	1.7513
19	0.4051	0.4552	0.5054	0.5555	0.6557	0.7560	0.8563	0.9565	1.0568	1.1570	1.2572	1.3575	1.4577	1.5579	1.6581
20	0.3265	0.3766	0.4267	0.4769	0.5771	0.6774	0.7776	0.8779	0.9782	1.0784	1.1786	1.2789	1.3791	1.4793	1.5796
21	0.2623	0.3125	0.3626	0.4128	0.5130	0.6133	0.7136	0.8138	0.9141	1.0143	1.1146	1.2148	1.3150	1.4153	1.5155
22	0.2127	0.2628	0.3129	0.3631	0.4634	0.5636	0.6639	0.7642	0.8644	0.9647	1.0649	1.1652	1.2654	1.3656	1.4659
23	0.1773	0.2274	0.2776	0.3277	0.4280	0.5283	0.6285	0.7288	0.8291	0.9293	1.0296	1.1298	1.2301	1.3303	1.4305
24	0.1561	0.2063	0.2564	0.3066	0.4068	0.5071	0.6074	0.7077	0.8079	0.9082	1.0084	1.1087	1.2089	1.3092	1.4094
25	0.1491	0.1993	0.2494	0.2995	0.3998	0.5001	0.6004	0.7006	0.8009	0.9012	1.0014	1.1017	1.2019	1.3021	1.4024
26	0.1561	0.2062	0.2564	0.3065	0.4068	0.5071	0.6074	0.7076	0.8079	0.9081	1.0084	1.1086	1.2089	1.3091	1.4094
27	0.1770	0.2272	0.2773	0.3275	0.4277	0.5280	0.6283	0.7285	0.8288	0.9291	1.0293	1.1296	1.2298	1.3300	1.4303
28	0.2118	0.2619	0.3121	0.3622	0.4625	0.5628	0.6630	0.7633	0.8635	0.9638	1.0640	1.1643	1.2645	1.3648	1.4650
29	0.2603	0.3104	0.3606	0.4107	0.5110	0.6112	0.7115	0.8118	0.9120	1.0123	1.1125	1.2127	1.3130	1.4132	1.5134
30	0.3224	0.3726	0.4227	0.4728	0.5731	0.6734	0.7736	0.8739	0.9741	1.0744	1.1746	1.2749	1.3751	1.4753	1.5755

TABLE 6.2 Thermo-mechanical exergy of liquid water from 2.5 to 15 bar. Dashed lines indicate that water is not a liquid at that temperature and pressure (it is a mixture of gas and liquid, or a gas).

| | | | | | | e^{tm} (kJ/kg) | | | | | | | | | |
| | | | | | | Pressure (bar) | | | | | | | | | |
T (°C)	2.5	3	3.5	4	5	6	7	8	9	10	11	12	13	14	15
31	0.3981	0.4482	0.4984	0.5485	0.6488	0.7490	0.8493	0.9495	1.0498	1.1500	1.2503	1.3505	1.4507	1.5510	1.6512
32	0.4873	0.5374	0.5875	0.6377	0.7379	0.8382	0.9384	1.0387	1.1389	1.2392	1.3394	1.4396	1.5398	1.6400	1.7403
33	0.5898	0.6399	0.6901	0.7402	0.8405	0.9407	1.0409	1.1412	1.2414	1.3416	1.4419	1.5421	1.6423	1.7425	1.8427
34	0.7056	0.7558	0.8059	0.8560	0.9563	1.0565	1.1567	1.2570	1.3572	1.4574	1.5576	1.6579	1.7581	1.8583	1.9585
35	0.8347	0.8848	0.9349	0.9850	1.0853	1.1855	1.2857	1.3860	1.4862	1.5864	1.6866	1.7868	1.8870	1.9872	2.0874
36	0.9768	1.0270	1.0771	1.1272	1.2274	1.3276	1.4279	1.5281	1.6283	1.7285	1.8287	1.9289	2.0291	2.1293	2.2295
37	1.1320	1.1822	1.2323	1.2824	1.3826	1.4828	1.5830	1.6832	1.7834	1.8836	1.9838	2.0840	2.1842	2.2844	2.3845
38	1.3002	1.3503	1.4004	1.4505	1.5507	1.6509	1.7511	1.8513	1.9515	2.0517	2.1519	2.2520	2.3522	2.4524	2.5526
39	1.4812	1.5313	1.5814	1.6315	1.7317	1.8319	1.9321	2.0323	2.1325	2.2326	2.3328	2.4330	2.5331	2.6333	2.7334
40	1.6750	1.7251	1.7752	1.8253	1.9255	2.0257	2.1258	2.2260	2.3262	2.4264	2.5265	2.6267	2.7268	2.8270	2.9271
41	1.8815	1.9316	1.9817	2.0318	2.1320	2.2321	2.3323	2.4325	2.5326	2.6328	2.7329	2.8331	2.9332	3.0333	3.1335
42	2.1007	2.1508	2.2009	2.2509	2.3511	2.4513	2.5514	2.6516	2.7517	2.8518	2.9520	3.0521	3.1522	3.2523	3.3525
43	2.3324	2.3825	2.4325	2.4826	2.5828	2.6829	2.7830	2.8832	2.9833	3.0834	3.1836	3.2837	3.3838	3.4839	3.5840
44	2.5766	2.6266	2.6767	2.7268	2.8269	2.9270	3.0272	3.1273	3.2274	3.3275	3.4276	3.5277	3.6278	3.7279	3.8280
45	2.8332	2.8832	2.9333	2.9833	3.0835	3.1836	3.2837	3.3838	3.4839	3.5840	3.6841	3.7842	3.8843	3.9843	4.0844
46	3.1021	3.1521	3.2022	3.2522	3.3523	3.4524	3.5525	3.6526	3.7527	3.8528	3.9529	4.0529	4.1530	4.2531	4.3531
47	3.3832	3.4333	3.4833	3.5334	3.6334	3.7335	3.8336	3.9337	4.0338	4.1338	4.2339	4.3339	4.4340	4.5340	4.6341
48	3.6765	3.7266	3.7766	3.8267	3.9267	4.0268	4.1269	4.2269	4.3270	4.4270	4.5271	4.6271	4.7271	4.8272	4.9272
49	3.9820	4.0320	4.0820	4.1321	4.2321	4.3322	4.4322	4.5323	4.6323	4.7323	4.8324	4.9324	5.0324	5.1324	5.2324
50	4.2994	4.3495	4.3995	4.4495	4.5495	4.6496	4.7496	4.8496	4.9497	5.0497	5.1497	5.2497	5.3497	5.4497	5.5497
51	4.6289	4.6789	4.7289	4.7789	4.8789	4.9789	5.0789	5.1790	5.2790	5.3790	5.4789	5.5789	5.6789	5.7789	5.8789
52	4.9702	5.0202	5.0702	5.1202	5.2202	5.3202	5.4202	5.5202	5.6201	5.7201	5.8201	5.9201	6.0200	6.1200	6.2200
53	5.3233	5.3733	5.4233	5.4733	5.5733	5.6733	5.7732	5.8732	5.9732	6.0731	6.1731	6.2730	6.3730	6.4729	6.5729
54	5.6882	5.7382	5.7882	5.8381	5.9381	6.0381	6.1380	6.2380	6.3379	6.4379	6.5378	6.6377	6.7376	6.8376	6.9375
55	6.0648	6.1147	6.1647	6.2147	6.3146	6.4146	6.5145	6.6144	6.7143	6.8143	6.9142	7.0141	7.1140	7.2139	7.3138
56	6.4529	6.5029	6.5529	6.6028	6.7027	6.8027	6.9026	7.0025	7.1024	7.2023	7.3022	7.4021	7.5020	7.6018	7.7017
57	6.8527	6.9026	6.9526	7.0025	7.1024	7.2023	7.3022	7.4021	7.5020	7.6018	7.7017	7.8016	7.9015	8.0013	8.1012
58	7.2639	7.3138	7.3638	7.4137	7.5136	7.6134	7.7133	7.8132	7.9130	8.0129	8.1127	8.2126	8.3124	8.4123	8.5121
59	7.6865	7.7364	7.7864	7.8363	7.9361	8.0360	8.1358	8.2357	8.3355	8.4353	8.5352	8.6350	8.7348	8.8346	8.9344
60	8.1205	8.1704	8.2203	8.2702	8.3700	8.4699	8.5697	8.6695	8.7693	8.8691	8.9689	9.0688	9.1685	9.2683	9.3681

TABLE 6.2 Thermo-mechanical exergy of liquid water from 2.5 to 15 bar. Dashed lines indicate that water is not a liquid at that temperature and pressure (it is a mixture of gas and liquid, or a gas). (Continued)

83

e^{tm} (kJ/kg)

T (°C)	Pressure (bar)														
	2.5	3	3.5	4	5	6	7	8	9	10	11	12	13	14	15
61	8.5657	8.6156	8.6655	8.7154	8.8153	8.9151	9.0149	9.1146	9.2144	9.3142	9.4140	9.5138	9.6136	9.7133	9.8131
62	9.0222	9.0721	9.1220	9.1719	9.2717	9.3715	9.4712	9.5710	9.6708	9.7705	9.8703	9.9700	10.070	10.170	10.269
63	9.4899	9.5397	9.5896	9.6395	9.7393	9.8390	9.9388	10.039	10.138	10.238	10.338	10.437	10.537	10.637	10.737
64	9.9686	10.018	10.068	10.118	10.218	10.318	10.417	10.517	10.617	10.717	10.816	10.916	11.016	11.115	11.215
65	10.458	10.508	10.558	10.608	10.708	10.807	10.907	11.007	11.106	11.206	11.306	11.406	11.505	11.605	11.705
66	10.959	11.009	11.059	11.109	11.208	11.308	11.408	11.507	11.607	11.707	11.806	11.906	12.006	12.105	12.205
67	11.471	11.521	11.570	11.620	11.720	11.820	11.919	12.019	12.119	12.218	12.318	12.417	12.517	12.617	12.716
68	11.993	12.043	12.093	12.143	12.242	12.342	12.442	12.541	12.641	12.741	12.840	12.940	13.039	13.139	13.238
69	12.527	12.577	12.626	12.676	12.776	12.875	12.975	13.074	13.174	13.274	13.373	13.473	13.572	13.672	13.771
70	13.071	13.121	13.170	13.220	13.320	13.419	13.519	13.618	13.718	13.818	13.917	14.017	14.116	14.216	14.315
71	13.626	13.675	13.725	13.775	13.874	13.974	14.073	14.173	14.273	14.372	14.472	14.571	14.671	14.770	14.870
72	14.191	14.241	14.290	14.340	14.440	14.539	14.639	14.738	14.838	14.937	15.037	15.136	15.236	15.335	15.435
73	14.767	14.817	14.866	14.916	15.016	15.115	15.215	15.314	15.414	15.513	15.612	15.712	15.811	15.911	16.010
74	15.353	15.403	15.453	15.503	15.602	15.701	15.801	15.900	16.000	16.099	16.199	16.298	16.398	16.497	16.596
75	15.950	16.000	16.050	16.099	16.199	16.298	16.398	16.497	16.597	16.696	16.795	16.895	16.994	17.094	17.193
76	16.558	16.607	16.657	16.707	16.806	16.906	17.005	17.104	17.204	17.303	17.402	17.502	17.601	17.701	17.800
77	17.175	17.225	17.275	17.324	17.424	17.523	17.623	17.722	17.821	17.921	18.020	18.119	18.219	18.318	18.417
78	17.803	17.853	17.903	17.952	18.052	18.151	18.250	18.350	18.449	18.548	18.648	18.747	18.846	18.946	19.045
79	18.442	18.491	18.541	18.591	18.690	18.789	18.889	18.988	19.087	19.186	19.286	19.385	19.484	19.583	19.683
80	19.090	19.140	19.189	19.239	19.338	19.438	19.537	19.636	19.735	19.835	19.934	20.033	20.132	20.232	20.331
81	19.749	19.798	19.848	19.898	19.997	20.096	20.195	20.295	20.394	20.493	20.592	20.691	20.791	20.890	20.989
82	20.418	20.467	20.517	20.566	20.666	20.765	20.864	20.963	21.062	21.161	21.261	21.360	21.459	21.558	21.657
83	21.096	21.146	21.195	21.245	21.344	21.443	21.543	21.642	21.741	21.840	21.939	22.038	22.137	22.236	22.336
84	21.785	21.835	21.884	21.934	22.033	22.132	22.231	22.330	22.429	22.528	22.628	22.727	22.826	22.925	23.024
85	22.484	22.533	22.583	22.632	22.732	22.831	22.930	23.029	23.128	23.227	23.326	23.425	23.524	23.623	23.722
86	23.192	23.242	23.292	23.341	23.440	23.539	23.638	23.737	23.836	23.935	24.034	24.133	24.232	24.331	24.430
87	23.911	23.960	24.010	24.059	24.158	24.257	24.356	24.455	24.554	24.653	24.752	24.851	24.950	25.049	25.148
88	24.639	24.689	24.738	24.788	24.887	24.986	25.085	25.184	25.283	25.381	25.480	25.579	25.678	25.777	25.876
89	25.377	25.427	25.476	25.526	25.625	25.724	25.822	25.921	26.020	26.119	26.218	26.317	26.416	26.515	26.614

TABLE 6.2 Thermo-mechanical exergy of liquid water from 2.5 to 15 bar. Dashed lines indicate that water is not a liquid at that temperature and pressure (it is a mixture of gas and liquid, or a gas). (Continued)

e^{tm} (kJ/kg)

T (°C)	Pressure (bar)														
	2.5	3	3.5	4	5	6	7	8	9	10	11	12	13	14	15
90	26.125	26.175	26.224	26.273	26.372	26.471	26.570	26.669	26.768	26.867	26.966	27.064	27.163	27.262	27.361
91	26.883	26.932	26.981	27.031	27.130	27.229	27.327	27.426	27.525	27.624	27.723	27.822	27.920	28.019	28.118
92	27.650	27.699	27.748	27.798	27.897	27.995	28.094	28.193	28.292	28.391	28.489	28.588	28.687	28.786	28.885
93	28.426	28.476	28.525	28.574	28.673	28.772	28.871	28.970	29.068	29.167	29.266	29.365	29.463	29.562	29.661
94	29.213	29.262	29.311	29.361	29.459	29.558	29.657	29.755	29.854	29.953	30.052	30.150	30.249	30.348	30.446
95	30.008	30.058	30.107	30.156	30.255	30.354	30.452	30.551	30.650	30.748	30.847	30.946	31.044	31.143	31.242
96	30.813	30.863	30.912	30.961	31.060	31.159	31.257	31.356	31.455	31.553	31.652	31.750	31.849	31.948	32.046
97	31.628	31.677	31.727	31.776	31.875	31.973	32.072	32.170	32.269	32.367	32.466	32.565	32.663	32.762	32.860
98	32.452	32.501	32.551	32.600	32.698	32.797	32.895	32.994	33.093	33.191	33.290	33.388	33.487	33.585	33.684
99	33.285	33.335	33.384	33.433	33.532	33.630	33.729	33.827	33.926	34.024	34.123	34.221	34.320	34.418	34.516
100	34.128	34.177	34.226	34.276	34.374	34.473	34.571	34.669	34.768	34.866	34.965	35.063	35.162	35.260	35.358
101	34.980	35.029	35.078	35.127	35.226	35.324	35.423	35.521	35.619	35.718	35.816	35.915	36.013	36.111	36.210
102	35.841	35.890	35.939	35.988	36.087	36.185	36.284	36.382	36.480	36.579	36.677	36.775	36.874	36.972	37.070
103	36.711	36.760	36.809	36.859	36.957	37.055	37.154	37.252	37.350	37.448	37.547	37.645	37.743	37.842	37.940
104	37.591	37.640	37.689	37.738	37.836	37.935	38.033	38.131	38.229	38.328	38.426	38.524	38.622	38.721	38.819
105	38.479	38.528	38.577	38.626	38.725	38.823	38.921	39.019	39.117	39.216	39.314	39.412	39.510	39.608	39.707
106	39.377	39.426	39.475	39.524	39.622	39.720	39.818	39.917	40.015	40.113	40.211	40.309	40.407	40.505	40.604
107	40.283	40.332	40.381	40.430	40.529	40.627	40.725	40.823	40.921	41.019	41.117	41.215	41.313	41.412	41.510
108	41.199	41.248	41.297	41.346	41.444	41.542	41.640	41.738	41.836	41.934	42.032	42.130	42.229	42.327	42.425
109	42.124	42.173	42.222	42.271	42.369	42.467	42.565	42.663	42.761	42.859	42.957	43.055	43.153	43.251	43.349
110	43.057	43.106	43.155	43.204	43.302	43.400	43.498	43.596	43.694	43.792	43.890	43.988	44.086	44.183	44.281
111	43.999	44.048	44.097	44.146	44.244	44.342	44.440	44.538	44.636	44.734	44.832	44.930	45.027	45.125	45.223
112	44.951	45.000	45.049	45.098	45.195	45.293	45.391	45.489	45.587	45.685	45.782	45.880	45.978	46.076	46.174
113	45.911	45.960	46.009	46.058	46.155	46.253	46.351	46.449	46.547	46.644	46.742	46.840	46.938	47.035	47.133
114	46.880	46.929	46.978	47.026	47.124	47.222	47.320	47.417	47.515	47.613	47.711	47.808	47.906	48.004	48.101
115	47.857	47.906	47.955	48.004	48.102	48.199	48.297	48.395	48.492	48.590	48.688	48.785	48.883	48.981	49.078
116	48.844	48.893	48.942	48.990	49.088	49.186	49.283	49.381	49.478	49.576	49.674	49.771	49.869	49.966	50.064
117	49.839	49.888	49.937	49.985	50.083	50.180	50.278	50.376	50.473	50.571	50.668	50.766	50.863	50.961	51.058
118	50.843	50.892	50.940	50.989	51.087	51.184	51.282	51.379	51.477	51.574	51.672	51.769	51.867	51.964	52.061
119	51.855	51.904	51.953	52.001	52.099	52.196	52.294	52.391	52.489	52.586	52.683	52.781	52.878	52.976	53.073

TABLE 6.2 Thermo-mechanical exergy of liquid water from 2.5 to 15 bar. Dashed lines indicate that water is not a liquid at that temperature and pressure (it is a mixture of gas and liquid, or a gas). (Continued)

e^tm (kJ/kg)

T (°C)	Pressure (bar)														
	2.5	3	3.5	4	5	6	7	8	9	10	11	12	13	14	15
120	52.876	52.925	52.974	53.022	53.120	53.217	53.315	53.412	53.509	53.607	53.704	53.801	53.899	53.996	54.093
121	53.906	53.955	54.003	54.052	54.149	54.247	54.344	54.441	54.539	54.636	54.733	54.830	54.928	55.025	55.122
122	54.944	54.993	55.041	55.090	55.187	55.285	55.382	55.479	55.576	55.674	55.771	55.868	55.965	56.063	56.160
123	55.991	56.040	56.088	56.137	56.234	56.331	56.428	56.525	56.623	56.720	56.817	56.914	57.011	57.109	57.206
124	57.046	57.095	57.143	57.192	57.289	57.386	57.483	57.580	57.678	57.775	57.872	57.969	58.066	58.163	58.260
125	58.110	58.158	58.207	58.256	58.353	58.450	58.547	58.644	58.741	58.838	58.935	59.032	59.129	59.226	59.323
126	59.182	59.231	59.279	59.328	59.425	59.522	59.619	59.716	59.813	59.910	60.007	60.104	60.200	60.297	60.394
127	60.263	60.311	60.360	60.408	60.505	60.602	60.699	60.796	60.893	60.990	61.087	61.183	61.280	61.377	61.474
128	—	61.400	61.449	61.497	61.594	61.691	61.788	61.884	61.981	62.078	62.175	62.272	62.369	62.466	62.562
129	—	62.498	62.546	62.594	62.691	62.788	62.885	62.981	63.078	63.175	63.272	63.369	63.465	63.562	63.659
130	—	63.603	63.652	63.700	63.797	63.893	63.990	64.087	64.184	64.280	64.377	64.474	64.570	64.667	64.764
131	—	64.717	64.766	64.814	64.911	65.007	65.104	65.200	65.297	65.394	65.490	65.587	65.684	65.780	65.877
132	—	65.840	65.888	65.936	66.033	66.129	66.226	66.322	66.419	66.516	66.612	66.709	66.805	66.902	66.998
133	—	66.970	67.018	67.067	67.163	67.260	67.356	67.453	67.549	67.646	67.742	67.839	67.935	68.032	68.128
134	—	—	68.157	68.206	68.302	68.398	68.495	68.591	68.688	68.784	68.881	68.977	69.073	69.170	69.266
135	—	—	69.304	69.353	69.449	69.545	69.642	69.738	69.834	69.931	70.027	70.123	70.220	70.316	70.412
136	—	—	70.460	70.508	70.604	70.700	70.797	70.893	70.989	71.086	71.182	71.278	71.374	71.471	71.567
137	—	—	71.623	71.671	71.768	71.864	71.960	72.056	72.152	72.249	72.345	72.441	72.537	72.633	72.729
138	—	—	72.795	72.843	72.939	73.035	73.131	73.227	73.324	73.420	73.516	73.612	73.708	73.804	73.900
139	—	—	—	74.023	74.119	74.215	74.311	74.407	74.503	74.599	74.695	74.791	74.887	74.983	75.079
140	—	—	—	75.211	75.307	75.403	75.499	75.595	75.691	75.787	75.882	75.978	76.074	76.170	76.266
141	—	—	—	76.407	76.503	76.599	76.695	76.790	76.886	76.982	77.078	77.174	77.270	77.366	77.461
142	—	—	—	77.611	77.707	77.803	77.898	77.994	78.090	78.186	78.282	78.377	78.473	78.569	78.665
143	—	—	—	78.823	78.919	79.015	79.110	79.206	79.302	79.398	79.493	79.589	79.685	79.780	79.876
144	—	—	—	—	80.139	80.235	80.331	80.426	80.522	80.617	80.713	80.809	80.904	81.000	81.095
145	—	—	—	—	81.368	81.463	81.559	81.654	81.750	81.845	81.941	82.036	82.132	82.227	82.323
146	—	—	—	—	82.604	82.699	82.795	82.890	82.986	83.081	83.176	83.272	83.367	83.463	83.558
147	—	—	—	—	83.848	83.944	84.039	84.134	84.230	84.325	84.420	84.516	84.611	84.706	84.802
148	—	—	—	—	85.101	85.196	85.291	85.386	85.482	85.577	85.672	85.767	85.863	85.958	86.053

TABLE 6.2 Thermo-mechanical exergy of liquid water from 2.5 to 15 bar. Dashed lines indicate that water is not a liquid at that temperature and pressure (it is a mixture of gas and liquid, or a gas). (Continued)

e^{tm} (kJ/kg)

T (°C)	2.5	3	3.5	4	5	6	7	8	9	10	11	12	13	14	15
										Pressure (bar)					
149	—	—	—	—	86.361	86.456	86.551	86.646	86.742	86.837	86.932	87.027	87.122	87.217	87.312
150	—	—	—	—	87.629	87.724	87.819	87.914	88.010	88.105	88.200	88.295	88.390	88.485	88.580
151	—	—	—	—	88.906	89.001	89.096	89.190	89.285	89.380	89.475	89.570	89.665	89.760	89.855
152	—	—	—	—	—	90.285	90.380	90.474	90.569	90.664	90.759	90.854	90.949	91.044	91.138
153	—	—	—	—	—	91.577	91.672	91.766	91.861	91.956	92.051	92.145	92.240	92.335	92.430
154	—	—	—	—	—	92.877	92.971	93.066	93.161	93.255	93.350	93.445	93.539	93.634	93.729
155	—	—	—	—	—	94.185	94.279	94.374	94.468	94.563	94.657	94.752	94.847	94.941	95.036
156	—	—	—	—	—	95.500	95.595	95.689	95.784	95.878	95.973	96.067	96.162	96.256	96.350
157	—	—	—	—	—	96.824	96.918	97.013	97.107	97.202	97.296	97.390	97.485	97.579	97.673
158	—	—	—	—	—	98.156	98.250	98.344	98.438	98.533	98.627	98.721	98.815	98.910	99.004
159	—	—	—	—	—	—	99.589	99.683	99.778	99.872	99.966	100.06	100.15	100.25	100.34
160	—	—	—	—	—	—	100.94	101.03	101.12	101.22	101.31	101.41	101.50	101.59	101.69
161	—	—	—	—	—	—	102.29	102.39	102.48	102.57	102.67	102.76	102.85	102.95	103.04
162	—	—	—	—	—	—	103.65	103.75	103.84	103.94	104.03	104.12	104.22	104.31	104.40
163	—	—	—	—	—	—	105.03	105.12	105.21	105.31	105.40	105.49	105.59	105.68	105.77
164	—	—	—	—	—	—	106.40	106.50	106.59	106.68	106.78	106.87	106.96	107.06	107.15
165	—	—	—	—	—	—	—	107.88	107.98	108.07	108.16	108.26	108.35	108.44	108.54
166	—	—	—	—	—	—	—	109.28	109.37	109.46	109.56	109.65	109.74	109.84	109.93
167	—	—	—	—	—	—	—	110.68	110.77	110.87	110.96	111.05	111.15	111.24	111.33
168	—	—	—	—	—	—	—	112.09	112.18	112.28	112.37	112.46	112.55	112.65	112.74
169	—	—	—	—	—	—	—	113.51	113.60	113.69	113.79	113.88	113.97	114.06	114.16
170	—	—	—	—	—	—	—	114.93	115.02	115.12	115.21	115.30	115.40	115.49	115.58
171	—	—	—	—	—	—	—	—	116.46	116.55	116.64	116.74	116.83	116.92	117.01
172	—	—	—	—	—	—	—	—	117.90	117.99	118.08	118.18	118.27	118.36	118.45
173	—	—	—	—	—	—	—	—	119.35	119.44	119.53	119.62	119.72	119.81	119.90
174	—	—	—	—	—	—	—	—	120.80	120.90	120.99	121.08	121.17	121.26	121.36
175	—	—	—	—	—	—	—	—	122.27	122.36	122.45	122.54	122.64	122.73	122.82
176	—	—	—	—	—	—	—	—	—	123.83	123.92	124.01	124.11	124.20	124.29
177	—	—	—	—	—	—	—	—	—	125.31	125.40	125.49	125.59	125.68	125.77
178	—	—	—	—	—	—	—	—	—	126.80	126.89	126.98	127.07	127.16	127.26

TABLE 6.2 Thermo-mechanical exergy of liquid water from 2.5 to 15 bar. Dashed lines indicate that water is not a liquid at that temperature and pressure (it is a mixture of gas and liquid, or a gas). (*Continued*)

T (°C)	2.5	3	3.5	4	5	6	7	8	9	10	11	12	13	14	15
											e^{tm} (kJ/kg) Pressure (bar)				
179	—	—	—	—	—	—	—	—	—	128.29	128.38	128.48	128.57	128.66	128.75
180	—	—	—	—	—	—	—	—	—	—	129.89	129.98	130.07	130.16	130.25
181	—	—	—	—	—	—	—	—	—	—	131.40	131.49	131.58	131.67	131.76
182	—	—	—	—	—	—	—	—	—	—	132.92	133.01	133.10	133.19	133.28
183	—	—	—	—	—	—	—	—	—	—	134.44	134.53	134.62	134.71	134.81
184	—	—	—	—	—	—	—	—	—	—	135.98	136.07	136.16	136.25	136.34
185	—	—	—	—	—	—	—	—	—	—	—	137.61	137.70	137.79	137.88
186	—	—	—	—	—	—	—	—	—	—	—	139.16	139.25	139.34	139.43
187	—	—	—	—	—	—	—	—	—	—	—	140.71	140.80	140.89	140.98
188	—	—	—	—	—	—	—	—	—	—	—	—	142.37	142.46	142.55
189	—	—	—	—	—	—	—	—	—	—	—	—	143.94	144.03	144.12
190	—	—	—	—	—	—	—	—	—	—	—	—	145.52	145.61	145.70
191	—	—	—	—	—	—	—	—	—	—	—	—	147.11	147.20	147.29
192	—	—	—	—	—	—	—	—	—	—	—	—	—	148.80	148.89
193	—	—	—	—	—	—	—	—	—	—	—	—	—	150.40	150.49
194	—	—	—	—	—	—	—	—	—	—	—	—	—	152.01	152.10
195	—	—	—	—	—	—	—	—	—	—	—	—	—	153.63	153.72
196	—	—	—	—	—	—	—	—	—	—	—	—	—	—	155.35
197	—	—	—	—	—	—	—	—	—	—	—	—	—	—	156.98
198	—	—	—	—	—	—	—	—	—	—	—	—	—	—	158.62

TABLE 6.2 Thermo-mechanical exergy of liquid water from 2.5 to 15 bar. Dashed lines indicate that water is not a liquid at that temperature and pressure (it is a mixture of gas and liquid, or a gas). (Continued)

88

e^{tm} (kJ/kg)

T (°C)	Pressure (bar)														
	16	17	18	19	20	21	22	23	24	25	26	27	28	29	30
0	6.1614	6.2612	6.3610	6.4607	6.5605	6.6603	6.7600	6.8598	6.9595	7.0593	7.1590	7.2588	7.3585	7.4583	7.5580
1	5.7844	5.8843	5.9841	6.0839	6.1837	6.2835	6.3833	6.4831	6.5829	6.6827	6.7825	6.8823	6.9821	7.0818	7.1816
2	5.4244	5.5243	5.6241	5.7240	5.8238	5.9237	6.0235	6.1234	6.2232	6.3231	6.4229	6.5227	6.6225	6.7223	6.8222
3	5.0812	5.1811	5.2810	5.3809	5.4808	5.5807	5.6805	5.7804	5.8803	5.9802	6.0800	6.1799	6.2798	6.3796	6.4795
4	4.7546	4.8546	4.9545	5.0544	5.1543	5.2543	5.3542	5.4541	5.5540	5.6539	5.7538	5.8537	5.9536	6.0535	6.1533
5	4.4445	4.5445	4.6445	4.7444	4.8444	4.9443	5.0443	5.1442	5.2442	5.3441	5.4440	5.5439	5.6439	5.7438	5.8437
6	4.1508	4.2508	4.3508	4.4508	4.5507	4.6507	4.7507	4.8507	4.9506	5.0506	5.1506	5.2505	5.3505	5.4504	5.5504
7	3.8732	3.9732	4.0733	4.1733	4.2733	4.3733	4.4733	4.5733	4.6733	4.7733	4.8733	4.9733	5.0732	5.1732	5.2732
8	3.6117	3.7118	3.8118	3.9119	4.0119	4.1119	4.2119	4.3120	4.4120	4.5120	4.6120	4.7120	4.8120	4.9120	5.0120
9	3.3661	3.4662	3.5663	3.6663	3.7664	3.8665	3.9665	4.0666	4.1666	4.2666	4.3667	4.4667	4.5667	4.6667	4.7668
10	3.1363	3.2364	3.3365	3.4366	3.5367	3.6368	3.7368	3.8369	3.9370	4.0370	4.1371	4.2371	4.3372	4.4372	4.5372
11	2.9222	3.0223	3.1224	3.2225	3.3226	3.4227	3.5228	3.6229	3.7230	3.8231	3.9231	4.0232	4.1233	4.2233	4.3234
12	2.7236	2.8237	2.9239	3.0240	3.1241	3.2242	3.3243	3.4244	3.5245	3.6246	3.7247	3.8248	3.9249	4.0249	4.1250
13	2.5404	2.6406	2.7407	2.8408	2.9410	3.0411	3.1412	3.2413	3.3415	3.4416	3.5417	3.6418	3.7419	3.8420	3.9420
14	2.3725	2.4727	2.5728	2.6730	2.7731	2.8733	2.9734	3.0735	3.1737	3.2738	3.3739	3.4740	3.5741	3.6742	3.7743
15	2.2198	2.3200	2.4201	2.5203	2.6205	2.7206	2.8208	2.9209	3.0210	3.1212	3.2213	3.3214	3.4216	3.5217	3.6218
16	2.0821	2.1823	2.2825	2.3827	2.4828	2.5830	2.6832	2.7833	2.8835	2.9836	3.0837	3.1839	3.2840	3.3841	3.4843
17	1.9594	2.0596	2.1598	2.2600	2.3601	2.4603	2.5605	2.6606	2.7608	2.8610	2.9611	3.0613	3.1614	3.2615	3.3617
18	1.8515	1.9517	2.0519	2.1521	2.2523	2.3525	2.4526	2.5528	2.6530	2.7531	2.8533	2.9535	3.0536	3.1538	3.2539
19	1.7583	1.8585	1.9587	2.0589	2.1591	2.2593	2.3595	2.4597	2.5599	2.6600	2.7602	2.8604	2.9605	3.0607	3.1608
20	1.6798	1.7800	1.8802	1.9804	2.0806	2.1808	2.2810	2.3812	2.4814	2.5815	2.6817	2.7819	2.8820	2.9822	3.0823
21	1.6157	1.7159	1.8162	1.9164	2.0166	2.1168	2.2170	2.3172	2.4173	2.5175	2.6177	2.7179	2.8180	2.9182	3.0184
22	1.5661	1.6663	1.7665	1.8667	1.9669	2.0671	2.1673	2.2675	2.3677	2.4679	2.5681	2.6683	2.7684	2.8686	2.9688
23	1.5308	1.6310	1.7312	1.8314	1.9316	2.0318	2.1320	2.2322	2.3324	2.4326	2.5328	2.6330	2.7331	2.8333	2.9335
24	1.5096	1.6099	1.7101	1.8103	1.9105	2.0107	2.1109	2.2111	2.3113	2.4115	2.5117	2.6119	2.7120	2.8122	2.9124
25	1.5026	1.6028	1.7031	1.8033	1.9035	2.0037	2.1039	2.2041	2.3043	2.4045	2.5047	2.6048	2.7050	2.8052	2.9054
26	1.5096	1.6098	1.7100	1.8103	1.9105	2.0107	2.1109	2.2111	2.3113	2.4115	2.5116	2.6118	2.7120	2.8122	2.9123
27	1.5305	1.6307	1.7309	1.8312	1.9314	2.0316	2.1318	2.2320	2.3322	2.4324	2.5325	2.6327	2.7329	2.8331	2.9332
28	1.5652	1.6654	1.7657	1.8659	1.9661	2.0663	2.1665	2.2667	2.3669	2.4671	2.5672	2.6674	2.7676	2.8678	2.9679
29	1.6137	1.7139	1.8141	1.9143	2.0145	2.1147	2.2149	2.3151	2.4153	2.5155	2.6157	2.7158	2.8160	2.9162	3.0163

TABLE 6.3 Thermo-mechanical exergy of liquid water from 16 to 30 bar. Dashed lines indicate that water is not a liquid at that temperature and pressure (it is a mixture of gas and liquid, or a gas).

e^{tm} (kJ/kg)

T (°C)	Pressure (bar)														
	16	17	18	19	20	21	22	23	24	25	26	27	28	29	30
30	1.6758	1.7760	1.8762	1.9764	2.0766	2.1768	2.2770	2.3772	2.4773	2.5775	2.6777	2.7779	2.8780	2.9782	3.0783
31	1.7514	1.8516	1.9518	2.0520	2.1522	2.2524	2.3526	2.4528	2.5529	2.6531	2.7533	2.8534	2.9536	3.0538	3.1539
32	1.8405	1.9407	2.0409	2.1411	2.2413	2.3414	2.4416	2.5418	2.6420	2.7421	2.8423	2.9425	3.0426	3.1428	3.2429
33	1.9429	2.0431	2.1433	2.2435	2.3437	2.4439	2.5440	2.6442	2.7444	2.8445	2.9447	3.0449	3.1450	3.2451	3.3453
34	2.0587	2.1589	2.2590	2.3592	2.4594	2.5596	2.6597	2.7599	2.8601	2.9602	3.0604	3.1605	3.2607	3.3608	3.4609
35	2.1876	2.2878	2.3880	2.4881	2.5883	2.6885	2.7886	2.8888	2.9889	3.0891	3.1892	3.2894	3.3895	3.4896	3.5897
36	2.3296	2.4298	2.5300	2.6301	2.7303	2.8305	2.9306	3.0308	3.1309	3.2310	3.3312	3.4313	3.5314	3.6315	3.7317
37	2.4847	2.5849	2.6850	2.7852	2.8853	2.9855	3.0856	3.1858	3.2859	3.3860	3.4861	3.5863	3.6864	3.7865	3.8866
38	2.6527	2.7529	2.8530	2.9532	3.0533	3.1534	3.2536	3.3537	3.4538	3.5539	3.6541	3.7542	3.8543	3.9544	4.0545
39	2.8336	2.9337	3.0339	3.1340	3.2341	3.3343	3.4344	3.5345	3.6346	3.7347	3.8348	3.9349	4.0350	4.1351	4.2352
40	3.0272	3.1274	3.2275	3.3276	3.4277	3.5279	3.6280	3.7281	3.8282	3.9283	4.0284	4.1285	4.2285	4.3286	4.4287
41	3.2336	3.3337	3.4338	3.5339	3.6341	3.7342	3.8343	3.9343	4.0344	4.1345	4.2346	4.3347	4.4347	4.5348	4.6349
42	3.4526	3.5527	3.6528	3.7529	3.8530	3.9531	4.0532	4.1532	4.2533	4.3534	4.4535	4.5535	4.6536	4.7536	4.8537
43	3.6841	3.7842	3.8843	3.9844	4.0845	4.1845	4.2846	4.3847	4.4847	4.5848	4.6848	4.7849	4.8849	4.9850	5.0850
44	3.9281	4.0282	4.1283	4.2283	4.3284	4.4285	4.5285	4.6286	4.7286	4.8287	4.9287	5.0287	5.1288	5.2288	5.3288
45	4.1845	4.2845	4.3846	4.4847	4.5847	4.6848	4.7848	4.8848	4.9849	5.0849	5.1849	5.2850	5.3850	5.4850	5.5850
46	4.4532	4.5532	4.6533	4.7533	4.8534	4.9534	5.0534	5.1534	5.2535	5.3535	5.4535	5.5535	5.6535	5.7535	5.8535
47	4.7341	4.8342	4.9342	5.0342	5.1342	5.2343	5.3343	5.4343	5.5343	5.6343	5.7343	5.8343	5.9343	6.0342	6.1342
48	5.0272	5.1273	5.2273	5.3273	5.4273	5.5273	5.6273	5.7273	5.8273	5.9272	6.0272	6.1272	6.2272	6.3271	6.4271
49	5.3324	5.4324	5.5324	5.6324	5.7324	5.8324	5.9324	6.0324	6.1323	6.2323	6.3323	6.4322	6.5322	6.6321	6.7321
50	5.6497	5.7497	5.8496	5.9496	6.0496	6.1495	6.2495	6.3495	6.4494	6.5494	6.6493	6.7493	6.8492	6.9491	7.0491
51	5.9788	6.0788	6.1788	6.2787	6.3787	6.4786	6.5786	6.6785	6.7785	6.8784	6.9783	7.0782	7.1782	7.2781	7.3780
52	6.3199	6.4199	6.5198	6.6197	6.7197	6.8196	6.9195	7.0195	7.1194	7.2193	7.3192	7.4191	7.5190	7.6189	7.7188
53	6.6728	6.7727	6.8726	6.9726	7.0725	7.1724	7.2723	7.3722	7.4721	7.5720	7.6719	7.7718	7.8717	7.9715	8.0714
54	7.0374	7.1373	7.2372	7.3371	7.4370	7.5369	7.6368	7.7367	7.8366	7.9364	8.0363	8.1362	8.2360	8.3359	8.4357
55	7.4137	7.5136	7.6135	7.7133	7.8132	7.9131	8.0130	8.1128	8.2127	8.3125	8.4124	8.5122	8.6121	8.7119	8.8118
56	7.8016	7.9015	8.0013	8.1012	8.2010	8.3009	8.4007	8.5006	8.6004	8.7003	8.8001	8.8999	8.9997	9.0995	9.1994
57	8.2010	8.3009	8.4007	8.5006	8.6004	8.7002	8.8000	8.8999	8.9997	9.0995	9.1993	9.2991	9.3989	9.4987	9.5985
58	8.6119	8.7118	8.8116	8.9114	9.0112	9.1110	9.2108	9.3106	9.4104	9.5102	9.6100	9.7098	9.8096	9.9093	10.009

TABLE 6.3 Thermo-mechanical exergy of liquid water from 16 to 30 bar. Dashed lines indicate that water is not a liquid at that temperature and pressure (it is a mixture of gas and liquid, or a gas). (Continued)

e^{tm} (kJ/kg)

T (°C)	Pressure (bar)														
	16	17	18	19	20	21	22	23	24	25	26	27	28	29	30
59	9.0343	9.1341	9.2339	9.3337	9.4334	9.5332	9.6330	9.7328	9.8326	9.9323	10.032	10.132	10.232	10.331	10.431
60	9.4679	9.5677	9.6675	9.7672	9.8670	9.9668	10.067	10.166	10.266	10.366	10.466	10.565	10.665	10.765	10.864
61	9.9129	10.013	10.112	10.212	10.312	10.412	10.511	10.611	10.711	10.811	10.910	11.010	11.110	11.209	11.309
62	10.369	10.469	10.568	10.668	10.768	10.868	10.967	11.067	11.167	11.266	11.366	11.466	11.566	11.665	11.765
63	10.836	10.936	11.036	11.135	11.235	11.335	11.435	11.534	11.634	11.734	11.833	11.933	12.033	12.132	12.232
64	11.315	11.414	11.514	11.614	11.713	11.813	11.913	12.012	12.112	12.212	12.311	12.411	12.511	12.610	12.710
65	11.804	11.904	12.003	12.103	12.203	12.302	12.402	12.502	12.601	12.701	12.801	12.900	13.000	13.099	13.199
66	12.305	12.404	12.504	12.603	12.703	12.803	12.902	13.002	13.102	13.201	13.301	13.400	13.500	13.599	13.699
67	12.816	12.916	13.015	13.115	13.214	13.314	13.413	13.513	13.613	13.712	13.812	13.911	14.011	14.110	14.210
68	13.338	13.438	13.537	13.637	13.736	13.836	13.935	14.035	14.135	14.234	14.334	14.433	14.533	14.632	14.732
69	13.871	13.971	14.070	14.170	14.269	14.369	14.468	14.568	14.667	14.767	14.866	14.966	15.065	15.165	15.264
70	14.415	14.514	14.614	14.713	14.813	14.912	15.012	15.111	15.211	15.310	15.410	15.509	15.609	15.708	15.808
71	14.969	15.069	15.168	15.268	15.367	15.467	15.566	15.665	15.765	15.864	15.964	16.063	16.163	16.262	16.362
72	15.534	15.634	15.733	15.833	15.932	16.031	16.131	16.230	16.330	16.429	16.529	16.628	16.727	16.827	16.926
73	16.110	16.209	16.309	16.408	16.507	16.607	16.706	16.806	16.905	17.004	17.104	17.203	17.303	17.402	17.501
74	16.696	16.795	16.895	16.994	17.093	17.193	17.292	17.392	17.491	17.590	17.690	17.789	17.888	17.988	18.087
75	17.292	17.392	17.491	17.590	17.690	17.789	17.888	17.988	18.087	18.187	18.286	18.385	18.485	18.584	18.683
76	17.899	17.999	18.098	18.197	18.297	18.396	18.495	18.595	18.694	18.793	18.892	18.992	19.091	19.190	19.290
77	18.517	18.616	18.715	18.814	18.914	19.013	19.112	19.212	19.311	19.410	19.509	19.609	19.708	19.807	19.907
78	19.144	19.243	19.343	19.442	19.541	19.640	19.740	19.839	19.938	20.037	20.137	20.236	20.335	20.434	20.534
79	19.782	19.881	19.980	20.080	20.179	20.278	20.377	20.477	20.576	20.675	20.774	20.873	20.973	21.072	21.171
80	20.430	20.529	20.628	20.728	20.827	20.926	21.025	21.124	21.224	21.323	21.422	21.521	21.620	21.719	21.819
81	21.088	21.187	21.286	21.386	21.485	21.584	21.683	21.782	21.881	21.981	22.080	22.179	22.278	22.377	22.476
82	21.756	21.856	21.955	22.054	22.153	22.252	22.351	22.450	22.549	22.649	22.748	22.847	22.946	23.045	23.144
83	22.435	22.534	22.633	22.732	22.831	22.930	23.029	23.128	23.227	23.326	23.426	23.525	23.624	23.723	23.822
84	23.123	23.222	23.321	23.420	23.519	23.618	23.717	23.816	23.915	24.014	24.113	24.213	24.312	24.411	24.510
85	23.821	23.920	24.019	24.118	24.217	24.316	24.415	24.514	24.613	24.712	24.811	24.910	25.009	25.108	25.207
86	24.529	24.628	24.727	24.826	24.925	25.024	25.123	25.222	25.321	25.420	25.519	25.618	25.717	25.816	25.915
87	25.247	25.346	25.445	25.544	25.643	25.742	25.841	25.940	26.039	26.138	26.237	26.336	26.435	26.533	26.632
88	25.975	26.074	26.173	26.272	26.371	26.470	26.569	26.667	26.766	26.865	26.964	27.063	27.162	27.261	27.360
89	26.713	26.811	26.910	27.009	27.108	27.207	27.306	27.405	27.504	27.602	27.701	27.800	27.899	27.998	28.097

TABLE 6.3 Thermo-mechanical exergy of liquid water from 16 to 30 bar. Dashed lines indicate that water is not a liquid at that temperature and pressure (it is a mixture of gas and liquid, or a gas). (Continued)

e^{tm} (kJ/kg)

T (°C)	16	17	18	19	20	21	22	23	24	25	26	27	28	29	30
							Pressure (bar)								
90	27.460	27.559	27.658	27.756	27.855	27.954	28.053	28.152	28.251	28.349	28.448	28.547	28.646	28.745	28.843
91	28.217	28.316	28.414	28.513	28.612	28.711	28.810	28.908	29.007	29.106	29.205	29.303	29.402	29.501	29.600
92	28.983	29.082	29.181	29.280	29.378	29.477	29.576	29.675	29.773	29.872	29.971	30.070	30.168	30.267	30.366
93	29.759	29.858	29.957	30.056	30.154	30.253	30.352	30.450	30.549	30.648	30.747	30.845	30.944	31.043	31.141
94	30.545	30.644	30.742	30.841	30.940	31.038	31.137	31.236	31.334	31.433	31.532	31.630	31.729	31.828	31.926
95	31.340	31.439	31.538	31.636	31.735	31.833	31.932	32.031	32.129	32.228	32.326	32.425	32.524	32.622	32.721
96	32.145	32.243	32.342	32.441	32.539	32.638	32.736	32.835	32.934	33.032	33.131	33.229	33.328	33.426	33.525
97	32.959	33.057	33.156	33.254	33.353	33.452	33.550	33.649	33.747	33.846	33.944	34.043	34.141	34.240	34.338
98	33.782	33.881	33.979	34.078	34.176	34.275	34.373	34.472	34.570	34.669	34.767	34.866	34.964	35.063	35.161
99	34.615	34.713	34.812	34.910	35.009	35.107	35.206	35.304	35.403	35.501	35.599	35.698	35.796	35.895	35.993
100	35.457	35.555	35.654	35.752	35.851	35.949	36.047	36.146	36.244	36.343	36.441	36.539	36.638	36.736	36.834
101	36.308	36.407	36.505	36.603	36.702	36.800	36.898	36.997	37.095	37.193	37.292	37.390	37.488	37.587	37.685
102	37.169	37.267	37.365	37.464	37.562	37.660	37.758	37.857	37.955	38.053	38.152	38.250	38.348	38.447	38.545
103	38.038	38.136	38.235	38.333	38.431	38.530	38.628	38.726	38.824	38.923	39.021	39.119	39.217	39.316	39.414
104	38.917	39.015	39.113	39.212	39.310	39.408	39.506	39.605	39.703	39.801	39.899	39.997	40.096	40.194	40.292
105	39.805	39.903	40.001	40.099	40.198	40.296	40.394	40.492	40.590	40.688	40.786	40.885	40.983	41.081	41.179
106	40.702	40.800	40.898	40.996	41.094	41.192	41.290	41.389	41.487	41.585	41.683	41.781	41.879	41.977	42.075
107	41.608	41.706	41.804	41.902	42.000	42.098	42.196	42.294	42.392	42.490	42.588	42.686	42.784	42.882	42.981
108	42.523	42.621	42.719	42.817	42.915	43.013	43.111	43.209	43.307	43.405	43.503	43.601	43.699	43.797	43.895
109	43.447	43.544	43.642	43.740	43.838	43.936	44.034	44.132	44.230	44.328	44.426	44.524	44.622	44.720	44.818
110	44.379	44.477	44.575	44.673	44.771	44.869	44.967	45.065	45.163	45.261	45.358	45.456	45.554	45.652	45.750
111	45.321	45.419	45.517	45.615	45.712	45.810	45.908	46.006	46.104	46.202	46.300	46.397	46.495	46.593	46.691
112	46.272	46.369	46.467	46.565	46.663	46.761	46.858	46.956	47.054	47.152	47.250	47.347	47.445	47.543	47.641
113	47.231	47.329	47.426	47.524	47.622	47.720	47.817	47.915	48.013	48.111	48.208	48.306	48.404	48.502	48.599
114	48.199	48.297	48.394	48.492	48.590	48.688	48.785	48.883	48.981	49.078	49.176	49.274	49.371	49.469	49.567
115	49.176	49.274	49.371	49.469	49.567	49.664	49.762	49.859	49.957	50.055	50.152	50.250	50.348	50.445	50.543
116	50.162	50.259	50.357	50.454	50.552	50.650	50.747	50.845	50.942	51.040	51.137	51.235	51.333	51.430	51.528
117	51.156	51.254	51.351	51.449	51.546	51.644	51.741	51.839	51.936	52.034	52.131	52.229	52.326	52.424	52.521
118	52.159	52.256	52.354	52.451	52.549	52.646	52.744	52.841	52.939	53.036	53.134	53.231	53.328	53.426	53.523

TABLE 6.3 Thermo-mechanical exergy of liquid water from 16 to 30 bar. Dashed lines indicate that water is not a liquid at that temperature and pressure (it is a mixture of gas and liquid, or a gas). (Continued)

e^{tm} (kJ/kg)

T (°C)	Pressure (bar)														
	16	17	18	19	20	21	22	23	24	25	26	27	28	29	30
119	53.171	53.268	53.365	53.463	53.560	53.658	53.755	53.852	53.950	54.047	54.145	54.242	54.339	54.437	54.534
120	54.191	54.288	54.386	54.483	54.580	54.678	54.775	54.872	54.970	55.067	55.164	55.262	55.359	55.456	55.554
121	55.220	55.317	55.414	55.512	55.609	55.706	55.803	55.901	55.998	56.095	56.192	56.290	56.387	56.484	56.581
122	56.257	56.354	56.451	56.549	56.646	56.743	56.840	56.938	57.035	57.132	57.229	57.326	57.424	57.521	57.618
123	57.303	57.400	57.497	57.594	57.692	57.789	57.886	57.983	58.080	58.177	58.274	58.372	58.469	58.566	58.663
124	58.357	58.454	58.551	58.649	58.746	58.843	58.940	59.037	59.134	59.231	59.328	59.425	59.522	59.619	59.716
125	59.420	59.517	59.614	59.711	59.808	59.905	60.002	60.099	60.196	60.293	60.390	60.487	60.584	60.681	60.778
126	60.491	60.588	60.685	60.782	60.879	60.976	61.073	61.170	61.267	61.364	61.461	61.558	61.655	61.752	61.849
127	61.571	61.668	61.765	61.862	61.959	62.056	62.152	62.249	62.346	62.443	62.540	62.637	62.734	62.831	62.927
128	62.659	62.756	62.853	62.950	63.046	63.143	63.240	63.337	63.434	63.531	63.627	63.724	63.821	63.918	64.015
129	63.756	63.852	63.949	64.046	64.143	64.239	64.336	64.433	64.530	64.626	64.723	64.820	64.917	65.013	65.110
130	64.860	64.957	65.054	65.151	65.247	65.344	65.441	65.537	65.634	65.731	65.827	65.924	66.021	66.117	66.214
131	65.974	66.070	66.167	66.263	66.360	66.457	66.553	66.650	66.746	66.843	66.940	67.036	67.133	67.229	67.326
132	67.095	67.191	67.288	67.385	67.481	67.578	67.674	67.771	67.867	67.964	68.060	68.157	68.253	68.350	68.446
133	68.225	68.321	68.418	68.514	68.611	68.707	68.803	68.900	68.996	69.093	69.189	69.286	69.382	69.479	69.575
134	69.363	69.459	69.555	69.652	69.748	69.845	69.941	70.037	70.134	70.230	70.326	70.423	70.519	70.616	70.712
135	70.509	70.605	70.701	70.798	70.894	70.990	71.087	71.183	71.279	71.376	71.472	71.568	71.664	71.761	71.857
136	71.663	71.759	71.856	71.952	72.048	72.144	72.241	72.337	72.433	72.529	72.625	72.722	72.818	72.914	73.010
137	72.826	72.922	73.018	73.114	73.210	73.306	73.403	73.499	73.595	73.691	73.787	73.883	73.980	74.076	74.172
138	73.996	74.092	74.188	74.285	74.381	74.477	74.573	74.669	74.765	74.861	74.957	75.053	75.149	75.245	75.341
139	75.175	75.271	75.367	75.463	75.559	75.655	75.751	75.847	75.943	76.039	76.135	76.231	76.327	76.423	76.519
140	76.362	76.458	76.554	76.650	76.746	76.842	76.938	77.034	77.130	77.225	77.321	77.417	77.513	77.609	77.705
141	77.557	77.653	77.749	77.845	77.941	78.036	78.132	78.228	78.324	78.420	78.516	78.611	78.707	78.803	78.899
142	78.760	78.856	78.952	79.048	79.143	79.239	79.335	79.431	79.526	79.622	79.718	79.814	79.909	80.005	80.101
143	79.972	80.067	80.163	80.259	80.354	80.450	80.546	80.641	80.737	80.833	80.928	81.024	81.120	81.215	81.311
144	81.191	81.287	81.382	81.478	81.573	81.669	81.764	81.860	81.956	82.051	82.147	82.242	82.338	82.433	82.529
145	82.418	82.514	82.609	82.705	82.800	82.896	82.991	83.087	83.182	83.278	83.373	83.469	83.564	83.660	83.755
146	83.654	83.749	83.844	83.940	84.035	84.131	84.226	84.321	84.417	84.512	84.608	84.703	84.798	84.894	84.989
147	84.897	84.992	85.088	85.183	85.278	85.374	85.469	85.564	85.660	85.755	85.850	85.945	86.041	86.136	86.231
148	86.148	86.244	86.339	86.434	86.529	86.624	86.720	86.815	86.910	87.005	87.101	87.196	87.291	87.386	87.481
149	87.408	87.503	87.598	87.693	87.788	87.883	87.978	88.074	88.169	88.264	88.359	88.454	88.549	88.644	88.739

TABLE 6.3 Thermo-mechanical exergy of liquid water from 16 to 30 bar. Dashed lines indicate that water is not a liquid at that temperature and pressure (it is a mixture of gas and liquid, or a gas). (Continued)

93

T (°C)															
	16	17	18	19	20	21	22	23	24	25	26	27	28	29	30
150	88.675	88.770	88.865	88.960	89.055	89.150	89.245	89.340	89.435	89.530	89.625	89.720	89.815	89.910	90.005
151	89.950	90.045	90.140	90.235	90.330	90.425	90.520	90.615	90.710	90.805	90.900	90.995	91.089	91.184	91.279
152	91.233	91.328	91.423	91.518	91.613	91.708	91.802	91.897	91.992	92.087	92.182	92.277	92.371	92.466	92.561
153	92.524	92.619	92.714	92.809	92.903	92.998	93.093	93.188	93.282	93.377	93.472	93.567	93.661	93.756	93.851
154	93.823	93.918	94.013	94.107	94.202	94.297	94.391	94.486	94.581	94.675	94.770	94.864	94.959	95.054	95.148
155	95.130	95.225	95.319	95.414	95.508	95.603	95.697	95.792	95.887	95.981	96.076	96.170	96.265	96.359	96.454
156	96.445	96.539	96.634	96.728	96.823	96.917	97.012	97.106	97.201	97.295	97.389	97.484	97.578	97.673	97.767
157	97.768	97.862	97.956	98.051	98.145	98.239	98.334	98.428	98.522	98.617	98.711	98.805	98.900	98.994	99.088
158	99.098	99.192	99.287	99.381	99.475	99.569	99.663	99.758	99.852	99.946	100.04	100.13	100.23	100.32	100.42
159	100.44	100.53	100.62	100.72	100.81	100.91	101.00	101.10	101.19	101.28	101.38	101.47	101.57	101.66	101.75
160	101.78	101.88	101.97	102.06	102.16	102.25	102.35	102.44	102.53	102.63	102.72	102.82	102.91	103.00	103.10
161	103.14	103.23	103.32	103.42	103.51	103.61	103.70	103.79	103.89	103.98	104.08	104.17	104.26	104.36	104.45
162	104.50	104.59	104.69	104.78	104.87	104.97	105.06	105.16	105.25	105.34	105.44	105.53	105.62	105.72	105.81
163	105.87	105.96	106.06	106.15	106.24	106.34	106.43	106.52	106.62	106.71	106.80	106.90	106.99	107.09	107.18
164	107.25	107.34	107.43	107.53	107.62	107.71	107.81	107.90	107.99	108.09	108.18	108.27	108.37	108.46	108.56
165	108.63	108.72	108.82	108.91	109.00	109.10	109.19	109.28	109.38	109.47	109.57	109.66	109.75	109.85	109.94
166	110.02	110.12	110.21	110.30	110.40	110.49	110.58	110.68	110.77	110.86	110.96	111.05	111.14	111.24	111.33
167	111.42	111.52	111.61	111.70	111.80	111.89	111.98	112.08	112.17	112.26	112.36	112.45	112.54	112.64	112.73
168	112.83	112.93	113.02	113.11	113.21	113.30	113.39	113.49	113.58	113.67	113.76	113.86	113.95	114.04	114.14
169	114.25	114.34	114.44	114.53	114.62	114.71	114.81	114.90	114.99	115.09	115.18	115.27	115.37	115.46	115.55
170	115.67	115.77	115.86	115.95	116.05	116.14	116.23	116.32	116.42	116.51	116.60	116.70	116.79	116.88	116.97
171	117.11	117.20	117.29	117.38	117.48	117.57	117.66	117.75	117.85	117.94	118.03	118.13	118.22	118.31	118.40
172	118.55	118.64	118.73	118.82	118.92	119.01	119.10	119.19	119.29	119.38	119.47	119.56	119.66	119.75	119.84
173	119.99	120.09	120.18	120.27	120.36	120.46	120.55	120.64	120.73	120.83	120.92	121.01	121.10	121.19	121.29
174	121.45	121.54	121.63	121.73	121.82	121.91	122.00	122.09	122.19	122.28	122.37	122.46	122.56	122.65	122.74
175	122.91	123.00	123.10	123.19	123.28	123.37	123.46	123.56	123.65	123.74	123.83	123.93	124.02	124.11	124.20
176	124.38	124.47	124.57	124.66	124.75	124.84	124.93	125.03	125.12	125.21	125.30	125.39	125.49	125.58	125.67
177	125.86	125.95	126.04	126.14	126.23	126.32	126.41	126.50	126.60	126.69	126.78	126.87	126.96	127.06	127.15
178	127.35	127.44	127.53	127.62	127.71	127.81	127.90	127.99	128.08	128.17	128.26	128.36	128.45	128.54	128.63

e^{tm} (kJ/kg), Pressure (bar)

TABLE 6.3 Thermo-mechanical exergy of liquid water from 16 to 30 bar. Dashed lines indicate that water is not a liquid at that temperature and pressure (it is a mixture of gas and liquid, or a gas). (Continued)

e^{tm} (kJ/kg)

T (°C)	Pressure (bar)														
	16	17	18	19	20	21	22	23	24	25	26	27	28	29	30
179	128.84	128.93	129.02	129.12	129.21	129.30	129.39	129.48	129.57	129.67	129.76	129.85	129.94	130.03	130.12
180	130.34	130.44	130.53	130.62	130.71	130.80	130.89	130.98	131.08	131.17	131.26	131.35	131.44	131.53	131.62
181	131.85	131.94	132.04	132.13	132.22	132.31	132.40	132.49	132.58	132.67	132.77	132.86	132.95	133.04	133.13
182	133.37	133.46	133.55	133.64	133.74	133.83	133.92	134.01	134.10	134.19	134.28	134.37	134.46	134.56	134.65
183	134.90	134.99	135.08	135.17	135.26	135.35	135.44	135.53	135.62	135.71	135.81	135.90	135.99	136.08	136.17
184	136.43	136.52	136.61	136.70	136.79	136.88	136.97	137.06	137.16	137.25	137.34	137.43	137.52	137.61	137.70
185	137.97	138.06	138.15	138.24	138.33	138.42	138.51	138.60	138.69	138.79	138.88	138.97	139.06	139.15	139.24
186	139.52	139.61	139.70	139.79	139.88	139.97	140.06	140.15	140.24	140.33	140.42	140.51	140.60	140.69	140.79
187	141.08	141.17	141.26	141.35	141.44	141.53	141.62	141.71	141.80	141.89	141.98	142.07	142.16	142.25	142.34
188	142.64	142.73	142.82	142.91	143.00	143.09	143.18	143.27	143.36	143.45	143.54	143.63	143.72	143.81	143.90
189	144.21	144.30	144.39	144.48	144.57	144.66	144.75	144.84	144.93	145.02	145.11	145.20	145.29	145.38	145.47
190	145.79	145.88	145.97	146.06	146.15	146.24	146.33	146.42	146.51	146.60	146.69	146.78	146.87	146.96	147.05
191	147.38	147.47	147.56	147.65	147.74	147.83	147.92	148.01	148.10	148.19	148.27	148.36	148.45	148.54	148.63
192	148.97	149.06	149.15	149.24	149.33	149.42	149.51	149.60	149.69	149.78	149.87	149.96	150.05	150.14	150.23
193	150.58	150.67	150.76	150.85	150.93	151.02	151.11	151.20	151.29	151.38	151.47	151.56	151.65	151.74	151.83
194	152.19	152.28	152.37	152.46	152.55	152.63	152.72	152.81	152.90	152.99	153.08	153.17	153.26	153.35	153.44
195	153.81	153.90	153.99	154.07	154.16	154.25	154.34	154.43	154.52	154.61	154.70	154.79	154.87	154.96	155.05
196	155.44	155.52	155.61	155.70	155.79	155.88	155.97	156.06	156.14	156.23	156.32	156.41	156.50	156.59	156.68
197	157.07	157.16	157.25	157.34	157.42	157.51	157.60	157.69	157.78	157.87	157.95	158.04	158.13	158.22	158.31
198	158.71	158.80	158.89	158.98	159.07	159.15	159.24	159.33	159.42	159.51	159.60	159.68	159.77	159.86	159.95
199	160.36	160.45	160.54	160.63	160.72	160.80	160.89	160.98	161.07	161.16	161.24	161.33	161.42	161.51	161.60
200	162.02	162.11	162.20	162.29	162.37	162.46	162.55	162.64	162.72	162.81	162.90	162.99	163.08	163.16	163.25
201	163.69	163.78	163.86	163.95	164.04	164.13	164.21	164.30	164.39	164.48	164.57	164.65	164.74	164.83	164.92
202	—	165.45	165.54	165.63	165.71	165.80	165.89	165.98	166.06	166.15	166.24	166.32	166.41	166.50	166.59
203	—	167.13	167.22	167.31	167.40	167.48	167.57	167.66	167.74	167.83	167.92	168.01	168.09	168.18	168.27
204	—	168.82	168.91	169.00	169.09	169.17	169.26	169.35	169.43	169.52	169.61	169.69	169.78	169.87	169.95
205	—	—	170.61	170.70	170.78	170.87	170.96	171.04	171.13	171.22	171.30	171.39	171.48	171.56	171.65
206	—	—	172.32	172.40	172.49	172.58	172.66	172.75	172.83	172.92	173.01	173.09	173.18	173.27	173.35
207	—	—	174.03	174.12	174.20	174.29	174.38	174.46	174.55	174.63	174.72	174.81	174.89	174.98	175.07
208	—	—	—	175.84	175.92	176.01	176.10	176.18	176.27	176.35	176.44	176.53	176.61	176.70	176.78

TABLE 6.3 Thermo-mechanical exergy of liquid water from 16 to 30 bar. Dashed lines indicate that water is not a liquid at that temperature and pressure (it is a mixture of gas and liquid, or a gas). (*Continued*)

e^{tm} (kJ/kg)

T (°C)	16	17	18	19	20	21	22	23	24	25	26	27	28	29	30
										Pressure (bar)					
209	—	—	—	177.57	177.66	177.74	177.83	177.91	178.00	178.08	178.17	178.26	178.34	178.43	178.51
210	—	—	—	—	179.39	179.48	179.56	179.65	179.74	179.82	179.91	179.99	180.08	180.16	180.25
211	—	—	—	—	181.14	181.23	181.31	181.40	181.48	181.57	181.65	181.74	181.82	181.91	181.99
212	—	—	—	—	182.89	182.98	183.06	183.15	183.23	183.32	183.40	183.49	183.57	183.66	183.74
213	—	—	—	—	—	184.74	184.83	184.91	185.00	185.08	185.17	185.25	185.33	185.42	185.50
214	—	—	—	—	—	186.51	186.60	186.68	186.77	186.85	186.94	187.02	187.10	187.19	187.27
215	—	—	—	—	—	—	188.38	188.46	188.54	188.63	188.71	188.80	188.88	188.97	189.05
216	—	—	—	—	—	—	190.16	190.25	190.33	190.42	190.50	190.58	190.67	190.75	190.83
217	—	—	—	—	—	—	191.96	192.04	192.13	192.21	192.29	192.38	192.46	192.54	192.63
218	—	—	—	—	—	—	—	193.85	193.93	194.01	194.10	194.18	194.26	194.35	194.43
219	—	—	—	—	—	—	—	195.66	195.74	195.82	195.91	195.99	196.07	196.16	196.24
220	—	—	—	—	—	—	—	—	197.56	197.64	197.73	197.81	197.89	197.97	198.06
221	—	—	—	—	—	—	—	—	199.39	199.47	199.56	199.64	199.72	199.80	199.88
222	—	—	—	—	—	—	—	—	—	201.31	201.39	201.47	201.56	201.64	201.72
223	—	—	—	—	—	—	—	—	—	203.16	203.24	203.32	203.40	203.48	203.56
224	—	—	—	—	—	—	—	—	—	—	205.09	205.17	205.25	205.33	205.42
225	—	—	—	—	—	—	—	—	—	—	206.95	207.03	207.11	207.20	207.28
226	—	—	—	—	—	—	—	—	—	—	208.82	208.90	208.99	209.07	209.15
227	—	—	—	—	—	—	—	—	—	—	—	210.78	210.86	210.94	211.02
228	—	—	—	—	—	—	—	—	—	—	—	212.67	212.75	212.83	212.91
229	—	—	—	—	—	—	—	—	—	—	—	—	214.65	214.73	214.81
230	—	—	—	—	—	—	—	—	—	—	—	—	216.55	216.63	216.71
231	—	—	—	—	—	—	—	—	—	—	—	—	—	218.55	218.63
232	—	—	—	—	—	—	—	—	—	—	—	—	—	—	220.55
233	—	—	—	—	—	—	—	—	—	—	—	—	—	—	222.48

TABLE 6.3 Thermo-mechanical exergy of liquid water from 16 to 30 bar. Dashed lines indicate that water is not a liquid at that temperature and pressure (it is a mixture of gas and liquid, or a gas). (Continued)

T (°C)	35	40	45	50	60	70	80	90	100	120	140	160	180	200	220
							e^{tm} (kJ/kg) Pressure (bar)								
0	8.0566	8.5550	9.0534	9.5517	10.548	11.544	12.539	13.534	14.529	16.517	18.503	20.488	22.472	24.453	26.433
2	7.3211	7.8200	8.3188	8.8175	9.8146	10.811	11.807	12.803	13.799	15.788	17.776	19.762	21.747	23.730	25.712
4	6.6527	7.1519	7.6511	8.1501	9.1478	10.145	11.142	12.139	13.135	15.125	17.115	19.102	21.088	23.072	25.055
6	6.0500	6.5495	7.0490	7.5483	8.5466	9.5445	10.542	11.539	12.536	14.528	16.518	18.507	20.494	22.479	24.463
8	5.5119	6.0117	6.5114	7.0110	8.0098	9.0082	10.006	11.004	12.001	13.994	15.985	17.975	19.962	21.949	23.933
10	5.0374	5.5374	6.0373	6.5371	7.5363	8.5351	9.5335	10.531	11.529	13.523	15.515	17.505	19.494	21.481	23.466
12	4.6253	5.1255	5.6256	6.1255	7.1251	8.1243	9.1230	10.121	11.119	13.114	15.106	17.097	19.086	21.074	23.060
14	4.2748	4.7751	5.2753	5.7754	6.7753	7.7748	8.7738	9.7723	10.770	12.765	14.759	16.750	18.740	20.728	22.714
16	3.9848	4.4853	4.9856	5.4858	6.4859	7.4856	8.4848	9.4836	10.482	12.477	14.471	16.463	18.453	20.442	22.428
18	3.7545	4.2551	4.7555	5.2558	6.2561	7.2559	8.2553	9.2543	10.253	12.248	14.242	16.235	18.225	20.214	22.201
20	3.5831	4.0836	4.5841	5.0845	6.0849	7.0849	8.0844	9.0835	10.082	12.078	14.072	16.065	18.055	20.044	22.032
22	3.4695	3.9702	4.4707	4.9711	5.9716	6.9716	7.9712	8.9704	9.9691	11.965	13.960	15.952	17.943	19.932	21.920
24	3.4131	3.9138	4.4143	4.9148	5.9153	6.9154	7.9150	8.9142	9.9130	11.909	13.904	15.896	17.887	19.876	21.864
26	3.4131	3.9138	4.4143	4.9147	5.9153	6.9153	7.9150	8.9142	9.9129	11.909	13.904	15.896	17.887	19.876	21.864
28	3.4687	3.9693	4.4698	4.9702	5.9707	6.9708	7.9704	8.9695	9.9682	11.964	13.959	15.951	17.942	19.931	21.919
30	3.5791	4.0797	4.5802	5.0805	6.0810	7.0809	8.0805	9.0795	10.078	12.074	14.068	16.061	18.052	20.041	22.028
32	3.7436	4.2441	4.7446	5.2449	6.2452	7.2451	8.2445	9.2435	10.242	12.238	14.232	16.224	18.215	20.204	22.191
34	3.9615	4.4620	4.9624	5.4626	6.4628	7.4626	8.4619	9.4607	10.459	12.455	14.448	16.440	18.431	20.419	22.406
36	4.2322	4.7326	5.2329	5.7330	6.7331	7.7326	8.7318	9.7305	10.729	12.724	14.717	16.709	18.699	20.688	22.674
38	4.5549	5.0552	5.5554	6.0555	7.0553	8.0547	9.0537	10.052	11.050	13.045	15.038	17.030	19.019	21.007	22.994
40	4.9290	5.4292	5.9293	6.4293	7.4289	8.4281	9.4268	10.425	11.423	13.417	15.410	17.401	19.390	21.378	23.364
42	5.3539	5.8539	6.3539	6.8538	7.8531	8.8521	9.8506	10.849	11.846	13.840	15.833	17.823	19.812	21.799	23.784
44	5.8289	6.3288	6.8286	7.3284	8.3275	9.3262	10.324	11.322	12.320	14.313	16.305	18.295	20.283	22.270	24.255
46	6.3534	6.8532	7.3529	7.8525	8.8513	9.8497	10.848	11.845	12.842	14.835	16.826	18.816	20.804	22.790	24.774
48	6.9268	7.4265	7.9260	8.4254	9.4239	10.422	11.420	12.417	13.414	15.406	17.397	19.385	21.373	23.358	25.342
50	7.5486	8.0481	8.5475	9.0467	10.045	11.043	12.040	13.037	14.033	16.025	18.015	20.003	21.990	23.975	25.958
52	8.2182	8.7175	9.2167	9.7157	10.714	11.711	12.708	13.704	14.701	16.692	18.681	20.668	22.654	24.639	26.621
54	8.9349	9.4340	9.9330	10.432	11.429	12.426	13.423	14.419	15.415	17.405	19.394	21.381	23.366	25.349	27.331
56	9.6984	10.197	10.696	11.195	12.192	13.188	14.185	15.180	16.176	18.165	20.153	22.139	24.124	26.106	28.088
58	10.508	11.007	11.505	12.004	13.000	13.996	14.992	15.988	16.983	18.971	20.958	22.943	24.927	26.909	28.890
60	11.363	11.861	12.360	12.858	13.854	14.850	15.845	16.840	17.835	19.823	21.809	23.793	25.776	27.757	29.737

TABLE 6.4 Thermo-mechanical exergy of liquid water from 35 to 220 bar. Dashed lines indicate that water is not a liquid at that temperature and pressure (it is a mixture of gas and liquid, or a gas).

e^{tm} (kJ/kg)

Pressure (bar)

T (°C)	35	40	45	50	60	70	80	90	100	120	140	160	180	200	220
62	12.263	12.761	13.260	13.758	14.753	15.749	16.743	17.738	18.732	20.719	22.704	24.688	26.670	28.650	30.629
64	13.208	13.706	14.204	14.702	15.697	16.692	17.686	18.680	19.674	21.660	23.644	25.627	27.608	29.587	31.565
66	14.197	14.695	15.192	15.690	16.684	17.679	18.673	19.666	20.659	22.644	24.628	26.609	28.589	30.568	32.545
68	15.229	15.727	16.224	16.721	17.716	18.709	19.703	20.696	21.688	23.672	25.655	27.635	29.615	31.592	33.568
70	16.305	16.802	17.299	17.796	18.790	19.783	20.776	21.768	22.760	24.743	26.725	28.704	30.682	32.659	34.634
72	17.423	17.920	18.417	18.914	19.907	20.899	21.892	22.884	23.875	25.857	27.837	29.816	31.793	33.768	35.742
74	18.584	19.080	19.577	20.073	21.066	22.058	23.050	24.041	25.032	27.012	28.992	30.969	32.945	34.919	36.892
76	19.786	20.282	20.779	21.275	22.267	23.258	24.249	25.240	26.230	28.210	30.188	32.164	34.139	36.112	38.084
78	21.030	21.526	22.022	22.517	23.509	24.500	25.490	26.480	27.470	29.448	31.425	33.400	35.374	37.346	39.316
80	22.314	22.810	23.306	23.801	24.792	25.782	26.772	27.761	28.750	30.727	32.703	34.677	36.649	38.620	40.589
82	23.639	24.135	24.630	25.125	26.115	27.105	28.094	29.083	30.071	32.047	34.021	35.993	37.965	39.934	41.902
84	25.005	25.500	25.994	26.489	27.479	28.467	29.456	30.444	31.432	33.406	35.379	37.350	39.320	41.288	43.255
86	26.410	26.904	27.399	27.893	28.882	29.870	30.858	31.845	32.832	34.805	36.776	38.746	40.715	42.682	44.648
88	27.854	28.348	28.842	29.336	30.324	31.312	32.299	33.285	34.272	36.243	38.213	40.182	42.149	44.115	46.079
90	29.337	29.831	30.325	30.819	31.806	32.792	33.779	34.764	35.750	37.720	39.689	41.656	43.621	45.586	47.549
92	30.859	31.353	31.846	32.339	33.326	34.311	35.297	36.282	37.267	39.235	41.202	43.168	45.132	47.095	49.056
94	32.419	32.912	33.405	33.898	34.884	35.869	36.853	37.838	38.822	40.789	42.754	44.718	46.681	48.642	50.602
96	34.018	34.510	35.003	35.495	36.480	37.464	38.448	39.431	40.415	42.380	44.344	46.306	48.267	50.227	52.186
98	35.653	36.145	36.638	37.130	38.113	39.097	40.080	41.062	42.045	44.008	45.971	47.932	49.891	51.849	53.806
100	37.326	37.818	38.310	38.801	39.784	40.767	41.749	42.731	43.712	45.674	47.635	49.594	51.552	53.508	55.463
102	39.036	39.527	40.019	40.510	41.492	42.473	43.455	44.435	45.416	47.376	49.335	51.293	53.249	55.204	57.157
104	40.783	41.274	41.764	42.255	43.236	44.217	45.197	46.177	47.157	49.115	51.072	53.028	54.982	56.935	58.887
106	42.566	43.056	43.546	44.036	45.016	45.996	46.975	47.955	48.933	50.890	52.845	54.799	56.752	58.703	60.653
108	44.385	44.874	45.364	45.854	46.833	47.812	48.790	49.768	50.746	52.700	54.654	56.606	58.557	60.506	62.454
110	46.239	46.729	47.218	47.707	48.685	49.663	50.640	51.617	52.594	54.547	56.498	58.448	60.397	62.345	64.291
112	48.130	48.618	49.107	49.596	50.573	51.549	52.526	53.502	54.477	56.428	58.377	60.325	62.272	64.218	66.162
114	50.055	50.543	51.031	51.519	52.495	53.471	54.446	55.421	56.396	58.344	60.292	62.238	64.183	66.126	68.069
116	52.015	52.503	52.991	53.478	54.453	55.427	56.402	57.375	58.349	60.295	62.241	64.185	66.127	68.069	70.010
118	54.010	54.498	54.985	55.471	56.445	57.418	58.392	59.364	60.337	62.281	64.224	66.166	68.106	70.046	71.984

TABLE 6.4 Thermo-mechanical exergy of liquid water from 35 to 220 bar. Dashed lines indicate that water is not a liquid at that temperature and pressure (it is a mixture of gas and liquid, or a gas). (Continued)

e^{tm} (kJ/kg)

T (°C)	\multicolumn{15}{c}{Pressure (bar)}
	35	40	45	50	60	70	80	90	100	120	140	160	180	200	220
120	56.040	56.527	57.013	57.499	58.472	59.444	60.416	61.387	62.359	64.301	66.241	68.181	70.120	72.057	73.993
122	58.104	58.590	59.076	59.561	60.532	61.503	62.474	63.445	64.415	66.354	68.293	70.230	72.166	74.102	76.036
124	60.202	60.687	61.172	61.657	62.627	63.597	64.566	65.536	66.505	68.442	70.378	72.313	74.247	76.180	78.112
126	62.333	62.818	63.302	63.787	64.756	65.724	66.692	67.660	68.628	70.563	72.496	74.429	76.360	78.291	80.221
128	64.499	64.982	65.466	65.950	66.918	67.885	68.852	69.818	70.785	72.717	74.648	76.578	78.507	80.435	82.363
130	66.697	67.180	67.664	68.147	69.113	70.079	71.044	72.009	72.975	74.904	76.833	78.760	80.687	82.613	84.537
132	68.929	69.412	69.894	70.376	71.341	72.305	73.270	74.234	75.197	77.124	79.050	80.975	82.899	84.822	86.745
134	71.194	71.676	72.157	72.639	73.602	74.565	75.528	76.491	77.453	79.377	81.300	83.222	85.144	87.064	88.984
136	73.491	73.972	74.453	74.934	75.896	76.858	77.819	78.780	79.741	81.662	83.583	85.502	87.421	89.339	91.256
138	75.822	76.302	76.782	77.262	78.223	79.183	80.142	81.102	82.061	83.980	85.897	87.814	89.730	91.645	93.559
140	78.184	78.664	79.143	79.623	80.581	81.540	82.498	83.456	84.414	86.329	88.244	90.158	92.071	93.983	95.895
142	80.580	81.058	81.537	82.015	82.972	83.929	84.886	85.843	86.799	88.711	90.623	92.533	94.444	96.353	98.262
144	83.007	83.485	83.962	84.440	85.396	86.351	87.306	88.261	89.216	91.125	93.033	94.941	96.848	98.754	100.66
146	85.466	85.943	86.420	86.897	87.851	88.804	89.758	90.711	91.664	93.570	95.475	97.379	99.283	101.19	103.09
148	87.958	88.434	88.910	89.386	90.337	91.289	92.241	93.192	94.144	96.046	97.948	99.849	101.75	103.65	105.55
150	90.481	90.956	91.431	91.906	92.856	93.806	94.756	95.706	96.655	98.554	100.45	102.35	104.25	106.14	108.04
152	93.035	93.509	93.984	94.458	95.406	96.354	97.302	98.250	99.198	101.09	102.99	104.88	106.78	108.67	110.56
154	95.622	96.095	96.568	97.041	97.987	98.934	99.880	100.83	101.77	103.66	105.55	107.45	109.34	111.23	113.12
156	98.239	98.712	99.184	99.656	100.60	101.54	102.49	103.43	104.38	106.26	108.15	110.04	111.93	113.81	115.70
158	100.89	101.36	101.83	102.30	103.24	104.19	105.13	106.07	107.01	108.90	110.78	112.66	114.55	116.43	118.31
160	103.57	104.04	104.51	104.98	105.92	106.86	107.80	108.74	109.68	111.56	113.44	115.32	117.20	119.08	120.96
162	106.28	106.75	107.22	107.69	108.63	109.56	110.50	111.44	112.38	114.25	116.13	118.00	119.88	121.76	123.63
164	109.02	109.49	109.96	110.43	111.36	112.30	113.23	114.17	115.11	116.98	118.85	120.72	122.59	124.46	126.33
166	111.80	112.26	112.73	113.20	114.13	115.06	116.00	116.93	117.86	119.73	121.60	123.47	125.33	127.20	129.07
168	114.60	115.07	115.53	116.00	116.93	117.86	118.79	119.72	120.65	122.52	124.38	126.24	128.11	129.97	131.83
170	117.44	117.90	118.37	118.83	119.76	120.69	121.62	122.55	123.47	125.33	127.19	129.05	130.91	132.77	134.63
172	120.30	120.77	121.23	121.69	122.62	123.55	124.47	125.40	126.32	128.18	130.03	131.89	133.74	135.59	137.45
174	123.20	123.66	124.13	124.59	125.51	126.43	127.36	128.28	129.21	131.05	132.90	134.75	136.60	138.45	140.30
176	126.13	126.59	127.05	127.51	128.43	129.35	130.27	131.20	132.12	133.96	135.80	137.65	139.49	141.34	143.18
178	129.09	129.55	130.01	130.47	131.38	132.30	133.22	134.14	135.06	136.90	138.74	140.57	142.41	144.25	146.10
180	132.08	132.54	133.00	133.45	134.37	135.28	136.20	137.11	138.03	139.86	141.70	143.53	145.37	147.20	149.04

TABLE 6.4 Thermo-mechanical exergy of liquid water from 35 to 220 bar. Dashed lines indicate that water is not a liquid at that temperature and pressure (it is a mixture of gas and liquid, or a gas). (Continued)

e^tm (kJ/kg)

T (°C)	Pressure (bar)														
	35	40	45	50	60	70	80	90	100	120	140	160	180	200	220
182	135.10	135.56	136.01	136.47	137.38	138.29	139.21	140.12	141.03	142.86	144.69	146.52	148.35	150.18	152.01
184	138.15	138.61	139.06	139.52	140.43	141.34	142.25	143.16	144.07	145.89	147.71	149.53	151.36	153.18	155.01
186	141.24	141.69	142.14	142.60	143.50	144.41	145.31	146.22	147.13	148.94	150.76	152.58	154.40	156.22	158.04
188	144.35	144.80	145.25	145.70	146.61	147.51	148.41	149.32	150.22	152.03	153.84	155.65	157.47	159.28	161.09
190	147.50	147.95	148.40	148.84	149.74	150.64	151.54	152.44	153.34	155.15	156.95	158.76	160.56	162.37	164.18
192	150.67	151.12	151.57	152.02	152.91	153.81	154.70	155.60	156.50	158.29	160.09	161.89	163.69	165.49	167.30
194	153.88	154.33	154.77	155.22	156.11	157.00	157.90	158.79	159.68	161.47	163.26	165.06	166.85	168.65	170.44
196	157.12	157.56	158.01	158.45	159.34	160.23	161.12	162.01	162.90	164.68	166.46	168.25	170.04	171.83	173.62
198	160.39	160.83	161.27	161.72	162.60	163.48	164.37	165.26	166.14	167.92	169.70	171.47	173.26	175.04	176.82
200	163.69	164.13	164.57	165.01	165.89	166.77	167.65	168.54	169.42	171.19	172.96	174.73	176.50	178.28	180.06
202	167.02	167.46	167.90	168.34	169.21	170.09	170.97	171.85	172.73	174.49	176.25	178.01	179.78	181.55	183.32
204	170.39	170.82	171.26	171.70	172.57	173.44	174.31	175.19	176.06	177.81	179.57	181.33	183.09	184.85	186.61
206	173.79	174.22	174.65	175.09	175.95	176.82	177.69	178.56	179.43	181.17	182.92	184.67	186.42	188.18	189.93
208	177.21	177.65	178.08	178.51	179.37	180.23	181.10	181.96	182.83	184.56	186.30	188.04	189.79	191.53	193.28
210	180.68	181.10	181.53	181.96	182.82	183.68	184.54	185.40	186.26	187.99	189.72	191.45	193.18	194.92	196.66
212	184.17	184.59	185.02	185.45	186.30	187.15	188.01	188.86	189.72	191.44	193.16	194.88	196.61	198.34	200.07
214	187.70	188.12	188.54	188.96	189.81	190.66	191.51	192.36	193.21	194.92	196.63	198.35	200.07	201.79	203.51
216	191.25	191.67	192.09	192.52	193.36	194.20	195.05	195.89	196.74	198.44	200.14	201.84	203.55	205.27	206.98
218	194.85	195.26	195.68	196.10	196.94	197.77	198.61	199.45	200.29	201.98	203.67	205.37	207.07	208.77	210.48
220	198.47	198.89	199.30	199.71	200.55	201.38	202.21	203.05	203.88	205.56	207.24	208.93	210.62	212.31	214.01
222	202.13	202.54	202.95	203.36	204.19	205.02	205.84	206.67	207.50	209.17	210.84	212.52	214.20	215.88	217.57
224	205.82	206.23	206.64	207.05	207.87	208.69	209.51	210.33	211.16	212.81	214.47	216.13	217.80	219.48	221.16
226	209.55	209.95	210.36	210.76	211.58	212.39	213.21	214.02	214.84	216.48	218.13	219.79	221.45	223.11	224.78
228	213.31	213.71	214.11	214.52	215.32	216.13	216.94	217.75	218.56	220.19	221.83	223.47	225.12	226.77	228.43
230	217.11	217.51	217.90	218.30	219.10	219.90	220.70	221.51	222.31	223.93	225.55	227.18	228.82	230.46	232.11
232	220.94	221.33	221.73	222.12	222.91	223.71	224.50	225.30	226.10	227.70	229.31	230.93	232.55	234.18	235.82
234	224.81	225.20	225.59	225.98	226.76	227.55	228.33	229.12	229.92	231.51	233.10	234.71	236.32	237.94	239.56
236	228.71	229.10	229.48	229.87	230.65	231.42	232.20	232.98	233.77	235.34	236.93	238.52	240.12	241.72	243.34
238	232.65	233.04	233.42	233.80	234.56	235.33	236.11	236.88	237.66	239.22	240.79	242.36	243.95	245.54	247.14

TABLE 6.4 Thermo-mechanical exergy of liquid water from 35 to 220 bar. Dashed lines indicate that water is not a liquid at that temperature and pressure (it is a mixture of gas and liquid, or a gas). (Continued)

e^{tm} (kJ/kg)

T (°C)	Pressure (bar)														
	35	40	45	50	60	70	80	90	100	120	140	160	180	200	220
240	236.63	237.01	237.39	237.76	238.52	239.28	240.04	240.81	241.58	243.12	244.68	246.24	247.81	249.39	250.98
242	240.65	241.02	241.39	241.77	242.51	243.26	244.02	244.78	245.54	247.06	248.60	250.15	251.71	253.27	254.85
244	—	245.07	245.44	245.81	246.54	247.29	248.03	248.78	249.53	251.04	252.56	254.10	255.64	257.19	258.75
246	—	249.16	249.52	249.88	250.61	251.34	252.08	252.82	253.56	255.05	256.56	258.07	259.60	261.14	262.68
248	—	253.29	253.64	254.00	254.72	255.44	256.17	256.89	257.63	259.10	260.59	262.09	263.60	265.12	266.65
250	—	257.46	257.81	258.16	258.86	259.58	260.29	261.01	261.73	263.19	264.66	266.14	267.63	269.13	270.65
252	—	—	262.01	262.36	263.05	263.75	264.45	265.16	265.87	267.31	268.76	270.22	271.70	273.18	274.68
254	—	—	266.25	266.59	267.28	267.96	268.66	269.35	270.05	271.47	272.90	274.34	275.80	277.27	278.75
256	—	—	270.54	270.88	271.55	272.22	272.90	273.59	274.27	275.67	277.07	278.50	279.93	281.38	282.85
258	—	—	—	275.20	275.86	276.52	277.19	277.86	278.53	279.90	281.29	282.69	284.11	285.54	286.98
260	—	—	—	279.57	280.21	280.86	281.51	282.17	282.84	284.18	285.54	286.92	288.32	289.73	291.15
262	—	—	—	283.98	284.61	285.24	285.88	286.53	287.18	288.50	289.83	291.19	292.56	293.95	295.36
264	—	—	—	—	289.05	289.67	290.29	290.93	291.56	292.85	294.17	295.50	296.85	298.22	299.60
266	—	—	—	—	293.54	294.14	294.75	295.37	295.99	297.25	298.54	299.85	301.17	302.52	303.88
268	—	—	—	—	298.08	298.66	299.26	299.86	300.46	301.70	302.95	304.23	305.54	306.86	308.20
270	—	—	—	—	302.67	303.23	303.81	304.39	304.98	306.18	307.41	308.66	309.94	311.23	312.55
272	—	—	—	—	307.30	307.85	308.41	308.97	309.54	310.71	311.91	313.13	314.38	315.65	316.94
274	—	—	—	—	311.99	312.52	313.06	313.60	314.16	315.29	316.45	317.65	318.87	320.11	321.37
276	—	—	—	—	—	317.24	317.76	318.28	318.82	319.91	321.04	322.20	323.39	324.61	325.84
278	—	—	—	—	—	322.02	322.51	323.02	323.53	324.59	325.68	326.81	327.96	329.15	330.36
280	—	—	—	—	—	326.85	327.32	327.80	328.29	329.31	330.36	331.45	332.58	333.73	334.91
282	—	—	—	—	—	331.74	332.19	332.64	333.11	334.08	335.10	336.15	337.23	338.35	339.50
284	—	—	—	—	—	336.69	337.11	337.54	337.99	338.91	339.88	340.89	341.94	343.02	344.14
286	—	—	—	—	—	—	342.10	342.50	342.92	343.79	344.71	345.68	346.69	347.74	348.82
288	—	—	—	—	—	—	347.15	347.52	347.91	348.73	349.60	350.53	351.50	352.50	353.55
290	—	—	—	—	—	—	352.26	352.60	352.96	353.73	354.55	355.42	356.35	357.31	358.32
292	—	—	—	—	—	—	357.45	357.75	358.08	358.78	359.55	360.38	361.25	362.17	363.14
294	—	—	—	—	—	—	362.71	362.98	363.27	363.90	364.61	365.38	366.21	367.08	368.01
296	—	—	—	—	—	—	—	368.27	368.52	369.09	369.74	370.45	371.22	372.05	372.92
298	—	—	—	—	—	—	—	373.65	373.86	374.35	374.92	375.57	376.29	377.07	377.89
300	—	—	—	—	—	—	—	379.10	379.27	379.68	380.18	380.76	381.42	382.14	382.92

TABLE 6.4 Thermo-mechanical exergy of liquid water from 35 to 220 bar. Dashed lines indicate that water is not a liquid at that temperature and pressure (it is a mixture of gas and liquid, or a gas). (Continued)

e^{tm} (kJ/kg)

T (°C)	35	40	45	50	60	70	80	90	100	120	140	160	180	200	220
302	—	—	—	—	—	—	—	384.64	384.76	385.08	385.50	386.02	386.61	387.27	387.99
304	—	—	—	—	—	—	—	—	390.34	390.57	390.90	391.34	391.86	392.46	393.13
306	—	—	—	—	—	—	—	—	396.02	396.13	396.38	396.73	397.18	397.71	398.32
308	—	—	—	—	—	—	—	—	401.79	401.79	401.94	402.20	402.57	403.03	403.57
310	—	—	—	—	—	—	—	—	407.67	407.55	407.58	407.75	408.03	408.41	408.88
312	—	—	—	—	—	—	—	—	—	413.40	413.31	413.38	413.57	413.87	414.26
314	—	—	—	—	—	—	—	—	—	419.37	419.15	419.09	419.18	419.39	419.71
316	—	—	—	—	—	—	—	—	—	425.45	425.08	424.90	424.88	425.00	425.23
318	—	—	—	—	—	—	—	—	—	431.66	431.13	430.81	430.67	430.68	430.82
320	—	—	—	—	—	—	—	—	—	438.01	437.30	436.83	436.55	436.45	436.49
322	—	—	—	—	—	—	—	—	—	444.51	443.59	442.95	442.54	442.31	442.24
324	—	—	—	—	—	—	—	—	—	451.18	450.03	449.20	448.63	448.27	448.08
326	—	—	—	—	—	—	—	—	—	—	456.62	455.58	454.83	454.32	454.01
328	—	—	—	—	—	—	—	—	—	—	463.39	462.10	461.16	460.48	460.03
330	—	—	—	—	—	—	—	—	—	—	470.34	468.78	467.62	466.76	466.16
332	—	—	—	—	—	—	—	—	—	—	477.51	475.63	474.22	473.17	472.39
334	—	—	—	—	—	—	—	—	—	—	484.93	482.68	480.98	479.70	478.74
336	—	—	—	—	—	—	—	—	—	—	492.63	489.94	487.92	486.38	485.22
338	—	—	—	—	—	—	—	—	—	—	—	497.45	495.05	493.22	491.83
340	—	—	—	—	—	—	—	—	—	—	—	505.25	502.40	500.24	498.58
342	—	—	—	—	—	—	—	—	—	—	—	513.38	509.99	507.45	505.50
344	—	—	—	—	—	—	—	—	—	—	—	521.92	517.87	514.87	512.58
346	—	—	—	—	—	—	—	—	—	—	—	530.97	526.08	522.54	519.87
348	—	—	—	—	—	—	—	—	—	—	—	—	534.69	530.49	527.36

Table 6.4 Thermo-mechanical exergy of liquid water from 35 to 220 bar. Dashed lines indicate that water is not a liquid at that temperature and pressure (it is a mixture of gas and liquid, or a gas). (*Continued*)

T (°C)	35	40	45	50	60	70	80	90	100	120	140	160	180	200	220
350	—	—	—	—	—	—	—	—	—	—	—	—	543.78	538.77	535.10
352	—	—	—	—	—	—	—	—	—	—	—	—	553.50	547.43	543.11
354	—	—	—	—	—	—	—	—	—	—	—	—	564.05	556.55	551.44
356	—	—	—	—	—	—	—	—	—	—	—	—	575.79	566.26	560.13
358	—	—	—	—	—	—	—	—	—	—	—	—	—	576.73	569.28
360	—	—	—	—	—	—	—	—	—	—	—	—	—	588.22	578.96
362	—	—	—	—	—	—	—	—	—	—	—	—	—	601.23	589.33
364	—	—	—	—	—	—	—	—	—	—	—	—	—	616.75	600.59
366	—	—	—	—	—	—	—	—	—	—	—	—	—	—	613.07
368	—	—	—	—	—	—	—	—	—	—	—	—	—	—	627.35
370	—	—	—	—	—	—	—	—	—	—	—	—	—	—	644.61
372	—	—	—	—	—	—	—	—	—	—	—	—	—	—	668.48

e^{tm} (kJ/kg)

Pressure (bar)

TABLE 6.4 Thermo-mechanical exergy of liquid water from 35 to 220 bar. Dashed lines indicate that water is not a liquid at that temperature and pressure (it is a mixture of gas and liquid, or a gas). (Continued)

T (°C)	1.01325	1.084	1.154	1.225	1.295	1.366	1.436	1.507	1.577	1.648	1.718	1.789	1.859	1.930	2.000
									e^{tm} (kJ/kg) Pressure (bar)						
100	487.38	—	—	—	—	—	—	—	—	—	—	—	—	—	—
101	487.80	—	—	—	—	—	—	—	—	—	—	—	—	—	—
102	488.22	497.14	—	—	—	—	—	—	—	—	—	—	—	—	—
103	488.65	497.57	—	—	—	—	—	—	—	—	—	—	—	—	—
104	489.08	498.00	506.33	—	—	—	—	—	—	—	—	—	—	—	—
105	489.51	498.44	506.77	—	—	—	—	—	—	—	—	—	—	—	—
106	489.95	498.88	507.22	515.04	—	—	—	—	—	—	—	—	—	—	—
107	490.39	499.32	507.66	515.49	—	—	—	—	—	—	—	—	—	—	—
108	490.84	499.77	508.11	515.94	523.32	—	—	—	—	—	—	—	—	—	—
109	491.28	500.22	508.57	516.40	523.78	530.74	—	—	—	—	—	—	—	—	—
110	491.73	500.67	509.03	516.86	524.24	531.21	—	—	—	—	—	—	—	—	—
111	492.19	501.13	509.49	517.33	524.71	531.68	538.28	—	—	—	—	—	—	—	—
112	492.65	501.59	509.95	517.79	525.18	532.15	538.75	545.02	—	—	—	—	—	—	—
113	493.11	502.06	510.42	518.26	525.65	532.62	539.23	545.50	551.47	—	—	—	—	—	—
114	493.58	502.52	510.89	518.74	526.12	533.10	539.71	545.99	551.96	—	—	—	—	—	—
115	494.05	503.00	511.36	519.21	526.60	533.59	540.20	546.48	552.45	558.15	—	—	—	—	—
116	494.52	503.47	511.84	519.69	527.09	534.07	540.69	546.97	552.95	558.65	564.10	—	—	—	—
117	494.99	503.95	512.32	520.18	527.57	534.56	541.18	547.46	553.45	559.15	564.60	569.82	—	—	—
118	495.47	504.43	512.81	520.66	528.06	535.05	541.67	547.96	553.95	559.65	565.11	570.33	575.33	—	—
119	495.96	504.92	513.29	521.15	528.56	535.55	542.17	548.46	554.45	560.16	565.62	570.84	575.85	—	—
120	496.44	505.41	513.78	521.65	529.05	536.05	542.67	548.97	554.96	560.67	566.13	571.35	576.36	581.17	—
121	496.93	505.90	514.28	522.14	529.55	536.55	543.18	549.47	555.47	561.18	566.64	571.87	576.89	581.70	586.32
122	497.43	506.39	514.78	522.64	530.05	537.05	543.69	549.98	555.98	561.70	567.16	572.39	577.41	582.23	586.86
123	497.92	506.89	515.28	523.15	530.56	537.56	544.20	550.50	556.49	562.22	567.68	572.92	577.94	582.75	587.39
124	498.42	507.40	515.78	523.66	531.07	538.07	544.71	551.01	557.01	562.74	568.21	573.44	578.47	583.29	587.92
125	498.93	507.90	516.29	524.17	531.58	538.59	545.23	551.53	557.54	563.26	568.73	573.97	579.00	583.82	588.46
126	499.44	508.41	516.80	524.68	532.10	539.11	545.75	552.06	558.06	563.79	569.26	574.51	579.53	584.36	589.00
127	499.95	508.92	517.32	525.20	532.62	539.63	546.27	552.58	558.59	564.32	569.80	575.04	580.07	584.90	589.55
128	500.46	509.44	517.84	525.72	533.14	540.15	546.80	553.11	559.12	564.85	570.33	575.58	580.61	585.45	590.09

TABLE 6.5 Thermo-mechanical exergy of steam from 1.01325 to 2 bar. Dashed lines indicate that water is not a gas at that temperature and pressure (it is a mixture of liquid and gas).

e^{tm} (kJ/kg)

T (°C)	Pressure (bar)														
	1.01325	1.084	1.154	1.225	1.295	1.366	1.436	1.507	1.577	1.648	1.718	1.789	1.859	1.930	2.000
129	500.98	509.96	518.36	526.24	533.66	540.68	547.33	553.64	559.65	565.39	570.87	576.12	581.16	585.99	590.64
130	501.50	510.48	518.88	526.77	534.19	541.21	547.86	554.18	560.19	565.93	571.41	576.67	581.70	586.54	591.19
131	502.02	511.01	519.41	527.30	534.73	541.75	548.40	554.72	560.73	566.47	571.96	577.21	582.25	587.09	591.75
132	502.55	511.54	519.94	527.83	535.26	542.28	548.94	555.26	561.28	567.02	572.51	577.76	582.81	587.65	592.31
133	503.08	512.07	520.47	528.37	535.80	542.82	549.48	555.80	561.82	567.57	573.06	578.32	583.36	588.21	592.87
134	503.61	512.60	521.01	528.90	536.34	543.37	550.02	556.35	562.37	568.12	573.61	578.87	583.92	588.77	593.43
135	504.15	513.14	521.55	529.45	536.88	543.91	550.57	556.90	562.93	568.67	574.17	579.43	584.48	589.33	594.00
136	504.69	513.69	522.10	529.99	537.43	544.46	551.13	557.45	563.48	569.23	574.73	580.00	585.05	589.90	594.56
137	505.23	514.23	522.64	530.54	537.98	545.02	551.68	558.01	564.04	569.79	575.29	580.56	585.61	590.47	595.14
138	505.78	514.78	523.19	531.10	538.54	545.57	552.24	558.57	564.60	570.36	575.86	581.13	586.18	591.04	595.71
139	506.33	515.33	523.75	531.65	539.10	546.13	552.80	559.14	565.17	570.93	576.43	581.70	586.76	591.61	596.29
140	506.88	515.89	524.31	532.21	539.66	546.69	553.36	559.70	565.74	571.50	577.00	582.27	587.33	592.19	596.87
141	507.44	516.44	524.87	532.77	540.22	547.26	553.93	560.27	566.31	572.07	577.58	582.85	587.91	592.77	597.45
142	508.00	517.01	525.43	533.34	540.79	547.83	554.50	560.84	566.88	572.64	578.15	583.43	588.49	593.36	598.04
143	508.56	517.57	526.00	533.90	541.36	548.40	555.08	561.42	567.46	573.22	578.74	584.01	589.08	593.94	598.63
144	509.13	518.14	526.56	534.48	541.93	548.97	555.65	562.00	568.04	573.81	579.32	584.60	589.67	594.53	599.22
145	509.70	518.71	527.14	535.05	542.51	549.55	556.23	562.58	568.62	574.39	579.91	585.19	590.26	595.13	599.81
146	510.27	519.28	527.71	535.63	543.09	550.13	556.81	563.16	569.21	574.98	580.50	585.78	590.85	595.72	600.41
147	510.85	519.86	528.29	536.21	543.67	550.72	557.40	563.75	569.80	575.57	581.09	586.38	591.45	596.32	601.01
148	511.42	520.44	528.87	536.79	544.25	551.30	557.99	564.34	570.39	576.16	581.69	586.97	592.05	596.92	601.61
149	512.01	521.03	529.46	537.38	544.84	551.90	558.58	564.93	570.99	576.76	582.28	587.57	592.65	597.53	602.22
150	512.59	521.61	530.05	537.97	545.43	552.49	559.18	565.53	571.58	577.36	582.89	588.18	593.25	598.13	602.83
151	513.18	522.20	530.64	538.56	546.03	553.08	559.77	566.13	572.19	577.96	583.49	588.78	593.86	598.74	603.44
152	513.77	522.79	531.23	539.16	546.63	553.68	560.37	566.73	572.79	578.57	584.10	589.39	594.47	599.36	604.05
153	514.37	523.39	531.83	539.76	547.23	554.29	560.98	567.34	573.40	579.18	584.71	590.01	595.09	599.97	604.67
154	514.96	523.99	532.43	540.36	547.83	554.89	561.59	567.95	574.01	579.79	585.32	590.62	595.70	600.59	605.29
155	515.56	524.59	533.03	540.96	548.44	555.50	562.19	568.56	574.62	580.40	585.94	591.24	596.32	601.21	605.91
156	516.17	525.20	533.64	541.57	549.05	556.11	562.81	569.17	575.24	581.02	586.56	591.86	596.95	601.84	606.54
157	516.77	525.80	534.25	542.18	549.66	556.72	563.42	569.79	575.85	581.64	587.18	592.48	597.57	602.46	607.17
158	517.38	526.42	534.86	542.80	550.27	557.34	564.04	570.41	576.48	582.27	587.80	593.11	598.20	603.09	607.80

TABLE 6.5 Thermo-mechanical exergy of steam from 1.01325 to 2 bar. Dashed lines indicate that water is not a gas at that temperature and pressure (it is a mixture of liquid and gas). (Continued)

e^{tm} (kJ/kg)

T (°C)	Pressure (bar)														
	1.01325	1.084	1.154	1.225	1.295	1.366	1.436	1.507	1.577	1.648	1.718	1.789	1.859	1.930	2.000
159	518.00	527.03	535.48	543.41	550.89	557.96	564.66	571.03	577.10	582.89	588.43	593.74	598.83	603.72	608.43
160	518.61	527.65	536.10	544.03	551.51	558.58	565.29	571.66	577.73	583.52	589.06	594.37	599.46	604.36	609.07
161	519.23	528.27	536.72	544.66	552.14	559.21	565.92	572.29	578.36	584.15	589.70	595.00	600.10	605.00	609.71
162	519.85	528.89	537.34	545.28	552.77	559.84	566.55	572.92	578.99	584.79	590.33	595.64	600.74	605.64	610.35
163	520.48	529.52	537.97	545.91	553.40	560.47	567.18	573.55	579.63	585.43	590.97	596.28	601.38	606.28	611.00
164	521.10	530.15	538.60	546.54	554.03	561.11	567.81	574.19	580.27	586.07	591.61	596.93	602.03	606.93	611.65
165	521.74	530.78	539.24	547.18	554.67	561.74	568.45	574.83	580.91	586.71	592.26	597.57	602.67	607.58	612.30
166	522.37	531.41	539.87	547.82	555.30	562.38	569.10	575.48	581.55	587.36	592.90	598.22	603.32	608.23	612.95
167	523.01	532.05	540.51	548.46	555.95	563.03	569.74	576.12	582.20	588.00	593.56	598.87	603.98	608.88	613.61
168	523.65	532.69	541.15	549.10	556.59	563.67	570.39	576.77	582.85	588.66	594.21	599.53	604.63	609.54	614.26
169	524.29	533.33	541.80	549.75	557.24	564.32	571.04	577.42	583.50	589.31	594.86	600.19	605.29	610.20	614.93
170	524.93	533.98	542.45	550.40	557.89	564.97	571.69	578.08	584.16	589.97	595.52	600.85	605.95	610.86	615.59
171	525.58	534.63	543.10	551.05	558.54	565.63	572.35	578.73	584.82	590.63	596.18	601.51	606.62	611.53	616.26
172	526.23	535.28	543.75	551.70	559.20	566.29	573.01	579.39	585.48	591.29	596.85	602.17	607.29	612.20	616.93
173	526.89	535.94	544.41	552.36	559.86	566.95	573.67	580.06	586.15	591.96	597.52	602.84	607.96	612.87	617.60
174	527.54	536.60	545.07	553.02	560.52	567.61	574.33	580.72	586.81	592.63	598.19	603.51	608.63	613.54	618.28
175	528.20	537.26	545.73	553.69	561.19	568.28	575.00	581.39	587.48	593.30	598.86	604.19	609.30	614.22	618.95
176	528.87	537.92	546.40	554.35	561.85	568.95	575.67	582.06	588.16	593.97	599.53	604.86	609.98	614.90	619.63
177	529.53	538.59	547.06	555.02	562.52	569.62	576.34	582.74	588.83	594.65	600.21	605.54	610.66	615.58	620.32
178	530.20	539.26	547.73	555.69	563.20	570.29	577.02	583.42	589.51	595.33	600.89	606.23	611.35	616.27	621.00
179	530.87	539.93	548.41	556.37	563.87	570.97	577.70	584.10	590.19	596.01	601.58	606.91	612.03	616.95	621.69
180	531.55	540.61	549.08	557.05	564.55	571.65	578.38	584.78	590.87	596.69	602.26	607.60	612.72	617.64	622.38
181	532.22	541.29	549.76	557.73	565.24	572.33	579.06	585.46	591.56	597.38	602.95	608.29	613.41	618.34	623.08
182	532.90	541.97	550.45	558.41	565.92	573.02	579.75	586.15	592.25	598.07	603.64	608.98	614.11	619.03	623.77
183	533.59	542.65	551.13	559.10	566.61	573.71	580.44	586.84	592.94	598.77	604.34	609.68	614.80	619.73	624.47
184	534.27	543.34	551.82	559.79	567.30	574.40	581.13	587.54	593.64	599.46	605.03	610.38	615.50	620.43	625.17
185	534.96	544.03	552.51	560.48	567.99	575.09	581.83	588.23	594.33	600.16	605.73	611.08	616.20	621.13	625.88
186	535.65	544.72	553.20	561.17	568.69	575.79	582.53	588.93	595.03	600.86	606.44	611.78	616.91	621.84	626.59
187	536.35	545.41	553.90	561.87	569.38	576.49	583.23	589.63	595.74	601.57	607.14	612.49	617.62	622.55	627.30

TABLE 6.5 Thermo-mechanical exergy of steam from 1.01325 to 2 bar. Dashed lines indicate that water is not a gas at that temperature and pressure (it is a mixture of liquid and gas). (Continued)

| T (°C) | | e^tm (kJ/kg) Pressure (bar) | | | | | | | | | | | | |
	1.01325	1.084	1.154	1.225	1.295	1.366	1.436	1.507	1.577	1.648	1.718	1.789	1.859	1.930	2.000
188	537.04	546.11	554.60	562.57	570.09	577.19	583.93	590.34	596.44	602.27	607.85	613.20	618.33	623.26	628.01
189	537.74	546.81	555.30	563.27	570.79	577.90	584.64	591.04	597.15	602.98	608.56	613.91	619.04	623.97	628.72
190	538.44	547.52	556.00	563.98	571.50	578.60	585.35	591.75	597.86	603.69	609.27	614.62	619.75	624.69	629.44
191	539.15	548.22	556.71	564.69	572.20	579.31	586.06	592.47	598.58	604.41	609.99	615.34	620.47	625.41	630.16
192	539.86	548.93	557.42	565.40	572.92	580.03	586.77	593.18	599.29	605.13	610.71	616.06	621.19	626.13	630.88
193	540.57	549.64	558.13	566.11	573.63	580.74	587.49	593.90	600.01	605.85	611.43	616.78	621.92	626.85	631.61
194	541.28	550.36	558.85	566.83	574.35	581.46	588.21	594.62	600.73	606.57	612.15	617.50	622.64	627.58	632.34
195	542.00	551.07	559.57	567.55	575.07	582.18	588.93	595.34	601.46	607.29	612.88	618.23	623.37	628.31	633.07
196	542.72	551.79	560.29	568.27	575.79	582.91	589.65	596.07	602.18	608.02	613.61	618.96	624.10	629.04	633.80
197	543.44	552.52	561.01	568.99	576.52	583.63	590.38	596.80	602.91	608.75	614.34	619.69	624.84	629.78	634.54
198	544.16	553.24	561.74	569.72	577.24	584.36	591.11	597.53	603.64	609.49	615.07	620.43	625.57	630.52	635.28
199	544.89	553.97	562.47	570.45	577.98	585.09	591.84	598.26	604.38	610.22	615.81	621.17	626.31	631.26	636.02
200	545.62	554.70	563.20	571.18	578.71	585.83	592.58	599.00	605.12	610.96	616.55	621.91	627.05	632.00	636.76
201	546.35	555.43	563.93	571.92	579.44	586.56	593.32	599.74	605.86	611.70	617.29	622.65	627.80	632.74	637.51
202	547.08	556.17	564.67	572.65	580.18	587.30	594.06	600.48	606.60	612.44	618.04	623.40	628.54	633.49	638.25
203	547.82	556.91	565.41	573.39	580.92	588.05	594.80	601.22	607.34	613.19	618.78	624.14	629.29	634.24	639.00
204	548.56	557.65	566.15	574.14	581.67	588.79	595.55	601.97	608.09	613.94	619.53	624.89	630.04	634.99	639.76
205	549.31	558.39	566.89	574.88	582.42	589.54	596.30	602.72	608.84	614.69	620.28	625.65	630.80	635.75	640.51
206	550.05	559.14	567.64	575.63	583.16	590.29	597.05	603.47	609.59	615.44	621.04	626.40	631.55	636.50	641.27
207	550.80	559.89	568.39	576.38	583.92	591.04	597.80	604.23	610.35	616.20	621.80	627.16	632.31	637.26	642.03
208	551.55	560.64	569.14	577.14	584.67	591.80	598.56	604.98	611.11	616.96	622.56	627.92	633.07	638.03	642.80
209	552.30	561.39	569.90	577.89	585.43	592.55	599.31	605.74	611.87	617.72	623.32	628.69	633.84	638.79	643.56
210	553.06	562.15	570.66	578.65	586.19	593.31	600.08	606.50	612.63	618.48	624.08	629.45	634.60	639.56	644.33
211	553.82	562.91	571.42	579.41	586.95	594.08	600.84	607.27	613.40	619.25	624.85	630.22	635.37	640.33	645.10
212	554.58	563.67	572.18	580.17	587.71	594.84	601.61	608.04	614.17	620.02	625.62	630.99	636.15	641.10	645.88
213	555.34	564.44	572.95	580.94	588.48	595.61	602.37	608.81	614.94	620.79	626.39	631.76	636.92	641.88	646.65
214	556.11	565.20	573.71	581.71	589.25	596.38	603.15	609.58	615.71	621.56	627.17	632.54	637.70	642.66	647.43
215	556.88	565.97	574.48	582.48	590.02	597.15	603.92	610.35	616.48	622.34	627.95	633.32	638.48	643.44	648.21
216	557.65	566.74	575.26	583.25	590.80	597.93	604.70	611.13	617.26	623.12	628.73	634.10	639.26	644.22	648.99
217	558.42	567.52	576.03	584.03	591.57	598.71	605.48	611.91	618.04	623.90	629.51	634.88	640.04	645.00	649.78

Table 6.5 Thermo-mechanical exergy of steam from 1.01325 to 2 bar. Dashed lines indicate that water is not a gas at that temperature and pressure (it is a mixture of liquid and gas). (*Continued*)

e^tm (kJ/kg)

T (°C)	1.01325	1.084	1.154	1.225	1.295	1.366	1.436	1.507	1.577	1.648	1.718	1.789	1.859	1.930	2.000
								Pressure (bar)							
218	559.20	568.30	576.81	584.81	592.35	599.49	606.26	612.69	618.83	624.69	630.29	635.67	640.83	645.79	650.57
219	559.98	569.08	577.59	585.59	593.14	600.27	607.04	613.48	619.61	625.47	631.08	636.46	641.62	646.58	651.36
220	560.76	569.86	578.37	586.38	593.92	601.06	607.83	614.26	620.40	626.26	631.87	637.25	642.41	647.37	652.15
221	561.54	570.64	579.16	587.16	594.71	601.85	608.62	615.05	621.19	627.05	632.66	638.04	643.20	648.17	652.95
222	562.33	571.43	579.95	587.95	595.50	602.64	609.41	615.85	621.98	627.85	633.46	638.83	644.00	648.96	653.75
223	563.12	572.22	580.74	588.74	596.29	603.43	610.20	616.64	622.78	628.64	634.25	639.63	644.80	649.76	654.55
224	563.91	573.01	581.53	589.54	597.08	604.22	611.00	617.44	623.58	629.44	635.05	640.43	645.60	650.57	655.35
225	564.70	573.81	582.33	590.33	597.88	605.02	611.80	618.24	624.38	630.24	635.85	641.24	646.40	651.37	656.15
226	565.50	574.61	583.13	591.13	598.68	605.82	612.60	619.04	625.18	631.05	636.66	642.04	647.21	652.18	656.96
227	566.30	575.40	583.93	591.93	599.48	606.63	613.40	619.84	625.99	631.85	637.47	642.85	648.02	652.99	657.77
228	567.10	576.21	584.73	592.74	600.29	607.43	614.21	620.65	626.79	632.66	638.28	643.66	648.83	653.80	658.58
229	567.91	577.01	585.53	593.54	601.10	608.24	615.02	621.46	627.60	633.47	639.09	644.47	649.64	654.61	659.40
230	568.71	577.82	586.34	594.35	601.91	609.05	615.83	622.27	628.42	634.28	639.90	645.29	650.46	655.43	660.22
231	569.52	578.63	587.15	595.16	602.72	609.86	616.64	623.09	629.23	635.10	640.72	646.10	651.27	656.25	661.04
232	570.33	579.44	587.97	595.98	603.53	610.68	617.46	623.90	630.05	635.92	641.54	646.92	652.09	657.07	661.86
233	571.15	580.26	588.78	596.79	604.35	611.49	618.27	624.72	630.87	636.74	642.36	647.74	652.92	657.89	662.68
234	571.96	581.07	589.60	597.61	605.17	612.31	619.10	625.54	631.69	637.56	643.18	648.57	653.74	658.72	663.51
235	572.78	581.89	590.42	598.43	605.99	613.14	619.92	626.37	632.51	638.39	644.01	649.40	654.57	659.55	664.34
236	573.60	582.71	591.24	599.26	606.81	613.96	620.74	627.19	633.34	639.21	644.84	650.22	655.40	660.38	665.17
237	574.43	583.54	592.07	600.08	607.64	614.79	621.57	628.02	634.17	640.04	645.67	651.06	656.23	661.21	666.00
238	575.25	584.37	592.89	600.91	608.47	615.62	622.40	628.85	635.00	640.88	646.50	651.89	657.07	662.04	666.84
239	576.08	585.19	593.72	601.74	609.30	616.45	623.23	629.69	635.84	641.71	647.33	652.73	657.90	662.88	667.68
240	576.91	586.03	594.56	602.57	610.13	617.28	624.07	630.52	636.67	642.55	648.17	653.56	658.74	663.72	668.52
241	577.75	586.86	595.39	603.41	610.97	618.12	624.91	631.36	637.51	643.39	649.01	654.40	659.58	664.56	669.36
242	578.58	587.70	596.23	604.25	611.81	618.96	625.75	632.20	638.35	644.23	649.85	655.25	660.43	665.41	670.21
243	579.42	588.53	597.07	605.09	612.65	619.80	626.59	633.04	639.20	645.07	650.70	656.09	661.27	666.25	671.05
244	580.26	589.38	597.91	605.93	613.49	620.65	627.43	633.89	640.04	645.92	651.55	656.94	662.12	667.10	671.90
245	581.10	590.22	598.75	606.77	614.34	621.49	628.28	634.74	640.89	646.77	652.40	657.79	662.97	667.96	672.75
246	581.95	591.07	599.60	607.62	615.18	622.34	629.13	635.59	641.74	647.62	653.25	658.64	663.83	668.81	673.61

TABLE 6.5 Thermo-mechanical exergy of steam from 1.01325 to 2 bar. Dashed lines indicate that water is not a gas at that temperature and pressure (it is a mixture of liquid and gas). (Continued)

e^{tm} (kJ/kg)

T (°C)	Pressure (bar)														
	1.01325	1.084	1.154	1.225	1.295	1.366	1.436	1.507	1.577	1.648	1.718	1.789	1.859	1.930	2.000
247	582.79	591.91	600.45	608.47	616.03	623.19	629.98	636.44	642.59	648.47	654.10	659.50	664.68	669.67	674.47
248	583.64	592.76	601.30	609.32	616.89	624.04	630.83	637.29	643.45	649.33	654.96	660.36	665.54	670.52	675.33
249	584.50	593.62	602.15	610.18	617.74	624.90	631.69	638.15	644.31	650.19	655.82	661.21	666.40	671.38	676.19
250	585.35	594.47	603.01	611.03	618.60	625.76	632.55	639.01	645.17	651.05	656.68	662.08	667.26	672.25	677.05
251	586.21	595.33	603.87	611.89	619.46	626.62	633.41	639.87	646.03	651.91	657.54	662.94	668.13	673.11	677.92
252	587.07	596.19	604.73	612.75	620.32	627.48	634.27	640.73	646.89	652.78	658.41	663.81	668.99	673.98	678.78
253	587.93	597.05	605.59	613.62	621.18	628.34	635.14	641.60	647.76	653.64	659.27	664.67	669.86	674.85	679.65
254	588.79	597.92	606.46	614.48	622.05	629.21	636.00	642.47	648.63	654.51	660.14	665.55	670.73	675.72	680.53
255	589.66	598.78	607.32	615.35	622.92	630.08	636.87	643.34	649.50	655.38	661.02	666.42	671.61	676.60	681.40
256	590.53	599.65	608.19	616.22	623.79	630.95	637.75	644.21	650.37	656.26	661.89	667.29	672.48	677.47	682.28
257	591.40	600.52	609.06	617.09	624.66	631.83	638.62	645.08	651.25	657.13	662.77	668.17	673.36	678.35	683.16
258	592.27	601.40	609.94	617.97	625.54	632.70	639.50	645.96	652.12	658.01	663.65	669.05	674.24	679.23	684.04
259	593.15	602.27	610.82	618.84	626.42	633.58	640.38	646.84	653.00	658.89	664.53	669.93	675.12	680.11	684.92
260	594.02	603.15	611.69	619.72	627.30	634.46	641.26	647.72	653.89	659.78	665.41	670.82	676.01	681.00	685.81
261	594.90	604.03	612.58	620.61	628.18	635.34	642.14	648.61	654.77	660.66	666.30	671.70	676.89	681.89	686.70
262	595.79	604.91	613.46	621.49	629.06	636.23	643.03	649.49	655.66	661.55	667.19	672.59	677.78	682.78	687.59
263	596.67	605.80	614.34	622.38	629.95	637.12	643.91	650.38	656.55	662.44	668.08	673.48	678.68	683.67	688.48
264	597.56	606.69	615.23	623.26	630.84	638.01	644.80	651.27	657.44	663.33	668.97	674.38	679.57	684.56	689.37
265	598.45	607.58	616.12	624.15	631.73	638.90	645.70	652.17	658.33	664.22	669.86	675.27	680.46	685.46	690.27
266	599.34	608.47	617.01	625.05	632.62	639.79	646.59	653.06	659.23	665.12	670.76	676.17	681.36	686.36	691.17
267	600.23	609.36	617.91	625.94	633.52	640.69	647.49	653.96	660.13	666.02	671.66	677.07	682.26	687.26	692.07
268	601.13	610.26	618.81	626.84	634.42	641.59	648.39	654.86	661.03	666.92	672.56	677.97	683.17	688.16	692.98
269	602.02	611.16	619.70	627.74	635.32	642.49	649.29	655.76	661.93	667.82	673.46	678.87	684.07	689.07	693.88
270	602.92	612.06	620.61	628.64	636.22	643.39	650.19	656.66	662.83	668.73	674.37	679.78	684.98	689.98	694.79
271	603.83	612.96	621.51	629.54	637.12	644.29	651.10	657.57	663.74	669.63	675.28	680.69	685.89	690.88	695.70
272	604.73	613.87	622.42	630.45	638.03	645.20	652.01	658.48	664.65	670.54	676.19	681.60	686.80	691.80	696.61
273	605.64	614.77	623.32	631.36	638.94	646.11	652.92	659.39	665.56	671.46	677.10	682.51	687.71	692.71	697.53
274	606.55	615.68	624.23	632.27	639.85	647.02	653.83	660.30	666.47	672.37	678.01	683.43	688.63	693.63	698.44
275	607.46	616.59	625.15	633.18	640.76	647.94	654.74	661.22	667.39	673.29	678.93	684.34	689.54	694.54	699.36
276	608.37	617.51	626.06	634.10	641.68	648.85	655.66	662.13	668.31	674.20	679.85	685.26	690.46	695.46	700.28

TABLE 6.5 Thermo-mechanical exergy of steam from 1.01325 to 2 bar. Dashed lines indicate that water is not a gas at that temperature and pressure (it is a mixture of liquid and gas). (Continued)

| | e^{tm} (kJ/kg) | | | | | | | | | | | | | | |
| T (°C) | Pressure (bar) | | | | | | | | | | | | | | |
	1.01325	1.084	1.154	1.225	1.295	1.366	1.436	1.507	1.577	1.648	1.718	1.789	1.859	1.930	2.000
277	609.29	618.42	626.98	635.02	642.60	649.77	656.58	663.05	669.23	675.12	680.77	686.18	691.38	696.39	701.21
278	610.21	619.34	627.90	635.93	643.52	650.69	657.50	663.97	670.15	676.05	681.69	687.11	692.31	697.31	702.13
279	611.13	620.26	628.82	636.86	644.44	651.61	658.42	664.90	671.07	676.97	682.62	688.03	693.23	698.24	703.06
280	612.05	621.19	629.74	637.78	645.36	652.54	659.35	665.82	672.00	677.90	683.55	688.96	694.16	699.17	703.99
281	612.97	622.11	630.66	638.71	646.29	653.47	660.27	666.75	672.93	678.83	684.47	689.89	695.09	700.10	704.92
282	613.90	623.04	631.59	639.63	647.22	654.39	661.20	667.68	673.86	679.76	685.41	690.82	696.03	701.03	705.85
283	614.83	623.97	632.52	640.56	648.15	655.33	662.14	668.61	674.79	680.69	686.34	691.76	696.96	701.97	706.79
284	615.76	624.90	633.45	641.50	649.08	656.26	663.07	669.55	675.72	681.63	687.28	692.69	697.90	702.90	707.73
285	616.69	625.83	634.39	642.43	650.02	657.19	664.01	670.48	676.66	682.56	688.21	693.63	698.84	703.84	708.67
286	617.63	626.77	635.32	643.37	650.95	658.13	664.94	671.42	677.60	683.50	689.15	694.57	699.78	704.78	709.61
287	618.56	627.70	636.26	644.31	651.89	659.07	665.88	672.36	678.54	684.44	690.10	695.52	700.72	705.73	710.55
288	619.50	628.64	637.20	645.25	652.83	660.01	666.83	673.31	679.48	685.39	691.04	696.46	701.67	706.67	711.50
289	620.44	629.59	638.14	646.19	653.78	660.96	667.77	674.25	680.43	686.33	691.99	697.41	702.61	707.62	712.45
290	621.39	630.53	639.09	647.13	654.72	661.90	668.72	675.20	681.38	687.28	692.93	698.36	703.56	708.57	713.40
291	622.33	631.48	640.04	648.08	655.67	662.85	669.67	676.15	682.33	688.23	693.89	699.31	704.51	709.52	714.35
292	623.28	632.42	640.98	649.03	656.62	663.80	670.62	677.10	683.28	689.18	694.84	700.26	705.47	710.48	715.30
293	624.23	633.38	641.94	649.98	657.57	664.75	671.57	678.05	684.23	690.14	695.79	701.21	706.42	711.43	716.26
294	625.18	634.33	642.89	650.94	658.53	665.71	672.52	679.01	685.19	691.09	696.75	702.17	707.38	712.39	717.22
295	626.14	635.28	643.84	651.89	659.48	666.66	673.48	679.96	686.15	692.05	697.71	703.13	708.34	713.35	718.18
296	627.09	636.24	644.80	652.85	660.44	667.62	674.44	680.92	687.11	693.01	698.67	704.09	709.30	714.31	719.14
297	628.05	637.20	645.76	653.81	661.40	668.58	675.40	681.88	688.07	693.98	699.63	705.06	710.27	715.28	720.11
298	629.01	638.16	646.72	654.77	662.36	669.55	676.36	682.85	689.03	694.94	700.60	706.02	711.23	716.24	721.07
299	629.98	639.12	647.69	655.73	663.33	670.51	677.33	683.81	690.00	695.91	701.56	706.99	712.20	717.21	722.04
300	630.94	640.09	648.65	656.70	664.29	671.48	678.30	684.78	690.97	696.88	702.53	707.96	713.17	718.18	723.01
305	635.79	644.94	653.51	661.56	669.16	676.34	683.16	689.65	695.84	701.75	707.41	712.84	718.05	723.07	727.90
310	640.70	649.85	658.42	666.47	674.07	681.26	688.08	694.57	700.76	706.67	712.34	717.77	722.98	728.00	732.83
315	645.66	654.81	663.38	671.43	679.03	686.23	693.05	699.54	705.73	711.65	717.31	722.74	727.96	732.98	737.82
320	650.66	659.82	668.39	676.45	684.05	691.24	698.07	704.56	710.75	716.67	722.34	727.77	732.99	738.01	742.85
325	655.72	664.87	673.45	681.51	689.11	696.31	703.13	709.63	715.83	721.75	727.41	732.85	738.07	743.09	747.93

TABLE 6.5 Thermo-mechanical exergy of steam from 1.01325 to 2 bar. Dashed lines indicate that water is not a gas at that temperature and pressure (it is a mixture of liquid and gas). (Continued)

T (°C)	e^{tm} (kJ/kg) Pressure (bar)														
	1.01325	1.084	1.154	1.225	1.295	1.366	1.436	1.507	1.577	1.648	1.718	1.789	1.859	1.930	2.000
330	660.82	669.98	678.56	686.62	694.22	701.42	708.25	714.75	720.95	726.87	732.54	737.97	743.20	748.22	753.07
335	665.98	675.14	683.71	691.78	699.38	706.58	713.42	719.92	726.11	732.04	737.71	743.15	748.37	753.40	758.24
340	671.18	680.34	688.92	696.98	704.59	711.79	718.63	725.13	731.33	737.25	742.93	748.37	753.60	758.63	763.47
345	676.43	685.59	694.17	702.24	709.85	717.05	723.89	730.39	736.59	742.52	748.19	753.64	758.87	763.90	768.74
350	681.72	690.89	699.47	707.54	715.15	722.36	729.19	735.70	741.90	747.83	753.51	758.95	764.18	769.21	774.06
355	687.07	696.23	704.82	712.89	720.50	727.71	734.55	741.05	747.26	753.19	758.87	764.31	769.54	774.58	779.43
360	692.45	701.62	710.21	718.28	725.90	733.10	739.94	746.45	752.66	758.59	764.27	769.72	774.95	779.99	784.84
365	697.89	707.06	715.65	723.72	731.34	738.55	745.39	751.90	758.11	764.04	769.72	775.17	780.41	785.44	790.30
370	703.37	712.54	721.13	729.21	736.82	744.03	750.88	757.39	763.60	769.53	775.22	780.67	785.90	790.94	795.80
375	708.90	718.07	726.66	734.74	742.36	749.57	756.41	762.92	769.14	775.07	780.76	786.21	791.45	796.49	801.34
380	714.47	723.64	732.23	740.31	747.93	755.14	761.99	768.50	774.72	780.65	786.34	791.79	797.03	802.07	806.93
385	720.08	729.26	737.85	745.93	753.55	760.76	767.61	774.13	780.34	786.28	791.97	797.42	802.66	807.71	812.57
390	725.74	734.91	743.51	751.59	759.21	766.43	773.28	779.79	786.01	791.95	797.64	803.09	808.34	813.38	818.24
395	731.44	740.62	749.21	757.29	764.92	772.14	778.99	785.50	791.72	797.66	803.35	808.81	814.05	819.10	823.96
400	737.18	746.36	754.96	763.04	770.67	777.89	784.74	791.26	797.48	803.42	809.11	814.57	819.81	824.86	829.72
405	742.97	752.15	760.75	768.83	776.46	783.68	790.53	797.05	803.27	809.22	814.91	820.37	825.62	830.66	835.53
410	748.80	757.98	766.58	774.67	782.29	789.52	796.37	802.89	809.11	815.06	820.75	826.21	831.46	836.51	841.37
415	754.67	763.85	772.45	780.54	788.17	795.39	802.25	808.77	814.99	820.94	826.63	832.10	837.34	842.39	847.26
420	760.58	769.77	778.37	786.46	794.09	801.31	808.17	814.69	820.91	826.86	832.56	838.02	843.27	848.32	853.19
425	766.54	775.72	784.33	792.41	800.05	807.27	814.13	820.65	826.88	832.83	838.52	843.99	849.24	854.29	859.16
430	772.53	781.72	790.32	798.41	806.05	813.27	820.13	826.66	832.88	838.83	844.53	849.99	855.25	860.30	865.17
435	778.57	787.76	796.36	804.45	812.09	819.31	826.17	832.70	838.92	844.88	850.57	856.04	861.29	866.35	871.22
440	784.64	793.83	802.44	810.53	818.17	825.39	832.25	838.78	845.01	850.96	856.66	862.13	867.38	872.44	877.31
445	790.76	799.95	808.56	816.65	824.29	831.51	838.38	844.91	851.13	857.09	862.79	868.26	873.51	878.57	883.44
450	796.92	806.11	814.71	822.81	830.45	837.68	844.54	851.07	857.30	863.25	868.95	874.42	879.68	884.74	889.61
455	803.11	812.30	820.91	829.01	836.64	843.88	850.74	857.27	863.50	869.45	875.16	880.63	885.89	890.94	895.82
460	809.34	818.54	827.15	835.24	842.88	850.11	856.98	863.51	869.74	875.70	881.40	886.87	892.13	897.19	902.07
465	815.62	824.81	833.42	841.52	849.16	856.39	863.26	869.79	876.02	881.98	887.68	893.16	898.42	903.48	908.35
470	821.93	831.13	839.74	847.83	855.48	862.71	869.57	876.11	882.34	888.30	894.00	899.48	904.74	909.80	914.68
475	828.28	837.48	846.09	854.19	861.83	869.06	875.93	882.47	888.70	894.66	900.36	905.84	911.10	916.16	921.04
480	834.67	843.87	852.48	860.58	868.22	875.46	882.32	888.86	895.09	901.05	906.76	912.24	917.50	922.56	927.44

TABLE 6.5 Thermo-mechanical exergy of steam from 1.01325 to 2 bar. Dashed lines indicate that water is not a gas at that temperature and pressure (it is a mixture of liquid and gas). (Continued)

| | e^tm (kJ/kg) | | | | | | | | | | | | | | |
| | Pressure (bar) | | | | | | | | | | | | | | |
T (°C)	1.01325	1.084	1.154	1.225	1.295	1.366	1.436	1.507	1.577	1.648	1.718	1.789	1.859	1.930	2.000
485	841.10	850.29	858.91	867.01	874.65	881.89	888.76	895.29	901.53	907.49	913.20	918.67	923.93	929.00	933.88
490	847.56	856.76	865.37	873.47	881.12	888.35	895.22	901.76	908.00	913.96	919.67	925.15	930.41	935.47	940.36
495	854.06	863.26	871.88	879.98	887.62	894.86	901.73	908.27	914.51	920.47	926.18	931.66	936.92	941.99	946.87
500	860.60	869.80	878.42	886.52	894.16	901.40	908.27	914.81	921.05	927.01	932.72	938.20	943.47	948.53	953.42
505	867.17	876.37	884.99	893.10	900.74	907.98	914.85	921.39	927.63	933.60	939.31	944.79	950.05	955.12	960.00
510	873.79	882.99	891.61	899.71	907.36	914.60	921.47	928.01	934.25	940.21	945.93	951.41	956.67	961.74	966.63
515	880.43	889.64	898.26	906.36	914.01	921.25	928.12	934.66	940.90	946.87	952.58	958.06	963.33	968.40	973.29
520	887.12	896.32	904.94	913.05	920.70	927.94	934.81	941.35	947.59	953.56	959.27	964.76	970.02	975.09	979.98
525	893.84	903.04	911.66	919.77	927.42	934.66	941.54	948.08	954.32	960.29	966.00	971.48	976.75	981.82	986.71
530	900.60	909.80	918.42	926.53	934.18	941.42	948.30	954.84	961.08	967.05	972.76	978.25	983.52	988.59	993.48
535	907.39	916.59	925.21	933.32	940.97	948.22	955.09	961.64	967.88	973.85	979.56	985.05	990.32	995.39	1000.3
540	914.21	923.42	932.04	940.15	947.80	955.05	961.92	968.47	974.71	980.68	986.40	991.88	997.15	1002.2	1007.1
545	921.08	930.28	938.91	947.02	954.67	961.91	968.79	975.34	981.58	987.55	993.27	998.75	1004.0	1009.1	1014.0
550	927.97	937.18	945.81	953.92	961.57	968.81	975.69	982.24	988.48	994.45	1000.2	1005.7	1010.9	1016.0	1020.9
555	934.91	944.11	952.74	960.85	968.50	975.75	982.63	989.17	995.42	1001.4	1007.1	1012.6	1017.9	1022.9	1027.8
560	941.87	951.08	959.71	967.82	975.47	982.72	989.60	996.15	1002.4	1008.4	1014.1	1019.6	1024.8	1029.9	1034.8
565	948.87	958.08	966.71	974.82	982.48	989.72	996.60	1003.2	1009.4	1015.4	1021.1	1026.6	1031.9	1036.9	1041.8
570	955.91	965.12	973.75	981.86	989.51	996.76	1003.6	1010.2	1016.4	1022.4	1028.1	1033.6	1038.9	1044.0	1048.9
575	962.98	972.19	980.82	988.93	996.59	1003.8	1010.7	1017.3	1023.5	1029.5	1035.2	1040.7	1046.0	1051.0	1055.9
580	970.08	979.29	987.92	996.03	1003.7	1010.9	1017.8	1024.4	1030.6	1036.6	1042.3	1047.8	1053.1	1058.2	1063.1
585	977.22	986.43	995.06	1003.2	1010.8	1018.1	1025.0	1031.5	1037.8	1043.7	1049.5	1054.9	1060.2	1065.3	1070.2
590	984.39	993.60	1002.2	1010.3	1018.0	1025.3	1032.1	1038.7	1044.9	1050.9	1056.6	1062.1	1067.4	1072.5	1077.4
595	991.59	1000.8	1009.4	1017.6	1025.2	1032.5	1039.3	1045.9	1052.1	1058.1	1063.8	1069.3	1074.6	1079.7	1084.6
600	998.83	1008.0	1016.7	1024.8	1032.4	1039.7	1046.6	1053.1	1059.4	1065.4	1071.1	1076.6	1081.9	1086.9	1091.8
605	1006.1	1015.3	1023.9	1032.1	1039.7	1047.0	1053.9	1060.4	1066.7	1072.6	1078.4	1083.9	1089.1	1094.2	1099.1
610	1013.4	1022.6	1031.2	1039.4	1047.0	1054.3	1061.2	1067.7	1074.0	1079.9	1085.7	1091.2	1096.4	1101.5	1106.4
615	1020.7	1030.0	1038.6	1046.7	1054.4	1061.6	1068.5	1075.1	1081.3	1087.3	1093.0	1098.5	1103.8	1108.9	1113.8
620	1028.1	1037.3	1046.0	1054.1	1061.7	1069.0	1075.9	1082.4	1088.7	1094.7	1100.4	1105.9	1111.2	1116.2	1121.1
625	1035.5	1044.7	1053.4	1061.5	1069.1	1076.4	1083.3	1089.8	1096.1	1102.1	1107.8	1113.3	1118.6	1123.6	1128.5

TABLE 6.5 Thermo-mechanical exergy of steam from 1.01325 to 2 bar. Dashed lines indicate that water is not a gas at that temperature and pressure (it is a mixture of liquid and gas). (Continued)

		e^{tm} (kJ/kg)													
T (°C)		**Pressure (bar)**													
	1.01325	**1.084**	**1.154**	**1.225**	**1.295**	**1.366**	**1.436**	**1.507**	**1.577**	**1.648**	**1.718**	**1.789**	**1.859**	**1.930**	**2.000**
630	1042.9	1052.2	1060.8	1068.9	1076.6	1083.8	1090.7	1097.3	1103.5	1109.5	1115.2	1120.7	1126.0	1131.1	1136.0
635	1050.4	1059.6	1068.3	1076.4	1084.0	1091.3	1098.2	1104.7	1111.0	1117.0	1122.7	1128.2	1133.5	1138.6	1143.5
640	1057.9	1067.1	1075.8	1083.9	1091.5	1098.8	1105.7	1112.2	1118.5	1124.5	1130.2	1135.7	1141.0	1146.1	1151.0
645	1065.4	1074.6	1083.3	1091.4	1099.1	1106.3	1113.2	1119.8	1126.0	1132.0	1137.7	1143.2	1148.5	1153.6	1158.5
650	1073.0	1082.2	1090.8	1099.0	1106.6	1113.9	1120.8	1127.3	1133.6	1139.6	1145.3	1150.8	1156.1	1161.2	1166.1
655	1080.6	1089.8	1098.4	1106.6	1114.2	1121.5	1128.4	1134.9	1141.2	1147.2	1152.9	1158.4	1163.7	1168.8	1173.7
660	1088.2	1097.4	1106.1	1114.2	1121.8	1129.1	1136.0	1142.5	1148.8	1154.8	1160.5	1166.0	1171.3	1176.4	1181.3
665	1095.9	1105.1	1113.7	1121.8	1129.5	1136.8	1143.6	1150.2	1156.5	1162.4	1168.2	1173.7	1179.0	1184.0	1188.9
670	1103.5	1112.8	1121.4	1129.5	1137.2	1144.4	1151.3	1157.9	1164.1	1170.1	1175.9	1181.4	1186.6	1191.7	1196.6
675	1111.3	1120.5	1129.1	1137.2	1144.9	1152.2	1159.1	1165.6	1171.9	1177.9	1183.6	1189.1	1194.4	1199.5	1204.4
680	1119.0	1128.2	1136.9	1145.0	1152.7	1159.9	1166.8	1173.4	1179.6	1185.6	1191.3	1196.8	1202.1	1207.2	1212.1
685	1126.8	1136.0	1144.6	1152.8	1160.4	1167.7	1174.6	1181.1	1187.4	1193.4	1199.1	1204.6	1209.9	1215.0	1219.9
690	1134.6	1143.8	1152.4	1160.6	1168.2	1175.5	1182.4	1189.0	1195.2	1201.2	1206.9	1212.4	1217.7	1222.8	1227.7
695	1142.4	1151.6	1160.3	1168.4	1176.1	1183.3	1190.2	1196.8	1203.1	1209.0	1214.8	1220.3	1225.6	1230.6	1235.5
700	1150.3	1159.5	1168.2	1176.3	1183.9	1191.2	1198.1	1204.7	1210.9	1216.9	1222.6	1228.1	1233.4	1238.5	1243.4
705	1158.2	1167.4	1176.1	1184.2	1191.9	1199.1	1206.0	1212.6	1218.8	1224.8	1230.5	1236.0	1241.3	1246.4	1251.3
710	1166.1	1175.3	1184.0	1192.1	1199.8	1207.0	1213.9	1220.5	1226.8	1232.7	1238.5	1244.0	1249.3	1254.4	1259.3
715	1174.1	1183.3	1191.9	1200.1	1207.7	1215.0	1221.9	1228.5	1234.7	1240.7	1246.4	1251.9	1257.2	1262.3	1267.2
720	1182.1	1191.3	1199.9	1208.1	1215.7	1223.0	1229.9	1236.4	1242.7	1248.7	1254.4	1259.9	1265.2	1270.3	1275.2
725	1190.1	1199.3	1208.0	1216.1	1223.7	1231.0	1237.9	1244.5	1250.7	1256.7	1262.5	1268.0	1273.2	1278.3	1283.2
730	1198.1	1207.4	1216.0	1224.1	1231.8	1239.1	1246.0	1252.5	1258.8	1264.8	1270.5	1276.0	1281.3	1286.4	1291.3
735	1206.2	1215.4	1224.1	1232.2	1239.9	1247.1	1254.0	1260.6	1266.9	1272.8	1278.6	1284.1	1289.4	1294.5	1299.4
740	1214.3	1223.5	1232.2	1240.3	1248.0	1255.2	1262.1	1268.7	1275.0	1281.0	1286.7	1292.2	1297.5	1302.6	1307.5
745	1222.5	1231.7	1240.3	1248.4	1256.1	1263.4	1270.3	1276.8	1283.1	1289.1	1294.8	1300.3	1305.6	1310.7	1315.6
750	1230.6	1239.8	1248.5	1256.6	1264.3	1271.5	1278.4	1285.0	1291.3	1297.3	1303.0	1308.5	1313.8	1318.9	1323.8
755	1238.8	1248.0	1256.7	1264.8	1272.5	1279.7	1286.6	1293.2	1299.5	1305.5	1311.2	1316.7	1322.0	1327.1	1332.0
760	1247.0	1256.3	1264.9	1273.0	1280.7	1288.0	1294.9	1301.4	1307.7	1313.7	1319.4	1324.9	1330.2	1335.3	1340.2
765	1255.3	1264.5	1273.2	1281.3	1289.0	1296.2	1303.1	1309.7	1315.9	1321.9	1327.7	1333.2	1338.5	1343.6	1348.5
770	1263.6	1272.8	1281.4	1289.6	1297.2	1304.5	1311.4	1318.0	1324.2	1330.2	1336.0	1341.5	1346.8	1351.8	1356.8
775	1271.9	1281.1	1289.7	1297.9	1305.5	1312.8	1319.7	1326.3	1332.5	1338.5	1344.3	1349.8	1355.1	1360.2	1365.1
780	1280.2	1289.4	1298.1	1306.2	1313.9	1321.1	1328.0	1334.6	1340.9	1346.9	1352.6	1358.1	1363.4	1368.5	1373.4

TABLE 6.5 Thermo-mechanical exergy of steam from 1.01325 to 2 bar. Dashed lines indicate that water is not a gas at that temperature and pressure (it is a mixture of liquid and gas). (Continued)

e^tm (kJ/kg)

T (°C)	1.01325	1.084	1.154	1.225	1.295	1.366	1.436	1.507	1.577	1.648	1.718	1.789	1.859	1.930	2.000
							Pressure (bar)								
785	1288.6	1297.8	1306.4	1314.6	1322.2	1329.5	1336.4	1343.0	1349.2	1355.2	1361.0	1366.5	1371.8	1376.9	1381.8
790	1297.0	1306.2	1314.8	1323.0	1330.6	1337.9	1344.8	1351.4	1357.6	1363.6	1369.4	1374.9	1380.2	1385.3	1390.2
795	1305.4	1314.6	1323.3	1331.4	1339.1	1346.3	1353.2	1359.8	1366.1	1372.1	1377.8	1383.3	1388.6	1393.7	1398.6
800	1313.8	1323.1	1331.7	1339.8	1347.5	1354.8	1361.7	1368.2	1374.5	1380.5	1386.2	1391.7	1397.0	1402.1	1407.0
805	1322.3	1331.5	1340.2	1348.3	1356.0	1363.3	1370.2	1376.7	1383.0	1389.0	1394.7	1400.2	1405.5	1410.6	1415.5
810	1330.8	1340.0	1348.7	1356.8	1364.5	1371.8	1378.7	1385.2	1391.5	1397.5	1403.2	1408.7	1414.0	1419.1	1424.0
815	1339.3	1348.6	1357.2	1365.3	1373.0	1380.3	1387.2	1393.8	1400.0	1406.0	1411.8	1417.3	1422.6	1427.7	1432.6
820	1347.9	1357.1	1365.8	1373.9	1381.6	1388.9	1395.8	1402.3	1408.6	1414.6	1420.3	1425.8	1431.1	1436.2	1441.1
825	1356.5	1365.7	1374.4	1382.5	1390.2	1397.4	1404.3	1410.9	1417.2	1423.2	1428.9	1434.4	1439.7	1444.8	1449.7
830	1365.1	1374.3	1383.0	1391.1	1398.8	1406.1	1413.0	1419.5	1425.8	1431.8	1437.5	1443.0	1448.3	1453.4	1458.3
835	1373.7	1383.0	1391.6	1399.7	1407.4	1414.7	1421.6	1428.2	1434.4	1440.4	1446.2	1451.7	1457.0	1462.1	1467.0
840	1382.4	1391.6	1400.3	1408.4	1416.1	1423.4	1430.3	1436.8	1443.1	1449.1	1454.8	1460.3	1465.6	1470.7	1475.7
845	1391.1	1400.3	1409.0	1417.1	1424.8	1432.1	1439.0	1445.5	1451.8	1457.8	1463.5	1469.0	1474.3	1479.4	1484.3
850	1399.8	1409.1	1417.7	1425.8	1433.5	1440.8	1447.7	1454.3	1460.5	1466.5	1472.3	1477.8	1483.1	1488.2	1493.1
855	1408.6	1417.8	1426.4	1434.6	1442.3	1449.5	1456.4	1463.0	1469.3	1475.3	1481.0	1486.5	1491.8	1496.9	1501.8
860	1417.3	1426.6	1435.2	1443.4	1451.0	1458.3	1465.2	1471.8	1478.0	1484.0	1489.8	1495.3	1500.6	1505.7	1510.6
865	1426.1	1435.4	1444.0	1452.2	1459.8	1467.1	1474.0	1480.6	1486.8	1492.8	1498.6	1504.1	1509.4	1514.5	1519.4
870	1435.0	1444.2	1452.8	1461.0	1468.7	1475.9	1482.8	1489.4	1495.7	1501.7	1507.4	1512.9	1518.2	1523.3	1528.2
875	1443.8	1453.1	1461.7	1469.8	1477.5	1484.8	1491.7	1498.3	1504.5	1510.5	1516.3	1521.8	1527.1	1532.2	1537.1
880	1452.7	1461.9	1470.6	1478.7	1486.4	1493.7	1500.6	1507.1	1513.4	1519.4	1525.2	1530.7	1536.0	1541.1	1546.0
885	1461.6	1470.8	1479.5	1487.6	1495.3	1502.6	1509.5	1516.0	1522.3	1528.3	1534.1	1539.6	1544.9	1550.0	1554.9
890	1470.5	1479.8	1488.4	1496.6	1504.2	1511.5	1518.4	1525.0	1531.3	1537.2	1543.0	1548.5	1553.8	1558.9	1563.8
895	1479.5	1488.7	1497.4	1505.5	1513.2	1520.5	1527.4	1533.9	1540.2	1546.2	1552.0	1557.5	1562.8	1567.9	1572.8
900	1488.5	1497.7	1506.4	1514.5	1522.2	1529.4	1536.4	1542.9	1549.2	1555.2	1560.9	1566.4	1571.7	1576.8	1581.8
905	1497.5	1506.7	1515.4	1523.5	1531.2	1538.5	1545.4	1551.9	1558.2	1564.2	1569.9	1575.5	1580.8	1585.9	1590.8
910	1506.5	1515.8	1524.4	1532.5	1540.2	1547.5	1554.4	1561.0	1567.2	1573.2	1579.0	1584.5	1589.8	1594.9	1599.8
915	1515.6	1524.8	1533.5	1541.6	1549.3	1556.6	1563.5	1570.0	1576.3	1582.3	1588.0	1593.6	1598.9	1604.0	1608.9
920	1524.7	1533.9	1542.6	1550.7	1558.4	1565.6	1572.5	1579.1	1585.4	1591.4	1597.1	1602.6	1607.9	1613.0	1618.0
925	1533.8	1543.0	1551.7	1559.8	1567.5	1574.8	1581.7	1588.2	1594.5	1600.5	1606.2	1611.8	1617.1	1622.2	1627.1
930	1542.9	1552.1	1560.8	1568.9	1576.6	1583.9	1590.8	1597.4	1603.6	1609.6	1615.4	1620.9	1626.2	1631.3	1636.2
935	1552.1	1561.3	1570.0	1578.1	1585.8	1593.0	1600.0	1606.5	1612.8	1618.8	1624.5	1630.1	1635.4	1640.5	1645.4

TABLE 6.5 Thermo-mechanical exergy of steam from 1.01325 to 2 bar. Dashed lines indicate that water is not a gas at that temperature and pressure (it is a mixture of liquid and gas). (Continued)

e^tm (kJ/kg)

T (°C)	Pressure (bar)														
	1.01325	1.084	1.154	1.225	1.295	1.366	1.436	1.507	1.577	1.648	1.718	1.789	1.859	1.930	2.000
940	1561.3	1570.5	1579.1	1587.3	1595.0	1602.2	1609.1	1615.7	1622.0	1628.0	1633.7	1639.2	1644.5	1649.6	1654.6
945	1570.5	1579.7	1588.4	1596.5	1604.2	1611.4	1618.4	1624.9	1631.2	1637.2	1642.9	1648.5	1653.8	1658.9	1663.8
950	1579.7	1588.9	1597.6	1605.7	1613.4	1620.7	1627.6	1634.2	1640.4	1646.4	1652.2	1657.7	1663.0	1668.1	1673.0
955	1589.0	1598.2	1606.8	1615.0	1622.7	1629.9	1636.9	1643.4	1649.7	1655.7	1661.4	1667.0	1672.3	1677.4	1682.3
960	1598.2	1607.5	1616.1	1624.3	1632.0	1639.2	1646.1	1652.7	1659.0	1665.0	1670.7	1676.2	1681.5	1686.6	1691.6
965	1607.6	1616.8	1625.4	1633.6	1641.3	1648.5	1655.4	1662.0	1668.3	1674.3	1680.0	1685.6	1690.9	1696.0	1700.9
970	1616.9	1626.1	1634.8	1642.9	1650.6	1657.9	1664.8	1671.4	1677.6	1683.6	1689.4	1694.9	1700.2	1705.3	1710.2
975	1626.2	1635.5	1644.1	1652.3	1660.0	1667.2	1674.1	1680.7	1687.0	1693.0	1698.7	1704.2	1709.6	1714.7	1719.6
980	1635.6	1644.9	1653.5	1661.7	1669.3	1676.6	1683.5	1690.1	1696.4	1702.4	1708.1	1713.6	1718.9	1724.0	1729.0
985	1645.0	1654.3	1662.9	1671.1	1678.7	1686.0	1692.9	1699.5	1705.8	1711.8	1717.5	1723.0	1728.3	1733.4	1738.4
990	1654.5	1663.7	1672.4	1680.5	1688.2	1695.5	1702.4	1708.9	1715.2	1721.2	1727.0	1732.5	1737.8	1742.9	1747.8
995	1663.9	1673.2	1681.8	1689.9	1697.6	1704.9	1711.8	1718.4	1724.7	1730.7	1736.4	1741.9	1747.2	1752.3	1757.3
1000	1673.4	1682.6	1691.3	1699.4	1707.1	1714.4	1721.3	1727.9	1734.1	1740.1	1745.9	1751.4	1756.7	1761.8	1766.7
1020	1711.5	1720.8	1729.4	1737.6	1745.3	1752.5	1759.4	1766.0	1772.3	1778.3	1784.0	1789.6	1794.9	1800.0	1804.9
1040	1750.1	1759.3	1768.0	1776.1	1783.8	1791.1	1798.0	1804.5	1810.8	1816.8	1822.6	1828.1	1833.4	1838.5	1843.4
1060	1788.9	1798.2	1806.8	1815.0	1822.7	1829.9	1836.9	1843.4	1849.7	1855.7	1861.5	1867.0	1872.3	1877.4	1882.3
1080	1828.2	1837.4	1846.1	1854.2	1861.9	1869.2	1876.1	1882.7	1889.0	1895.0	1900.7	1906.2	1911.5	1916.6	1921.6
1100	1867.8	1877.0	1885.7	1893.8	1901.5	1908.8	1915.7	1922.3	1928.6	1934.6	1940.3	1945.8	1951.2	1956.3	1961.2
1120	1907.8	1917.0	1925.7	1933.8	1941.5	1948.8	1955.7	1962.3	1968.5	1974.5	1980.3	1985.8	1991.1	1996.2	2001.1
1140	1948.1	1957.3	1966.0	1974.1	1981.8	1989.1	1996.0	2002.6	2008.8	2014.8	2020.6	2026.1	2031.4	2036.5	2041.5
1160	1988.7	1997.9	2006.6	2014.7	2022.4	2029.7	2036.6	2043.2	2049.5	2055.5	2061.2	2066.8	2072.1	2077.2	2082.1
1180	2029.7	2038.9	2047.6	2055.7	2063.4	2070.7	2077.6	2084.2	2090.5	2096.5	2102.2	2107.7	2113.0	2118.2	2123.1
1200	2071.0	2080.2	2088.9	2097.0	2104.7	2112.0	2118.9	2125.5	2131.8	2137.8	2143.5	2149.0	2154.4	2159.5	2164.4
1220	2112.6	2121.8	2130.5	2138.6	2146.3	2153.6	2160.5	2167.1	2173.4	2179.4	2185.2	2190.7	2196.0	2201.1	2206.0
1240	2154.5	2163.8	2172.4	2180.6	2188.3	2195.6	2202.5	2209.1	2215.3	2221.3	2227.1	2232.6	2237.9	2243.0	2248.0
1260	2196.8	2206.0	2214.7	2222.8	2230.5	2237.8	2244.7	2251.3	2257.6	2263.6	2269.4	2274.9	2280.2	2285.3	2290.2
1280	2239.4	2248.6	2257.3	2265.4	2273.1	2280.4	2287.3	2293.9	2300.2	2306.2	2311.9	2317.4	2322.8	2327.9	2332.8
1300	2282.2	2291.5	2300.1	2308.3	2316.0	2323.2	2330.2	2336.7	2343.0	2349.0	2354.8	2360.3	2365.6	2370.7	2375.7
1320	2325.4	2334.6	2343.3	2351.4	2359.1	2366.4	2373.3	2379.9	2386.2	2392.2	2398.0	2403.5	2408.8	2413.9	2418.8
1340	2368.8	2378.1	2386.7	2394.9	2402.6	2409.9	2416.8	2423.4	2429.7	2435.7	2441.4	2446.9	2452.2	2457.4	2462.3
1360	2412.6	2421.8	2430.5	2438.6	2446.3	2453.6	2460.5	2467.1	2473.4	2479.4	2485.2	2490.7	2496.0	2501.1	2506.0
1380	2456.6	2465.8	2474.5	2482.7	2490.4	2497.6	2504.6	2511.1	2517.4	2523.4	2529.2	2534.7	2540.0	2545.1	2550.1

TABLE 6.5 Thermo-mechanical exergy of steam from 1.01325 to 2 bar. Dashed lines indicate that water is not a gas at that temperature and pressure (it is a mixture of liquid and gas). (Continued)

115

| | | | | | | | e^{tm} (kJ/kg) | | | | | | | |
| | | | | | | | Pressure (bar) | | | | | | | |
T (°C)	1.01325	1.084	1.154	1.225	1.295	1.366	1.436	1.507	1.577	1.648	1.718	1.789	1.859	1.930	2.000
1400	2500.9	2510.2	2518.8	2527.0	2534.7	2541.9	2548.9	2555.4	2561.7	2567.7	2573.5	2579.0	2584.3	2589.4	2594.4
1420	2545.5	2554.7	2563.4	2571.5	2579.2	2586.5	2593.4	2600.0	2606.3	2612.3	2618.1	2623.6	2628.9	2634.0	2639.0
1440	2590.3	2599.6	2608.2	2616.4	2624.1	2631.4	2638.3	2644.9	2651.2	2657.2	2662.9	2668.4	2673.8	2678.9	2683.8
1460	2635.4	2644.7	2653.3	2661.5	2669.2	2676.5	2683.4	2690.0	2696.3	2702.3	2708.0	2713.6	2718.9	2724.0	2728.9
1480	2680.8	2690.0	2698.7	2706.9	2714.6	2721.8	2728.8	2735.3	2741.6	2747.6	2753.4	2758.9	2764.2	2769.4	2774.3
1500	2726.4	2735.7	2744.3	2752.5	2760.2	2767.5	2774.4	2781.0	2787.3	2793.3	2799.0	2804.6	2809.9	2815.0	2819.9
1520	2772.3	2781.6	2790.2	2798.4	2806.1	2813.3	2820.3	2826.9	2833.1	2839.1	2844.9	2850.4	2855.7	2860.9	2865.8
1540	2818.4	2827.7	2836.3	2844.5	2852.2	2859.5	2866.4	2873.0	2879.3	2885.3	2891.0	2896.6	2901.9	2907.0	2911.9
1560	2864.8	2874.0	2882.7	2890.9	2898.6	2905.8	2912.8	2919.3	2925.6	2931.6	2937.4	2942.9	2948.2	2953.4	2958.3
1580	2911.4	2920.6	2929.3	2937.5	2945.2	2952.4	2959.4	2966.0	2972.2	2978.3	2984.0	2989.5	2994.9	3000.0	3004.9
1600	2958.2	2967.5	2976.2	2984.3	2992.0	2999.3	3006.2	3012.8	3019.1	3025.1	3030.9	3036.4	3041.7	3046.8	3051.7
1620	3005.3	3014.6	3023.2	3031.4	3039.1	3046.4	3053.3	3059.9	3066.2	3072.2	3077.9	3083.5	3088.8	3093.9	3098.8
1640	3052.6	3061.9	3070.5	3078.7	3086.4	3093.7	3100.6	3107.2	3113.5	3119.5	3125.2	3130.8	3136.1	3141.2	3146.1
1660	3100.1	3109.4	3118.1	3126.2	3133.9	3141.2	3148.1	3154.7	3161.0	3167.0	3172.8	3178.3	3183.6	3188.7	3193.7
1680	3147.9	3157.1	3165.8	3174.0	3181.6	3188.9	3195.9	3202.4	3208.7	3214.7	3220.5	3226.0	3231.3	3236.5	3241.4
1700	3195.8	3205.1	3213.8	3221.9	3229.6	3236.9	3243.8	3250.4	3256.7	3262.7	3268.5	3274.0	3279.3	3284.4	3289.4
1720	3244.0	3253.3	3261.9	3270.1	3277.8	3285.1	3292.0	3298.6	3304.9	3310.9	3316.6	3322.2	3327.5	3332.6	3337.5
1740	3292.4	3301.7	3310.3	3318.5	3326.2	3333.5	3340.4	3347.0	3353.3	3359.3	3365.0	3370.6	3375.9	3381.0	3385.9
1760	3341.0	3350.2	3358.9	3367.1	3374.8	3382.0	3389.0	3395.6	3401.8	3407.9	3413.6	3419.2	3424.5	3429.6	3434.5
1780	3389.8	3399.0	3407.7	3415.9	3423.6	3430.8	3437.8	3444.4	3450.6	3456.7	3462.4	3467.9	3473.3	3478.4	3483.3
1800	3438.8	3448.0	3456.7	3464.8	3472.5	3479.8	3486.8	3493.3	3499.6	3505.6	3511.4	3516.9	3522.3	3527.4	3532.3
1820	3488.0	3497.2	3505.9	3514.0	3521.7	3529.0	3535.9	3542.5	3548.8	3554.8	3560.6	3566.1	3571.4	3576.6	3581.5
1840	3537.3	3546.6	3555.3	3563.4	3571.1	3578.4	3585.3	3591.9	3598.2	3604.2	3610.0	3615.5	3620.8	3625.9	3630.9
1860	3586.9	3596.2	3604.8	3613.0	3620.7	3628.0	3634.9	3641.5	3647.8	3653.8	3659.6	3665.1	3670.4	3675.5	3680.5
1880	3636.7	3645.9	3654.6	3662.8	3670.4	3677.7	3684.7	3691.3	3697.5	3703.6	3709.3	3714.8	3720.2	3725.3	3730.2
1900	3686.6	3695.9	3704.5	3712.7	3720.4	3727.7	3734.6	3741.2	3747.5	3753.5	3759.3	3764.8	3770.1	3775.2	3780.2
1920	3736.7	3746.0	3754.7	3762.8	3770.5	3777.8	3784.7	3791.3	3797.6	3803.6	3809.4	3814.9	3820.2	3825.4	3830.3
1940	3787.0	3796.3	3805.0	3813.1	3820.8	3828.1	3835.0	3841.6	3847.9	3853.9	3859.7	3865.2	3870.5	3875.7	3880.6
1960	3837.5	3846.8	3855.4	3863.6	3871.3	3878.6	3885.5	3892.1	3898.4	3904.4	3910.2	3915.7	3921.0	3926.1	3931.1
1980	3888.2	3897.4	3906.1	3914.3	3922.0	3929.2	3936.2	3942.8	3949.1	3955.1	3960.8	3966.4	3971.7	3976.8	3981.7
2000	3939.0	3948.3	3956.9	3965.1	3972.8	3980.1	3987.0	3993.6	3999.9	4005.9	4011.7	4017.2	4022.5	4027.6	4032.6

TABLE 6.5 Thermo-mechanical exergy of steam from 1.01325 to 2 bar. Dashed lines indicate that water is not a gas at that temperature and pressure (it is a mixture of liquid and gas). (Continued)

116

e^{tm} (kJ/kg)

Pressure (bar)

T (°C)	2.5	3	3.5	4	5	6	7	8	9	10	11	12	13	14	15
128	618.70	—	—	—	—	—	—	—	—	—	—	—	—	—	—
129	619.27	—	—	—	—	—	—	—	—	—	—	—	—	—	—
130	619.84	—	—	—	—	—	—	—	—	—	—	—	—	—	—
131	620.41	—	—	—	—	—	—	—	—	—	—	—	—	—	—
132	620.99	—	—	—	—	—	—	—	—	—	—	—	—	—	—
133	621.57	—	—	—	—	—	—	—	—	—	—	—	—	—	—
134	622.15	645.18	—	—	—	—	—	—	—	—	—	—	—	—	—
135	622.73	645.79	—	—	—	—	—	—	—	—	—	—	—	—	—
136	623.32	646.39	—	—	—	—	—	—	—	—	—	—	—	—	—
137	623.91	647.00	—	—	—	—	—	—	—	—	—	—	—	—	—
138	624.50	647.61	—	—	—	—	—	—	—	—	—	—	—	—	—
139	625.09	648.22	667.41	—	—	—	—	—	—	—	—	—	—	—	—
140	625.69	648.84	668.04	—	—	—	—	—	—	—	—	—	—	—	—
141	626.29	649.46	668.68	—	—	—	—	—	—	—	—	—	—	—	—
142	626.89	650.07	669.32	—	—	—	—	—	—	—	—	—	—	—	—
143	627.49	650.69	669.96	—	—	—	—	—	—	—	—	—	—	—	—
144	628.10	651.32	670.60	686.99	—	—	—	—	—	—	—	—	—	—	—
145	628.71	651.94	671.25	687.65	—	—	—	—	—	—	—	—	—	—	—
146	629.32	652.57	671.89	688.32	—	—	—	—	—	—	—	—	—	—	—
147	629.93	653.20	672.54	688.99	—	—	—	—	—	—	—	—	—	—	—
148	630.55	653.83	673.19	689.66	—	—	—	—	—	—	—	—	—	—	—
149	631.17	654.47	673.84	690.33	—	—	—	—	—	—	—	—	—	—	—
150	631.79	655.10	674.49	691.00	—	—	—	—	—	—	—	—	—	—	—
151	632.42	655.74	675.15	691.67	—	—	—	—	—	—	—	—	—	—	—
152	633.04	656.38	675.80	692.34	719.21	—	—	—	—	—	—	—	—	—	—
153	633.67	657.03	676.46	693.02	719.93	—	—	—	—	—	—	—	—	—	—
154	634.31	657.67	677.12	693.70	720.65	—	—	—	—	—	—	—	—	—	—
155	634.94	658.32	677.79	694.38	721.37	—	—	—	—	—	—	—	—	—	—
156	635.58	658.97	678.45	695.06	722.09	—	—	—	—	—	—	—	—	—	—
157	636.22	659.63	679.12	695.74	722.81	—	—	—	—	—	—	—	—	—	—

Table 6.6 Thermo-mechanical exergy of steam from 2.5 to 15 bar. Dashed lines indicate that water is not a pure gas at that temperature and pressure (it is a mixture of liquid and gas).

T (°C)	2.5	3	3.5	4	5	6	7	8	9	10	11	12	13	14	15
						e^{tm} (kJ/kg) Pressure (bar)									
158	636.86	660.28	679.79	696.42	723.53	—	—	—	—	—	—	—	—	—	—
159	637.51	660.94	680.46	697.11	724.25	745.57	—	—	—	—	—	—	—	—	—
160	638.16	661.60	681.13	697.80	724.97	746.33	—	—	—	—	—	—	—	—	—
161	638.81	662.26	681.81	698.49	725.69	747.10	—	—	—	—	—	—	—	—	—
162	639.46	662.93	682.49	699.18	726.41	747.86	—	—	—	—	—	—	—	—	—
163	640.12	663.60	683.17	699.87	727.14	748.63	—	—	—	—	—	—	—	—	—
164	640.78	664.27	683.85	700.57	727.87	749.39	—	—	—	—	—	—	—	—	—
165	641.44	664.94	684.54	701.27	728.59	750.15	767.65	—	—	—	—	—	—	—	—
166	642.10	665.62	685.23	701.97	729.32	750.92	768.46	—	—	—	—	—	—	—	—
167	642.77	666.29	685.92	702.67	730.05	751.68	769.26	—	—	—	—	—	—	—	—
168	643.44	666.97	686.61	703.38	730.78	752.45	770.07	—	—	—	—	—	—	—	—
169	644.11	667.66	687.30	704.08	731.52	753.21	770.88	—	—	—	—	—	—	—	—
170	644.79	668.34	688.00	704.79	732.25	753.98	771.68	—	—	—	—	—	—	—	—
171	645.46	669.03	688.70	705.50	732.99	754.74	772.48	787.21	—	—	—	—	—	—	—
172	646.14	669.72	689.40	706.22	733.73	755.51	773.29	788.06	—	—	—	—	—	—	—
173	646.83	670.41	690.10	706.93	734.47	756.28	774.09	788.90	—	—	—	—	—	—	—
174	647.51	671.11	690.81	707.65	735.21	757.05	774.89	789.75	—	—	—	—	—	—	—
175	648.20	671.81	691.52	708.37	735.95	757.82	775.69	790.59	—	—	—	—	—	—	—
176	648.89	672.51	692.23	709.09	736.70	758.59	776.50	791.43	804.03	—	—	—	—	—	—
177	649.58	673.21	692.94	709.81	737.45	759.37	777.30	792.27	804.92	—	—	—	—	—	—
178	650.28	673.92	693.66	710.54	738.19	760.14	778.10	793.11	805.80	—	—	—	—	—	—
179	650.98	674.62	694.37	711.27	738.95	760.92	778.91	793.95	806.68	—	—	—	—	—	—
180	651.68	675.33	695.09	712.00	739.70	761.69	779.71	794.79	807.56	818.46	—	—	—	—	—
181	652.38	676.05	695.82	712.73	740.45	762.47	780.52	795.62	808.44	819.38	—	—	—	—	—
182	653.08	676.76	696.54	713.46	741.21	763.25	781.33	796.46	809.31	820.30	—	—	—	—	—
183	653.79	677.48	697.27	714.20	741.97	764.03	782.14	797.30	810.18	821.22	—	—	—	—	—
184	654.50	678.20	698.00	714.94	742.73	764.82	782.94	798.14	811.05	822.13	—	—	—	—	—
185	655.22	678.92	698.73	715.68	743.49	765.60	783.75	798.98	811.93	823.04	832.63	—	—	—	—
186	655.93	679.64	699.46	716.43	744.25	766.39	784.56	799.81	812.80	823.95	833.58	—	—	—	—
187	656.65	680.37	700.20	717.17	745.02	767.18	785.38	800.65	813.66	824.86	834.53	—	—	—	—
188	657.37	681.10	700.94	717.92	745.79	767.97	786.19	801.49	814.53	825.76	835.48	843.91	—	—	—

TABLE 6.6 Thermo-mechanical exergy of steam from 2.5 to 15 bar. Dashed lines indicate that water is not a pure gas at that temperature and pressure (it is a mixture of liquid and gas). (Continued)

T (°C)	e^{tm} (kJ/kg) Pressure (bar)														
	2.5	3	3.5	4	5	6	7	8	9	10	11	12	13	14	15
189	658.09	681.83	701.68	718.67	746.56	768.76	787.00	802.33	815.40	826.66	836.42	844.90	—	—	—
190	658.82	682.57	702.42	719.42	747.33	769.55	787.82	803.17	816.27	827.56	837.36	845.88	—	—	—
191	659.55	683.30	703.17	720.18	748.10	770.34	788.64	804.01	817.14	828.46	838.30	846.87	—	—	—
192	660.28	684.04	703.91	720.93	748.88	771.14	789.45	804.86	818.01	829.36	839.23	847.84	855.35	—	—
193	661.01	684.78	704.66	721.69	749.66	771.94	790.27	805.70	818.88	830.26	840.17	848.82	856.37	—	—
194	661.75	685.53	705.42	722.45	750.44	772.74	791.09	806.54	819.75	831.16	841.10	849.78	857.39	—	—
195	662.49	686.27	706.17	723.22	751.22	773.54	791.92	807.39	820.62	832.06	842.03	850.75	858.40	—	—
196	663.23	687.02	706.93	723.98	752.00	774.34	792.74	808.23	821.49	832.96	842.96	851.71	859.40	866.14	—
197	663.97	687.77	707.69	724.75	752.79	775.15	793.57	809.08	822.36	833.86	843.88	852.68	860.40	867.19	—
198	664.72	688.53	708.45	725.52	753.57	775.95	794.39	809.93	823.23	834.75	844.81	853.64	861.40	868.23	—
199	665.47	689.28	709.21	726.29	754.36	776.76	795.22	810.78	824.11	835.65	845.74	854.59	862.40	869.27	875.32
200	666.22	690.04	709.98	727.07	755.15	777.57	796.05	811.63	824.98	836.55	846.66	855.55	863.39	870.30	876.39
201	666.97	690.80	710.75	727.84	755.95	778.38	796.88	812.48	825.85	837.45	847.59	856.50	864.37	871.33	877.47
202	667.73	691.57	711.52	728.62	756.74	779.19	797.71	813.33	826.73	838.35	848.51	857.46	865.36	872.35	878.53
203	668.48	692.33	712.29	729.40	757.54	780.01	798.55	814.19	827.60	839.24	849.43	858.41	866.34	873.37	879.59
204	669.24	693.10	713.07	730.19	758.34	780.83	799.38	815.04	828.48	840.14	850.36	859.36	867.33	874.39	880.65
205	670.01	693.87	713.85	730.97	759.14	781.65	800.22	815.90	829.36	841.04	851.28	860.31	868.31	875.40	881.70
206	670.77	694.64	714.63	731.76	759.95	782.47	801.06	816.76	830.24	841.94	852.21	861.26	869.28	876.41	882.75
207	671.54	695.42	715.41	732.55	760.75	783.29	801.90	817.62	831.11	842.84	853.13	862.21	870.26	877.42	883.79
208	672.31	696.20	716.19	733.34	761.56	784.11	802.74	818.48	832.00	843.75	854.06	863.16	871.24	878.43	884.83
209	673.08	696.98	716.98	734.14	762.37	784.94	803.58	819.34	832.88	844.65	854.98	864.11	872.21	879.43	885.87
210	673.86	697.76	717.77	734.93	763.18	785.77	804.43	820.20	833.76	845.55	855.91	865.06	873.19	880.43	886.90
211	674.64	698.54	718.56	735.73	764.00	786.60	805.28	821.07	834.64	846.46	856.83	866.01	874.16	881.43	887.93
212	675.42	699.33	719.35	736.53	764.81	787.43	806.12	821.93	835.53	847.36	857.76	866.96	875.13	882.43	888.96
213	676.20	700.12	720.15	737.34	765.63	788.26	806.97	822.80	836.41	848.27	858.68	867.90	876.11	883.43	889.99
214	676.98	700.91	720.95	738.14	766.45	789.10	807.82	823.67	837.30	849.17	859.61	868.85	877.08	884.43	891.01
215	677.77	701.70	721.75	738.95	767.27	789.93	808.68	824.54	838.19	850.08	860.54	869.80	878.05	885.43	892.04
216	678.56	702.50	722.55	739.76	768.09	790.77	809.53	825.41	839.08	850.99	861.47	870.75	879.02	886.42	893.06
217	679.35	703.30	723.36	740.57	768.92	791.61	810.39	826.28	839.97	851.90	862.39	871.70	879.99	887.42	894.08

TABLE 6.6 Thermo-mechanical exergy of steam from 2.5 to 15 bar. Dashed lines indicate that water is not a pure gas at that temperature and pressure (it is a mixture of liquid and gas). (Continued)

e^{tm} (kJ/kg)

T (°C)	Pressure (bar)														
	2.5	3	3.5	4	5	6	7	8	9	10	11	12	13	14	15
218	680.15	704.10	724.16	741.38	769.75	792.46	811.25	827.16	840.86	852.81	863.32	872.65	880.97	888.41	895.10
219	680.94	704.90	724.97	742.20	770.58	793.30	812.11	828.04	841.75	853.72	864.25	873.60	881.94	889.41	896.12
220	681.74	705.71	725.79	743.02	771.41	794.15	812.97	828.91	842.65	854.63	865.18	874.55	882.91	890.40	897.13
221	682.54	706.52	726.60	743.84	772.25	795.00	813.83	829.79	843.54	855.54	866.12	875.50	883.88	891.39	898.15
222	683.35	707.33	727.42	744.66	773.08	795.85	814.70	830.67	844.44	856.46	867.05	876.45	884.85	892.39	899.17
223	684.16	708.14	728.24	745.49	773.92	796.70	815.57	831.56	845.34	857.37	867.98	877.40	885.82	893.38	900.18
224	684.96	708.95	729.06	746.31	774.76	797.55	816.43	832.44	846.24	858.29	868.92	878.36	886.80	894.37	901.20
225	685.77	709.77	729.88	747.14	775.60	798.41	817.30	833.32	847.14	859.21	869.85	879.31	887.77	895.37	902.21
226	686.59	710.59	730.70	747.97	776.45	799.27	818.18	834.21	848.04	860.12	870.79	880.26	888.74	896.36	903.23
227	687.40	711.41	731.53	748.81	777.29	800.13	819.05	835.10	848.95	861.04	871.72	881.22	889.71	897.35	904.24
228	688.22	712.23	732.36	749.64	778.14	800.99	819.92	835.99	849.85	861.97	872.66	882.17	890.69	898.34	905.25
229	689.04	713.06	733.19	750.48	778.99	801.85	820.80	836.88	850.76	862.89	873.60	883.13	891.66	899.34	906.26
230	689.86	713.89	734.03	751.32	779.84	802.72	821.68	837.77	851.66	863.81	874.54	884.09	892.64	900.33	907.28
231	690.69	714.72	734.86	752.17	780.70	803.58	822.56	838.67	852.57	864.74	875.48	885.04	893.61	901.32	908.29
232	691.52	715.55	735.70	753.01	781.56	804.45	823.44	839.56	853.49	865.66	876.42	886.00	894.59	902.32	909.30
233	692.35	716.39	736.54	753.86	782.41	805.33	824.33	840.46	854.40	866.59	877.36	886.96	895.56	903.31	910.32
234	693.18	717.22	737.39	754.70	783.28	806.20	825.21	841.36	855.31	867.52	878.31	887.92	896.54	904.30	911.33
235	694.01	718.06	738.23	755.56	784.14	807.07	826.10	842.26	856.23	868.45	879.25	888.88	897.52	905.30	912.34
236	694.85	718.91	739.08	756.41	785.00	807.95	826.99	843.17	857.14	869.38	880.20	889.84	898.49	906.29	913.35
237	695.69	719.75	739.93	757.26	785.87	808.83	827.88	844.07	858.06	870.31	881.14	890.80	899.47	907.29	914.37
238	696.53	720.60	740.78	758.12	786.74	809.71	828.78	844.98	858.98	871.24	882.09	891.77	900.45	908.29	915.38
239	697.37	721.45	741.64	758.98	787.61	810.59	829.67	845.88	859.90	872.18	883.04	892.73	901.43	909.28	916.39
240	698.22	722.30	742.49	759.84	788.48	811.48	830.57	846.79	860.82	873.11	883.99	893.70	902.41	910.28	917.41
241	699.07	723.15	743.35	760.71	789.36	812.36	831.47	847.70	861.75	874.05	884.94	894.66	903.39	911.28	918.42
242	699.92	724.01	744.21	761.57	790.23	813.25	832.37	848.62	862.67	874.99	885.90	895.63	904.38	912.27	919.44
243	700.77	724.86	745.07	762.44	791.11	814.14	833.27	849.53	863.60	875.93	886.85	896.60	905.36	913.27	920.45
244	701.62	725.72	745.94	763.31	791.99	815.03	834.17	850.45	864.53	876.87	887.81	897.57	906.34	914.27	921.47
245	702.48	726.58	746.81	764.18	792.88	815.93	835.08	851.36	865.46	877.82	888.76	898.54	907.33	915.27	922.48
246	703.34	727.45	747.67	765.06	793.76	816.82	835.99	852.28	866.39	878.76	889.72	899.51	908.32	916.27	923.50
247	704.20	728.32	748.55	765.93	794.65	817.72	836.89	853.21	867.32	879.71	890.68	900.48	909.30	917.27	924.51
248	705.07	729.18	749.42	766.81	795.54	818.62	837.81	854.13	868.26	880.65	891.64	901.46	910.29	918.28	925.53

TABLE 6.6 Thermo-mechanical exergy of steam from 2.5 to 15 bar. Dashed lines indicate that water is not a pure gas at that temperature and pressure (it is a mixture of liquid and gas). (Continued)

	e^{tm} (kJ/kg) Pressure (bar)														
T (°C)	2.5	3	3.5	4	5	6	7	8	9	10	11	12	13	14	15
249	705.93	730.05	750.30	767.69	796.43	819.52	838.72	855.05	869.19	881.60	892.60	902.43	911.28	919.28	926.55
250	706.80	730.93	751.17	768.58	797.32	820.43	839.63	855.98	870.13	882.55	893.56	903.41	912.27	920.28	927.57
251	707.67	731.80	752.05	769.46	798.22	821.33	840.55	856.91	871.07	883.50	894.53	904.39	913.26	921.29	928.59
252	708.54	732.68	752.94	770.35	799.11	822.24	841.47	857.84	872.01	884.46	895.49	905.36	914.25	922.29	929.61
253	709.42	733.56	753.82	771.24	800.01	823.15	842.39	858.77	872.95	885.41	896.46	906.34	915.24	923.30	930.63
254	710.30	734.44	754.71	772.13	800.91	824.06	843.31	859.70	873.90	886.37	897.43	907.32	916.24	924.31	931.65
255	711.17	735.33	755.59	773.02	801.82	824.97	844.23	860.63	874.84	887.32	898.40	908.31	917.23	925.31	932.67
256	712.06	736.21	756.49	773.92	802.72	825.89	845.16	861.57	875.79	888.28	899.37	909.29	918.23	926.32	933.69
257	712.94	737.10	757.38	774.81	803.63	826.81	846.09	862.51	876.74	889.24	900.34	910.27	919.22	927.33	934.71
258	713.83	737.99	758.27	775.71	804.54	827.73	847.02	863.45	877.69	890.20	901.31	911.26	920.22	928.34	935.74
259	714.71	738.88	759.17	776.62	805.45	828.65	847.95	864.39	878.64	891.17	902.29	912.24	921.22	929.35	936.76
260	715.60	739.78	760.07	777.52	806.36	829.57	848.88	865.33	879.60	892.13	903.26	913.23	922.22	930.37	937.79
261	716.50	740.67	760.97	778.43	807.28	830.49	849.81	866.28	880.55	893.10	904.24	914.22	923.22	931.38	938.81
262	717.39	741.57	761.87	779.33	808.19	831.42	850.75	867.22	881.51	894.07	905.22	915.21	924.22	932.39	939.84
263	718.29	742.47	762.78	780.24	809.11	832.35	851.69	868.17	882.47	895.04	906.20	916.20	925.23	933.41	940.87
264	719.19	743.38	763.69	781.16	810.03	833.28	852.63	869.12	883.43	896.01	907.18	917.20	926.23	934.43	941.90
265	720.09	744.28	764.60	782.07	810.96	834.21	853.57	870.07	884.39	896.98	908.16	918.19	927.24	935.44	942.93
266	720.99	745.19	765.51	782.99	811.88	835.14	854.51	871.03	885.35	897.95	909.15	919.18	928.24	936.46	943.96
267	721.90	746.10	766.42	783.90	812.81	836.08	855.46	871.98	886.32	898.93	910.13	920.18	929.25	937.48	944.99
268	722.80	747.01	767.34	784.82	813.74	837.02	856.41	872.94	887.28	899.90	911.12	921.18	930.26	938.50	946.02
269	723.71	747.93	768.26	785.75	814.67	837.96	857.35	873.90	888.25	900.88	912.11	922.18	931.27	939.52	947.06
270	724.63	748.84	769.18	786.67	815.60	838.90	858.31	874.86	889.22	901.86	913.10	923.18	932.28	940.55	948.09
271	725.54	749.76	770.10	787.60	816.54	839.84	859.26	875.82	890.19	902.84	914.09	924.18	933.29	941.57	949.12
272	726.46	750.68	771.02	788.53	817.47	840.79	860.21	876.78	891.17	903.83	915.09	925.18	934.31	942.60	950.16
273	727.37	751.60	771.95	789.46	818.41	841.74	861.17	877.75	892.14	904.81	916.08	926.19	935.32	943.62	951.20
274	728.30	752.53	772.88	790.39	819.35	842.69	862.13	878.71	893.12	905.80	917.08	927.20	936.34	944.65	952.24
275	729.22	753.45	773.81	791.32	820.30	843.64	863.09	879.68	894.10	906.78	918.07	928.20	937.36	945.68	953.28
276	730.14	754.38	774.74	792.26	821.24	844.59	864.05	880.65	895.08	907.77	919.07	929.21	938.38	946.71	954.32
277	731.07	755.31	775.68	793.20	822.19	845.55	865.01	881.63	896.06	908.76	920.07	930.22	939.40	947.74	955.36
278	732.00	756.24	776.61	794.14	823.14	846.50	865.98	882.60	897.04	909.76	921.07	931.23	940.42	948.77	956.40
279	732.93	757.18	777.55	795.08	824.09	847.46	866.94	883.58	898.03	910.75	922.08	932.25	941.44	949.80	957.44

TABLE 6.6 Thermo-mechanical exergy of steam from 2.5 to 15 bar. Dashed lines indicate that water is not a pure gas at that temperature and pressure (it is a mixture of liquid and gas). (Continued)

121

e^{tm} (kJ/kg)

T (°C)	2.5	3	3.5	4	5	6	7	8	9	10	11	12	13	14	15
										Pressure (bar)					
280	733.86	758.12	778.49	796.03	825.04	848.42	867.91	884.55	899.01	911.75	923.08	933.26	942.46	950.84	958.49
281	734.80	759.06	779.43	796.97	825.99	849.38	868.88	885.53	900.00	912.74	924.09	934.28	943.49	951.87	959.53
282	735.73	760.00	780.38	797.92	826.95	850.35	869.86	886.51	900.99	913.74	925.10	935.29	944.52	952.91	960.58
283	736.67	760.94	781.33	798.87	827.91	851.31	870.83	887.50	901.98	914.74	926.10	936.31	945.54	953.95	961.63
284	737.62	761.88	782.27	799.82	828.87	852.28	871.81	888.48	902.97	915.74	927.12	937.33	946.57	954.98	962.68
285	738.56	762.83	783.22	800.78	829.83	853.25	872.79	889.47	903.97	916.75	928.13	938.35	947.60	956.02	963.73
286	739.50	763.78	784.18	801.73	830.79	854.22	873.77	890.46	904.96	917.75	929.14	939.37	948.64	957.07	964.78
287	740.45	764.73	785.13	802.69	831.76	855.20	874.75	891.44	905.96	918.76	930.16	940.40	949.67	958.11	965.83
288	741.40	765.69	786.09	803.65	832.73	856.17	875.73	892.44	906.96	919.77	931.17	941.42	950.70	959.15	966.88
289	742.35	766.64	787.05	804.62	833.70	857.15	876.72	893.43	907.96	920.78	932.19	942.45	951.74	960.20	967.94
290	743.31	767.60	788.01	805.58	834.67	858.13	877.70	894.42	908.97	921.79	933.21	943.48	952.78	961.24	968.99
291	744.26	768.56	788.97	806.55	835.64	859.11	878.69	895.42	909.97	922.80	934.23	944.51	953.82	962.29	970.05
292	745.22	769.52	789.94	807.52	836.62	860.09	879.68	896.42	910.98	923.81	935.25	945.54	954.86	963.34	971.11
293	746.18	770.48	790.90	808.49	837.60	861.08	880.67	897.42	911.99	924.83	936.28	946.57	955.90	964.39	972.17
294	747.14	771.45	791.87	809.46	838.58	862.07	881.67	898.42	912.99	925.85	937.30	947.61	956.94	965.44	973.23
295	748.11	772.41	792.84	810.43	839.56	863.05	882.66	899.42	914.01	926.87	938.33	948.64	957.98	966.49	974.29
296	749.07	773.38	793.82	811.41	840.54	864.05	883.66	900.43	915.02	927.89	939.36	949.68	959.03	967.55	975.35
297	750.04	774.36	794.79	812.39	841.53	865.04	884.66	901.44	916.03	928.91	940.39	950.72	960.08	968.60	976.42
298	751.01	775.33	795.77	813.37	842.51	866.03	885.66	902.45	917.05	929.93	941.42	951.76	961.12	969.66	977.48
299	751.98	776.30	796.75	814.35	843.50	867.03	886.67	903.46	918.07	930.96	942.46	952.80	962.17	970.72	978.55
300	752.96	777.28	797.73	815.33	844.49	868.03	887.67	904.47	919.09	931.99	943.49	953.84	963.22	971.78	979.61
305	757.86	782.20	802.66	820.28	849.48	873.04	892.72	909.55	924.21	937.15	948.69	959.08	968.50	977.09	984.97
310	762.81	787.16	807.64	825.28	854.50	878.10	897.82	914.68	929.37	942.35	953.93	964.35	973.81	982.44	990.36
315	767.81	792.18	812.67	830.32	859.58	883.21	902.96	919.86	934.58	947.59	959.20	969.66	979.16	987.83	995.78
320	772.86	797.24	817.75	835.42	864.70	888.36	908.14	925.07	939.83	952.87	964.51	975.01	984.54	993.25	1001.2
325	777.95	802.35	822.87	840.56	869.87	893.56	913.37	930.33	945.12	958.19	969.87	980.40	989.96	998.70	1006.7
330	783.10	807.51	828.05	845.74	875.08	898.80	918.64	935.63	950.45	963.55	975.26	985.82	995.42	1004.2	1012.2
335	788.29	812.72	833.26	850.97	880.34	904.09	923.95	940.97	955.82	968.95	980.69	991.28	1000.9	1009.7	1017.8
340	793.53	817.97	838.53	856.25	885.65	909.42	929.31	946.36	961.23	974.39	986.16	996.78	1006.4	1015.3	1023.4
345	798.81	823.27	843.84	861.57	891.00	914.79	934.71	951.79	966.69	979.87	991.67	1002.3	1012.0	1020.9	1029.0
350	804.15	828.61	849.19	866.94	896.39	920.21	940.15	957.25	972.18	985.39	997.22	1007.9	1017.6	1026.5	1034.7

TABLE 6.6 Thermo-mechanical exergy of steam from 2.5 to 15 bar. Dashed lines indicate that water is not a pure gas at that temperature and pressure (it is a mixture of liquid and gas). (Continued)

	e^{tm} (kJ/kg)														
T (°C)	Pressure (bar)														
	2.5	3	3.5	4	5	6	7	8	9	10	11	12	13	14	15
355	809.52	834.00	854.59	872.35	901.82	925.67	945.64	962.76	977.72	990.95	1002.8	1013.5	1023.2	1032.2	1040.4
360	814.94	839.43	860.04	877.81	907.30	931.18	951.17	968.32	983.29	996.55	1008.4	1019.2	1028.9	1037.9	1046.1
365	820.41	844.91	865.53	883.31	912.83	936.72	956.74	973.91	988.91	1002.2	1014.1	1024.8	1034.6	1043.6	1051.9
370	825.92	850.43	871.06	888.85	918.39	942.31	962.35	979.54	994.56	1007.9	1019.8	1030.6	1040.4	1049.4	1057.7
375	831.48	856.00	876.64	894.44	924.00	947.94	968.00	985.21	1000.3	1013.6	1025.5	1036.3	1046.2	1055.2	1063.5
380	837.08	861.61	882.26	900.07	929.65	953.61	973.69	990.93	1006.0	1019.3	1031.3	1042.1	1052.0	1061.0	1069.4
385	842.72	867.26	887.92	905.75	935.35	959.32	979.42	996.68	1011.8	1025.1	1037.1	1048.0	1057.9	1066.9	1075.3
390	848.41	872.96	893.63	911.46	941.08	965.08	985.20	1002.5	1017.6	1031.0	1043.0	1053.9	1063.8	1072.8	1081.2
395	854.14	878.69	899.37	917.22	946.86	970.87	991.01	1008.3	1023.4	1036.9	1048.9	1059.8	1069.7	1078.8	1087.2
400	859.91	884.47	905.16	923.02	952.67	976.71	996.87	1014.2	1029.3	1042.8	1054.8	1065.7	1075.7	1084.8	1093.2
405	865.72	890.30	910.99	928.85	958.53	982.58	1002.8	1020.1	1035.3	1048.7	1060.8	1071.7	1081.7	1090.8	1099.2
410	871.58	896.16	916.87	934.74	964.43	988.50	1008.7	1026.0	1041.2	1054.7	1066.8	1077.7	1087.7	1096.9	1105.3
415	877.47	902.06	922.78	940.66	970.37	994.46	1014.7	1032.0	1047.2	1060.7	1072.8	1083.8	1093.8	1103.0	1111.4
420	883.41	908.01	928.73	946.62	976.34	1000.5	1020.7	1038.1	1053.3	1066.8	1078.9	1089.9	1099.9	1109.1	1117.6
425	889.39	913.99	934.73	952.62	982.36	1006.5	1026.7	1044.1	1059.4	1072.9	1085.0	1096.0	1106.1	1115.3	1123.8
430	895.40	920.02	940.76	958.66	988.42	1012.6	1032.8	1050.2	1065.5	1079.0	1091.2	1102.2	1112.2	1121.5	1130.0
435	901.46	926.09	946.83	964.74	994.52	1018.7	1038.9	1056.4	1071.7	1085.2	1097.4	1108.4	1118.5	1127.7	1136.3
440	907.56	932.19	952.95	970.86	1000.7	1024.8	1045.1	1062.6	1077.8	1091.4	1103.6	1114.6	1124.7	1134.0	1142.6
445	913.70	938.34	959.10	977.02	1006.8	1031.0	1051.3	1068.8	1084.1	1097.7	1109.9	1120.9	1131.0	1140.3	1148.9
450	919.87	944.52	965.29	983.22	1013.0	1037.2	1057.6	1075.0	1090.4	1104.0	1116.2	1127.2	1137.4	1146.7	1155.2
455	926.09	950.74	971.52	989.46	1019.3	1043.5	1063.8	1081.3	1096.7	1110.3	1122.5	1133.6	1143.7	1153.0	1161.6
460	932.34	957.00	977.79	995.73	1025.6	1049.8	1070.2	1087.7	1103.0	1116.6	1128.9	1140.0	1150.1	1159.4	1168.1
465	938.64	963.30	984.09	1002.0	1031.9	1056.1	1076.5	1094.0	1109.4	1123.0	1135.3	1146.4	1156.6	1165.9	1174.5
470	944.97	969.64	990.44	1008.4	1038.3	1062.5	1082.9	1100.4	1115.8	1129.5	1141.7	1152.9	1163.0	1172.4	1181.0
475	951.34	976.02	996.82	1014.8	1044.7	1068.9	1089.3	1106.9	1122.3	1135.9	1148.2	1159.4	1169.5	1178.9	1187.6
480	957.74	982.43	1003.2	1021.2	1051.1	1075.4	1095.8	1113.3	1128.7	1142.4	1154.7	1165.9	1176.1	1185.5	1194.1
485	964.19	988.88	1009.7	1027.7	1057.6	1081.9	1102.3	1119.9	1135.3	1149.0	1161.3	1172.4	1182.6	1192.0	1200.7
490	970.67	995.37	1016.2	1034.2	1064.1	1088.4	1108.8	1126.4	1141.8	1155.5	1167.9	1179.0	1189.3	1198.7	1207.4
495	977.19	1001.9	1022.7	1040.7	1070.6	1095.0	1115.4	1133.0	1148.4	1162.1	1174.5	1185.7	1195.9	1205.3	1214.0

TABLE 6.6 Thermo-mechanical exergy of steam from 2.5 to 15 bar. Dashed lines indicate that water is not a pure gas at that temperature and pressure (it is a mixture of liquid and gas). *(Continued)*

e^{tm} (kJ/kg)

T (°C)	Pressure (bar)														
	2.5	3	3.5	4	5	6	7	8	9	10	11	12	13	14	15
500	983.74	1008.5	1029.3	1047.3	1077.2	1101.5	1122.0	1139.6	1155.0	1168.8	1181.1	1192.3	1202.6	1212.0	1220.7
505	990.34	1015.0	1035.9	1053.8	1083.8	1108.2	1128.6	1146.3	1161.7	1175.5	1187.8	1199.0	1209.3	1218.7	1227.5
510	996.96	1021.7	1042.5	1060.5	1090.5	1114.8	1135.3	1152.9	1168.4	1182.2	1194.5	1205.8	1216.0	1225.5	1234.2
515	1003.6	1028.4	1049.2	1067.2	1097.2	1121.5	1142.0	1159.7	1175.1	1188.9	1201.3	1212.5	1222.8	1232.3	1241.0
520	1010.3	1035.1	1055.9	1073.9	1103.9	1128.3	1148.8	1166.4	1181.9	1195.7	1208.1	1219.3	1229.6	1239.1	1247.9
525	1017.1	1041.8	1062.7	1080.7	1110.7	1135.1	1155.6	1173.2	1188.7	1202.5	1214.9	1226.2	1236.5	1245.9	1254.7
530	1023.8	1048.6	1069.4	1087.5	1117.5	1141.9	1162.4	1180.0	1195.6	1209.4	1221.8	1233.0	1243.3	1252.8	1261.6
535	1030.6	1055.4	1076.3	1094.3	1124.3	1148.7	1169.2	1186.9	1202.4	1216.2	1228.7	1239.9	1250.3	1259.8	1268.6
540	1037.5	1062.2	1083.1	1101.1	1131.2	1155.6	1176.1	1193.8	1209.3	1223.1	1235.6	1246.9	1257.2	1266.7	1275.5
545	1044.4	1069.1	1090.0	1108.0	1138.1	1162.5	1183.0	1200.7	1216.3	1230.1	1242.5	1253.8	1264.2	1273.7	1282.5
550	1051.3	1076.0	1096.9	1115.0	1145.0	1169.4	1190.0	1207.7	1223.2	1237.1	1249.5	1260.8	1271.2	1280.7	1289.5
555	1058.2	1083.0	1103.9	1121.9	1152.0	1176.4	1197.0	1214.7	1230.2	1244.1	1256.5	1267.9	1278.2	1287.8	1296.6
560	1065.2	1090.0	1110.9	1128.9	1159.0	1183.4	1204.0	1221.7	1237.3	1251.1	1263.6	1274.9	1285.3	1294.9	1303.7
565	1072.2	1097.0	1117.9	1135.9	1166.0	1190.5	1211.0	1228.8	1244.4	1258.2	1270.7	1282.0	1292.4	1302.0	1310.8
570	1079.3	1104.0	1124.9	1143.0	1173.1	1197.5	1218.1	1235.9	1251.5	1265.3	1277.8	1289.2	1299.5	1309.1	1318.0
575	1086.3	1111.1	1132.0	1150.1	1180.2	1204.7	1225.2	1243.0	1258.6	1272.5	1285.0	1296.3	1306.7	1316.3	1325.2
580	1093.5	1118.2	1139.1	1157.2	1187.3	1211.8	1232.4	1250.2	1265.8	1279.7	1292.2	1303.5	1313.9	1323.5	1332.4
585	1100.6	1125.4	1146.3	1164.4	1194.5	1219.0	1239.6	1257.4	1273.0	1286.9	1299.4	1310.7	1321.2	1330.7	1339.6
590	1107.8	1132.6	1153.5	1171.6	1201.7	1226.2	1246.8	1264.6	1280.2	1294.1	1306.6	1318.0	1328.4	1338.0	1346.9
595	1115.0	1139.8	1160.7	1178.8	1208.9	1233.4	1254.1	1271.8	1287.5	1301.4	1313.9	1325.3	1335.7	1345.3	1354.2
600	1122.2	1147.0	1168.0	1186.1	1216.2	1240.7	1261.3	1279.1	1294.8	1308.7	1321.2	1332.6	1343.0	1352.7	1361.6
605	1129.5	1154.3	1175.3	1193.4	1223.5	1248.0	1268.7	1286.5	1302.1	1316.0	1328.6	1340.0	1350.4	1360.0	1369.0
610	1136.8	1161.7	1182.6	1200.7	1230.8	1255.3	1276.0	1293.8	1309.5	1323.4	1335.9	1347.4	1357.8	1367.4	1376.4
615	1144.2	1169.0	1189.9	1208.0	1238.2	1262.7	1283.4	1301.2	1316.8	1330.8	1343.3	1354.8	1365.2	1374.9	1383.8
620	1151.6	1176.4	1197.3	1215.4	1245.6	1270.1	1290.8	1308.6	1324.3	1338.2	1350.8	1362.2	1372.7	1382.3	1391.3
625	1159.0	1183.8	1204.7	1222.8	1253.0	1277.6	1298.2	1316.1	1331.7	1345.7	1358.3	1369.7	1380.2	1389.8	1398.8
630	1166.4	1191.2	1212.2	1230.3	1260.5	1285.0	1305.7	1323.5	1339.2	1353.2	1365.8	1377.2	1387.7	1397.3	1406.3
635	1173.9	1198.7	1219.7	1237.8	1268.0	1292.5	1313.2	1331.1	1346.7	1360.7	1373.3	1384.7	1395.2	1404.9	1413.9
640	1181.4	1206.2	1227.2	1245.3	1275.5	1300.1	1320.7	1338.6	1354.3	1368.3	1380.9	1392.3	1402.8	1412.5	1421.4
645	1188.9	1213.8	1234.7	1252.8	1283.0	1307.6	1328.3	1346.2	1361.9	1375.8	1388.4	1399.9	1410.4	1420.1	1429.1
650	1196.5	1221.3	1242.3	1260.4	1290.6	1315.2	1335.9	1353.8	1369.5	1383.5	1396.1	1407.5	1418.0	1427.7	1436.7

TABLE 6.6 Thermo-mechanical exergy of steam from 2.5 to 15 bar. Dashed lines indicate that water is not a pure gas at that temperature and pressure (it is a mixture of liquid and gas). (*Continued*)

T (°C)	\multicolumn{15}{c}{e^{tm} (kJ/kg) — Pressure (bar)}

T (°C)	2.5	3	3.5	4	5	6	7	8	9	10	11	12	13	14	15
655	1204.1	1228.9	1249.9	1268.0	1298.2	1322.8	1343.5	1361.4	1377.1	1391.1	1403.7	1415.2	1425.7	1435.4	1444.4
660	1211.7	1236.6	1257.5	1275.7	1305.9	1330.5	1351.2	1369.1	1384.8	1398.8	1411.4	1422.9	1433.4	1443.1	1452.1
665	1219.4	1244.2	1265.2	1283.3	1313.6	1338.2	1358.9	1376.8	1392.5	1406.5	1419.1	1430.6	1441.1	1450.8	1459.8
670	1227.1	1251.9	1272.9	1291.1	1321.3	1345.9	1366.6	1384.5	1400.2	1414.2	1426.9	1438.4	1448.9	1458.6	1467.6
675	1234.8	1259.7	1280.6	1298.8	1329.0	1353.6	1374.4	1392.3	1408.0	1422.0	1434.6	1446.1	1456.7	1466.4	1475.4
680	1242.6	1267.4	1288.4	1306.5	1336.8	1361.4	1382.1	1400.0	1415.8	1429.8	1442.5	1454.0	1464.5	1474.2	1483.2
685	1250.4	1275.2	1296.2	1314.3	1344.6	1369.2	1390.0	1407.9	1423.6	1437.6	1450.3	1461.8	1472.3	1482.1	1491.1
690	1258.2	1283.0	1304.0	1322.2	1352.4	1377.0	1397.8	1415.7	1431.5	1445.5	1458.2	1469.7	1480.2	1490.0	1499.0
695	1266.0	1290.9	1311.9	1330.0	1360.3	1384.9	1405.7	1423.6	1439.3	1453.4	1466.0	1477.6	1488.1	1497.9	1506.9
700	1273.9	1298.8	1319.8	1337.9	1368.2	1392.8	1413.6	1431.5	1447.3	1461.3	1474.0	1485.5	1496.1	1505.8	1514.8
705	1281.8	1306.7	1327.7	1345.8	1376.1	1400.7	1421.5	1439.4	1455.2	1469.2	1481.9	1493.4	1504.0	1513.8	1522.8
710	1289.7	1314.6	1335.6	1353.8	1384.0	1408.7	1429.5	1447.4	1463.2	1477.2	1489.9	1501.4	1512.0	1521.8	1530.8
715	1297.7	1322.6	1343.6	1361.7	1392.0	1416.7	1437.4	1455.4	1471.2	1485.2	1497.9	1509.5	1520.0	1529.8	1538.9
720	1305.7	1330.6	1351.6	1369.7	1400.0	1424.7	1445.5	1463.4	1479.2	1493.3	1506.0	1517.5	1528.1	1537.9	1546.9
725	1313.7	1338.6	1359.6	1377.8	1408.1	1432.7	1453.5	1471.5	1487.3	1501.3	1514.0	1525.6	1536.2	1545.9	1555.0
730	1321.8	1346.7	1367.7	1385.8	1416.1	1440.8	1461.6	1479.5	1495.3	1509.4	1522.1	1533.7	1544.3	1554.0	1563.1
735	1329.9	1354.8	1375.8	1393.9	1424.2	1448.9	1469.7	1487.7	1503.5	1517.5	1530.2	1541.8	1552.4	1562.2	1571.3
740	1338.0	1362.9	1383.9	1402.0	1432.3	1457.0	1477.8	1495.8	1511.6	1525.7	1538.4	1550.0	1560.6	1570.4	1579.4
745	1346.1	1371.0	1392.0	1410.2	1440.5	1465.2	1486.0	1504.0	1519.8	1533.9	1546.6	1558.1	1568.8	1578.6	1587.6
750	1354.3	1379.2	1400.2	1418.4	1448.7	1473.4	1494.2	1512.2	1528.0	1542.1	1554.8	1566.4	1577.0	1586.8	1595.9
755	1362.5	1387.4	1408.4	1426.6	1456.9	1481.6	1502.4	1520.4	1536.2	1550.3	1563.0	1574.6	1585.2	1595.0	1604.1
760	1370.7	1395.6	1416.6	1434.8	1465.1	1489.8	1510.7	1528.6	1544.5	1558.6	1571.3	1582.9	1593.5	1603.3	1612.4
765	1379.0	1403.9	1424.9	1443.1	1473.4	1498.1	1518.9	1536.9	1552.7	1566.9	1579.6	1591.2	1601.8	1611.6	1620.7
770	1387.3	1412.2	1433.2	1451.4	1481.7	1506.4	1527.2	1545.2	1561.1	1575.2	1587.9	1599.5	1610.1	1620.0	1629.1
775	1395.6	1420.5	1441.5	1459.7	1490.0	1514.7	1535.6	1553.6	1569.4	1583.5	1596.3	1607.9	1618.5	1628.3	1637.4
780	1403.9	1428.8	1449.9	1468.0	1498.4	1523.1	1543.9	1561.9	1577.8	1591.9	1604.6	1616.2	1626.9	1636.7	1645.8
785	1412.3	1437.2	1458.2	1476.4	1506.8	1531.5	1552.3	1570.3	1586.2	1600.3	1613.0	1624.7	1635.3	1645.1	1654.3
790	1420.7	1445.6	1466.6	1484.8	1515.2	1539.9	1560.7	1578.8	1594.6	1608.7	1621.5	1633.1	1643.7	1653.6	1662.7
795	1429.1	1454.0	1475.1	1493.3	1523.6	1548.3	1569.2	1587.2	1603.0	1617.2	1629.9	1641.6	1652.2	1662.0	1671.2
800	1437.6	1462.5	1483.5	1501.7	1532.1	1556.8	1577.7	1595.7	1611.5	1625.7	1638.4	1650.0	1660.7	1670.5	1679.7
805	1446.1	1471.0	1492.0	1510.2	1540.6	1565.3	1586.2	1604.2	1620.0	1634.2	1646.9	1658.6	1669.1	1679.1	1688.2

TABLE 6.6 Thermo-mechanical exergy of steam from 2.5 to 15 bar. Dashed lines indicate that water is not a pure gas at that temperature and pressure (it is a mixture of liquid and gas). (Continued)

125

T (°C)	e^tm (kJ/kg) Pressure (bar)														
	2.5	3	3.5	4	5	6	7	8	9	10	11	12	13	14	15
810	1454.6	1479.5	1500.5	1518.7	1549.1	1573.8	1594.7	1612.7	1628.6	1642.7	1655.5	1667.1	1677.8	1687.6	1696.8
815	1463.1	1488.0	1509.1	1527.3	1557.6	1582.4	1603.2	1621.3	1637.1	1651.3	1664.1	1675.7	1686.4	1696.2	1705.4
820	1471.7	1496.6	1517.6	1535.8	1566.2	1590.9	1611.8	1629.9	1645.7	1659.9	1672.7	1684.3	1695.0	1704.8	1714.0
825	1480.3	1505.2	1526.2	1544.4	1574.8	1599.5	1620.4	1638.5	1654.3	1668.5	1681.3	1692.9	1703.6	1713.5	1722.6
830	1488.9	1513.8	1534.8	1553.1	1583.4	1608.2	1629.1	1647.1	1663.0	1677.1	1689.9	1701.6	1712.2	1722.1	1731.3
835	1497.5	1522.4	1543.5	1561.7	1592.1	1616.8	1637.7	1655.8	1671.6	1685.8	1698.6	1710.2	1720.9	1730.8	1740.0
840	1506.2	1531.1	1552.2	1570.4	1600.8	1625.5	1646.4	1664.5	1680.3	1694.5	1707.3	1719.0	1729.6	1739.5	1748.7
845	1514.9	1539.8	1560.9	1579.1	1609.5	1634.2	1655.1	1673.2	1689.1	1703.2	1716.0	1727.7	1738.4	1748.3	1757.4
850	1523.6	1548.5	1569.6	1587.8	1618.2	1643.0	1663.9	1681.9	1697.8	1712.0	1724.8	1736.4	1747.1	1757.0	1766.2
855	1532.4	1557.3	1578.4	1596.6	1627.0	1651.7	1672.6	1690.7	1706.6	1720.8	1733.6	1745.2	1755.9	1765.8	1775.0
860	1541.2	1566.1	1587.1	1605.4	1635.8	1660.5	1681.4	1699.5	1715.4	1729.6	1742.4	1754.0	1764.7	1774.6	1783.8
865	1550.0	1574.9	1595.9	1614.2	1644.6	1669.3	1690.3	1708.3	1724.2	1738.4	1751.2	1762.9	1773.6	1783.5	1792.7
870	1558.8	1583.7	1604.8	1623.0	1653.4	1678.2	1699.1	1717.2	1733.1	1747.3	1760.1	1771.7	1782.5	1792.3	1801.5
875	1567.6	1592.6	1613.6	1631.9	1662.3	1687.1	1708.0	1726.0	1741.9	1756.1	1769.0	1780.6	1791.3	1801.2	1810.4
880	1576.5	1601.5	1622.5	1640.8	1671.2	1696.0	1716.9	1734.9	1750.9	1765.1	1777.9	1789.5	1800.3	1810.2	1819.4
885	1585.4	1610.4	1631.4	1649.7	1680.1	1704.9	1725.8	1743.9	1759.8	1774.0	1786.8	1798.5	1809.2	1819.1	1828.3
890	1594.4	1619.3	1640.4	1658.6	1689.0	1713.8	1734.7	1752.8	1768.7	1782.9	1795.8	1807.5	1818.2	1828.1	1837.3
895	1603.3	1628.3	1649.3	1667.6	1698.0	1722.8	1743.7	1761.8	1777.7	1791.9	1804.8	1816.4	1827.2	1837.1	1846.3
900	1612.3	1637.3	1658.3	1676.6	1707.0	1731.8	1752.7	1770.8	1786.7	1800.9	1813.8	1825.5	1836.2	1846.1	1855.3
905	1621.3	1646.3	1667.4	1685.6	1716.0	1740.8	1761.7	1779.8	1795.8	1810.0	1822.8	1834.5	1845.2	1855.2	1864.4
910	1630.4	1655.3	1676.4	1694.6	1725.1	1749.9	1770.8	1788.9	1804.8	1819.0	1831.9	1843.6	1854.3	1864.2	1873.4
915	1639.4	1664.4	1685.5	1703.7	1734.1	1758.9	1779.9	1798.0	1813.9	1828.1	1841.0	1852.7	1863.4	1873.3	1882.5
920	1648.5	1673.5	1694.6	1712.8	1743.2	1768.0	1789.0	1807.1	1823.0	1837.2	1850.1	1861.8	1872.5	1882.4	1891.7
925	1657.6	1682.6	1703.7	1721.9	1752.3	1777.2	1798.1	1816.2	1832.1	1846.4	1859.2	1870.9	1881.7	1891.6	1900.8
930	1666.8	1691.7	1712.8	1731.1	1761.5	1786.3	1807.3	1825.4	1841.3	1855.5	1868.4	1880.1	1890.8	1900.8	1910.0
935	1676.0	1700.9	1722.0	1740.2	1770.7	1795.5	1816.4	1834.5	1850.5	1864.7	1877.6	1889.3	1900.0	1910.0	1919.2
940	1685.1	1710.1	1731.2	1749.4	1779.9	1804.7	1825.6	1843.8	1859.7	1873.9	1886.8	1898.5	1909.2	1919.2	1928.4
945	1694.4	1719.3	1740.4	1758.6	1789.1	1813.9	1834.9	1853.0	1868.9	1883.2	1896.0	1907.7	1918.5	1928.4	1937.7
950	1703.6	1728.6	1749.6	1767.9	1798.3	1823.2	1844.1	1862.2	1878.2	1892.4	1905.3	1917.0	1927.8	1937.7	1946.9
955	1712.9	1737.8	1758.9	1777.1	1807.6	1832.4	1853.4	1871.5	1887.5	1901.7	1914.6	1926.3	1937.1	1947.0	1956.2
960	1722.1	1747.1	1768.2	1786.4	1816.9	1841.7	1862.7	1880.8	1896.8	1911.0	1923.9	1935.6	1946.4	1956.3	1965.6

TABLE 6.6 Thermo-mechanical exergy of steam from 2.5 to 15 bar. Dashed lines indicate that water is not a pure gas at that temperature and pressure (it is a mixture of liquid and gas). (Continued)

e^tm (kJ/kg)

T (°C)	Pressure (bar)														
	2.5	3	3.5	4	5	6	7	8	9	10	11	12	13	14	15
965	1731.5	1756.4	1777.5	1795.8	1826.2	1851.1	1872.0	1890.1	1906.1	1920.4	1933.2	1944.9	1955.7	1965.7	1974.9
970	1740.8	1765.8	1786.8	1805.1	1835.6	1860.4	1881.4	1899.5	1915.5	1929.7	1942.6	1954.3	1965.1	1975.0	1984.3
975	1750.2	1775.1	1796.2	1814.5	1844.9	1869.8	1890.7	1908.9	1924.8	1939.1	1952.0	1963.7	1974.5	1984.4	1993.7
980	1759.5	1784.5	1805.6	1823.9	1854.3	1879.2	1900.1	1918.3	1934.2	1948.5	1961.4	1973.1	1983.9	1993.8	2003.1
985	1769.0	1793.9	1815.0	1833.3	1863.7	1888.6	1909.6	1927.7	1943.7	1957.9	1970.8	1982.5	1993.3	2003.3	2012.5
990	1778.4	1803.4	1824.4	1842.7	1873.2	1898.0	1919.0	1937.1	1953.1	1967.4	1980.3	1992.0	2002.8	2012.7	2022.0
995	1787.8	1812.8	1833.9	1852.2	1882.6	1907.5	1928.5	1946.6	1962.6	1976.8	1989.7	2001.5	2012.3	2022.2	2031.5
1000	1797.3	1822.3	1843.4	1861.7	1892.1	1917.0	1938.0	1956.1	1972.1	1986.3	1999.2	2011.0	2021.8	2031.7	2041.0
1020	1835.5	1860.5	1881.6	1899.8	1930.3	1955.2	1976.2	1994.3	2010.3	2024.6	2037.5	2049.2	2060.0	2070.0	2079.3
1040	1874.0	1899.0	1920.1	1938.4	1968.9	1993.7	2014.7	2032.9	2048.9	2063.2	2076.1	2087.8	2098.6	2108.6	2117.9
1060	1912.9	1937.9	1959.0	1977.3	2007.8	2032.7	2053.7	2071.8	2087.8	2102.1	2115.1	2126.8	2137.6	2147.6	2156.9
1080	1952.2	1977.2	1998.3	2016.6	2047.1	2072.0	2093.0	2111.2	2127.2	2141.5	2154.4	2166.2	2177.0	2187.0	2196.3
1100	1991.8	2016.8	2037.9	2056.2	2086.7	2111.6	2132.6	2150.8	2166.8	2181.1	2194.1	2205.9	2216.7	2226.7	2236.0
1120	2031.8	2056.8	2077.9	2096.2	2126.7	2151.6	2172.6	2190.8	2206.8	2221.2	2234.1	2245.9	2256.7	2266.7	2276.1
1140	2072.1	2097.1	2118.2	2136.5	2167.0	2191.9	2213.0	2231.2	2247.2	2261.5	2274.5	2286.3	2297.1	2307.1	2316.5
1160	2112.7	2137.7	2158.9	2177.1	2207.7	2232.6	2253.7	2271.9	2287.9	2302.2	2315.2	2327.0	2337.8	2347.9	2357.2
1180	2153.7	2178.7	2199.8	2218.1	2248.7	2273.6	2294.7	2312.9	2328.9	2343.3	2356.2	2368.0	2378.9	2388.9	2398.3
1200	2195.0	2220.0	2241.2	2259.5	2290.0	2314.9	2336.0	2354.2	2370.3	2384.6	2397.6	2409.4	2420.3	2430.3	2439.7
1220	2236.7	2261.7	2282.8	2301.1	2331.7	2356.6	2377.7	2395.9	2412.0	2426.3	2439.3	2451.1	2462.0	2472.0	2481.4
1240	2278.6	2303.6	2324.8	2343.1	2373.6	2398.6	2419.6	2437.9	2453.9	2468.3	2481.3	2493.1	2504.0	2514.0	2523.4
1260	2320.9	2345.9	2367.0	2385.3	2415.9	2440.9	2461.9	2480.2	2496.2	2510.6	2523.6	2535.4	2546.3	2556.4	2565.7
1280	2363.4	2388.5	2409.6	2427.9	2458.5	2483.4	2504.5	2522.8	2538.9	2553.2	2566.2	2578.0	2588.9	2599.0	2608.4
1300	2406.3	2431.3	2452.5	2470.8	2501.4	2526.3	2547.4	2565.7	2581.8	2596.1	2609.1	2621.0	2631.9	2641.9	2651.3
1320	2449.5	2474.5	2495.7	2514.0	2544.6	2569.5	2590.6	2608.9	2625.0	2639.3	2652.3	2664.2	2675.1	2685.2	2694.5
1340	2492.9	2518.0	2539.1	2557.4	2588.0	2613.0	2634.1	2652.4	2668.5	2682.8	2695.8	2707.7	2718.6	2728.7	2738.1
1360	2536.7	2561.7	2582.9	2601.2	2631.8	2656.8	2677.9	2696.1	2712.2	2726.6	2739.6	2751.5	2762.4	2772.5	2781.9
1380	2580.7	2605.8	2626.9	2645.2	2675.8	2700.8	2721.9	2740.2	2756.3	2770.7	2783.7	2795.6	2806.5	2816.6	2826.0
1400	2625.0	2650.1	2671.2	2689.5	2720.2	2745.1	2766.3	2784.5	2800.6	2815.0	2828.0	2839.9	2850.8	2860.9	2870.3
1420	2669.6	2694.6	2715.8	2734.1	2764.7	2789.7	2810.9	2829.1	2845.2	2859.6	2872.7	2884.5	2895.5	2905.6	2915.0
1440	2714.5	2739.5	2760.7	2779.0	2809.6	2834.6	2855.7	2874.0	2890.1	2904.5	2917.5	2929.4	2940.3	2950.5	2959.9
1460	2759.6	2784.6	2805.8	2824.1	2854.7	2879.7	2900.9	2919.1	2935.3	2949.7	2962.7	2974.6	2985.5	2995.6	3005.0

TABLE 6.6 Thermo-mechanical exergy of steam from 2.5 to 15 bar. Dashed lines indicate that water is not a pure gas at that temperature and pressure (it is a mixture of liquid and gas). (*Continued*)

127

e^{tm} (kJ/kg)

T (°C)	Pressure (bar)														
	2.5	3	3.5	4	5	6	7	8	9	10	11	12	13	14	15
1480	2804.9	2830.0	2851.2	2869.5	2900.1	2925.1	2946.2	2964.5	2980.7	2995.1	3008.1	3020.0	3030.9	3041.0	3050.4
1500	2850.6	2875.6	2896.8	2915.1	2945.8	2970.8	2991.9	3010.2	3026.3	3040.7	3053.8	3065.7	3076.6	3086.7	3096.1
1520	2896.5	2921.5	2942.7	2961.0	2991.6	3016.7	3037.8	3056.1	3072.2	3086.6	3099.7	3111.6	3122.5	3132.6	3142.0
1540	2942.6	2967.6	2988.8	3007.2	3037.8	3062.8	3083.9	3102.2	3118.4	3132.8	3145.8	3157.7	3168.7	3178.8	3188.2
1560	2989.0	3014.0	3035.2	3053.5	3084.2	3109.2	3130.3	3148.6	3164.8	3179.2	3192.2	3204.1	3215.1	3225.2	3234.6
1580	3035.6	3060.6	3081.8	3100.1	3130.8	3155.8	3177.0	3195.3	3211.4	3225.8	3238.9	3250.8	3261.7	3271.9	3281.3
1600	3082.4	3107.5	3128.7	3147.0	3177.6	3202.7	3223.8	3242.1	3258.3	3272.7	3285.8	3297.7	3308.6	3318.7	3328.2
1620	3129.5	3154.6	3175.7	3194.1	3224.7	3249.8	3270.9	3289.2	3305.4	3319.8	3332.9	3344.8	3355.7	3365.9	3375.3
1640	3176.8	3201.9	3223.0	3241.4	3272.0	3297.1	3318.2	3336.5	3352.7	3367.1	3380.2	3392.1	3403.1	3413.2	3422.6
1660	3224.3	3249.4	3270.6	3288.9	3319.6	3344.6	3365.8	3384.1	3400.2	3414.7	3427.7	3439.7	3450.6	3460.8	3470.2
1680	3272.1	3297.1	3318.3	3336.7	3367.3	3392.4	3413.5	3431.8	3448.0	3462.4	3475.5	3487.4	3498.4	3508.5	3518.0
1700	3320.0	3345.1	3366.3	3384.6	3415.3	3440.3	3461.5	3479.8	3496.0	3510.4	3523.5	3535.4	3546.4	3556.5	3566.0
1720	3368.2	3393.3	3414.5	3432.8	3463.5	3488.5	3509.7	3528.0	3544.2	3558.6	3571.7	3583.6	3594.6	3604.7	3614.2
1740	3416.6	3441.7	3462.9	3481.2	3511.9	3536.9	3558.1	3576.4	3592.6	3607.0	3620.1	3632.0	3643.0	3653.2	3662.6
1760	3465.2	3490.3	3511.5	3529.8	3560.5	3585.5	3606.7	3625.0	3641.2	3655.7	3668.7	3680.7	3691.6	3701.8	3711.2
1780	3514.0	3539.1	3560.3	3578.6	3609.3	3634.3	3655.5	3673.8	3690.0	3704.5	3717.5	3729.5	3740.5	3750.6	3760.1
1800	3563.0	3588.1	3609.3	3627.6	3658.3	3683.3	3704.5	3722.8	3739.0	3753.5	3766.6	3778.5	3789.5	3799.6	3809.1
1820	3612.2	3637.3	3658.5	3676.8	3707.5	3732.5	3753.7	3772.1	3788.2	3802.7	3815.8	3827.7	3838.7	3848.9	3858.3
1840	3661.6	3686.6	3707.8	3726.2	3756.9	3781.9	3803.1	3821.5	3837.6	3852.1	3865.2	3877.1	3888.1	3898.3	3907.7
1860	3711.1	3736.2	3757.4	3775.8	3806.5	3831.5	3852.7	3871.0	3887.2	3901.7	3914.8	3926.7	3937.7	3947.9	3957.3
1880	3760.9	3786.0	3807.2	3825.5	3856.2	3881.3	3902.5	3920.8	3937.0	3951.5	3964.6	3976.5	3987.5	3997.7	4007.1
1900	3810.9	3835.9	3857.1	3875.5	3906.2	3931.2	3952.4	3970.8	3987.0	4001.4	4014.5	4026.5	4037.5	4047.6	4057.1
1920	3861.0	3886.1	3907.3	3925.6	3956.3	3981.4	4002.6	4020.9	4037.1	4051.6	4064.7	4076.6	4087.6	4097.8	4107.3
1940	3911.3	3936.4	3957.6	3975.9	4006.6	4031.7	4052.9	4071.2	4087.4	4101.9	4115.0	4126.9	4137.9	4148.1	4157.6
1960	3961.8	3986.9	4008.1	4026.4	4057.1	4082.2	4103.4	4121.7	4137.9	4152.4	4165.5	4177.4	4188.4	4198.6	4208.1
1980	4012.4	4037.5	4058.7	4077.1	4107.8	4132.8	4154.0	4172.4	4188.6	4203.1	4216.2	4228.1	4239.1	4249.3	4258.8
2000	4063.3	4088.3	4109.5	4127.9	4158.6	4183.7	4204.9	4223.2	4239.4	4253.9	4267.0	4279.0	4290.0	4300.1	4309.6

TABLE 6.6   Thermo-mechanical exergy of steam from 2.5 to 15 bar. Dashed lines indicate that water is not a pure gas at that temperature and pressure (it is a mixture of liquid and gas). (Continued)

e^tm (kJ/kg)

Pressure (bar)

T (°C)	16	17	18	19	20	21	22	23	24	25	26	27	28	29	30
202	883.98	—	—	—	—	—	—	—	—	—	—	—	—	—	—
203	885.09	—	—	—	—	—	—	—	—	—	—	—	—	—	—
204	886.19	—	—	—	—	—	—	—	—	—	—	—	—	—	—
205	887.28	892.21	—	—	—	—	—	—	—	—	—	—	—	—	—
206	888.37	893.34	—	—	—	—	—	—	—	—	—	—	—	—	—
207	889.45	894.47	—	—	—	—	—	—	—	—	—	—	—	—	—
208	890.53	895.59	900.07	—	—	—	—	—	—	—	—	—	—	—	—
209	891.60	896.71	901.23	—	—	—	—	—	—	—	—	—	—	—	—
210	892.67	897.82	902.38	906.41	—	—	—	—	—	—	—	—	—	—	—
211	893.74	898.92	903.53	907.60	—	—	—	—	—	—	—	—	—	—	—
212	894.80	900.02	904.67	908.79	—	—	—	—	—	—	—	—	—	—	—
213	895.86	901.11	905.80	909.97	913.64	—	—	—	—	—	—	—	—	—	—
214	896.91	902.20	906.93	911.14	914.86	—	—	—	—	—	—	—	—	—	—
215	897.97	903.29	908.05	912.30	916.07	919.39	—	—	—	—	—	—	—	—	—
216	899.02	904.37	909.17	913.45	917.27	920.64	—	—	—	—	—	—	—	—	—
217	900.06	905.45	910.28	914.60	918.46	921.88	—	—	—	—	—	—	—	—	—
218	901.11	906.52	911.39	915.75	919.65	923.11	926.15	—	—	—	—	—	—	—	—
219	902.16	907.59	912.49	916.89	920.82	924.33	927.42	—	—	—	—	—	—	—	—
220	903.20	908.67	913.59	918.02	921.99	925.54	928.68	931.44	—	—	—	—	—	—	—
221	904.24	909.73	914.69	919.15	923.16	926.74	929.93	932.73	—	—	—	—	—	—	—
222	905.28	910.80	915.78	920.28	924.32	927.94	931.17	934.02	936.51	—	—	—	—	—	—
223	906.32	911.87	916.88	921.40	925.47	929.13	932.40	935.30	937.84	—	—	—	—	—	—
224	907.36	912.93	917.97	922.52	926.63	930.32	933.62	936.56	939.15	941.40	—	—	—	—	—
225	908.40	913.99	919.05	923.63	927.77	931.50	934.84	937.82	940.46	942.76	—	—	—	—	—
226	909.43	915.05	920.14	924.75	928.91	932.67	936.05	939.07	941.75	944.10	—	—	—	—	—
227	910.47	916.11	921.22	925.86	930.05	933.84	937.26	940.31	943.03	945.43	947.51	—	—	—	—
228	911.50	917.17	922.30	926.97	931.19	935.01	938.45	941.55	944.31	946.75	948.88	—	—	—	—
229	912.54	918.22	923.38	928.07	932.32	936.17	939.65	942.77	945.57	948.05	950.23	952.11	—	—	—
230	913.57	919.28	924.46	929.17	933.45	937.33	940.84	944.00	946.83	949.35	951.57	953.51	—	—	—

Table 6.7 Thermo-mechanical exergy of steam from 16 to 30 bar. Dashed lines indicate that water is not a pure gas at that temperature and pressure (it is a mixture of liquid and gas).

T (°C)	16	17	18	19	20	21	22	23	24	25	26	27	28	29	30
										e^{tm} (kJ/kg)					
										Pressure (bar)					
231	914.60	920.33	925.54	930.28	934.58	938.48	942.02	945.21	948.08	950.64	952.91	954.88	956.58	—	—
232	915.64	921.39	926.62	931.38	935.70	939.64	943.20	946.42	949.33	951.92	954.23	956.25	958.00	959.48	—
233	916.67	922.44	927.69	932.47	936.83	940.78	944.38	947.63	950.56	953.20	955.54	957.61	959.40	960.93	—
234	917.70	923.49	928.77	933.57	937.95	941.93	945.55	948.83	951.80	954.46	956.84	958.95	960.79	962.37	963.70
235	918.73	924.54	929.84	934.66	939.06	943.07	946.72	950.03	953.03	955.72	958.14	960.28	962.17	963.80	965.18
236	919.76	925.60	930.91	935.76	940.18	944.21	947.89	951.22	954.25	956.98	959.43	961.61	963.54	965.21	966.64
237	920.79	926.65	931.98	936.85	941.30	945.35	949.05	952.42	955.47	958.23	960.71	962.93	964.89	966.61	968.08
238	921.83	927.70	933.05	937.94	942.41	946.49	950.21	953.60	956.68	959.47	961.99	964.24	966.24	968.00	969.52
239	922.86	928.75	934.12	939.03	943.52	947.62	951.37	954.79	957.89	960.71	963.26	965.54	967.58	969.37	970.94
240	923.89	929.80	935.19	940.12	944.63	948.76	952.53	955.97	959.10	961.95	964.52	966.84	968.91	970.74	972.34
241	924.92	930.85	936.26	941.21	945.74	949.89	953.68	957.14	960.30	963.18	965.78	968.13	970.23	972.10	973.74
242	925.95	931.89	937.33	942.30	946.85	951.02	954.83	958.32	961.50	964.40	967.04	969.41	971.55	973.45	975.13
243	926.98	932.94	938.39	943.38	947.95	952.14	955.98	959.49	962.70	965.63	968.29	970.69	972.86	974.79	976.51
244	928.02	933.99	939.46	944.47	949.06	953.27	957.13	960.66	963.89	966.85	969.53	971.97	974.16	976.13	977.88
245	929.05	935.04	940.53	945.55	950.16	954.39	958.27	961.83	965.09	968.06	970.77	973.24	975.46	977.46	979.24
246	930.08	936.09	941.59	946.64	951.27	955.52	959.42	963.00	966.27	969.27	972.01	974.50	976.75	978.78	980.60
247	931.11	937.14	942.66	947.72	952.37	956.64	960.56	964.16	967.46	970.48	973.24	975.76	978.04	980.10	981.94
248	932.14	938.19	943.73	948.81	953.47	957.76	961.70	965.32	968.64	971.69	974.48	977.02	979.32	981.41	983.28
249	933.18	939.24	944.79	949.89	954.57	958.88	962.84	966.48	969.83	972.89	975.70	978.27	980.60	982.72	984.62
250	934.21	940.29	945.86	950.97	955.67	960.00	963.98	967.64	971.01	974.10	976.93	979.52	981.88	984.02	985.95
251	935.25	941.34	946.92	952.06	956.77	961.12	965.12	968.80	972.18	975.29	978.15	980.76	983.14	985.31	987.27
252	936.28	942.39	947.99	953.14	957.87	962.23	966.25	969.95	973.36	976.49	979.37	982.00	984.41	986.60	988.59
253	937.32	943.44	949.06	954.22	958.97	963.35	967.39	971.11	974.53	977.69	980.58	983.24	985.67	987.89	989.90
254	938.35	944.49	950.12	955.30	960.07	964.47	968.52	972.26	975.70	978.88	981.80	984.48	986.93	989.17	991.21
255	939.39	945.54	951.19	956.38	961.17	965.58	969.65	973.41	976.87	980.07	983.01	985.71	988.19	990.45	992.51
256	940.42	946.59	952.25	957.46	962.27	966.70	970.79	974.56	978.04	981.26	984.22	986.94	989.44	991.73	993.81
257	941.46	947.64	953.32	958.55	963.36	967.81	971.92	975.71	979.21	982.44	985.42	988.17	990.69	993.00	995.11
258	942.50	948.69	954.39	959.63	964.46	968.92	973.05	976.86	980.38	983.63	986.63	989.39	991.93	994.27	996.40
259	943.54	949.74	955.45	960.71	965.56	970.04	974.18	978.00	981.54	984.81	987.83	990.61	993.18	995.53	997.69
260	944.57	950.80	956.52	961.79	966.66	971.15	975.31	979.15	982.71	985.99	989.03	991.83	994.42	996.79	998.97
261	945.61	951.85	957.59	962.87	967.75	972.26	976.43	980.30	983.87	987.17	990.23	993.05	995.66	998.05	1000.3

TABLE 6.7 Thermo-mechanical exergy of steam from 16 to 30 bar. Dashed lines indicate that water is not a pure gas at that temperature and pressure (it is a mixture of liquid and gas). (Continued)

T (°C)	e^tm (kJ/kg) Pressure (bar)														
	16	17	18	19	20	21	22	23	24	25	26	27	28	29	30
262	946.65	952.90	958.65	963.95	968.85	973.37	977.56	981.44	985.03	988.35	991.43	994.27	996.89	999.31	1001.5
263	947.69	953.96	959.72	965.04	969.95	974.49	978.69	982.58	986.19	989.53	992.62	995.48	998.13	1000.6	1002.8
264	948.74	955.01	960.79	966.12	971.04	975.60	979.82	983.73	987.35	990.71	993.82	996.70	999.36	1001.8	1004.1
265	949.78	956.07	961.86	967.20	972.14	976.71	980.94	984.87	988.51	991.88	995.01	997.91	1000.6	1003.1	1005.3
266	950.82	957.12	962.93	968.28	973.24	977.82	982.07	986.01	989.67	993.06	996.20	999.12	1001.8	1004.3	1006.6
267	951.86	958.18	964.00	969.37	974.33	978.93	983.20	987.15	990.82	994.23	997.39	1000.3	1003.0	1005.6	1007.9
268	952.91	959.24	965.07	970.45	975.43	980.04	984.32	988.29	991.98	995.40	998.58	1001.5	1004.3	1006.8	1009.1
269	953.95	960.29	966.14	971.53	976.53	981.15	985.45	989.43	993.13	996.57	999.77	1002.7	1005.5	1008.0	1010.4
270	955.00	961.35	967.21	972.62	977.62	982.26	986.57	990.57	994.29	997.74	1001.0	1003.9	1006.7	1009.3	1011.6
271	956.05	962.41	968.28	973.70	978.72	983.38	987.70	991.71	995.44	998.91	1002.1	1005.1	1007.9	1010.5	1012.9
272	957.10	963.47	969.35	974.79	979.82	984.49	988.82	992.85	996.60	1000.1	1003.3	1006.3	1009.1	1011.7	1014.2
273	958.14	964.53	970.42	975.87	980.92	985.60	989.95	993.99	997.75	1001.2	1004.5	1007.5	1010.4	1013.0	1015.4
274	959.19	965.59	971.50	976.96	982.02	986.71	991.07	995.13	998.90	1002.4	1005.7	1008.7	1011.6	1014.2	1016.7
275	960.24	966.65	972.57	978.04	983.11	987.82	992.20	996.27	1000.1	1003.6	1006.9	1009.9	1012.8	1015.4	1017.9
276	961.29	967.72	973.65	979.13	984.21	988.93	993.32	997.40	1001.2	1004.7	1008.1	1011.1	1014.0	1016.7	1019.1
277	962.35	968.78	974.72	980.22	985.31	990.04	994.44	998.54	1002.4	1005.9	1009.2	1012.3	1015.2	1017.9	1020.4
278	963.40	969.85	975.80	981.30	986.41	991.16	995.57	999.68	1003.5	1007.1	1010.4	1013.5	1016.4	1019.1	1021.6
279	964.45	970.91	976.87	982.39	987.51	992.27	996.69	1000.8	1004.7	1008.2	1011.6	1014.7	1017.6	1020.3	1022.9
280	965.51	971.98	977.95	983.48	988.61	993.38	997.82	1002.0	1005.8	1009.4	1012.8	1015.9	1018.8	1021.6	1024.1
281	966.57	973.04	979.03	984.57	989.71	994.49	998.94	1003.1	1007.0	1010.6	1013.9	1017.1	1020.0	1022.8	1025.3
282	967.62	974.11	980.11	985.66	990.81	995.61	1000.1	1004.2	1008.1	1011.7	1015.1	1018.3	1021.2	1024.0	1026.6
283	968.68	975.18	981.19	986.75	991.92	996.72	1001.2	1005.4	1009.3	1012.9	1016.3	1019.5	1022.4	1025.2	1027.8
284	969.74	976.25	982.27	987.84	993.02	997.83	1002.3	1006.5	1010.4	1014.1	1017.5	1020.7	1023.6	1026.4	1029.0
285	970.80	977.32	983.35	988.93	994.12	998.95	1003.4	1007.6	1011.6	1015.2	1018.7	1021.9	1024.9	1027.7	1030.3
286	971.86	978.39	984.43	990.02	995.22	1000.1	1004.6	1008.8	1012.7	1016.4	1019.8	1023.0	1026.1	1028.9	1031.5
287	972.92	979.46	985.51	991.12	996.33	1001.2	1005.7	1009.9	1013.9	1017.6	1021.0	1024.2	1027.3	1030.1	1032.7
288	973.99	980.54	986.59	992.21	997.43	1002.3	1006.8	1011.1	1015.0	1018.7	1022.2	1025.4	1028.5	1031.3	1034.0
289	975.05	981.61	987.68	993.31	998.54	1003.4	1008.0	1012.2	1016.2	1019.9	1023.4	1026.6	1029.7	1032.5	1035.2
290	976.11	982.68	988.76	994.40	999.64	1004.5	1009.1	1013.3	1017.3	1021.0	1024.5	1027.8	1030.9	1033.7	1036.4
291	977.18	983.76	989.85	995.50	1000.7	1005.6	1010.2	1014.5	1018.5	1022.2	1025.7	1029.0	1032.1	1034.9	1037.6
292	978.25	984.84	990.94	996.59	1001.9	1006.8	1011.3	1015.6	1019.6	1023.4	1026.9	1030.2	1033.3	1036.1	1038.9

TABLE 6.7 Thermo-mechanical exergy of steam from 16 to 30 bar. Dashed lines indicate that water is not a pure gas at that temperature and pressure (it is a mixture of liquid and gas). (Continued)

e^{tm} (kJ/kg)

T (°C)	Pressure (bar)														
	16	17	18	19	20	21	22	23	24	25	26	27	28	29	30
293	979.32	985.92	992.02	997.69	1003.0	1007.9	1012.5	1016.8	1020.8	1024.5	1028.0	1031.4	1034.5	1037.4	1040.1
294	980.39	986.99	993.11	998.79	1004.1	1009.0	1013.6	1017.9	1021.9	1025.7	1029.2	1032.5	1035.7	1038.6	1041.3
295	981.46	988.07	994.20	999.89	1005.2	1010.1	1014.7	1019.0	1023.1	1026.9	1030.4	1033.7	1036.8	1039.8	1042.5
296	982.53	989.16	995.29	1001.0	1006.3	1011.2	1015.9	1020.2	1024.2	1028.0	1031.6	1034.9	1038.0	1041.0	1043.8
297	983.60	990.24	996.38	1002.1	1007.4	1012.4	1017.0	1021.3	1025.4	1029.2	1032.7	1036.1	1039.2	1042.2	1045.0
298	984.68	991.32	997.48	1003.2	1008.5	1013.5	1018.1	1022.5	1026.5	1030.3	1033.9	1037.3	1040.4	1043.4	1046.2
299	985.75	992.40	998.57	1004.3	1009.6	1014.6	1019.2	1023.6	1027.7	1031.5	1035.1	1038.5	1041.6	1044.6	1047.4
300	986.83	993.49	999.66	1005.4	1010.7	1015.7	1020.4	1024.7	1028.8	1032.7	1036.3	1039.7	1042.8	1045.8	1048.6
305	992.23	998.93	1005.1	1010.9	1016.3	1021.3	1026.1	1030.5	1034.6	1038.5	1042.1	1045.6	1048.8	1051.9	1054.7
310	997.66	1004.4	1010.7	1016.5	1021.9	1027.0	1031.7	1036.2	1040.4	1044.3	1048.0	1051.5	1054.8	1057.9	1060.8
315	1003.1	1009.9	1016.2	1022.1	1027.5	1032.7	1037.4	1041.9	1046.2	1050.2	1053.9	1057.4	1060.8	1063.9	1066.9
320	1008.6	1015.4	1021.8	1027.7	1033.2	1038.3	1043.2	1047.7	1052.0	1056.0	1059.8	1063.4	1066.8	1070.0	1073.0
325	1014.1	1021.0	1027.4	1033.3	1038.8	1044.0	1048.9	1053.5	1057.8	1061.9	1065.7	1069.3	1072.8	1076.0	1079.1
330	1019.7	1026.6	1033.0	1039.0	1044.5	1049.8	1054.7	1059.3	1063.7	1067.8	1071.6	1075.3	1078.8	1082.0	1085.2
335	1025.3	1032.2	1038.6	1044.6	1050.3	1055.5	1060.5	1065.1	1069.5	1073.7	1077.6	1081.3	1084.8	1088.1	1091.3
340	1030.9	1037.8	1044.3	1050.4	1056.0	1061.3	1066.3	1071.0	1075.4	1079.6	1083.5	1087.3	1090.8	1094.2	1097.4
345	1036.5	1043.5	1050.0	1056.1	1061.8	1067.1	1072.1	1076.9	1081.3	1085.5	1089.5	1093.3	1096.9	1100.3	1103.5
350	1042.2	1049.3	1055.8	1061.9	1067.6	1073.0	1078.0	1082.8	1087.3	1091.5	1095.5	1099.3	1102.9	1106.4	1109.6
355	1048.0	1055.0	1061.6	1067.7	1073.4	1078.8	1083.9	1088.7	1093.2	1097.5	1101.5	1105.4	1109.0	1112.5	1115.8
360	1053.7	1060.8	1067.4	1073.5	1079.3	1084.7	1089.8	1094.6	1099.2	1103.5	1107.6	1111.4	1115.1	1118.6	1121.9
365	1059.5	1066.6	1073.2	1079.4	1085.2	1090.7	1095.8	1100.6	1105.2	1109.5	1113.6	1117.5	1121.2	1124.8	1128.1
370	1065.3	1072.4	1079.1	1085.3	1091.1	1096.6	1101.8	1106.6	1111.2	1115.6	1119.7	1123.7	1127.4	1130.9	1134.3
375	1071.2	1078.3	1085.0	1091.2	1097.1	1102.6	1107.8	1112.7	1117.3	1121.7	1125.8	1129.8	1133.6	1137.1	1140.6
380	1077.1	1084.2	1090.9	1097.2	1103.1	1108.6	1113.8	1118.7	1123.4	1127.8	1132.0	1136.0	1139.8	1143.4	1146.8
385	1083.0	1090.2	1096.9	1103.2	1109.1	1114.6	1119.9	1124.8	1129.5	1133.9	1138.2	1142.2	1146.0	1149.6	1153.1
390	1089.0	1096.2	1102.9	1109.2	1115.1	1120.7	1126.0	1130.9	1135.6	1140.1	1144.3	1148.4	1152.2	1155.9	1159.4
395	1095.0	1102.2	1109.0	1115.3	1121.2	1126.8	1132.1	1137.1	1141.8	1146.3	1150.6	1154.6	1158.5	1162.2	1165.7
400	1101.0	1108.3	1115.0	1121.4	1127.3	1133.0	1138.3	1143.3	1148.0	1152.5	1156.8	1160.9	1164.8	1168.5	1172.0
405	1107.1	1114.3	1121.1	1127.5	1133.5	1139.1	1144.5	1149.5	1154.3	1158.8	1163.1	1167.2	1171.1	1174.8	1178.4
410	1113.2	1120.5	1127.3	1133.7	1139.7	1145.3	1150.7	1155.7	1160.5	1165.1	1169.4	1173.5	1177.4	1181.2	1184.8
415	1119.3	1126.6	1133.5	1139.9	1145.9	1151.6	1156.9	1162.0	1166.8	1171.4	1175.7	1179.9	1183.8	1187.6	1191.2

TABLE 6.7 Thermo-mechanical exergy of steam from 16 to 30 bar. Dashed lines indicate that water is not a pure gas at that temperature and pressure (it is a mixture of liquid and gas). (Continued)

T (°C)	e^tm (kJ/kg) Pressure (bar)														
	16	17	18	19	20	21	22	23	24	25	26	27	28	29	30
420	1125.5	1132.8	1139.7	1146.1	1152.1	1157.8	1163.2	1168.3	1173.1	1177.7	1182.1	1186.2	1190.2	1194.0	1197.6
425	1131.7	1139.0	1145.9	1152.3	1158.4	1164.1	1169.5	1174.6	1179.5	1184.1	1188.5	1192.7	1196.6	1200.4	1204.1
430	1137.9	1145.3	1152.2	1158.6	1164.7	1170.4	1175.9	1181.0	1185.9	1190.5	1194.9	1199.1	1203.1	1206.9	1210.6
435	1144.2	1151.6	1158.5	1165.0	1171.1	1176.8	1182.2	1187.4	1192.3	1196.9	1201.3	1205.6	1209.6	1213.4	1217.1
440	1150.5	1157.9	1164.8	1171.3	1177.4	1183.2	1188.6	1193.8	1198.7	1203.4	1207.8	1212.0	1216.1	1220.0	1223.7
445	1156.8	1164.3	1171.2	1177.7	1183.8	1189.6	1195.1	1200.3	1205.2	1209.9	1214.3	1218.6	1222.6	1226.5	1230.2
450	1163.2	1170.6	1177.6	1184.1	1190.3	1196.1	1201.6	1206.8	1211.7	1216.4	1220.9	1225.1	1229.2	1233.1	1236.8
455	1169.6	1177.1	1184.0	1190.6	1196.7	1202.6	1208.1	1213.3	1218.2	1222.9	1227.4	1231.7	1235.8	1239.7	1243.5
460	1176.1	1183.5	1190.5	1197.1	1203.2	1209.1	1214.6	1219.8	1224.8	1229.5	1234.0	1238.3	1242.4	1246.4	1250.1
465	1182.5	1190.0	1197.0	1203.6	1209.8	1215.6	1221.2	1226.4	1231.4	1236.1	1240.6	1245.0	1249.1	1253.0	1256.8
470	1189.1	1196.6	1203.6	1210.2	1216.4	1222.2	1227.8	1233.0	1238.0	1242.8	1247.3	1251.6	1255.8	1259.7	1263.5
475	1195.6	1203.1	1210.1	1216.7	1223.0	1228.8	1234.4	1239.7	1244.7	1249.4	1254.0	1258.3	1262.5	1266.4	1270.3
480	1202.2	1209.7	1216.7	1223.4	1229.6	1235.5	1241.0	1246.3	1251.4	1256.1	1260.7	1265.1	1269.2	1273.2	1277.0
485	1208.8	1216.3	1223.4	1230.0	1236.3	1242.1	1247.7	1253.0	1258.1	1262.9	1267.4	1271.8	1276.0	1280.0	1283.8
490	1215.4	1223.0	1230.1	1236.7	1242.9	1248.9	1254.5	1259.8	1264.8	1269.6	1274.2	1278.6	1282.8	1286.8	1290.7
495	1222.1	1229.7	1236.8	1243.4	1249.7	1255.6	1261.2	1266.5	1271.6	1276.4	1281.0	1285.4	1289.6	1293.7	1297.5
500	1228.8	1236.4	1243.5	1250.2	1256.4	1262.4	1268.0	1273.3	1278.4	1283.2	1287.9	1292.3	1296.5	1300.5	1304.4
505	1235.6	1243.2	1250.3	1256.9	1263.2	1269.2	1274.8	1280.2	1285.2	1290.1	1294.7	1299.1	1303.4	1307.4	1311.3
510	1242.4	1249.9	1257.1	1263.8	1270.1	1276.0	1281.7	1287.0	1292.1	1297.0	1301.6	1306.0	1310.3	1314.4	1318.3
515	1249.2	1256.8	1263.9	1270.6	1276.9	1282.9	1288.5	1293.9	1299.0	1303.9	1308.5	1313.0	1317.2	1321.3	1325.2
520	1256.0	1263.6	1270.8	1277.5	1283.8	1289.8	1295.5	1300.8	1306.0	1310.8	1315.5	1320.0	1324.2	1328.3	1332.2
525	1262.9	1270.5	1277.7	1284.4	1290.7	1296.7	1302.4	1307.8	1312.9	1317.8	1322.5	1327.0	1331.2	1335.3	1339.3
530	1269.8	1277.4	1284.6	1291.3	1297.7	1303.7	1309.4	1314.8	1319.9	1324.8	1329.5	1334.0	1338.3	1342.4	1346.3
535	1276.7	1284.4	1291.6	1298.3	1304.7	1310.7	1316.4	1321.8	1326.9	1331.9	1336.5	1341.0	1345.3	1349.5	1353.4
540	1283.7	1291.4	1298.5	1305.3	1311.7	1317.7	1323.4	1328.8	1334.0	1338.9	1343.6	1348.1	1352.4	1356.6	1360.5
545	1290.7	1298.4	1305.6	1312.3	1318.7	1324.7	1330.5	1335.9	1341.1	1346.0	1350.7	1355.2	1359.6	1363.7	1367.7
550	1297.8	1305.4	1312.6	1319.4	1325.8	1331.8	1337.6	1343.0	1348.2	1353.1	1357.9	1362.4	1366.7	1370.9	1374.9
555	1304.8	1312.5	1319.7	1326.5	1332.9	1339.0	1344.7	1350.1	1355.3	1360.3	1365.0	1369.6	1373.9	1378.1	1382.1
560	1311.9	1319.6	1326.8	1333.6	1340.0	1346.1	1351.9	1357.3	1362.5	1367.5	1372.2	1376.8	1381.1	1385.3	1389.3

TABLE 6.7 Thermo-mechanical exergy of steam from 16 to 30 bar. Dashed lines indicate that water is not a pure gas at that temperature and pressure (it is a mixture of liquid and gas). (Continued)

T (°C)	16	17	18	19	20	21	22	23	24	25	26	27	28	29	30
								e^{tm} (kJ/kg) Pressure (bar)							
565	1319.1	1326.8	1334.0	1340.8	1347.2	1353.3	1359.0	1364.5	1369.7	1374.7	1379.5	1384.0	1388.4	1392.5	1396.6
570	1326.2	1333.9	1341.2	1348.0	1354.4	1360.5	1366.3	1371.7	1377.0	1381.9	1386.7	1391.3	1395.6	1399.8	1403.9
575	1333.4	1341.1	1348.4	1355.2	1361.6	1367.7	1373.5	1379.0	1384.2	1389.2	1394.0	1398.6	1402.9	1407.1	1411.2
580	1340.7	1348.4	1355.6	1362.5	1368.9	1375.0	1380.8	1386.3	1391.5	1396.5	1401.3	1405.9	1410.3	1414.5	1418.5
585	1347.9	1355.7	1362.9	1369.7	1376.2	1382.3	1388.1	1393.6	1398.9	1403.9	1408.7	1413.2	1417.6	1421.8	1425.9
590	1355.2	1363.0	1370.2	1377.1	1383.5	1389.6	1395.4	1401.0	1406.2	1411.2	1416.0	1420.6	1425.0	1429.2	1433.3
595	1362.5	1370.3	1377.6	1384.4	1390.9	1397.0	1402.8	1408.3	1413.6	1418.6	1423.4	1428.0	1432.4	1436.7	1440.7
600	1369.9	1377.6	1384.9	1391.8	1398.3	1404.4	1410.2	1415.7	1421.0	1426.0	1430.9	1435.5	1439.9	1444.1	1448.2
605	1377.3	1385.0	1392.3	1399.2	1405.7	1411.8	1417.6	1423.2	1428.5	1433.5	1438.3	1442.9	1447.4	1451.6	1455.7
610	1384.7	1392.5	1399.8	1406.6	1413.1	1419.3	1425.1	1430.6	1435.9	1441.0	1445.8	1450.4	1454.9	1459.1	1463.2
615	1392.1	1399.9	1407.2	1414.1	1420.6	1426.7	1432.6	1438.1	1443.4	1448.5	1453.3	1458.0	1462.4	1466.7	1470.8
620	1399.6	1407.4	1414.7	1421.6	1428.1	1434.3	1440.1	1445.7	1451.0	1456.0	1460.9	1465.5	1470.0	1474.2	1478.4
625	1407.1	1414.9	1422.2	1429.1	1435.6	1441.8	1447.7	1453.2	1458.5	1463.6	1468.5	1473.1	1477.6	1481.8	1486.0
630	1414.6	1422.4	1429.8	1436.7	1443.2	1449.4	1455.2	1460.8	1466.1	1471.2	1476.1	1480.7	1485.2	1489.5	1493.6
635	1422.2	1430.0	1437.4	1444.3	1450.8	1457.0	1462.8	1468.4	1473.8	1478.8	1483.7	1488.4	1492.8	1497.1	1501.3
640	1429.8	1437.6	1445.0	1451.9	1458.4	1464.6	1470.5	1476.1	1481.4	1486.5	1491.4	1496.0	1500.5	1504.8	1509.0
645	1437.4	1445.3	1452.6	1459.5	1466.1	1472.3	1478.2	1483.7	1489.1	1494.2	1499.1	1503.7	1508.2	1512.5	1516.7
650	1445.1	1452.9	1460.3	1467.2	1473.8	1480.0	1485.8	1491.5	1496.8	1501.9	1506.8	1511.5	1516.0	1520.3	1524.4
655	1452.8	1460.6	1468.0	1474.9	1481.5	1487.7	1493.6	1499.2	1504.5	1509.6	1514.5	1519.2	1523.7	1528.0	1532.2
660	1460.5	1468.3	1475.7	1482.6	1489.2	1495.4	1501.3	1506.9	1512.3	1517.4	1522.3	1527.0	1531.5	1535.8	1540.0
665	1468.2	1476.1	1483.5	1490.4	1497.0	1503.2	1509.1	1514.7	1520.1	1525.2	1530.1	1534.8	1539.3	1543.7	1547.8
670	1476.0	1483.9	1491.3	1498.2	1504.8	1511.0	1516.9	1522.6	1527.9	1533.1	1538.0	1542.7	1547.2	1551.5	1555.7
675	1483.8	1491.7	1499.1	1506.0	1512.6	1518.8	1524.8	1530.4	1535.8	1540.9	1545.8	1550.5	1555.1	1559.4	1563.6
680	1491.7	1499.5	1506.9	1513.9	1520.5	1526.7	1532.6	1538.3	1543.7	1548.8	1553.7	1558.4	1563.0	1567.3	1571.5
685	1499.5	1507.4	1514.8	1521.8	1528.4	1534.6	1540.5	1546.2	1551.6	1556.7	1561.6	1566.4	1570.9	1575.3	1579.5
690	1507.4	1515.3	1522.7	1529.7	1536.3	1542.5	1548.5	1554.1	1559.5	1564.7	1569.6	1574.3	1578.9	1583.2	1587.4
695	1515.3	1523.2	1530.6	1537.6	1544.2	1550.5	1556.4	1562.1	1567.5	1572.6	1577.6	1582.3	1586.9	1591.2	1595.4
700	1523.3	1531.2	1538.6	1545.6	1552.2	1558.4	1564.4	1570.1	1575.5	1580.6	1585.6	1590.3	1594.9	1599.2	1603.5
705	1531.3	1539.2	1546.6	1553.6	1560.2	1566.5	1572.4	1578.1	1583.5	1588.7	1593.6	1598.4	1602.9	1607.3	1611.5
710	1539.3	1547.2	1554.6	1561.6	1568.2	1574.5	1580.4	1586.1	1591.5	1596.7	1601.7	1606.4	1611.0	1615.4	1619.6
715	1547.3	1555.2	1562.6	1569.6	1576.3	1582.6	1588.5	1594.2	1599.6	1604.8	1609.8	1614.5	1619.1	1623.5	1627.7

TABLE 6.7 Thermo-mechanical exergy of steam from 16 to 30 bar. Dashed lines indicate that water is not a pure gas at that temperature and pressure (it is a mixture of liquid and gas). (*Continued*)

e^{tm} (kJ/kg)

T (°C)	Pressure (bar)														
	16	17	18	19	20	21	22	23	24	25	26	27	28	29	30
720	1555.4	1563.3	1570.7	1577.7	1584.4	1590.6	1596.6	1602.3	1607.7	1612.9	1617.9	1622.6	1627.2	1631.6	1635.8
725	1563.5	1571.4	1578.8	1585.8	1592.5	1598.8	1604.7	1610.4	1615.9	1621.1	1626.0	1630.8	1635.4	1639.8	1644.0
730	1571.6	1579.5	1587.0	1594.0	1600.6	1606.9	1612.9	1618.6	1624.0	1629.2	1634.2	1639.0	1643.6	1648.0	1652.2
735	1579.7	1587.7	1595.1	1602.1	1608.8	1615.1	1621.1	1626.8	1632.2	1637.4	1642.4	1647.2	1651.8	1656.2	1660.4
740	1587.9	1595.8	1603.3	1610.3	1617.0	1623.3	1629.3	1635.0	1640.4	1645.6	1650.6	1655.4	1660.0	1664.4	1668.7
745	1596.1	1604.1	1611.5	1618.6	1625.2	1631.5	1637.5	1643.2	1648.7	1653.9	1658.9	1663.7	1668.3	1672.7	1676.9
750	1604.4	1612.3	1619.8	1626.8	1633.5	1639.8	1645.8	1651.5	1656.9	1662.2	1667.2	1672.0	1676.6	1681.0	1685.2
755	1612.6	1620.6	1628.0	1635.1	1641.7	1648.1	1654.1	1659.8	1665.2	1670.5	1675.5	1680.3	1684.9	1689.3	1693.6
760	1620.9	1628.9	1636.3	1643.4	1650.0	1656.4	1662.4	1668.1	1673.6	1678.8	1683.8	1688.6	1693.2	1697.7	1701.9
765	1629.2	1637.2	1644.7	1651.7	1658.4	1664.7	1670.7	1676.5	1681.9	1687.2	1692.2	1697.0	1701.6	1706.0	1710.3
770	1637.6	1645.5	1653.0	1660.1	1666.7	1673.1	1679.1	1684.8	1690.3	1695.5	1700.6	1705.4	1710.0	1714.4	1718.7
775	1645.9	1653.9	1661.4	1668.5	1675.1	1681.5	1687.5	1693.2	1698.7	1704.0	1709.0	1713.8	1718.4	1722.9	1727.2
780	1654.3	1662.3	1669.8	1676.9	1683.6	1689.9	1695.9	1701.7	1707.1	1712.4	1717.4	1722.2	1726.9	1731.3	1735.6
785	1662.8	1670.7	1678.2	1685.3	1692.0	1698.3	1704.4	1710.1	1715.6	1720.9	1725.9	1730.7	1735.3	1739.8	1744.1
790	1671.2	1679.2	1686.7	1693.8	1700.5	1706.8	1712.9	1718.6	1724.1	1729.4	1734.4	1739.2	1743.8	1748.3	1752.6
795	1679.7	1687.7	1695.2	1702.3	1709.0	1715.3	1721.4	1727.1	1732.6	1737.9	1742.9	1747.7	1752.4	1756.8	1761.2
800	1688.2	1696.2	1703.7	1710.8	1717.5	1723.8	1729.9	1735.7	1741.2	1746.4	1751.5	1756.3	1760.9	1765.4	1769.7
805	1696.7	1704.7	1712.2	1719.3	1726.0	1732.4	1738.5	1744.2	1749.7	1755.0	1760.0	1764.9	1769.5	1774.0	1778.3
810	1705.3	1713.3	1720.8	1727.9	1734.6	1741.0	1747.0	1752.8	1758.3	1763.6	1768.6	1773.5	1778.1	1782.6	1786.9
815	1713.9	1721.9	1729.4	1736.5	1743.2	1749.6	1755.7	1761.4	1766.9	1772.2	1777.3	1782.1	1786.8	1791.3	1795.6
820	1722.5	1730.5	1738.0	1745.1	1751.9	1758.2	1764.3	1770.1	1775.6	1780.9	1785.9	1790.8	1795.4	1799.9	1804.2
825	1731.2	1739.2	1746.7	1753.8	1760.5	1766.9	1773.0	1778.7	1784.3	1789.5	1794.6	1799.4	1804.1	1808.6	1812.9
830	1739.8	1747.8	1755.4	1762.5	1769.2	1775.6	1781.6	1787.4	1793.0	1798.2	1803.3	1808.2	1812.8	1817.3	1821.7
835	1748.5	1756.5	1764.1	1771.2	1777.9	1784.3	1790.4	1796.1	1801.7	1807.0	1812.0	1816.9	1821.6	1826.1	1830.4
840	1757.2	1765.3	1772.8	1779.9	1786.6	1793.0	1799.1	1804.9	1810.4	1815.7	1820.8	1825.7	1830.3	1834.8	1839.2
845	1766.0	1774.0	1781.5	1788.7	1795.4	1801.8	1807.9	1813.7	1819.2	1824.5	1829.6	1834.4	1839.1	1843.6	1848.0
850	1774.8	1782.8	1790.3	1797.5	1804.2	1810.6	1816.7	1822.5	1828.0	1833.3	1838.4	1843.3	1847.9	1852.4	1856.8
855	1783.6	1791.6	1799.1	1806.3	1813.0	1819.4	1825.5	1831.3	1836.8	1842.1	1847.2	1852.1	1856.8	1861.3	1865.6
860	1792.4	1800.4	1808.0	1815.1	1821.8	1828.2	1834.3	1840.1	1845.7	1851.0	1856.1	1861.0	1865.6	1870.2	1874.5

TABLE 6.7 Thermo-mechanical exergy of steam from 16 to 30 bar. Dashed lines indicate that water is not a pure gas at that temperature and pressure (it is a mixture of liquid and gas). (Continued)

T (°C)	e^{tm} (kJ/kg) Pressure (bar)														
	16	17	18	19	20	21	22	23	24	25	26	27	28	29	30
865	1801.2	1809.3	1816.8	1824.0	1830.7	1837.1	1843.2	1849.0	1854.6	1859.9	1865.0	1869.8	1874.5	1879.1	1883.4
870	1810.1	1818.1	1825.7	1832.8	1839.6	1846.0	1852.1	1857.9	1863.5	1868.8	1873.9	1878.8	1883.5	1888.0	1892.3
875	1819.0	1827.1	1834.6	1841.8	1848.5	1854.9	1861.0	1866.8	1872.4	1877.7	1882.8	1887.7	1892.4	1896.9	1901.3
880	1827.9	1836.0	1843.6	1850.7	1857.5	1863.9	1870.0	1875.8	1881.3	1886.7	1891.8	1896.7	1901.4	1905.9	1910.3
885	1836.9	1844.9	1852.5	1859.7	1866.4	1872.8	1878.9	1884.8	1890.3	1895.6	1900.8	1905.6	1910.4	1914.9	1919.3
890	1845.9	1853.9	1861.5	1868.6	1875.4	1881.8	1887.9	1893.8	1899.3	1904.7	1909.8	1914.7	1919.4	1923.9	1928.3
895	1854.9	1862.9	1870.5	1877.7	1884.4	1890.9	1897.0	1902.8	1908.4	1913.7	1918.8	1923.7	1928.4	1933.0	1937.3
900	1863.9	1872.0	1879.5	1886.7	1893.5	1899.9	1906.0	1911.8	1917.4	1922.7	1927.9	1932.8	1937.5	1942.0	1946.4
905	1873.0	1881.0	1888.6	1895.8	1902.5	1909.0	1915.1	1920.9	1926.5	1931.8	1936.9	1941.9	1946.6	1951.1	1955.5
910	1882.0	1890.1	1897.7	1904.8	1911.6	1918.1	1924.2	1930.0	1935.6	1940.9	1946.1	1951.0	1955.7	1960.2	1964.6
915	1891.1	1899.2	1906.8	1914.0	1920.7	1927.2	1933.3	1939.1	1944.7	1950.1	1955.2	1960.1	1964.8	1969.4	1973.8
920	1900.3	1908.3	1915.9	1923.1	1929.9	1936.3	1942.5	1948.3	1953.9	1959.2	1964.3	1969.3	1974.0	1978.5	1982.9
925	1909.4	1917.5	1925.1	1932.3	1939.0	1945.5	1951.6	1957.5	1963.1	1968.4	1973.5	1978.4	1983.2	1987.7	1992.1
930	1918.6	1926.7	1934.3	1941.4	1948.2	1954.7	1960.8	1966.7	1972.3	1977.6	1982.7	1987.7	1992.4	1997.0	2001.4
935	1927.8	1935.9	1943.5	1950.7	1957.5	1963.9	1970.0	1975.9	1981.5	1986.8	1992.0	1996.9	2001.6	2006.2	2010.6
940	1937.0	1945.1	1952.7	1959.9	1966.7	1973.1	1979.3	1985.1	1990.7	1996.1	2001.2	2006.1	2010.9	2015.5	2019.9
945	1946.3	1954.4	1962.0	1969.2	1976.0	1982.4	1988.5	1994.4	2000.0	2005.4	2010.5	2015.4	2020.2	2024.7	2029.2
950	1955.6	1963.6	1971.3	1978.4	1985.2	1991.7	1997.8	2003.7	2009.3	2014.7	2019.8	2024.7	2029.5	2034.1	2038.5
955	1964.9	1972.9	1980.6	1987.7	1994.6	2001.0	2007.2	2013.0	2018.6	2024.0	2029.1	2034.1	2038.8	2043.4	2047.8
960	1974.2	1982.3	1989.9	1997.1	2003.9	2010.3	2016.5	2022.4	2028.0	2033.3	2038.5	2043.4	2048.2	2052.8	2057.2
965	1983.5	1991.6	1999.2	2006.4	2013.2	2019.7	2025.9	2031.7	2037.3	2042.7	2047.9	2052.8	2057.6	2062.1	2066.5
970	1992.9	2001.0	2008.6	2015.8	2022.6	2029.1	2035.2	2041.1	2046.7	2052.1	2057.3	2062.2	2067.0	2071.5	2076.0
975	2002.3	2010.4	2018.0	2025.2	2032.0	2038.5	2044.7	2050.5	2056.1	2061.5	2066.7	2071.6	2076.4	2081.0	2085.4
980	2011.7	2019.8	2027.4	2034.6	2041.5	2047.9	2054.1	2060.0	2065.6	2071.0	2076.1	2081.1	2085.8	2090.4	2094.8
985	2021.2	2029.3	2036.9	2044.1	2050.9	2057.4	2063.6	2069.4	2075.0	2080.4	2085.6	2090.5	2095.3	2099.9	2104.3
990	2030.6	2038.7	2046.4	2053.6	2060.4	2066.9	2073.0	2078.9	2084.5	2089.9	2095.1	2100.0	2104.8	2109.4	2113.8
995	2040.1	2048.2	2055.9	2063.1	2069.9	2076.4	2082.5	2088.4	2094.0	2099.4	2104.6	2109.6	2114.3	2118.9	2123.3
1000	2049.6	2057.7	2065.4	2072.6	2079.4	2085.9	2092.1	2098.0	2103.6	2109.0	2114.1	2119.1	2123.9	2128.5	2132.9
1020	2087.9	2096.0	2103.7	2110.9	2117.7	2124.2	2130.4	2136.3	2141.9	2147.3	2152.5	2157.5	2162.3	2166.9	2171.3
1040	2126.6	2134.7	2142.4	2149.6	2156.4	2162.9	2169.1	2175.0	2180.7	2186.1	2191.3	2196.2	2201.0	2205.6	2210.1
1060	2165.6	2173.7	2181.4	2188.6	2195.5	2202.0	2208.2	2214.1	2219.8	2225.2	2230.4	2235.3	2240.1	2244.8	2249.2

TABLE 6.7 Thermo-mechanical exergy of steam from 16 to 30 bar. Dashed lines indicate that water is not a pure gas at that temperature and pressure (it is a mixture of liquid and gas). (*Continued*)

T (°C)															

e^tm (kJ/kg)

Pressure (bar)

T (°C)	16	17	18	19	20	21	22	23	24	25	26	27	28	29	30
1080	2205.0	2213.1	2220.8	2228.0	2234.9	2241.4	2247.6	2253.5	2259.2	2264.6	2269.8	2274.8	2279.6	2284.3	2288.7
1100	2244.7	2252.8	2260.5	2267.8	2274.6	2281.2	2287.4	2293.3	2299.0	2304.4	2309.6	2314.6	2319.4	2324.1	2328.6
1120	2284.8	2292.9	2300.6	2307.9	2314.7	2321.3	2327.5	2333.4	2339.1	2344.5	2349.8	2354.8	2359.6	2364.2	2368.7
1140	2325.2	2333.3	2341.0	2348.3	2355.2	2361.7	2367.9	2373.9	2379.6	2385.0	2390.2	2395.3	2400.1	2404.7	2409.2
1160	2365.9	2374.1	2381.8	2389.1	2396.0	2402.5	2408.7	2414.7	2420.4	2425.8	2431.1	2436.1	2440.9	2445.6	2450.1
1180	2407.0	2415.2	2422.9	2430.2	2437.1	2443.6	2449.8	2455.8	2461.5	2467.0	2472.2	2477.2	2482.1	2486.7	2491.2
1200	2448.4	2456.6	2464.3	2471.6	2478.5	2485.0	2491.3	2497.2	2502.9	2508.4	2513.7	2518.7	2523.5	2528.2	2532.7
1220	2490.1	2498.3	2506.0	2513.3	2520.2	2526.8	2533.0	2539.0	2544.7	2550.2	2555.4	2560.5	2565.3	2570.0	2574.5
1240	2532.1	2540.3	2548.1	2555.4	2562.3	2568.8	2575.1	2581.1	2586.8	2592.3	2597.5	2602.6	2607.4	2612.1	2616.7
1260	2574.5	2582.7	2590.4	2597.7	2604.6	2611.2	2617.5	2623.5	2629.2	2634.7	2639.9	2645.0	2649.9	2654.5	2659.1
1280	2617.1	2625.3	2633.1	2640.4	2647.3	2653.9	2660.2	2666.1	2671.9	2677.4	2682.6	2687.7	2692.6	2697.3	2701.8
1300	2660.1	2668.3	2676.0	2683.3	2690.3	2696.9	2703.1	2709.1	2714.9	2720.3	2725.6	2730.7	2735.6	2740.3	2744.8
1320	2703.3	2711.5	2719.3	2726.6	2733.5	2740.1	2746.4	2752.4	2758.1	2763.6	2768.9	2774.0	2778.9	2783.6	2788.1
1340	2746.8	2755.1	2762.8	2770.1	2777.1	2783.7	2790.0	2796.0	2801.7	2807.2	2812.5	2817.6	2822.5	2827.2	2831.7
1360	2790.6	2798.9	2806.6	2814.0	2820.9	2827.5	2833.8	2839.8	2845.6	2851.1	2856.3	2861.4	2866.3	2871.0	2875.6
1380	2834.7	2843.0	2850.7	2858.1	2865.0	2871.6	2877.9	2883.9	2889.7	2895.2	2900.5	2905.6	2910.5	2915.2	2919.8
1400	2879.1	2887.3	2895.1	2902.4	2909.4	2916.0	2922.3	2928.3	2934.1	2939.6	2944.9	2950.0	2954.9	2959.6	2964.2
1420	2923.7	2932.0	2939.8	2947.1	2954.1	2960.7	2967.0	2973.0	2978.8	2984.3	2989.6	2994.7	2999.6	3004.3	3008.9
1440	2968.6	2976.9	2984.7	2992.0	2999.0	3005.6	3011.9	3017.9	3023.7	3029.2	3034.5	3039.6	3044.5	3049.3	3053.9
1460	3013.8	3022.1	3029.8	3037.2	3044.2	3050.8	3057.1	3063.1	3068.9	3074.4	3079.7	3084.8	3089.8	3094.5	3099.1
1480	3059.2	3067.5	3075.3	3082.6	3089.6	3096.2	3102.6	3108.6	3114.4	3119.9	3125.2	3130.3	3135.2	3140.0	3144.6
1500	3104.9	3113.2	3121.0	3128.3	3135.3	3141.9	3148.3	3154.3	3160.1	3165.6	3170.9	3176.0	3181.0	3185.7	3190.3
1520	3150.9	3159.1	3166.9	3174.3	3181.3	3187.9	3194.2	3200.3	3206.0	3211.6	3216.9	3222.0	3226.9	3231.7	3236.3
1540	3197.0	3205.3	3213.1	3220.5	3227.4	3234.1	3240.4	3246.5	3252.2	3257.8	3263.1	3268.2	3273.2	3277.9	3282.5
1560	3243.4	3251.7	3259.5	3266.9	3273.9	3280.5	3286.8	3292.9	3298.7	3304.2	3309.6	3314.7	3319.6	3324.4	3329.0
1580	3290.1	3298.4	3306.2	3313.6	3320.5	3327.2	3333.5	3339.6	3345.4	3350.9	3356.2	3361.4	3366.3	3371.1	3375.7
1600	3337.0	3345.3	3353.1	3360.5	3367.4	3374.1	3380.4	3386.5	3392.3	3397.8	3403.2	3408.3	3413.2	3418.0	3422.6
1620	3384.1	3392.4	3400.2	3407.6	3414.6	3421.2	3427.6	3433.6	3439.4	3445.0	3450.3	3455.5	3460.4	3465.2	3469.8
1640	3431.5	3439.7	3447.6	3454.9	3461.9	3468.6	3474.9	3481.0	3486.8	3492.4	3497.7	3502.8	3507.8	3512.6	3517.2

TABLE 6.7 Thermo-mechanical exergy of steam from 16 to 30 bar. Dashed lines indicate that water is not a pure gas at that temperature and pressure (it is a mixture of liquid and gas). (Continued)

	e^tm (kJ/kg)														
	Pressure (bar)														
T (°C)	16	17	18	19	20	21	22	23	24	25	26	27	28	29	30
1660	3479.0	3487.3	3495.1	3502.5	3509.5	3516.2	3522.5	3528.6	3534.4	3540.0	3545.3	3550.4	3555.4	3560.2	3564.8
1680	3526.8	3535.1	3542.9	3550.3	3557.3	3564.0	3570.3	3576.4	3582.2	3587.8	3593.1	3598.3	3603.2	3608.0	3612.6
1700	3574.8	3583.1	3590.9	3598.3	3605.3	3612.0	3618.3	3624.4	3630.2	3635.8	3641.1	3646.3	3651.2	3656.0	3660.6
1720	3623.0	3631.3	3639.1	3646.5	3653.6	3660.2	3666.6	3672.6	3678.5	3684.0	3689.4	3694.5	3699.5	3704.3	3708.9
1740	3671.5	3679.8	3687.6	3695.0	3702.0	3708.7	3715.0	3721.1	3726.9	3732.5	3737.8	3743.0	3747.9	3752.7	3757.3
1760	3720.1	3728.4	3736.2	3743.6	3750.6	3757.3	3763.7	3769.7	3775.6	3781.1	3786.5	3791.6	3796.6	3801.4	3806.0
1780	3768.9	3777.2	3785.0	3792.4	3799.5	3806.1	3812.5	3818.6	3824.4	3830.0	3835.3	3840.5	3845.5	3850.3	3854.9
1800	3817.9	3826.2	3834.1	3841.5	3848.5	3855.2	3861.5	3867.6	3873.4	3879.0	3884.4	3889.5	3894.5	3899.3	3903.9
1820	3867.2	3875.5	3883.3	3890.7	3897.7	3904.4	3910.8	3916.9	3922.7	3928.3	3933.6	3938.8	3943.8	3948.6	3953.2
1840	3916.6	3924.9	3932.7	3940.1	3947.2	3953.8	3960.2	3966.3	3972.1	3977.7	3983.1	3988.2	3993.2	3998.0	4002.6
1860	3966.2	3974.5	3982.3	3989.7	3996.8	4003.5	4009.8	4015.9	4021.8	4027.3	4032.7	4037.9	4042.8	4047.6	4052.3
1880	4016.0	4024.3	4032.1	4039.5	4046.6	4053.3	4059.6	4065.7	4071.6	4077.1	4082.5	4087.7	4092.7	4097.5	4102.1
1900	4066.0	4074.3	4082.1	4089.5	4096.6	4103.3	4109.6	4115.7	4121.6	4127.1	4132.5	4137.7	4142.7	4147.5	4152.1
1920	4116.1	4124.4	4132.3	4139.7	4146.7	4153.4	4159.8	4165.9	4171.7	4177.3	4182.7	4187.9	4192.8	4197.7	4202.3
1940	4166.4	4174.8	4182.6	4190.0	4197.1	4203.8	4210.1	4216.2	4222.1	4227.7	4233.0	4238.2	4243.2	4248.0	4252.7
1960	4217.0	4225.3	4233.1	4240.5	4247.6	4254.3	4260.7	4266.8	4272.6	4278.2	4283.6	4288.7	4293.7	4298.5	4303.2
1980	4267.6	4276.0	4283.8	4291.2	4298.3	4305.0	4311.4	4317.5	4323.3	4328.9	4334.3	4339.5	4344.4	4349.3	4353.9
2000	4318.5	4326.8	4334.7	4342.1	4349.1	4355.8	4362.2	4368.3	4374.2	4379.8	4385.1	4390.3	4395.3	4400.1	4404.8

TABLE 6.7 Thermo-mechanical exergy of steam from 16 to 30 bar. Dashed lines indicate that water is not a pure gas at that temperature and pressure (it is a mixture of liquid and gas). (*Continued*)

e^{tm} (kJ/kg)

T (°C)	35	40	45	50	60	70	80	90	100	120	140	160	180	200	220
243	981.95	—	—	—	—	—	—	—	—	—	—	—	—	—	—
244	983.54	—	—	—	—	—	—	—	—	—	—	—	—	—	—
245	985.11	—	—	—	—	—	—	—	—	—	—	—	—	—	—
246	986.67	—	—	—	—	—	—	—	—	—	—	—	—	—	—
247	988.21	—	—	—	—	—	—	—	—	—	—	—	—	—	—
248	989.73	—	—	—	—	—	—	—	—	—	—	—	—	—	—
249	991.24	—	—	—	—	—	—	—	—	—	—	—	—	—	—
250	992.74	—	—	—	—	—	—	—	—	—	—	—	—	—	—
251	994.23	996.85	—	—	—	—	—	—	—	—	—	—	—	—	—
252	995.71	998.56	—	—	—	—	—	—	—	—	—	—	—	—	—
253	997.18	1000.2	—	—	—	—	—	—	—	—	—	—	—	—	—
254	998.63	1001.9	—	—	—	—	—	—	—	—	—	—	—	—	—
255	1000.1	1003.6	—	—	—	—	—	—	—	—	—	—	—	—	—
256	1001.5	1005.2	—	—	—	—	—	—	—	—	—	—	—	—	—
257	1003.0	1006.8	—	—	—	—	—	—	—	—	—	—	—	—	—
258	1004.4	1008.4	1008.8	—	—	—	—	—	—	—	—	—	—	—	—
259	1005.8	1010.0	1010.6	—	—	—	—	—	—	—	—	—	—	—	—
260	1007.2	1011.6	1012.4	—	—	—	—	—	—	—	—	—	—	—	—
261	1008.6	1013.1	1014.2	—	—	—	—	—	—	—	—	—	—	—	—
262	1010.0	1014.7	1015.9	—	—	—	—	—	—	—	—	—	—	—	—
263	1011.4	1016.2	1017.7	—	—	—	—	—	—	—	—	—	—	—	—
264	1012.8	1017.7	1019.4	1017.8	—	—	—	—	—	—	—	—	—	—	—
265	1014.2	1019.3	1021.1	1019.8	—	—	—	—	—	—	—	—	—	—	—
266	1015.5	1020.8	1022.8	1021.7	—	—	—	—	—	—	—	—	—	—	—
267	1016.9	1022.3	1024.5	1023.6	—	—	—	—	—	—	—	—	—	—	—
268	1018.3	1023.8	1026.1	1025.5	—	—	—	—	—	—	—	—	—	—	—
269	1019.6	1025.3	1027.8	1027.4	—	—	—	—	—	—	—	—	—	—	—
270	1021.0	1026.7	1029.4	1029.2	—	—	—	—	—	—	—	—	—	—	—
271	1022.4	1028.2	1031.0	1031.0	—	—	—	—	—	—	—	—	—	—	—
272	1023.7	1029.7	1032.6	1032.8	—	—	—	—	—	—	—	—	—	—	—

Pressure (bar)

TABLE 6.8 Thermo-mechanical exergy of steam from 35 to 220 bar. Dashed lines indicate that water is not a pure gas at that temperature and pressure (It is a mixture of liquid and gas).

T (°C)	35	40	45	50	60	70	80	90	100	120	140	160	180	200	220
273	1025.1	1031.1	1034.2	1034.6	—	—	—	—	—	—	—	—	—	—	—
274	1026.4	1032.6	1035.8	1036.4	—	—	—	—	—	—	—	—	—	—	—
275	1027.7	1034.1	1037.4	1038.1	—	—	—	—	—	—	—	—	—	—	—
276	1029.1	1035.5	1039.0	1039.8	1033.9	—	—	—	—	—	—	—	—	—	—
277	1030.4	1036.9	1040.5	1041.6	1036.1	—	—	—	—	—	—	—	—	—	—
278	1031.7	1038.4	1042.1	1043.3	1038.3	—	—	—	—	—	—	—	—	—	—
279	1033.1	1039.8	1043.6	1045.0	1040.4	—	—	—	—	—	—	—	—	—	—
280	1034.4	1041.2	1045.2	1046.6	1042.5	—	—	—	—	—	—	—	—	—	—
281	1035.7	1042.6	1046.7	1048.3	1044.6	—	—	—	—	—	—	—	—	—	—
282	1037.0	1044.0	1048.2	1050.0	1046.6	—	—	—	—	—	—	—	—	—	—
283	1038.3	1045.4	1049.8	1051.6	1048.6	—	—	—	—	—	—	—	—	—	—
284	1039.6	1046.9	1051.3	1053.3	1050.6	—	—	—	—	—	—	—	—	—	—
285	1041.0	1048.3	1052.8	1054.9	1052.6	—	—	—	—	—	—	—	—	—	—
286	1042.3	1049.7	1054.3	1056.5	1054.6	1043.9	—	—	—	—	—	—	—	—	—
287	1043.6	1051.0	1055.8	1058.1	1056.5	1046.4	—	—	—	—	—	—	—	—	—
288	1044.9	1052.4	1057.3	1059.7	1058.4	1048.8	—	—	—	—	—	—	—	—	—
289	1046.2	1053.8	1058.7	1061.3	1060.3	1051.2	—	—	—	—	—	—	—	—	—
290	1047.5	1055.2	1060.2	1062.9	1062.2	1053.6	—	—	—	—	—	—	—	—	—
291	1048.8	1056.6	1061.7	1064.5	1064.1	1055.9	—	—	—	—	—	—	—	—	—
292	1050.1	1058.0	1063.2	1066.1	1065.9	1058.1	—	—	—	—	—	—	—	—	—
293	1051.4	1059.3	1064.6	1067.6	1067.7	1060.4	—	—	—	—	—	—	—	—	—
294	1052.6	1060.7	1066.1	1069.2	1069.6	1062.6	—	—	—	—	—	—	—	—	—
295	1053.9	1062.1	1067.5	1070.8	1071.4	1064.8	—	—	—	—	—	—	—	—	—
296	1055.2	1063.4	1069.0	1072.3	1073.2	1066.9	1053.1	—	—	—	—	—	—	—	—
297	1056.5	1064.8	1070.4	1073.8	1075.0	1069.1	1055.8	—	—	—	—	—	—	—	—
298	1057.8	1066.1	1071.9	1075.4	1076.7	1071.2	1058.5	—	—	—	—	—	—	—	—
299	1059.1	1067.5	1073.3	1076.9	1078.5	1073.3	1061.1	—	—	—	—	—	—	—	—
300	1060.4	1068.9	1074.7	1078.4	1080.2	1075.3	1063.7	—	—	—	—	—	—	—	—
305	1066.8	1075.6	1081.9	1086.0	1088.8	1085.4	1075.9	1059.3	—	—	—	—	—	—	—
310	1073.1	1082.3	1088.9	1093.4	1097.2	1095.1	1087.3	1073.6	—	—	—	—	—	—	—
315	1079.5	1088.9	1095.9	1100.8	1105.4	1104.4	1098.2	1086.7	1068.9	—	—	—	—	—	—

e^{tm} (kJ/kg) — Pressure (bar)

TABLE 6.8 Thermo-mechanical exergy of steam from 35 to 220 bar. Dashed lines indicate that water is not a pure gas at that temperature and pressure (It is a mixture of liquid and gas). (Continued)

e^{tm} (kJ/kg)

T (°C)	Pressure (bar)														
	35	40	45	50	60	70	80	90	100	120	140	160	180	200	220
320	1085.8	1095.5	1102.8	1108.0	1113.5	1113.4	1108.5	1098.9	1083.9	—	—	—	—	—	—
325	1092.1	1102.1	1109.6	1115.2	1121.4	1122.3	1118.5	1110.4	1097.7	1053.4	—	—	—	—	—
330	1098.4	1108.7	1116.5	1122.3	1129.2	1130.9	1128.2	1121.4	1110.5	1073.5	—	—	—	—	—
335	1104.8	1115.2	1123.3	1129.4	1136.9	1139.3	1137.6	1132.0	1122.6	1091.0	—	—	—	—	—
340	1111.1	1121.7	1130.0	1136.4	1144.5	1147.6	1146.7	1142.2	1134.1	1106.8	1058.2	—	—	—	—
345	1117.4	1128.3	1136.8	1143.4	1152.0	1155.8	1155.6	1152.0	1145.2	1121.4	1080.6	—	—	—	—
350	1123.7	1134.8	1143.5	1150.3	1159.5	1163.9	1164.4	1161.6	1155.8	1135.0	1100.0	1040.0	—	—	—
355	1130.0	1141.3	1150.2	1157.3	1166.9	1171.8	1173.0	1171.0	1166.1	1147.9	1117.4	1068.5	—	—	—
360	1136.4	1147.8	1156.9	1164.2	1174.2	1179.7	1181.5	1180.2	1176.1	1160.2	1133.3	1091.9	1021.8	—	—
365	1142.7	1154.3	1163.6	1171.1	1181.6	1187.5	1189.8	1189.2	1185.9	1171.9	1148.1	1112.3	1057.0	—	—
370	1149.1	1160.8	1170.3	1177.9	1188.8	1195.2	1198.1	1198.0	1195.4	1183.2	1162.0	1130.5	1084.5	1007.6	—
375	1155.4	1167.4	1177.0	1184.8	1196.1	1202.9	1206.2	1206.7	1204.7	1194.2	1175.3	1147.3	1107.7	1048.6	921.35
380	1161.8	1173.9	1183.7	1191.6	1203.3	1210.5	1214.3	1215.3	1213.9	1204.8	1187.9	1162.9	1128.2	1079.4	1001.7
385	1168.2	1180.4	1190.4	1198.5	1210.5	1218.1	1222.3	1223.8	1222.9	1215.2	1200.0	1177.6	1146.8	1105.0	1045.2
390	1174.7	1187.0	1197.1	1205.3	1217.7	1225.6	1230.3	1232.2	1231.8	1225.3	1211.7	1191.5	1163.9	1127.4	1077.9
395	1181.1	1193.6	1203.8	1212.2	1224.8	1233.1	1238.1	1240.5	1240.6	1235.2	1223.1	1204.8	1179.9	1147.5	1105.1
400	1187.5	1200.2	1210.5	1219.1	1232.0	1240.6	1246.0	1248.7	1249.3	1244.9	1234.1	1217.5	1195.0	1165.9	1128.8
405	1194.0	1206.8	1217.2	1225.9	1239.1	1248.0	1253.8	1256.9	1257.9	1254.4	1244.9	1229.8	1209.3	1183.1	1150.1
410	1200.5	1213.4	1224.0	1232.8	1246.2	1255.5	1261.5	1265.0	1266.3	1263.8	1255.4	1241.7	1223.0	1199.2	1169.5
415	1207.0	1220.0	1230.7	1239.6	1253.4	1262.9	1269.2	1273.0	1274.8	1273.1	1265.7	1253.3	1236.2	1214.4	1187.5
420	1213.6	1226.7	1237.5	1246.5	1260.5	1270.3	1276.9	1281.1	1283.1	1282.2	1275.8	1264.5	1248.9	1228.9	1204.4
425	1220.2	1233.3	1244.3	1253.4	1267.6	1277.7	1284.6	1289.0	1291.4	1291.3	1285.7	1275.5	1261.2	1242.8	1220.4
430	1226.7	1240.0	1251.1	1260.3	1274.8	1285.0	1292.2	1297.0	1299.7	1300.2	1295.4	1286.3	1273.1	1256.2	1235.6
435	1233.4	1246.7	1257.9	1267.2	1281.9	1292.4	1299.9	1304.9	1307.9	1309.1	1305.1	1296.8	1284.8	1269.2	1250.1
440	1240.0	1253.5	1264.7	1274.2	1289.0	1299.8	1307.5	1312.7	1316.0	1317.8	1314.6	1307.2	1296.1	1281.7	1264.1
445	1246.7	1260.2	1271.5	1281.1	1296.2	1307.2	1315.1	1320.6	1324.1	1326.6	1324.0	1317.4	1307.2	1293.9	1277.6
450	1253.4	1267.0	1278.4	1288.1	1303.3	1314.5	1322.7	1328.4	1332.2	1335.2	1333.3	1327.4	1318.2	1305.9	1290.7
455	1260.1	1273.8	1285.3	1295.1	1310.5	1321.9	1330.3	1336.2	1340.3	1343.8	1342.5	1337.3	1328.9	1317.5	1303.4
460	1266.8	1280.6	1292.2	1302.0	1317.7	1329.3	1337.8	1344.0	1348.3	1352.4	1351.6	1347.1	1339.4	1328.9	1315.8
465	1273.6	1287.5	1299.1	1309.1	1324.9	1336.7	1345.4	1351.8	1356.3	1360.9	1360.7	1356.8	1349.8	1340.1	1327.9
470	1280.4	1294.3	1306.1	1316.1	1332.1	1344.0	1353.0	1359.6	1364.3	1369.3	1369.7	1366.4	1360.0	1351.0	1339.7

TABLE 6.8 Thermo-mechanical exergy of steam from 35 to 220 bar. Dashed lines indicate that water is not a pure gas at that temperature and pressure (It is a mixture of liquid and gas). (Continued)

141

| | e^{tm} (kJ/kg) | | | | | | | | | | | | | | |
| | Pressure (bar) | | | | | | | | | | | | | | |
T (°C)	35	40	45	50	60	70	80	90	100	120	140	160	180	200	220
475	1287.2	1301.2	1313.1	1323.1	1339.3	1351.4	1360.6	1367.4	1372.3	1377.7	1378.6	1375.8	1370.1	1361.8	1351.3
480	1294.0	1308.1	1320.0	1330.2	1346.5	1358.8	1368.1	1375.1	1380.3	1386.1	1387.5	1385.2	1380.1	1372.5	1362.7
485	1300.9	1315.1	1327.1	1337.3	1353.8	1366.2	1375.7	1382.9	1388.2	1394.5	1396.3	1394.6	1390.0	1383.0	1373.8
490	1307.8	1322.0	1334.1	1344.4	1361.0	1373.7	1383.3	1390.7	1396.2	1402.9	1405.1	1403.8	1399.8	1393.3	1384.9
495	1314.7	1329.0	1341.2	1351.5	1368.3	1381.1	1390.9	1398.4	1404.1	1411.2	1413.8	1413.0	1409.5	1403.6	1395.7
500	1321.7	1336.0	1348.2	1358.7	1375.6	1388.5	1398.5	1406.2	1412.1	1419.5	1422.5	1422.1	1419.1	1413.7	1406.4
505	1328.6	1343.1	1355.4	1365.9	1382.9	1396.0	1406.1	1414.0	1420.0	1427.8	1431.2	1431.2	1428.6	1423.8	1417.0
510	1335.6	1350.2	1362.5	1373.1	1390.3	1403.5	1413.8	1421.7	1427.9	1436.0	1439.8	1440.3	1438.1	1433.7	1427.4
515	1342.7	1357.3	1369.6	1380.3	1397.6	1411.0	1421.4	1429.5	1435.9	1444.3	1448.4	1449.3	1447.5	1443.6	1437.8
520	1349.7	1364.4	1376.8	1387.5	1405.0	1418.5	1429.0	1437.3	1443.8	1452.5	1457.0	1458.2	1456.9	1453.3	1448.0
525	1356.8	1371.5	1384.0	1394.8	1412.4	1426.0	1436.7	1445.1	1451.7	1460.8	1465.6	1467.1	1466.2	1463.1	1458.2
530	1363.9	1378.7	1391.3	1402.1	1419.8	1433.5	1444.3	1452.9	1459.7	1469.0	1474.1	1476.0	1475.4	1472.7	1468.2
535	1371.1	1385.9	1398.5	1409.4	1427.2	1441.1	1452.0	1460.7	1467.6	1477.2	1482.7	1484.9	1484.6	1482.3	1478.2
540	1378.3	1393.1	1405.8	1416.7	1434.7	1448.6	1459.7	1468.5	1475.6	1485.5	1491.2	1493.7	1493.8	1491.8	1488.1
545	1385.5	1400.4	1413.1	1424.1	1442.1	1456.2	1467.4	1476.4	1483.5	1493.7	1499.7	1502.6	1502.9	1501.3	1498.0
550	1392.7	1407.6	1420.4	1431.5	1449.6	1463.8	1475.1	1484.2	1491.5	1501.9	1508.2	1511.4	1512.0	1510.7	1507.8
555	1399.9	1415.0	1427.8	1438.9	1457.2	1471.5	1482.9	1492.1	1499.5	1510.2	1516.7	1520.1	1521.1	1520.1	1517.5
560	1407.2	1422.3	1435.2	1446.3	1464.7	1479.1	1490.6	1499.9	1507.4	1518.4	1525.2	1528.9	1530.2	1529.5	1527.2
565	1414.5	1429.6	1442.6	1453.8	1472.2	1486.8	1498.4	1507.8	1515.4	1526.6	1533.7	1537.6	1539.2	1538.8	1536.8
570	1421.9	1437.0	1450.0	1461.3	1479.8	1494.5	1506.2	1515.7	1523.4	1534.9	1542.2	1546.4	1548.2	1548.1	1546.4
575	1429.2	1444.4	1457.5	1468.8	1487.4	1502.2	1514.0	1523.6	1531.5	1543.1	1550.6	1555.1	1557.2	1557.4	1556.0
580	1436.6	1451.9	1465.0	1476.3	1495.1	1509.9	1521.8	1531.5	1539.5	1551.3	1559.1	1563.8	1566.2	1566.6	1565.5
585	1444.0	1459.3	1472.5	1483.9	1502.7	1517.6	1529.7	1539.5	1547.5	1559.6	1567.6	1572.5	1575.1	1575.8	1575.0
590	1451.5	1466.8	1480.0	1491.4	1510.4	1525.4	1537.5	1547.4	1555.6	1567.9	1576.1	1581.3	1584.1	1585.0	1584.4
595	1459.0	1474.4	1487.6	1499.0	1518.1	1533.2	1545.4	1555.4	1563.7	1576.1	1584.5	1590.0	1593.0	1594.2	1593.9
600	1466.5	1481.9	1495.2	1506.7	1525.8	1541.0	1553.3	1563.4	1571.7	1584.4	1593.0	1598.7	1601.9	1603.4	1603.3
605	1474.0	1489.5	1502.8	1514.3	1533.5	1548.8	1561.2	1571.4	1579.8	1592.7	1601.5	1607.4	1610.9	1612.5	1612.6
610	1481.6	1497.1	1510.4	1522.0	1541.3	1556.7	1569.1	1579.4	1588.0	1601.0	1610.0	1616.1	1619.8	1621.6	1622.0
615	1489.2	1504.7	1518.1	1529.7	1549.1	1564.5	1577.1	1587.5	1596.1	1609.3	1618.5	1624.8	1628.7	1630.8	1631.3
620	1496.8	1512.4	1525.8	1537.5	1556.9	1572.4	1585.1	1595.5	1604.2	1617.6	1627.0	1633.5	1637.6	1639.9	1640.6
625	1504.4	1520.0	1533.5	1545.2	1564.7	1580.3	1593.3	1603.6	1612.4	1626.0	1635.6	1642.5	1646.5	1649.0	1650.0

Table 6.8 Thermo-mechanical exergy of steam from 35 to 220 bar. Dashed lines indicate that water is not a pure gas at that temperature and pressure (It is a mixture of liquid and gas). (*Continued*)

e^{tm} (kJ/kg)

T (°C)	Pressure (bar)														
	35	40	45	50	60	70	80	90	100	120	140	160	180	200	220
630	1512.1	1527.8	1541.2	1553.0	1572.6	1588.3	1601.1	1611.7	1620.6	1634.3	1644.1	1650.9	1655.4	1658.1	1659.2
635	1519.8	1535.5	1549.0	1560.8	1580.5	1596.2	1609.1	1619.8	1628.8	1642.7	1652.6	1659.6	1664.3	1667.1	1668.5
640	1527.5	1543.3	1556.8	1568.6	1588.4	1604.2	1617.2	1628.0	1637.0	1651.1	1661.2	1668.3	1673.2	1676.2	1677.8
645	1535.3	1551.0	1564.6	1576.5	1596.3	1612.2	1625.3	1636.1	1645.2	1659.5	1669.7	1677.0	1682.1	1685.3	1687.1
650	1543.1	1558.9	1572.5	1584.4	1604.3	1620.3	1633.4	1644.3	1653.5	1667.9	1678.3	1685.7	1691.0	1694.4	1696.3
655	1550.9	1566.7	1580.4	1592.3	1612.3	1628.3	1641.5	1652.5	1661.7	1676.3	1686.8	1694.5	1699.9	1703.4	1705.6
660	1558.7	1574.6	1588.3	1600.2	1620.3	1636.4	1649.6	1660.7	1670.0	1684.7	1695.4	1703.2	1708.8	1712.5	1714.8
665	1566.6	1582.5	1596.2	1608.2	1628.3	1644.5	1657.8	1668.9	1678.3	1693.2	1704.0	1712.0	1717.7	1721.6	1724.1
670	1574.5	1590.4	1604.2	1616.2	1636.3	1652.6	1666.0	1677.2	1686.7	1701.6	1712.6	1720.7	1726.6	1730.7	1733.3
675	1582.4	1598.4	1612.1	1624.2	1644.4	1660.7	1674.2	1685.5	1695.0	1710.1	1721.3	1729.5	1735.5	1739.7	1742.5
680	1590.3	1606.3	1620.1	1632.2	1652.5	1668.9	1682.4	1693.8	1703.4	1718.6	1729.9	1738.3	1744.4	1748.8	1751.8
685	1598.3	1614.3	1628.2	1640.3	1660.7	1677.1	1690.7	1702.1	1711.8	1727.1	1738.6	1747.1	1753.4	1757.9	1761.0
690	1606.3	1622.4	1636.2	1648.4	1668.8	1685.3	1699.0	1710.4	1720.2	1735.7	1747.2	1755.9	1762.3	1767.0	1770.2
695	1614.4	1630.4	1644.3	1656.5	1677.0	1693.5	1707.3	1718.8	1728.6	1744.2	1755.9	1764.7	1771.3	1776.1	1779.5
700	1622.4	1638.5	1652.4	1664.7	1685.2	1701.8	1715.6	1727.2	1737.0	1752.8	1764.6	1773.5	1780.2	1785.2	1788.7
705	1630.5	1646.6	1660.6	1672.8	1693.4	1710.1	1723.9	1735.6	1745.5	1761.4	1773.3	1782.4	1789.2	1794.3	1797.9
710	1638.6	1654.8	1668.8	1681.0	1701.7	1718.4	1732.3	1744.0	1754.0	1770.0	1782.0	1791.2	1798.2	1803.4	1807.2
715	1646.7	1662.9	1676.9	1689.2	1709.9	1726.7	1740.7	1752.4	1762.5	1778.6	1790.8	1800.1	1807.2	1812.5	1816.4
720	1654.9	1671.1	1685.2	1697.5	1718.2	1735.1	1749.1	1760.9	1771.0	1787.2	1799.5	1809.0	1816.2	1821.7	1825.7
725	1663.1	1679.3	1693.4	1705.8	1726.6	1743.5	1757.5	1769.4	1779.5	1795.9	1808.3	1817.9	1825.2	1830.8	1835.0
730	1671.3	1687.6	1701.7	1714.1	1734.9	1751.9	1766.0	1777.9	1788.1	1804.6	1817.1	1826.8	1834.2	1839.9	1844.2
735	1679.6	1695.9	1710.0	1722.4	1743.3	1760.3	1774.4	1786.4	1796.7	1813.3	1825.9	1835.7	1843.3	1849.1	1853.5
740	1687.8	1704.1	1718.3	1730.7	1751.7	1768.7	1782.9	1795.0	1805.3	1822.0	1834.7	1844.6	1852.3	1858.3	1862.8
745	1696.1	1712.5	1726.6	1739.1	1760.1	1777.2	1791.5	1803.6	1813.9	1830.7	1843.6	1853.6	1861.4	1867.4	1872.1
750	1704.5	1720.8	1735.0	1747.5	1768.5	1785.7	1800.0	1812.2	1822.6	1839.5	1852.4	1862.5	1870.4	1876.6	1881.4
755	1712.8	1729.2	1743.4	1755.9	1777.0	1794.2	1808.6	1820.8	1831.2	1848.2	1861.3	1871.5	1879.5	1885.8	1890.7
760	1721.2	1737.6	1751.8	1764.4	1785.5	1802.8	1817.2	1829.4	1839.9	1857.0	1870.2	1880.5	1888.6	1895.0	1900.0
765	1729.6	1746.0	1760.3	1772.8	1794.0	1811.3	1825.8	1838.1	1848.6	1865.8	1879.1	1889.5	1897.8	1904.2	1909.3
770	1738.0	1754.5	1768.8	1781.3	1802.6	1819.9	1834.4	1846.8	1857.4	1874.7	1888.0	1898.6	1906.9	1913.5	1918.7
775	1746.5	1763.0	1777.3	1789.9	1811.1	1828.5	1843.1	1855.5	1866.1	1883.5	1897.0	1907.6	1916.0	1922.7	1928.0
780	1755.0	1771.5	1785.8	1798.4	1819.7	1837.2	1851.8	1864.2	1874.9	1892.4	1905.9	1916.7	1925.9	1932.7	1937.4

TABLE 6.8 Thermo-mechanical exergy of steam from 35 to 220 bar. Dashed lines indicate that water is not a pure gas at that temperature and pressure (It is a mixture of liquid and gas). (Continued)

T (°C)	e^tm (kJ/kg) Pressure (bar)														
	35	40	45	50	60	70	80	90	100	120	140	160	180	200	220
785	1763.5	1780.0	1794.3	1807.0	1828.4	1845.8	1860.5	1872.9	1883.7	1901.3	1914.9	1925.7	1934.4	1941.2	1946.7
790	1772.0	1788.5	1802.9	1815.6	1837.0	1854.5	1869.2	1881.7	1892.5	1910.2	1923.9	1934.8	1943.5	1950.5	1956.1
795	1780.6	1797.1	1811.5	1824.2	1845.7	1863.2	1878.0	1890.5	1901.4	1919.1	1933.0	1943.9	1952.7	1959.8	1965.5
800	1789.1	1805.7	1820.1	1832.9	1854.4	1872.0	1886.7	1899.3	1910.2	1928.1	1942.0	1953.1	1962.0	1969.1	1974.9
805	1797.8	1814.4	1828.8	1841.5	1863.1	1880.7	1895.5	1908.2	1919.1	1937.0	1951.0	1962.2	1971.2	1978.5	1984.3
810	1806.4	1823.0	1837.5	1850.2	1871.8	1889.5	1904.4	1917.0	1928.0	1946.0	1960.1	1971.4	1980.5	1987.8	1993.8
815	1815.1	1831.7	1846.2	1858.9	1880.6	1898.3	1913.2	1925.9	1936.9	1955.0	1969.2	1980.6	1989.7	1997.2	2003.2
820	1823.7	1840.4	1854.9	1867.7	1889.4	1907.1	1922.1	1934.8	1945.9	1964.0	1978.3	1989.8	1999.0	2006.5	2012.7
825	1832.5	1849.1	1863.7	1876.5	1898.2	1916.0	1931.0	1943.8	1954.8	1973.1	1987.5	1999.0	2008.3	2015.9	2022.1
830	1841.2	1857.9	1872.4	1885.3	1907.0	1924.9	1939.9	1952.7	1963.8	1982.2	1996.6	2008.2	2017.6	2025.3	2031.6
835	1850.0	1866.7	1881.2	1894.1	1915.9	1933.8	1948.8	1961.7	1972.9	1991.3	2005.8	2017.5	2026.9	2034.7	2041.1
840	1858.8	1875.5	1890.1	1902.9	1924.7	1942.7	1957.8	1970.7	1981.9	2000.4	2015.0	2026.7	2036.3	2044.2	2050.6
845	1867.6	1884.3	1898.9	1911.8	1933.7	1951.6	1966.7	1979.7	1990.9	2009.5	2024.2	2036.0	2045.7	2053.6	2060.1
850	1876.4	1893.2	1907.8	1920.7	1942.6	1960.6	1975.7	1988.7	2000.0	2018.7	2033.4	2045.3	2055.0	2063.1	2069.7
855	1885.3	1902.1	1916.7	1929.6	1951.5	1969.6	1984.8	1997.8	2009.1	2027.8	2042.7	2054.6	2064.4	2072.5	2079.2
860	1894.2	1911.0	1925.6	1938.6	1960.5	1978.6	1993.8	2006.9	2018.2	2037.0	2051.9	2064.0	2073.9	2082.0	2088.8
865	1903.1	1919.9	1934.6	1947.5	1969.5	1987.6	2002.9	2016.0	2027.4	2046.2	2061.2	2073.3	2083.3	2091.5	2098.4
870	1912.0	1928.9	1943.5	1956.5	1978.5	1996.7	2012.0	2025.1	2036.5	2055.5	2070.5	2082.7	2092.7	2101.1	2108.0
875	1921.0	1937.9	1952.5	1965.5	1987.6	2005.8	2021.1	2034.3	2045.7	2064.7	2079.8	2092.1	2102.2	2110.6	2117.6
880	1930.0	1946.9	1961.6	1974.6	1996.7	2014.9	2030.2	2043.5	2054.9	2074.0	2089.2	2101.5	2111.7	2120.2	2127.2
885	1939.0	1955.9	1970.6	1983.6	2005.8	2024.0	2039.4	2052.6	2064.2	2083.3	2098.6	2111.0	2121.2	2129.7	2136.9
890	1948.0	1964.9	1979.7	1992.7	2014.9	2033.2	2048.6	2061.9	2073.4	2092.6	2107.9	2120.4	2130.7	2139.3	2146.5
895	1957.1	1974.0	1988.8	2001.8	2024.0	2042.3	2057.8	2071.1	2082.7	2102.0	2117.3	2129.9	2140.3	2148.9	2156.2
900	1966.2	1983.1	1997.9	2011.0	2033.2	2051.5	2067.0	2080.4	2092.0	2111.3	2126.8	2139.4	2149.8	2158.5	2165.9
905	1975.3	1992.3	2007.0	2020.1	2042.4	2060.8	2076.3	2089.7	2101.3	2120.7	2136.2	2148.9	2159.4	2168.2	2175.6
910	1984.4	2001.4	2016.2	2029.3	2051.6	2070.0	2085.6	2099.0	2110.6	2130.1	2145.7	2158.4	2169.0	2177.8	2185.3
915	1993.6	2010.6	2025.4	2038.5	2060.8	2079.3	2094.9	2108.3	2120.0	2139.5	2155.2	2168.0	2178.6	2187.5	2195.1
920	2002.8	2019.8	2034.6	2047.7	2070.1	2088.6	2104.2	2117.6	2129.4	2149.0	2164.7	2177.5	2188.2	2197.2	2204.8
925	2012.0	2029.0	2043.9	2057.0	2079.4	2097.9	2113.5	2127.0	2138.8	2158.4	2174.2	2187.1	2197.9	2206.9	2214.6
930	2021.2	2038.3	2053.1	2066.3	2088.7	2107.2	2122.9	2136.4	2148.2	2167.9	2183.7	2196.7	2207.5	2216.6	2224.4
935	2030.5	2047.5	2062.4	2075.6	2098.0	2116.6	2132.3	2145.8	2157.7	2177.4	2193.3	2206.3	2217.2	2226.4	2234.2

TABLE 6.8 Thermo-mechanical exergy of steam from 35 to 220 bar. Dashed lines indicate that water is not a pure gas at that temperature and pressure (It is a mixture of liquid and gas). (Continued)

e^{tm} (kJ/kg)

Pressure (bar)

T (°C)	35	40	45	50	60	70	80	90	100	120	140	160	180	200	220
940	2039.8	2056.8	2071.7	2084.9	2107.4	2125.9	2141.7	2155.3	2167.1	2186.9	2202.9	2216.0	2226.9	2236.1	2244.0
945	2049.1	2066.1	2081.0	2094.2	2116.7	2135.3	2151.1	2164.7	2176.6	2196.5	2212.5	2225.6	2236.6	2245.9	2253.8
950	2058.4	2075.5	2090.4	2103.6	2126.1	2144.8	2160.6	2174.2	2186.1	2206.0	2222.1	2235.3	2246.4	2255.7	2263.7
955	2067.7	2084.8	2099.8	2113.0	2135.5	2154.2	2170.0	2183.7	2195.6	2215.6	2231.7	2245.0	2256.1	2265.5	2273.5
960	2077.1	2094.2	2109.2	2122.4	2145.0	2163.7	2179.5	2193.2	2205.2	2225.2	2241.4	2254.7	2265.9	2275.4	2283.4
965	2086.5	2103.6	2118.6	2131.9	2154.5	2173.2	2189.1	2202.8	2214.8	2234.9	2251.1	2264.5	2275.7	2285.2	2293.3
970	2095.9	2113.1	2128.1	2141.3	2163.9	2182.7	2198.6	2212.3	2224.4	2244.5	2260.8	2274.2	2285.5	2295.1	2303.3
975	2105.4	2122.5	2137.5	2150.8	2173.5	2192.2	2208.2	2221.9	2234.0	2254.2	2270.5	2284.0	2295.3	2304.9	2313.2
980	2114.9	2132.0	2147.0	2160.3	2183.0	2201.8	2217.7	2231.5	2243.6	2263.9	2280.2	2293.8	2305.2	2314.8	2323.1
985	2124.3	2141.5	2156.5	2169.8	2192.5	2211.4	2227.3	2241.2	2253.3	2273.6	2290.0	2303.6	2315.0	2324.8	2333.1
990	2133.9	2151.0	2166.1	2179.4	2202.1	2221.0	2237.0	2250.8	2262.9	2283.3	2299.8	2313.4	2324.9	2334.7	2343.1
995	2143.4	2160.6	2175.6	2189.0	2211.7	2230.6	2246.6	2260.5	2272.6	2293.1	2309.6	2323.3	2334.8	2344.7	2353.1
1000	2153.0	2170.2	2185.2	2198.6	2221.3	2240.2	2256.3	2270.2	2282.4	2302.8	2319.4	2333.2	2344.7	2354.6	2363.1
1020	2191.4	2208.7	2223.8	2237.2	2260.0	2279.0	2295.2	2309.2	2321.5	2342.1	2358.9	2372.8	2384.6	2394.7	2403.4
1040	2230.2	2247.6	2262.7	2276.1	2299.1	2318.2	2334.4	2348.5	2360.9	2381.7	2398.7	2412.8	2424.8	2435.0	2443.9
1060	2269.4	2286.8	2302.0	2315.4	2338.5	2357.7	2374.0	2388.2	2400.6	2421.6	2438.8	2453.1	2465.2	2475.7	2484.7
1080	2309.0	2326.3	2341.6	2355.1	2378.2	2397.5	2413.9	2428.1	2440.7	2461.9	2479.2	2493.6	2506.0	2516.6	2525.8
1100	2348.8	2366.3	2381.5	2395.1	2418.3	2437.6	2454.1	2468.5	2481.1	2502.4	2519.9	2534.5	2547.0	2557.8	2567.2
1120	2389.0	2406.5	2421.8	2435.4	2458.7	2478.1	2494.7	2509.1	2521.8	2543.3	2560.9	2575.7	2588.3	2599.2	2608.8
1140	2429.6	2447.1	2462.4	2476.1	2499.4	2518.9	2535.6	2550.0	2562.8	2584.4	2602.2	2617.1	2629.9	2641.0	2650.7
1160	2470.5	2488.0	2503.4	2517.0	2540.5	2560.0	2576.7	2591.3	2604.1	2625.9	2643.8	2658.9	2671.8	2683.0	2692.9
1180	2511.7	2529.2	2544.6	2558.4	2581.8	2601.5	2618.2	2632.9	2645.8	2667.7	2685.7	2700.9	2714.0	2725.3	2735.3
1200	2553.2	2570.8	2586.2	2600.0	2623.5	2643.2	2660.1	2674.7	2687.7	2709.8	2727.9	2743.3	2756.4	2767.9	2778.0
1220	2595.0	2612.7	2628.1	2641.9	2665.5	2685.3	2702.2	2716.9	2729.9	2752.1	2770.4	2785.9	2799.2	2810.8	2821.0
1240	2637.2	2654.8	2670.3	2684.1	2707.8	2727.6	2744.6	2759.4	2772.5	2794.8	2813.2	2828.8	2842.2	2853.9	2864.3
1260	2679.6	2697.3	2712.8	2726.7	2750.4	2770.3	2787.3	2802.1	2815.3	2837.7	2856.2	2871.9	2885.5	2897.3	2907.8
1280	2722.4	2740.1	2755.6	2769.5	2793.3	2813.2	2830.3	2845.2	2858.4	2880.9	2899.6	2915.4	2929.0	2941.0	2951.5
1300	2765.4	2783.2	2798.7	2812.6	2836.5	2856.4	2873.6	2888.5	2901.8	2924.4	2943.2	2959.1	2972.8	2984.9	2995.6
1320	2808.7	2826.5	2842.1	2856.0	2879.9	2900.0	2917.1	2932.2	2945.5	2968.2	2987.0	3003.1	3016.9	3029.1	3039.9
1340	2852.4	2870.2	2885.8	2899.7	2923.7	2943.7	2961.0	2976.1	2989.4	3012.2	3031.2	3047.3	3061.3	3073.5	3084.4
1360	2896.3	2914.1	2929.8	2943.7	2967.7	2987.8	3005.1	3020.2	3033.6	3056.5	3075.6	3091.8	3105.9	3118.2	3129.2

TABLE 6.8 Thermo-mechanical exergy of steam from 35 to 220 bar. Dashed lines indicate that water is not a pure gas at that temperature and pressure (It is a mixture of liquid and gas).
(Continued)

e^{tm} (kJ/kg)

T (°C)	Pressure (bar)														
	35	40	45	50	60	70	80	90	100	120	140	160	180	200	220
1380	2940.4	2958.3	2974.0	2988.0	3012.0	3032.2	3049.5	3064.7	3078.1	3101.1	3120.3	3136.6	3150.7	3163.1	3174.2
1400	2984.9	3002.8	3018.5	3032.5	3056.6	3076.8	3094.1	3109.3	3122.8	3145.9	3165.2	3181.6	3195.8	3208.3	3219.5
1420	3029.6	3047.5	3063.2	3077.3	3101.4	3121.6	3139.1	3154.3	3167.9	3191.0	3210.3	3226.8	3241.1	3253.8	3265.0
1440	3074.6	3092.5	3108.3	3122.3	3146.5	3166.8	3184.2	3199.5	3213.1	3236.4	3255.8	3272.3	3286.7	3299.4	3310.8
1460	3119.9	3137.8	3153.6	3167.6	3191.8	3212.2	3229.7	3245.0	3258.6	3282.0	3301.4	3318.1	3332.6	3345.3	3356.7
1480	3165.4	3183.3	3199.1	3213.2	3237.4	3257.8	3275.3	3290.7	3304.4	3327.8	3347.3	3364.1	3378.6	3391.5	3403.0
1500	3211.1	3229.1	3244.9	3259.0	3283.3	3303.7	3321.3	3336.7	3350.4	3373.9	3393.5	3410.3	3424.9	3437.9	3449.4
1520	3257.1	3275.1	3290.9	3305.0	3329.4	3349.8	3367.4	3382.9	3396.6	3420.2	3439.9	3456.7	3471.5	3484.5	3496.1
1540	3303.3	3321.4	3337.2	3351.3	3375.7	3396.2	3413.8	3429.3	3443.1	3466.7	3486.5	3503.4	3518.2	3531.3	3543.0
1560	3349.8	3367.9	3383.7	3397.9	3422.3	3442.8	3460.5	3476.0	3489.8	3513.5	3533.3	3550.3	3565.2	3578.3	3590.1
1580	3396.6	3414.6	3430.5	3444.6	3469.1	3489.6	3507.3	3522.9	3536.7	3560.5	3580.4	3597.5	3612.4	3625.6	3637.4
1600	3443.5	3461.6	3477.5	3491.6	3516.1	3536.7	3554.4	3570.0	3583.9	3607.7	3627.7	3644.8	3659.8	3673.1	3685.0
1620	3490.7	3508.8	3524.7	3538.9	3563.4	3584.0	3601.7	3617.4	3631.3	3655.2	3675.2	3692.4	3707.4	3720.8	3732.7
1640	3538.1	3556.2	3572.1	3586.3	3610.8	3631.5	3649.3	3664.9	3678.9	3702.8	3722.9	3740.2	3755.3	3768.7	3780.7
1660	3585.7	3603.8	3619.8	3634.0	3658.5	3679.2	3697.0	3712.7	3726.7	3750.7	3770.8	3788.2	3803.3	3816.8	3828.9
1680	3633.6	3651.7	3667.6	3681.9	3706.4	3727.1	3745.0	3760.7	3774.7	3798.8	3819.0	3836.4	3851.6	3865.1	3877.2
1700	3681.6	3699.7	3715.7	3730.0	3754.6	3775.3	3793.2	3808.9	3822.9	3847.1	3867.3	3884.8	3900.0	3913.6	3925.8
1720	3729.9	3748.0	3764.0	3778.3	3802.9	3823.7	3841.6	3857.3	3871.4	3895.6	3915.9	3933.4	3948.7	3962.3	3974.5
1740	3778.3	3796.5	3812.5	3826.8	3851.4	3872.2	3890.2	3906.0	3920.0	3944.3	3964.6	3982.2	3997.5	4011.2	4023.5
1760	3827.0	3845.2	3861.2	3875.5	3900.2	3921.0	3939.0	3954.8	3968.9	3993.2	4013.6	4031.2	4046.6	4060.3	4072.7
1780	3875.9	3894.1	3910.1	3924.4	3949.1	3970.0	3988.0	4003.8	4017.9	4042.2	4062.7	4080.4	4095.8	4109.6	4122.0
1800	3925.0	3943.2	3959.2	3973.5	3998.3	4019.1	4037.1	4053.0	4067.1	4091.5	4112.0	4129.7	4145.2	4159.1	4171.5
1820	3974.2	3992.5	4008.5	4022.8	4047.6	4068.5	4086.5	4102.4	4116.6	4141.0	4161.6	4179.3	4194.9	4208.7	4221.2
1840	4023.7	4041.9	4058.0	4072.3	4097.1	4118.0	4136.1	4152.0	4166.2	4190.6	4211.3	4229.0	4244.6	4258.6	4271.1
1860	4073.4	4091.6	4107.6	4122.0	4146.8	4167.7	4185.8	4201.7	4216.0	4240.5	4261.1	4279.0	4294.6	4308.6	4321.1
1880	4123.2	4141.4	4157.5	4171.9	4196.7	4217.6	4235.7	4251.7	4265.9	4290.5	4311.2	4329.1	4344.8	4358.8	4371.4
1900	4173.2	4191.5	4207.5	4221.9	4246.8	4267.7	4285.9	4301.8	4316.1	4340.7	4361.4	4379.3	4395.1	4409.1	4421.8
1920	4223.4	4241.7	4257.8	4272.1	4297.0	4318.0	4336.1	4352.1	4366.4	4391.1	4411.8	4429.8	4445.6	4459.6	4472.3
1940	4273.8	4292.0	4308.1	4322.5	4347.4	4368.4	4386.6	4402.6	4416.9	4441.6	4462.4	4480.4	4496.2	4510.3	4523.1
1960	4324.3	4342.6	4358.7	4373.1	4398.0	4419.0	4437.2	4453.3	4467.6	4492.3	4513.2	4531.2	4547.1	4561.2	4574.0
1980	4375.0	4393.3	4409.4	4423.9	4448.8	4469.8	4488.0	4504.1	4518.4	4543.2	4564.1	4582.2	4598.1	4612.2	4625.1
2000	4425.9	4444.2	4460.3	4474.8	4499.7	4520.8	4539.0	4555.1	4569.4	4594.2	4615.2	4633.3	4649.2	4663.4	4676.3

TABLE 6.8 Thermo-mechanical exergy of steam from 35 to 220 bar. Dashed lines indicate that water is not a pure gas at that temperature and pressure (It is a mixture of liquid and gas). (Continued)

T (°C)	0.001	0.072	0.144	0.215	0.286	0.358	0.429	0.500	0.572	0.643	0.715	0.786	0.857	0.929	1.000
									e^{tm} (kJ/kg)						
									Pressure (bar)						
1	3870.7	—	—	—	—	—	—	—	—	—	—	—	—	—	—
2	3884.3	—	—	—	—	—	—	—	—	—	—	—	—	—	—
3	3897.8	—	—	—	—	—	—	—	—	—	—	—	—	—	—
4	3911.4	—	—	—	—	—	—	—	—	—	—	—	—	—	—
5	3924.9	—	—	—	—	—	—	—	—	—	—	—	—	—	—
6	3938.5	—	—	—	—	—	—	—	—	—	—	—	—	—	—
7	3952.0	—	—	—	—	—	—	—	—	—	—	—	—	—	—
8	3965.6	—	—	—	—	—	—	—	—	—	—	—	—	—	—
9	3979.1	—	—	—	—	—	—	—	—	—	—	—	—	—	—
10	3992.6	—	—	—	—	—	—	—	—	—	—	—	—	—	—
11	4006.1	—	—	—	—	—	—	—	—	—	—	—	—	—	—
12	4019.6	—	—	—	—	—	—	—	—	—	—	—	—	—	—
13	4033.1	—	—	—	—	—	—	—	—	—	—	—	—	—	—
14	4046.6	—	—	—	—	—	—	—	—	—	—	—	—	—	—
15	4060.1	—	—	—	—	—	—	—	—	—	—	—	—	—	—
16	4073.5	—	—	—	—	—	—	—	—	—	—	—	—	—	—
17	4087.0	—	—	—	—	—	—	—	—	—	—	—	—	—	—
18	4100.4	—	—	—	—	—	—	—	—	—	—	—	—	—	—
19	4113.9	—	—	—	—	—	—	—	—	—	—	—	—	—	—
20	4127.3	—	—	—	—	—	—	—	—	—	—	—	—	—	—
21	4140.8	—	—	—	—	—	—	—	—	—	—	—	—	—	—
22	4154.2	—	—	—	—	—	—	—	—	—	—	—	—	—	—
23	4167.6	—	—	—	—	—	—	—	—	—	—	—	—	—	—
24	4181.0	—	—	—	—	—	—	—	—	—	—	—	—	—	—
25	4194.4	—	—	—	—	—	—	—	—	—	—	—	—	—	—
26	4207.8	—	—	—	—	—	—	—	—	—	—	—	—	—	—
27	4221.2	—	—	—	—	—	—	—	—	—	—	—	—	—	—
28	4234.6	—	—	—	—	—	—	—	—	—	—	—	—	—	—
29	4247.9	—	—	—	—	—	—	—	—	—	—	—	—	—	—
30	4261.3	—	—	—	—	—	—	—	—	—	—	—	—	—	—
31	4274.7	—	—	—	—	—	—	—	—	—	—	—	—	—	—

TABLE 6.9 Thermo-mechanical exergy of water vapour at vacuum pressures. Dashed lines indicate that water is not a pure gas at that temperature and pressure (it is a mixture of liquid and gas).

147

T (°C)	e^tm (kJ/kg) Pressure (bar)														
	0.001	0.072	0.144	0.215	0.286	0.358	0.429	0.500	0.572	0.643	0.715	0.786	0.857	0.929	1.000
32	4288.0	—	—	—	—	—	—	—	—	—	—	—	—	—	—
33	4301.4	—	—	—	—	—	—	—	—	—	—	—	—	—	—
34	4314.7	—	—	—	—	—	—	—	—	—	—	—	—	—	—
35	4328.0	—	—	—	—	—	—	—	—	—	—	—	—	—	—
36	4341.4	—	—	—	—	—	—	—	—	—	—	—	—	—	—
37	4354.7	—	—	—	—	—	—	—	—	—	—	—	—	—	—
38	4368.0	—	—	—	—	—	—	—	—	—	—	—	—	—	—
39	4381.3	—	—	—	—	—	—	—	—	—	—	—	—	—	—
40	4394.6	1178.6	—	—	—	—	—	—	—	—	—	—	—	—	—
41	4407.9	1182.1	—	—	—	—	—	—	—	—	—	—	—	—	—
42	4421.1	1185.6	—	—	—	—	—	—	—	—	—	—	—	—	—
43	4434.4	1189.0	—	—	—	—	—	—	—	—	—	—	—	—	—
44	4447.7	1192.5	—	—	—	—	—	—	—	—	—	—	—	—	—
45	4460.9	1196.0	—	—	—	—	—	—	—	—	—	—	—	—	—
46	4474.2	1199.5	—	—	—	—	—	—	—	—	—	—	—	—	—
47	4487.4	1203.0	—	—	—	—	—	—	—	—	—	—	—	—	—
48	4500.7	1206.5	—	—	—	—	—	—	—	—	—	—	—	—	—
49	4513.9	1210.0	—	—	—	—	—	—	—	—	—	—	—	—	—
50	4527.1	1213.5	—	—	—	—	—	—	—	—	—	—	—	—	—
51	4540.3	1217.0	—	—	—	—	—	—	—	—	—	—	—	—	—
52	4553.5	1220.5	—	—	—	—	—	—	—	—	—	—	—	—	—
53	4566.7	1224.0	—	—	—	—	—	—	—	—	—	—	—	—	—
54	4579.9	1227.5	960.57	—	—	—	—	—	—	—	—	—	—	—	—
55	4593.1	1231.0	963.07	—	—	—	—	—	—	—	—	—	—	—	—
56	4606.3	1234.5	965.58	—	—	—	—	—	—	—	—	—	—	—	—
57	4619.5	1238.0	968.09	—	—	—	—	—	—	—	—	—	—	—	—
58	4632.6	1241.5	970.60	—	—	—	—	—	—	—	—	—	—	—	—
59	4645.8	1245.0	973.11	—	—	—	—	—	—	—	—	—	—	—	—
60	4659.0	1248.6	975.62	—	—	—	—	—	—	—	—	—	—	—	—
61	4672.1	1252.1	978.14	—	—	—	—	—	—	—	—	—	—	—	—
62	4685.2	1255.6	980.66	844.88	—	—	—	—	—	—	—	—	—	—	—

TABLE 6.9 Thermo-mechanical exergy of water vapour at vacuum pressures. Dashed lines indicate that water is not a pure gas at that temperature and pressure (it is a mixture of liquid and gas). (Continued)

e^{tm} (kJ/kg)

T (°C)	0.001	0.072	0.144	0.215	0.286	0.358	0.429	0.500	0.572	0.643	0.715	0.786	0.857	0.929	1.000
								Pressure (bar)							
63	4698.4	1259.1	983.18	846.88	—	—	—	—	—	—	—	—	—	—	—
64	4711.5	1262.7	985.71	848.87	—	—	—	—	—	—	—	—	—	—	—
65	4724.6	1266.2	988.23	850.87	—	—	—	—	—	—	—	—	—	—	—
66	4737.7	1269.8	990.76	852.87	—	—	—	—	—	—	—	—	—	—	—
67	4750.9	1273.3	993.30	854.87	—	—	—	—	—	—	—	—	—	—	—
68	4764.0	1276.8	995.83	856.88	—	—	—	—	—	—	—	—	—	—	—
69	4777.0	1280.4	998.37	858.89	770.55	—	—	—	—	—	—	—	—	—	—
70	4790.1	1283.9	1000.9	860.90	772.21	—	—	—	—	—	—	—	—	—	—
71	4803.2	1287.5	1003.5	862.91	773.87	—	—	—	—	—	—	—	—	—	—
72	4816.3	1291.0	1006.0	864.93	775.54	—	—	—	—	—	—	—	—	—	—
73	4829.4	1294.6	1008.6	866.95	777.21	—	—	—	—	—	—	—	—	—	—
74	4842.4	1298.2	1011.1	868.97	778.88	714.84	—	—	—	—	—	—	—	—	—
75	4855.5	1301.7	1013.7	871.00	780.55	716.25	—	—	—	—	—	—	—	—	—
76	4868.5	1305.3	1016.2	873.03	782.23	717.67	—	—	—	—	—	—	—	—	—
77	4881.6	1308.9	1018.8	875.06	783.91	719.09	—	—	—	—	—	—	—	—	—
78	4894.6	1312.5	1021.4	877.09	785.59	720.51	671.02	—	—	—	—	—	—	—	—
79	4907.6	1316.0	1023.9	879.13	787.27	721.94	672.24	—	—	—	—	—	—	—	—
80	4920.7	1319.6	1026.5	881.17	788.96	723.37	673.47	—	—	—	—	—	—	—	—
81	4933.7	1323.2	1029.1	883.22	790.66	724.80	674.70	—	—	—	—	—	—	—	—
82	4946.7	1326.8	1031.7	885.27	792.35	726.24	675.93	635.91	—	—	—	—	—	—	—
83	4959.7	1330.4	1034.3	887.32	794.05	727.67	677.16	636.98	—	—	—	—	—	—	—
84	4972.7	1334.0	1036.8	889.38	795.75	729.12	678.40	638.05	—	—	—	—	—	—	—
85	4985.7	1337.6	1039.4	891.44	797.46	730.56	679.64	639.12	605.85	—	—	—	—	—	—
86	4998.7	1341.2	1042.0	893.50	799.17	732.01	680.88	640.20	606.79	—	—	—	—	—	—
87	5011.6	1344.8	1044.6	895.56	800.88	733.46	682.13	641.28	607.73	—	—	—	—	—	—
88	5024.6	1348.4	1047.2	897.63	802.60	734.92	683.38	642.36	608.67	580.34	—	—	—	—	—
89	5037.6	1352.0	1049.8	899.70	804.32	736.37	684.63	643.45	609.61	581.16	—	—	—	—	—
90	5050.6	1355.6	1052.4	901.78	806.04	737.84	685.89	644.54	610.56	581.99	—	—	—	—	—
91	5063.5	1359.2	1055.0	903.86	807.76	739.30	687.15	645.63	611.51	582.82	558.24	—	—	—	—
92	5076.5	1362.8	1057.7	905.94	809.49	740.77	688.41	646.72	612.47	583.65	558.96	—	—	—	—
93	5089.4	1366.4	1060.3	908.03	811.23	742.24	689.68	647.82	613.43	584.49	559.69	—	—	—	—

TABLE 6.9 Thermo-mechanical exergy of water vapour at vacuum pressures. Dashed lines indicate that water is not a pure gas at that temperature and pressure (it is a mixture of liquid and gas). (Continued)

T (°C)	0.001	0.072	0.144	0.215	0.286	0.358	0.429	0.500	0.572	0.643	0.715	0.786	0.857	0.929	1.000
															e^{tm} (kJ/kg)
															Pressure (bar)
94	5102.3	1370.0	1062.9	910.11	812.96	743.72	690.95	648.93	614.39	585.33	560.43	538.77	—	—	—
95	5115.3	1373.7	1065.5	912.21	814.70	745.20	692.22	650.03	615.36	586.17	561.16	539.42	—	—	—
96	5128.2	1377.3	1068.1	914.30	816.44	746.68	693.50	651.14	616.32	587.02	561.91	540.06	520.83	—	—
97	5141.1	1380.9	1070.8	916.40	818.19	748.16	694.78	652.26	617.30	587.87	562.65	540.71	521.40	—	—
98	5154.0	1384.6	1073.4	918.50	819.94	749.65	696.07	653.37	618.27	588.73	563.40	541.36	521.97	504.72	—
99	5166.9	1388.2	1076.0	920.61	821.69	751.14	697.36	654.49	619.25	589.58	564.15	542.02	522.54	505.21	—
100	5179.8	1391.8	1078.7	922.72	823.45	752.64	698.65	655.62	620.23	590.44	564.90	542.68	523.11	505.71	490.10
101	5192.7	1395.5	1081.3	924.83	825.21	754.14	699.94	656.74	621.22	591.31	565.66	543.34	523.69	506.21	490.54
102	5205.6	1399.1	1084.0	926.94	826.97	755.64	701.24	657.87	622.21	592.18	566.42	544.01	524.28	506.72	490.97
103	5218.5	1402.8	1086.6	929.06	828.74	757.15	702.54	659.01	623.20	593.05	567.19	544.68	524.86	507.23	491.41
104	5231.4	1406.4	1089.3	931.18	830.50	758.66	703.84	660.15	624.20	593.92	567.96	545.36	525.45	507.74	491.85
105	5244.3	1410.1	1091.9	933.31	832.28	760.17	705.15	661.29	625.20	594.80	568.73	546.03	526.04	508.26	492.30
106	5257.1	1413.7	1094.6	935.44	834.05	761.69	706.46	662.43	626.20	595.68	569.50	546.71	526.64	508.78	492.75
107	5270.0	1417.4	1097.2	937.57	835.83	763.20	707.78	663.58	627.21	596.57	570.28	547.40	527.24	509.30	493.20
108	5282.8	1421.0	1099.9	939.70	837.62	764.73	709.10	664.73	628.22	597.46	571.06	548.09	527.84	509.83	493.66
109	5295.7	1424.7	1102.6	941.84	839.40	766.25	710.42	665.88	629.23	598.35	571.85	548.78	528.45	510.36	494.12
110	5308.5	1428.4	1105.2	943.98	841.19	767.78	711.74	667.04	630.25	599.24	572.64	549.47	529.06	510.89	494.58
111	5321.4	1432.0	1107.9	946.13	842.98	769.32	713.07	668.20	631.27	600.14	573.43	550.17	529.67	511.43	495.05
112	5334.2	1435.7	1110.6	948.27	844.78	770.85	714.40	669.37	632.29	601.05	574.23	550.87	530.29	511.97	495.52
113	5347.0	1439.4	1113.3	950.42	846.58	772.39	715.74	670.54	633.32	601.95	575.03	551.58	530.91	512.51	496.00
114	5359.9	1443.1	1115.9	952.58	848.38	773.93	717.08	671.71	634.35	602.86	575.83	552.29	531.54	513.06	496.47
115	5372.7	1446.7	1118.6	954.73	850.19	775.48	718.42	672.88	635.38	603.77	576.64	553.00	532.17	513.61	496.96
116	5385.5	1450.4	1121.3	956.89	852.00	777.03	719.77	674.06	636.42	604.69	577.45	553.72	532.80	514.17	497.44
117	5398.3	1454.1	1124.0	959.06	853.81	778.58	721.11	675.24	637.46	605.61	578.26	554.44	533.43	514.73	497.93
118	5411.1	1457.8	1126.7	961.22	855.62	780.14	722.47	676.43	638.51	606.53	579.08	555.16	534.07	515.29	498.42
119	5423.9	1461.5	1129.4	963.39	857.44	781.70	723.82	677.61	639.55	607.46	579.90	555.89	534.71	515.85	498.92
120	5436.7	1465.2	1132.1	965.57	859.26	783.26	725.18	678.81	640.60	608.39	580.72	556.61	535.36	516.42	499.42
121	5449.4	1468.9	1134.8	967.74	861.09	784.82	726.54	680.00	641.66	609.32	581.55	557.35	536.00	516.99	499.92
122	5462.2	1472.6	1137.5	969.92	862.92	786.39	727.90	681.20	642.72	610.26	582.38	558.08	536.66	517.57	500.42
123	5475.0	1476.3	1140.2	972.11	864.75	787.96	729.27	682.40	643.78	611.20	583.21	558.82	537.31	518.15	500.93

TABLE 6.9 Thermo-mechanical exergy of water vapour at vacuum pressures. Dashed lines indicate that water is not a pure gas at that temperature and pressure (it is a mixture of liquid and gas). (Continued)

T (°C)	e^{tm} (kJ/kg) Pressure (bar)														
	1.000	0.929	0.857	0.786	0.715	0.643	0.572	0.500	0.429	0.358	0.286	0.215	0.144	0.072	0.001
124	501.45	518.73	537.97	559.57	584.05	612.14	644.84	683.60	730.64	789.54	866.58	974.29	1142.9	1480.0	5487.8
125	501.96	519.32	538.63	560.31	584.89	613.09	645.91	684.81	732.02	791.12	868.42	976.48	1145.7	1483.7	5500.5
126	502.48	519.90	539.30	561.06	585.73	614.04	646.98	686.02	733.40	792.70	870.26	978.67	1148.4	1487.4	5513.3
127	503.00	520.50	539.97	561.82	586.58	614.99	648.05	687.24	734.78	794.28	872.10	980.87	1151.1	1491.1	5526.0
128	503.53	521.09	540.64	562.57	587.43	615.94	649.13	688.45	736.16	795.87	873.95	983.06	1153.8	1494.8	5538.8
129	504.06	521.69	541.31	563.33	588.28	616.90	650.21	689.68	737.55	797.46	875.80	985.26	1156.5	1498.5	5551.5
130	504.59	522.29	541.99	564.09	589.14	617.87	651.29	690.90	738.94	799.06	877.66	987.47	1159.3	1502.3	5564.3
131	505.13	522.90	542.67	564.86	590.00	618.83	652.38	692.13	740.33	800.65	879.51	989.67	1162.0	1506.0	5577.0
132	505.67	523.51	543.36	565.63	590.86	619.80	653.47	693.36	741.73	802.25	881.37	991.88	1164.7	1509.7	5589.7
133	506.21	524.12	544.05	566.40	591.73	620.77	654.56	694.59	743.13	803.86	883.23	994.10	1167.5	1513.4	5602.4
134	506.76	524.74	544.74	567.18	592.60	621.75	655.66	695.83	744.53	805.47	885.10	996.31	1170.2	1517.2	5615.1
135	507.31	525.36	545.43	567.96	593.47	622.73	656.76	697.07	745.94	807.08	886.97	998.53	1173.0	1520.9	5627.8
136	507.86	525.98	546.13	568.74	594.35	623.71	657.86	698.31	747.35	808.69	888.84	1000.8	1175.7	1524.6	5640.6
137	508.41	526.60	546.83	569.53	595.23	624.70	658.97	699.56	748.76	810.31	890.72	1003.0	1178.5	1528.4	5653.3
138	508.97	527.23	547.54	570.31	596.11	625.68	660.08	700.81	750.18	811.92	892.59	1005.2	1181.2	1532.1	5665.9
139	509.53	527.86	548.25	571.11	597.00	626.68	661.19	702.06	751.60	813.55	894.48	1007.4	1184.0	1535.8	5678.6
140	510.10	528.50	548.96	571.90	597.89	627.67	662.30	703.31	753.02	815.17	896.36	1009.7	1186.7	1539.6	5691.3
141	510.67	529.14	549.67	572.70	598.78	628.67	663.42	704.57	754.45	816.80	898.25	1011.9	1189.5	1543.3	5704.0
142	511.24	529.78	550.39	573.50	599.67	629.67	664.54	705.83	755.87	818.43	900.14	1014.1	1192.3	1547.1	5716.7
143	511.82	530.42	551.11	574.31	600.57	630.67	665.67	707.10	757.31	820.07	902.03	1016.4	1195.0	1550.8	5729.3
144	512.39	531.07	551.83	575.11	601.47	631.68	666.80	708.37	758.74	821.71	903.93	1018.6	1197.8	1554.6	5742.0
145	512.98	531.72	552.56	575.92	602.38	632.69	667.93	709.64	760.18	823.35	905.83	1020.9	1200.6	1558.3	5754.7
146	513.56	532.37	553.29	576.74	603.29	633.70	669.06	710.91	761.62	824.99	907.73	1023.1	1203.3	1562.1	5767.3
147	514.15	533.03	554.02	577.56	604.20	634.72	670.20	712.19	763.06	826.64	909.63	1025.4	1206.1	1565.9	5780.0
148	514.74	533.69	554.76	578.38	605.11	635.74	671.34	713.47	764.51	828.29	911.54	1027.6	1208.9	1569.6	5792.6
149	515.33	534.36	555.50	579.20	606.03	636.76	672.48	714.75	765.96	829.94	913.45	1029.9	1211.7	1573.4	5805.2
150	515.93	535.02	556.24	580.03	606.95	637.79	673.63	716.04	767.41	831.60	915.37	1032.2	1214.5	1577.1	5817.9
151	516.53	535.69	556.99	580.86	607.87	638.82	674.78	717.33	768.87	833.25	917.28	1034.4	1217.3	1580.9	5830.5
152	517.13	536.36	557.74	581.69	608.80	639.85	675.93	718.62	770.33	834.92	919.20	1036.7	1220.1	1584.7	5843.1
153	517.74	537.04	558.49	582.53	609.73	640.89	677.09	719.92	771.79	836.58	921.13	1039.0	1222.8	1588.5	5855.8
154	518.35	537.72	559.24	583.36	610.66	641.92	678.25	721.22	773.25	838.25	923.05	1041.3	1225.6	1592.2	5868.4

TABLE 6.9 Thermo-mechanical exergy of water vapour at vacuum pressures. Dashed lines indicate that water is not a pure gas at that temperature and pressure (it is a mixture of liquid and gas). (Continued)

e^{tm} (kJ/kg)

T (°C)	Pressure (bar)														
	0.001	0.072	0.144	0.215	0.286	0.358	0.429	0.500	0.572	0.643	0.715	0.786	0.857	0.929	1.000
155	5881.0	1596.0	1228.4	1043.5	924.98	839.92	774.72	722.52	679.41	642.96	611.60	584.21	560.00	538.40	518.96
156	5893.6	1599.8	1231.2	1045.8	926.91	841.59	776.19	723.82	680.57	644.01	612.54	585.05	560.76	539.09	519.58
157	5906.2	1603.6	1234.1	1048.1	928.85	843.27	777.67	725.13	681.74	645.06	613.48	585.90	561.53	539.77	520.20
158	5918.8	1607.4	1236.9	1050.4	930.78	844.95	779.14	726.44	682.91	646.11	614.42	586.75	562.29	540.47	520.82
159	5931.4	1611.2	1239.7	1052.7	932.72	846.63	780.62	727.76	684.08	647.16	615.37	587.60	563.06	541.16	521.44
160	5944.0	1615.0	1242.5	1055.0	934.67	848.32	782.10	729.07	685.26	648.21	616.32	588.46	563.84	541.86	522.07
161	5956.5	1618.7	1245.3	1057.3	936.61	850.01	783.59	730.39	686.44	649.27	617.27	589.32	564.61	542.56	522.70
162	5969.1	1622.5	1248.1	1059.6	938.56	851.70	785.08	731.71	687.62	650.34	618.23	590.18	565.39	543.26	523.34
163	5981.7	1626.3	1250.9	1061.9	940.51	853.39	786.57	733.04	688.81	651.40	619.19	591.05	566.17	543.97	523.97
164	5994.3	1630.1	1253.8	1064.2	942.47	855.09	788.06	734.37	690.00	652.47	620.15	591.92	566.96	544.68	524.61
165	6006.8	1633.9	1256.6	1066.5	944.43	856.79	789.56	735.70	691.19	653.54	621.12	592.79	567.75	545.39	525.26
166	6019.4	1637.7	1259.4	1068.8	946.39	858.49	791.06	737.03	692.38	654.61	622.09	593.67	568.54	546.10	525.90
167	6031.9	1641.5	1262.3	1071.1	948.35	860.20	792.56	738.37	693.58	655.69	623.06	594.54	569.33	546.82	526.55
168	6044.5	1645.4	1265.1	1073.4	950.31	861.90	794.07	739.71	694.78	656.77	624.03	595.42	570.13	547.54	527.20
169	6057.0	1649.2	1267.9	1075.7	952.28	863.61	795.58	741.05	695.98	657.85	625.01	596.31	570.93	548.27	527.86
170	6069.6	1653.0	1270.8	1078.0	954.25	865.33	797.09	742.40	697.19	658.94	625.99	597.20	571.73	548.99	528.52
171	6082.1	1656.8	1273.6	1080.3	956.23	867.05	798.60	743.75	698.40	660.03	626.97	598.08	572.54	549.72	529.18
172	6094.6	1660.6	1276.5	1082.7	958.21	868.77	800.12	745.10	699.61	661.12	627.96	598.98	573.35	550.46	529.84
173	6107.2	1664.4	1279.3	1085.0	960.19	870.49	801.64	746.45	700.82	662.21	628.95	599.87	574.16	551.19	530.51
174	6119.7	1668.3	1282.1	1087.3	962.17	872.21	803.16	747.81	702.04	663.31	629.94	600.77	574.97	551.93	531.18
175	6132.2	1672.1	1285.0	1089.7	964.15	873.94	804.69	749.17	703.26	664.41	630.93	601.67	575.79	552.67	531.85
176	6144.7	1675.9	1287.9	1092.0	966.14	875.67	806.22	750.53	704.48	665.51	631.93	602.58	576.61	553.42	532.52
177	6157.2	1679.7	1290.7	1094.3	968.13	877.41	807.75	751.90	705.71	666.62	632.93	603.48	577.43	554.16	533.20
178	6169.7	1683.6	1293.6	1096.7	970.12	879.14	809.28	753.27	706.94	667.73	633.93	604.39	578.26	554.91	533.88
179	6182.2	1687.4	1296.4	1099.0	972.12	880.88	810.82	754.64	708.17	668.84	634.94	605.30	579.09	555.67	534.57
180	6194.7	1691.2	1299.3	1101.4	974.12	882.62	812.36	756.01	709.40	669.95	635.95	606.22	579.92	556.42	535.25
181	6207.2	1695.1	1302.2	1103.7	976.12	884.37	813.90	757.39	710.64	671.07	636.96	607.14	580.75	557.18	535.94
182	6219.7	1698.9	1305.0	1106.1	978.12	886.11	815.45	758.77	711.88	672.19	637.97	608.06	581.59	557.94	536.63
183	6232.2	1702.7	1307.9	1108.4	980.13	887.86	817.00	760.15	713.12	673.31	638.99	608.98	582.43	558.71	537.33
184	6244.7	1706.6	1310.8	1110.8	982.14	889.62	818.55	761.53	714.37	674.44	640.01	609.91	583.27	559.47	538.03
185	6257.2	1710.4	1313.7	1113.1	984.15	891.37	820.10	762.92	715.62	675.56	641.03	610.84	584.12	560.24	538.73

TABLE 6.9 Thermo-mechanical exergy of water vapour at vacuum pressures. Dashed lines indicate that water is not a pure gas at that temperature and pressure (it is a mixture of liquid and gas). (Continued)

e^{tm} (kJ/kg)

T (°C)	Pressure (bar)														
	0.001	0.072	0.144	0.215	0.286	0.358	0.429	0.500	0.572	0.643	0.715	0.786	0.857	0.929	1.000
186	6269.6	1714.3	1316.5	1115.5	986.17	893.13	821.66	764.31	716.87	676.69	642.06	611.77	584.97	561.01	539.43
187	6282.1	1718.1	1319.4	1117.9	988.19	894.89	823.21	765.71	718.12	677.83	643.09	612.71	585.82	561.79	540.14
188	6294.6	1722.0	1322.3	1120.2	990.21	896.65	824.78	767.10	719.38	678.96	644.12	613.64	586.67	562.57	540.85
189	6307.0	1725.8	1325.2	1122.6	992.23	898.42	826.34	768.50	720.64	680.10	645.15	614.58	587.53	563.35	541.56
190	6319.5	1729.7	1328.1	1125.0	994.25	900.19	827.91	769.90	721.90	681.24	646.19	615.53	588.39	564.13	542.27
191	6331.9	1733.5	1331.0	1127.3	996.28	901.96	829.48	771.31	723.16	682.39	647.23	616.47	589.25	564.92	542.99
192	6344.4	1737.4	1333.9	1129.7	998.31	903.73	831.05	772.71	724.43	683.54	648.27	617.42	590.12	565.71	543.71
193	6356.8	1741.3	1336.8	1132.1	1000.3	905.51	832.62	774.12	725.70	684.69	649.32	618.37	590.98	566.50	544.43
194	6369.2	1745.1	1339.7	1134.5	1002.4	907.29	834.20	775.54	726.97	685.84	650.36	619.33	591.85	567.29	545.16
195	6381.7	1749.0	1342.6	1136.9	1004.4	909.07	835.78	776.95	728.25	686.99	651.41	620.28	592.73	568.09	545.89
196	6394.1	1752.9	1345.5	1139.2	1006.5	910.86	837.36	778.37	729.53	688.15	652.47	621.24	593.60	568.89	546.62
197	6406.5	1756.7	1348.4	1141.6	1008.5	912.64	838.95	779.79	730.81	689.31	653.52	622.21	594.48	569.69	547.35
198	6419.0	1760.6	1351.3	1144.0	1010.5	914.43	840.54	781.21	732.09	690.48	654.58	623.17	595.36	570.50	548.09
199	6431.4	1764.5	1354.2	1146.4	1012.6	916.23	842.13	782.64	733.38	691.64	655.64	624.14	596.25	571.31	548.83
200	6443.8	1768.3	1357.1	1148.8	1014.6	918.02	843.72	784.06	734.67	692.81	656.70	625.11	597.13	572.12	549.57
201	6456.2	1772.2	1360.0	1151.2	1016.7	919.82	845.32	785.49	735.96	693.98	657.77	626.08	598.02	572.93	550.31
202	6468.6	1776.1	1362.9	1153.6	1018.8	921.62	846.92	786.93	737.25	695.16	658.84	627.06	598.91	573.75	551.06
203	6481.0	1780.0	1365.8	1156.0	1020.8	923.42	848.52	788.36	738.55	696.33	659.91	628.03	599.81	574.57	551.81
204	6493.4	1783.9	1368.8	1158.4	1022.9	925.23	850.12	789.80	739.85	697.51	660.99	629.02	600.71	575.39	552.56
205	6505.8	1787.7	1371.7	1160.8	1024.9	927.04	851.73	791.24	741.15	698.69	662.06	630.00	601.61	576.21	553.32
206	6518.2	1791.6	1374.6	1163.2	1027.0	928.85	853.34	792.69	742.45	699.88	663.14	630.99	602.51	577.04	554.07
207	6530.6	1795.5	1377.5	1165.7	1029.1	930.66	854.95	794.13	743.76	701.07	664.22	631.97	603.41	577.87	554.83
208	6543.0	1799.4	1380.5	1168.1	1031.1	932.47	856.56	795.58	745.07	702.25	665.31	632.97	604.32	578.70	555.60
209	6555.3	1803.3	1383.4	1170.5	1033.2	934.29	858.18	797.03	746.38	703.45	666.39	633.96	605.23	579.53	556.36
210	6567.7	1807.2	1386.3	1172.9	1035.3	936.11	859.80	798.49	747.70	704.64	667.48	634.96	606.14	580.37	557.13
211	6580.1	1811.1	1389.3	1175.3	1037.4	937.94	861.42	799.95	749.02	705.84	668.58	635.96	607.06	581.21	557.90
212	6592.5	1815.0	1392.2	1177.8	1039.5	939.76	863.04	801.40	750.34	707.04	669.67	636.96	607.98	582.05	558.67
213	6604.8	1818.9	1395.2	1180.2	1041.5	941.59	864.67	802.87	751.66	708.24	670.77	637.96	608.90	582.90	559.45
214	6617.2	1822.8	1398.1	1182.6	1043.6	943.42	866.30	804.33	752.98	709.45	671.87	638.97	609.82	583.75	560.23
215	6629.5	1826.7	1401.1	1185.0	1045.7	945.25	867.93	805.80	754.31	710.65	672.97	639.98	610.75	584.60	561.01
216	6641.9	1830.6	1404.0	1187.5	1047.8	947.09	869.57	807.27	755.64	711.87	674.08	640.99	611.67	585.45	561.79

TABLE 6.9 Thermo-mechanical exergy of water vapour at vacuum pressures. Dashed lines indicate that water is not a pure gas at that temperature and pressure (it is a mixture of liquid and gas). (Continued)

e^tm (kJ/kg)

T (°C)	Pressure (bar)														
	0.001	0.072	0.144	0.215	0.286	0.358	0.429	0.500	0.572	0.643	0.715	0.786	0.857	0.929	1.000
217	6654.2	1834.5	1407.0	1189.9	1049.9	948.93	871.20	808.74	756.97	713.08	675.19	642.00	612.61	586.30	562.58
218	6666.6	1838.4	1409.9	1192.4	1052.0	950.77	872.84	810.21	758.31	714.29	676.30	643.02	613.54	587.16	563.37
219	6678.9	1842.3	1412.9	1194.8	1054.1	952.61	874.48	811.69	759.65	715.51	677.41	644.04	614.48	588.02	564.16
220	6691.3	1846.2	1415.8	1197.2	1056.2	954.46	876.13	813.17	760.99	716.73	678.52	645.06	615.41	588.89	564.95
221	6703.6	1850.1	1418.8	1199.7	1058.3	956.30	877.78	814.65	762.33	717.95	679.64	646.09	616.35	589.75	565.75
222	6715.9	1854.1	1421.7	1202.1	1060.4	958.15	879.42	816.13	763.67	719.18	680.76	647.11	617.30	590.62	566.55
223	6728.3	1858.0	1424.7	1204.6	1062.5	960.01	881.08	817.62	765.02	720.41	681.88	648.14	618.24	591.49	567.35
224	6740.6	1861.9	1427.7	1207.0	1064.6	961.86	882.73	819.11	766.37	721.64	683.01	649.18	619.19	592.36	568.15
225	6752.9	1865.8	1430.6	1209.5	1066.7	963.72	884.39	820.60	767.72	722.87	684.14	650.21	620.14	593.24	568.96
226	6765.2	1869.7	1433.6	1212.0	1068.8	965.58	886.05	822.10	769.08	724.11	685.27	651.25	621.10	594.11	569.77
227	6777.5	1873.7	1436.6	1214.4	1071.0	967.44	887.71	823.59	770.44	725.34	686.40	652.29	622.05	595.00	570.58
228	6789.8	1877.6	1439.6	1216.9	1073.1	969.31	889.37	825.09	771.80	726.58	687.54	653.33	623.01	595.88	571.39
229	6802.1	1881.5	1442.5	1219.3	1075.2	971.17	891.04	826.59	773.16	727.83	688.67	654.37	623.97	596.76	572.21
230	6814.5	1885.5	1445.5	1221.8	1077.3	973.04	892.71	828.10	774.52	729.07	689.81	655.42	624.94	597.65	573.03
231	6826.7	1889.4	1448.5	1224.3	1079.5	974.91	894.38	829.60	775.89	730.32	690.96	656.47	625.90	598.54	573.85
232	6839.0	1893.3	1451.5	1226.8	1081.6	976.79	896.05	831.11	777.26	731.57	692.10	657.52	626.87	599.43	574.67
233	6851.3	1897.3	1454.5	1229.2	1083.7	978.66	897.73	832.62	778.63	732.82	693.25	658.58	627.84	600.33	575.50
234	6863.6	1901.2	1457.5	1231.7	1085.8	980.54	899.40	834.14	780.01	734.08	694.40	659.63	628.81	601.23	576.33
235	6875.9	1905.1	1460.5	1234.2	1088.0	982.42	901.08	835.65	781.39	735.33	695.55	660.69	629.79	602.13	577.16
236	6888.2	1909.1	1463.4	1236.7	1090.1	984.31	902.77	837.17	782.77	736.59	696.71	661.76	630.77	603.03	577.99
237	6900.5	1913.0	1466.4	1239.2	1092.3	986.19	904.45	838.69	784.15	737.86	697.86	662.82	631.75	603.93	578.83
238	6912.7	1917.0	1469.4	1241.6	1094.4	988.08	906.14	840.22	785.53	739.12	699.02	663.89	632.73	604.84	579.66
239	6925.0	1920.9	1472.4	1244.1	1096.6	989.97	907.83	841.74	786.92	740.39	700.19	664.95	633.72	605.75	580.51
240	6937.3	1924.9	1475.4	1246.6	1098.7	991.87	909.52	843.27	788.31	741.66	701.35	666.03	634.71	606.66	581.35
241	6949.5	1928.8	1478.4	1249.1	1100.8	993.76	911.22	844.80	789.70	742.93	702.52	667.10	635.70	607.58	582.19
242	6961.8	1932.8	1481.4	1251.6	1103.0	995.66	912.91	846.33	791.09	744.20	703.69	668.18	636.69	608.49	583.04
243	6974.1	1936.7	1484.4	1254.1	1105.2	997.56	914.61	847.87	792.49	745.48	704.86	669.25	637.68	609.41	583.89
244	6986.3	1940.7	1487.5	1256.6	1107.3	999.46	916.32	849.40	793.89	746.76	706.03	670.33	638.68	610.33	584.74
245	6998.6	1944.6	1490.5	1259.1	1109.5	1001.4	918.02	850.94	795.29	748.04	707.21	671.42	639.68	611.26	585.60
246	7010.8	1948.6	1493.5	1261.6	1111.6	1003.3	919.73	852.49	796.69	749.32	708.39	672.50	640.68	612.19	586.46
247	7023.1	1952.5	1496.5	1264.1	1113.8	1005.2	921.43	854.03	798.10	750.61	709.57	673.59	641.69	613.11	587.32

TABLE 6.9 Thermo-mechanical exergy of water vapour at vacuum pressures. Dashed lines indicate that water is not a pure gas at that temperature and pressure (it is a mixture of liquid and gas). (Continued)

154

e^{tm} (kJ/kg)

T (°C)	Pressure (bar)														
	0.001	0.072	0.144	0.215	0.286	0.358	0.429	0.500	0.572	0.643	0.715	0.786	0.857	0.929	1.000
248	7035.3	1956.5	1499.5	1266.6	1116.0	1007.1	923.15	855.58	799.50	751.90	710.75	674.68	642.69	614.04	588.18
249	7047.5	1960.5	1502.5	1269.1	1118.1	1009.0	924.86	857.12	800.91	753.19	711.94	675.77	643.70	614.98	589.04
250	7059.8	1964.4	1505.5	1271.6	1120.3	1010.9	926.57	858.68	802.33	754.48	713.12	676.87	644.71	615.91	589.91
251	7072.0	1968.4	1508.6	1274.1	1122.5	1012.8	928.29	860.23	803.74	755.77	714.31	677.97	645.73	616.85	590.78
252	7084.2	1972.4	1511.6	1276.7	1124.6	1014.8	930.01	861.79	805.16	757.07	715.51	679.07	646.74	617.79	591.65
253	7096.5	1976.3	1514.6	1279.2	1126.8	1016.7	931.73	863.34	806.58	758.37	716.70	680.17	647.76	618.74	592.52
254	7108.7	1980.3	1517.6	1281.7	1129.0	1018.6	933.46	864.90	808.00	759.67	717.90	681.27	648.78	619.68	593.40
255	7120.9	1984.3	1520.7	1284.2	1131.2	1020.5	935.19	866.47	809.42	760.98	719.10	682.38	649.80	620.63	594.28
256	7133.1	1988.2	1523.7	1286.7	1133.4	1022.5	936.91	868.03	810.85	762.28	720.30	683.49	650.83	621.58	595.16
257	7145.3	1992.2	1526.7	1289.3	1135.5	1024.4	938.65	869.60	812.28	763.59	721.50	684.60	651.86	622.53	596.04
258	7157.5	1996.2	1529.8	1291.8	1137.7	1026.3	940.38	871.17	813.71	764.90	722.71	685.71	652.89	623.48	596.93
259	7169.7	2000.2	1532.8	1294.3	1139.9	1028.3	942.12	872.74	815.14	766.22	723.92	686.83	653.92	624.44	597.81
260	7182.0	2004.2	1535.9	1296.9	1142.1	1030.2	943.85	874.31	816.58	767.53	725.13	687.95	654.95	625.40	598.70
261	7194.2	2008.1	1538.9	1299.4	1144.3	1032.1	945.59	875.89	818.01	768.85	726.34	689.07	655.99	626.36	599.60
262	7206.3	2012.1	1541.9	1301.9	1146.5	1034.1	947.34	877.47	819.45	770.17	727.56	690.19	657.03	627.32	600.49
263	7218.5	2016.1	1545.0	1304.5	1148.7	1036.0	949.08	879.05	820.90	771.49	728.78	691.31	658.07	628.29	601.39
264	7230.7	2020.1	1548.0	1307.0	1150.9	1038.0	950.83	880.63	822.34	772.82	730.00	692.44	659.11	629.26	602.29
265	7242.9	2024.1	1551.1	1309.6	1153.1	1039.9	952.58	882.22	823.79	774.15	731.22	693.57	660.16	630.23	603.19
266	7255.1	2028.1	1554.1	1312.1	1155.3	1041.9	954.33	883.80	825.23	775.48	732.44	694.70	661.21	631.20	604.09
267	7267.3	2032.1	1557.2	1314.6	1157.5	1043.8	956.08	885.39	826.69	776.81	733.67	695.83	662.26	632.17	605.00
268	7279.5	2036.1	1560.2	1317.2	1159.7	1045.8	957.84	886.98	828.14	778.14	734.90	696.97	663.31	633.15	605.90
269	7291.6	2040.1	1563.3	1319.7	1161.9	1047.7	959.60	888.58	829.59	779.48	736.13	698.11	664.37	634.13	606.81
270	7303.8	2044.1	1566.4	1322.3	1164.1	1049.7	961.36	890.17	831.05	780.81	737.36	699.25	665.42	635.11	607.73
271	7316.0	2048.1	1569.4	1324.9	1166.4	1051.7	963.12	891.77	832.51	782.15	738.60	700.39	666.48	636.09	608.64
272	7328.2	2052.1	1572.5	1327.4	1168.6	1053.6	964.88	893.37	833.97	783.50	739.84	701.54	667.54	637.08	609.56
273	7340.3	2056.1	1575.6	1330.0	1170.8	1055.6	966.65	894.97	835.44	784.84	741.08	702.68	668.61	638.07	610.48
274	7352.5	2060.1	1578.6	1332.5	1173.0	1057.6	968.42	896.58	836.90	786.19	742.32	703.83	669.67	639.06	611.40
275	7364.6	2064.1	1581.7	1335.1	1175.2	1059.5	970.19	898.19	838.37	787.54	743.56	704.98	670.74	640.05	612.32
276	7376.8	2068.1	1584.8	1337.7	1177.5	1061.5	971.96	899.80	839.84	788.89	744.81	706.14	671.81	641.05	613.25
277	7388.9	2072.1	1587.8	1340.2	1179.7	1063.5	973.74	901.41	841.32	790.24	746.06	707.29	672.88	642.04	614.17
278	7401.1	2076.1	1590.9	1342.8	1181.9	1065.5	975.51	903.02	842.79	791.60	747.31	708.45	673.96	643.04	615.10

TABLE 6.9 Thermo-mechanical exergy of water vapour at vacuum pressures. Dashed lines indicate that water is not a pure gas at that temperature and pressure (it is a mixture of liquid and gas). *(Continued)*

e^{tm} (kJ/kg)

Pressure (bar)

T (°C)	0.001	0.072	0.144	0.215	0.286	0.358	0.429	0.500	0.572	0.643	0.715	0.786	0.857	0.929	1.000
279	7413.2	2080.1	1594.0	1345.4	1184.2	1067.4	977.29	904.64	844.27	792.96	748.56	709.61	675.03	644.04	616.04
280	7425.4	2084.1	1597.1	1347.9	1186.4	1069.4	979.07	906.25	845.75	794.32	749.82	710.77	676.11	645.05	616.97
281	7437.5	2088.1	1600.1	1350.5	1188.6	1071.4	980.86	907.87	847.23	795.68	751.08	711.94	677.19	646.05	617.91
282	7449.7	2092.1	1603.2	1353.1	1190.9	1073.4	982.64	909.50	848.71	797.04	752.34	713.11	678.28	647.06	618.85
283	7461.8	2096.2	1606.3	1355.7	1193.1	1075.4	984.43	911.12	850.20	798.41	753.60	714.27	679.36	648.07	619.79
284	7473.9	2100.2	1609.4	1358.2	1195.3	1077.4	986.22	912.75	851.69	799.78	754.86	715.45	680.45	649.08	620.73
285	7486.1	2104.2	1612.5	1360.8	1197.6	1079.4	988.01	914.37	853.18	801.15	756.13	716.62	681.54	650.09	621.67
286	7498.2	2108.2	1615.6	1363.4	1199.8	1081.3	989.81	916.00	854.67	802.52	757.40	717.79	682.63	651.11	622.62
287	7510.3	2112.2	1618.7	1366.0	1202.1	1083.3	991.60	917.64	856.16	803.90	758.67	718.97	683.73	652.13	623.57
288	7522.4	2116.3	1621.8	1368.6	1204.3	1085.3	993.40	919.27	857.66	805.27	759.94	720.15	684.82	653.15	624.52
289	7534.5	2120.3	1624.8	1371.2	1206.6	1087.3	995.20	920.91	859.16	806.65	761.21	721.33	685.92	654.17	625.48
290	7546.7	2124.3	1627.9	1373.8	1208.8	1089.3	997.00	922.55	860.66	808.03	762.49	722.52	687.02	655.20	626.43
291	7558.8	2128.3	1631.0	1376.4	1211.1	1091.3	998.81	924.19	862.16	809.42	763.77	723.70	688.12	656.22	627.39
292	7570.9	2132.4	1634.1	1379.0	1213.3	1093.3	1000.6	925.83	863.66	810.80	765.05	724.89	689.23	657.25	628.35
293	7583.0	2136.4	1637.2	1381.6	1215.6	1095.4	1002.4	927.48	865.17	812.19	766.33	726.08	690.33	658.28	629.31
294	7595.1	2140.4	1640.3	1384.2	1217.9	1097.4	1004.2	929.12	866.68	813.58	767.62	727.27	691.44	659.32	630.28
295	7607.2	2144.5	1643.5	1386.8	1220.1	1099.4	1006.0	930.77	868.19	814.97	768.91	728.47	692.55	660.35	631.24
296	7619.3	2148.5	1646.6	1389.4	1222.4	1101.4	1007.9	932.42	869.70	816.36	770.19	729.66	693.67	661.39	632.21
297	7631.4	2152.5	1649.7	1392.0	1224.7	1103.4	1009.7	934.08	871.22	817.76	771.49	730.86	694.78	662.43	633.18
298	7643.5	2156.6	1652.8	1394.6	1226.9	1105.4	1011.5	935.73	872.74	819.16	772.78	732.06	695.90	663.47	634.15
299	7655.6	2160.6	1655.9	1397.2	1229.2	1107.4	1013.3	937.39	874.26	820.56	774.08	733.26	697.02	664.52	635.13
300	7667.7	2164.7	1659.0	1399.8	1231.5	1109.5	1015.1	939.05	875.78	821.96	775.37	734.47	698.14	665.56	636.11
305	7728.1	2184.9	1674.6	1412.9	1242.9	1119.6	1024.3	947.38	883.41	829.00	781.89	740.52	703.78	670.82	641.02
310	7788.4	2205.2	1690.2	1426.0	1254.3	1129.8	1033.5	955.75	891.10	836.09	788.46	746.63	709.47	676.13	645.99
315	7848.7	2225.5	1705.9	1439.2	1265.8	1140.0	1042.7	964.18	898.84	843.24	795.08	752.79	715.21	681.49	651.00
320	7908.9	2245.8	1721.6	1452.4	1277.4	1150.3	1052.0	972.66	906.63	850.43	801.75	758.99	721.00	686.91	656.07
325	7969.1	2266.2	1737.4	1465.7	1288.9	1160.6	1061.4	981.18	914.46	857.67	808.47	765.25	726.83	692.37	661.19
330	8029.2	2286.6	1753.2	1479.0	1300.6	1171.0	1070.7	989.76	922.35	864.96	815.24	771.55	732.72	697.88	666.35
335	8089.2	2307.0	1769.1	1492.4	1312.3	1181.5	1080.2	998.38	930.28	872.30	822.06	777.91	738.66	703.44	671.57
340	8149.2	2327.5	1785.0	1505.8	1324.0	1191.9	1089.7	1007.0	938.26	879.69	828.92	784.31	744.64	709.04	676.83
345	8209.1	2348.0	1800.9	1519.2	1335.7	1202.4	1099.2	1015.8	946.29	887.12	835.83	790.76	750.68	714.70	682.14

Table 6.9 Thermo-mechanical exergy of water vapour at vacuum pressures. Dashed lines indicate that water is not a pure gas at that temperature and pressure (it is a mixture of liquid and gas). (*Continued*)

e^tm (kJ/kg)

T (°C)	Pressure (bar)														
	0.001	0.072	0.144	0.215	0.286	0.358	0.429	0.500	0.572	0.643	0.715	0.786	0.857	0.929	1.000
350	8269.0	2368.5	1816.9	1532.7	1347.6	1213.0	1108.8	1024.5	954.36	894.60	842.79	797.25	756.76	720.40	687.50
355	8328.8	2389.1	1832.9	1546.2	1359.4	1223.6	1118.4	1033.3	962.48	902.13	849.80	803.79	762.88	726.15	692.90
360	8388.6	2409.7	1848.9	1559.8	1371.3	1234.3	1128.1	1042.2	970.65	909.70	856.85	810.38	769.05	731.94	698.35
365	8448.3	2430.3	1865.0	1573.4	1383.3	1245.0	1137.8	1051.1	978.86	917.32	863.95	817.02	775.27	737.78	703.85
370	8508.0	2451.0	1881.1	1587.1	1395.2	1255.7	1147.5	1060.0	987.12	924.98	871.09	823.70	781.53	743.67	709.39
375	8567.7	2471.7	1897.3	1600.8	1407.3	1266.5	1157.3	1069.0	995.42	932.69	878.28	830.42	787.84	749.60	714.97
380	8627.2	2492.4	1913.5	1614.5	1419.3	1277.3	1167.2	1078.0	1003.8	940.44	885.51	837.19	794.19	755.58	720.60
385	8686.8	2513.1	1929.7	1628.3	1431.4	1288.2	1177.0	1087.1	1012.2	948.24	892.78	844.00	800.59	761.60	726.28
390	8746.3	2533.9	1946.0	1642.1	1443.6	1299.1	1187.0	1096.2	1020.6	956.08	900.10	850.86	807.03	767.66	732.00
395	8805.7	2554.7	1962.3	1655.9	1455.8	1310.0	1196.9	1105.4	1029.1	963.96	907.47	857.76	813.52	773.77	737.76
400	8865.2	2575.6	1978.6	1669.8	1468.0	1321.0	1206.9	1114.6	1037.6	971.88	914.87	864.70	820.04	779.91	743.57
405	8924.5	2596.4	1995.0	1683.7	1480.3	1332.1	1217.0	1123.8	1046.1	979.85	922.32	871.68	826.61	786.11	749.41
410	8983.9	2617.3	2011.4	1697.7	1492.6	1343.1	1227.1	1133.1	1054.7	987.86	929.81	878.71	833.22	792.34	755.30
415	9043.2	2638.2	2027.9	1711.7	1504.9	1354.3	1237.2	1142.5	1063.4	995.91	937.34	885.78	839.88	798.62	761.23
420	9102.4	2659.2	2044.3	1725.8	1517.3	1365.4	1247.4	1151.8	1072.1	1004.0	944.91	892.89	846.57	804.94	767.21
425	9161.7	2680.2	2060.9	1739.8	1529.8	1376.6	1257.6	1161.2	1080.8	1012.1	952.53	900.04	853.31	811.30	773.22
430	9220.9	2701.2	2077.4	1753.9	1542.2	1387.8	1267.9	1170.7	1089.5	1020.3	960.18	907.24	860.09	817.70	779.28
435	9280.0	2722.2	2094.0	1768.1	1554.7	1399.1	1278.2	1180.2	1098.4	1028.5	967.88	914.47	866.90	824.14	785.38
440	9339.2	2743.3	2110.6	1782.3	1567.3	1410.4	1288.5	1189.7	1107.2	1036.8	975.61	921.74	873.76	830.62	791.51
445	9398.3	2764.4	2127.3	1796.5	1579.9	1421.8	1298.9	1199.3	1116.1	1045.1	983.39	929.06	880.66	837.14	797.69
450	9457.3	2785.5	2143.9	1810.8	1592.5	1433.1	1309.3	1208.9	1125.0	1053.4	991.20	936.41	887.60	843.70	803.91
455	9516.4	2806.7	2160.7	1825.1	1605.1	1444.6	1319.7	1218.5	1134.0	1061.8	999.06	943.80	894.58	850.30	810.16
460	9575.4	2827.9	2177.4	1839.4	1617.8	1456.0	1330.2	1228.2	1143.0	1070.2	1006.9	951.23	901.59	856.94	816.46
465	9634.4	2849.1	2194.2	1853.8	1630.5	1467.5	1340.7	1237.9	1152.0	1078.6	1014.9	958.70	908.65	863.62	822.79
470	9693.3	2870.3	2211.0	1868.2	1643.3	1479.1	1351.3	1247.7	1161.1	1087.1	1022.9	966.21	915.74	870.34	829.16
475	9752.2	2891.5	2227.9	1882.6	1656.1	1490.6	1361.9	1257.5	1170.2	1095.7	1030.9	973.76	922.87	877.09	835.58
480	9811.2	2912.8	2244.7	1897.1	1668.9	1502.2	1372.5	1267.3	1179.4	1104.2	1038.9	981.35	930.04	883.89	842.02
485	9870.0	2934.1	2261.6	1911.6	1681.8	1513.9	1383.2	1277.2	1188.5	1112.8	1047.0	988.97	937.25	890.72	848.51
490	9928.9	2955.5	2278.6	1926.1	1694.7	1525.6	1393.9	1287.1	1197.8	1121.5	1055.1	996.63	944.50	897.59	855.04
495	9987.7	2976.8	2295.6	1940.7	1707.6	1537.3	1404.7	1297.0	1207.0	1130.1	1063.3	1004.3	951.78	904.49	861.60
500	10046.6	2998.2	2312.6	1955.3	1720.6	1549.0	1415.4	1307.0	1216.3	1138.8	1071.5	1012.1	959.10	911.44	868.20

TABLE 6.9 Thermo-mechanical exergy of water vapour at vacuum pressures. Dashed lines indicate that water is not a pure gas at that temperature and pressure (it is a mixture of liquid and gas). (Continued)

T (°C)	0.001	0.072	0.144	0.215	0.286	0.358	0.429	0.500	0.572	0.643	0.715	0.786	0.857	0.929	1.000
															e^{tm} (kJ/kg)
															Pressure (bar)
505	10105.4	3019.7	2329.6	1969.9	1733.6	1560.8	1426.3	1317.0	1225.7	1147.6	1079.7	1019.8	966.46	918.42	874.83
510	10164.1	3041.1	2346.7	1984.6	1746.7	1572.6	1437.1	1327.1	1235.0	1156.4	1088.0	1027.6	973.85	925.43	881.51
515	10222.9	3062.6	2363.8	1999.3	1759.7	1584.5	1448.0	1337.2	1244.5	1165.2	1096.3	1035.5	981.28	932.49	888.22
520	10281.7	3084.1	2380.9	2014.0	1772.8	1596.4	1458.9	1347.3	1253.9	1174.1	1104.6	1043.4	988.75	939.58	894.96
525	10340.4	3105.6	2398.1	2028.8	1786.0	1608.3	1469.9	1357.4	1263.4	1183.0	1113.0	1051.3	996.25	946.70	901.74
530	10399.1	3127.1	2415.2	2043.6	1799.2	1620.3	1480.9	1367.6	1272.9	1191.9	1121.4	1059.2	1003.8	953.86	908.56
535	10457.8	3148.7	2432.5	2058.4	1812.4	1632.3	1491.9	1377.9	1282.5	1200.8	1129.8	1067.2	1011.4	961.06	915.41
540	10516.5	3170.3	2449.7	2073.3	1825.6	1644.3	1503.0	1388.1	1292.0	1209.8	1138.3	1075.2	1019.0	968.29	922.30
545	10575.2	3191.9	2467.0	2088.2	1838.9	1656.4	1514.1	1398.4	1301.7	1218.9	1146.9	1083.3	1026.6	975.56	929.22
550	10633.8	3213.6	2484.3	2103.1	1852.2	1668.5	1525.2	1408.8	1311.3	1228.0	1155.4	1091.4	1034.3	982.86	936.18
555	10692.5	3235.3	2501.6	2118.0	1865.5	1680.6	1536.3	1419.1	1321.0	1237.1	1164.0	1099.5	1042.0	990.20	943.17
560	10751.1	3257.0	2519.0	2133.0	1878.9	1692.7	1547.5	1429.5	1330.7	1246.2	1172.6	1107.7	1049.8	997.57	950.20
565	10809.8	3278.7	2536.4	2148.1	1892.3	1704.9	1558.8	1440.0	1340.5	1255.4	1181.3	1115.9	1057.5	1005.0	957.26
570	10868.4	3300.4	2553.8	2163.1	1905.8	1717.2	1570.0	1450.4	1350.3	1264.6	1190.0	1124.1	1065.4	1012.4	964.36
575	10927.0	3322.2	2571.3	2178.2	1919.2	1729.4	1581.3	1460.9	1360.1	1273.8	1198.7	1132.4	1073.2	1019.9	971.49
580	10985.6	3344.0	2588.8	2193.3	1932.7	1741.7	1592.7	1471.5	1370.0	1283.1	1207.4	1140.7	1081.1	1027.4	978.65
585	11044.2	3365.8	2606.3	2208.4	1946.3	1754.0	1604.0	1482.0	1379.9	1292.4	1216.2	1149.0	1089.0	1034.9	985.85
590	11102.8	3387.7	2623.8	2223.6	1959.8	1766.4	1615.4	1492.6	1389.8	1301.7	1225.1	1157.4	1097.0	1042.5	993.08
595	11161.4	3409.6	2641.4	2238.8	1973.4	1778.8	1626.8	1503.3	1399.7	1311.1	1233.9	1165.8	1104.9	1050.1	1000.3
600	11220.0	3431.5	2659.0	2254.1	1987.1	1791.2	1638.3	1513.9	1409.7	1320.5	1242.8	1174.2	1113.0	1057.8	1007.6
605	11278.6	3453.4	2676.6	2269.3	2000.7	1803.7	1649.8	1524.6	1419.7	1329.9	1251.7	1182.7	1121.0	1065.4	1015.0
610	11337.2	3475.4	2694.3	2284.6	2014.4	1816.1	1661.3	1535.3	1429.8	1339.4	1260.7	1191.2	1129.1	1073.2	1022.3
615	11395.7	3497.3	2711.9	2299.9	2028.1	1828.7	1672.9	1546.1	1439.9	1348.9	1269.7	1199.7	1137.2	1080.9	1029.7
620	11454.3	3519.3	2729.7	2315.3	2041.9	1841.2	1684.5	1556.9	1450.0	1358.5	1278.7	1208.3	1145.4	1088.7	1037.2
625	11512.9	3541.4	2747.4	2330.7	2055.7	1853.8	1696.1	1567.7	1460.2	1368.0	1287.8	1216.9	1153.5	1096.5	1044.6
630	11571.5	3563.4	2765.2	2346.1	2069.5	1866.4	1707.7	1578.6	1470.3	1377.6	1296.8	1225.5	1161.8	1104.3	1052.1
635	11630.0	3585.5	2783.0	2361.5	2083.3	1879.0	1719.4	1589.5	1480.6	1387.3	1306.0	1234.1	1170.0	1112.2	1059.6
640	—	3607.6	2800.8	2377.0	2097.2	1891.7	1731.1	1600.4	1490.8	1396.9	1315.1	1242.8	1178.3	1120.1	1067.2
645	—	3629.7	2818.6	2392.5	2111.1	1904.4	1742.9	1611.3	1501.1	1406.6	1324.3	1251.6	1186.6	1128.0	1074.8
650	—	3651.9	2836.5	2408.0	2125.0	1917.1	1754.6	1622.3	1511.4	1416.3	1333.5	1260.3	1194.9	1136.0	1082.4
655	—	3674.1	2854.4	2423.6	2139.0	1929.9	1766.4	1633.3	1521.7	1426.1	1342.7	1269.1	1203.3	1144.0	1090.1

TABLE 6.9 Thermo-mechanical exergy of water vapour at vacuum pressures. Dashed lines indicate that water is not a pure gas at that temperature and pressure (it is a mixture of liquid and gas). (Continued)

e^{tm} (kJ/kg)

Pressure (bar)

T (°C)	0.001	0.072	0.144	0.215	0.286	0.358	0.429	0.500	0.572	0.643	0.715	0.786	0.857	0.929	1.000
660	—	3696.3	2872.4	2439.2	2153.0	1942.7	1778.3	1644.4	1532.1	1435.9	1352.0	1277.9	1211.7	1152.0	1097.7
665	—	3718.5	2890.3	2454.8	2167.0	1955.5	1790.1	1655.5	1542.5	1445.7	1361.3	1286.8	1220.1	1160.1	1105.5
670	—	3740.8	2908.3	2470.5	2181.0	1968.4	1802.0	1666.6	1552.9	1455.6	1370.7	1295.6	1228.6	1168.2	1113.2
675	—	3763.0	2926.3	2486.1	2195.1	1981.2	1814.0	1677.7	1563.4	1465.4	1380.0	1304.5	1237.1	1176.3	1121.0
680	—	3785.3	2944.4	2501.8	2209.2	1994.1	1825.9	1688.9	1573.9	1475.4	1389.4	1313.5	1245.6	1184.4	1128.8
685	—	3807.7	2962.4	2517.6	2223.4	2007.1	1837.9	1700.1	1584.4	1485.3	1398.9	1322.5	1254.2	1192.6	1136.6
690	—	3830.0	2980.5	2533.3	2237.5	2020.1	1849.9	1711.3	1595.0	1495.3	1408.3	1331.5	1262.8	1200.8	1144.5
695	—	3852.4	2998.7	2549.1	2251.7	2033.1	1862.0	1722.6	1605.6	1505.3	1417.8	1340.5	1271.4	1209.1	1152.4
700	—	3874.8	3016.8	2564.9	2266.0	2046.1	1874.1	1733.8	1616.2	1515.3	1427.3	1349.6	1280.0	1217.3	1160.3
705	—	3897.2	3035.0	2580.8	2280.2	2059.1	1886.2	1745.2	1626.8	1525.4	1436.9	1358.6	1288.7	1225.6	1168.3
710	—	3919.7	3053.2	2596.6	2294.5	2072.2	1898.3	1756.5	1637.5	1535.5	1446.5	1367.8	1297.4	1234.0	1176.3
715	—	3942.2	3071.4	2612.5	2308.8	2085.3	1910.5	1767.9	1648.2	1545.6	1456.1	1376.9	1306.2	1242.3	1184.3
720	—	3964.7	3089.7	2628.5	2323.1	2098.5	1922.6	1779.3	1658.9	1555.7	1465.7	1386.1	1314.9	1250.7	1192.3
725	—	3987.2	3108.0	2644.4	2337.5	2111.7	1934.9	1790.7	1669.7	1565.9	1475.4	1395.3	1323.7	1259.2	1200.4
730	—	4009.7	3126.3	2660.4	2351.9	2124.9	1947.1	1802.2	1680.5	1576.1	1485.1	1404.5	1332.6	1267.6	1208.5
735	—	4032.3	3144.6	2676.4	2366.3	2138.1	1959.4	1813.7	1691.3	1586.4	1494.8	1413.8	1341.4	1276.1	1216.7
740	—	4054.9	3163.0	2692.5	2380.8	2151.3	1971.7	1825.2	1702.2	1596.6	1504.6	1423.1	1350.3	1284.6	1224.8
745	—	4077.5	3181.4	2708.5	2395.3	2164.6	1984.0	1836.7	1713.1	1606.9	1514.3	1432.4	1359.2	1293.1	1233.0
750	—	4100.2	3199.8	2724.6	2409.8	2178.0	1996.4	1848.3	1724.0	1617.3	1524.2	1441.8	1368.2	1301.7	1241.3
755	—	4122.8	3218.3	2740.7	2424.3	2191.3	2008.8	1859.9	1734.9	1627.6	1534.0	1451.2	1377.1	1310.3	1249.5
760	—	4145.5	3236.7	2756.9	2438.9	2204.7	2021.2	1871.6	1745.9	1638.0	1543.9	1460.6	1386.1	1318.9	1257.8
765	—	4168.3	3255.2	2773.1	2453.5	2218.1	2033.7	1883.2	1756.9	1648.4	1553.8	1470.0	1395.2	1327.6	1266.1
770	—	4191.0	3273.8	2789.3	2468.1	2231.5	2046.1	1894.9	1767.9	1658.9	1563.7	1479.5	1404.2	1336.3	1274.4
775	—	4213.8	3292.3	2805.5	2482.7	2244.9	2058.6	1906.6	1778.9	1669.3	1573.7	1489.0	1413.3	1345.0	1282.8
780	—	4236.6	3310.9	2821.7	2497.4	2258.4	2071.2	1918.4	1790.0	1679.8	1583.6	1498.5	1422.4	1353.7	1291.2
785	—	4259.4	3329.5	2838.0	2512.1	2271.9	2083.7	1930.1	1801.1	1690.3	1593.7	1508.1	1431.6	1362.5	1299.6
790	—	4282.2	3348.1	2854.3	2526.8	2285.5	2096.3	1941.9	1812.2	1700.9	1603.7	1517.7	1440.7	1371.3	1308.1
795	—	4305.1	3366.8	2870.7	2541.6	2299.0	2108.9	1953.8	1823.4	1711.5	1613.8	1527.3	1449.9	1380.1	1316.6
800	—	4328.0	3385.4	2887.0	2556.3	2312.6	2121.6	1965.6	1834.6	1722.1	1623.9	1536.9	1459.2	1389.0	1325.1
805	—	4350.9	3404.1	2903.4	2571.2	2326.2	2134.3	1977.5	1845.8	1732.7	1634.0	1546.6	1468.4	1397.8	1333.6
810	—	4373.8	3422.9	2919.8	2586.0	2339.9	2147.0	1989.4	1857.1	1743.4	1644.1	1556.3	1477.7	1406.8	1342.2

TABLE 6.9 Thermo-mechanical exergy of water vapour at vacuum pressures. Dashed lines indicate that water is not a pure gas at that temperature and pressure (it is a mixture of liquid and gas). (Continued)

e^{tm} (kJ/kg)

T (°C)	Pressure (bar)														
	0.001	0.072	0.144	0.215	0.286	0.358	0.429	0.500	0.572	0.643	0.715	0.786	0.857	0.929	1.000
815	—	4396.8	3441.6	2936.3	2600.8	2353.6	2159.7	2001.4	1868.3	1754.1	1654.3	1566.0	1487.0	1415.7	1350.8
820	—	4419.8	3460.4	2952.7	2615.7	2367.3	2172.4	2013.3	1879.6	1764.8	1664.5	1575.8	1496.4	1424.7	1359.4
825	—	4442.8	3479.2	2969.2	2630.6	2381.0	2185.2	2025.3	1891.0	1775.5	1674.8	1585.6	1505.7	1433.6	1368.0
830	—	4465.8	3498.0	2985.7	2645.6	2394.7	2198.0	2037.4	1902.3	1786.3	1685.0	1595.4	1515.1	1442.7	1376.7
835	—	4488.9	3516.9	3002.3	2660.6	2408.5	2210.8	2049.4	1913.7	1797.1	1695.3	1605.2	1524.5	1451.7	1385.4
840	—	4512.0	3535.8	3018.8	2675.5	2422.3	2223.7	2061.5	1925.1	1807.9	1705.6	1615.1	1534.0	1460.8	1394.1
845	—	4535.1	3554.7	3035.4	2690.6	2436.2	2236.6	2073.6	1936.5	1818.8	1716.0	1624.9	1543.5	1469.9	1402.9
850	—	4558.2	3573.6	3052.1	2705.6	2450.0	2249.5	2085.7	1948.0	1829.7	1726.3	1634.9	1553.0	1479.0	1411.7
855	—	4581.4	3592.6	3068.7	2720.7	2463.9	2262.4	2097.9	1959.5	1840.6	1736.7	1644.8	1562.5	1488.2	1420.5
860	—	4604.6	3611.5	3085.4	2735.8	2477.8	2275.4	2110.0	1971.0	1851.5	1747.2	1654.8	1572.0	1497.3	1429.3
865	—	4627.8	3630.6	3102.1	2750.9	2491.7	2288.4	2122.2	1982.5	1862.5	1757.6	1664.7	1581.6	1506.5	1438.2
870	—	4651.0	3649.6	3118.8	2766.0	2505.7	2301.4	2134.5	1994.1	1873.5	1768.1	1674.8	1591.2	1515.8	1447.1
875	—	4674.2	3668.6	3135.5	2781.2	2519.7	2314.4	2146.7	2005.7	1884.5	1778.6	1684.8	1600.9	1525.0	1456.0
880	—	4697.5	3687.7	3152.3	2796.4	2533.7	2327.5	2159.0	2017.3	1895.5	1789.1	1694.9	1610.5	1534.3	1464.9
885	—	4720.8	3706.8	3169.1	2811.6	2547.7	2340.6	2171.3	2028.9	1906.6	1799.7	1705.0	1620.2	1543.6	1473.9
890	—	4744.1	3725.9	3185.9	2826.9	2561.8	2353.7	2183.7	2040.6	1917.7	1810.2	1715.1	1629.9	1553.0	1482.9
895	—	4767.5	3745.1	3202.8	2842.1	2575.9	2366.9	2196.0	2052.3	1928.8	1820.9	1725.3	1639.7	1562.3	1491.9
900	—	4790.9	3764.3	3219.6	2857.4	2590.0	2380.0	2208.4	2064.0	1939.9	1831.5	1735.4	1649.4	1571.7	1500.9
905	—	4814.3	3783.5	3236.5	2872.8	2604.1	2393.2	2220.8	2075.8	1951.1	1842.1	1745.6	1659.2	1581.1	1510.0
910	—	4837.7	3802.7	3253.5	2888.1	2618.3	2406.5	2233.3	2087.5	1962.3	1852.8	1755.9	1669.0	1590.6	1519.1
915	—	4861.1	3822.0	3270.4	2903.5	2632.5	2419.7	2245.7	2099.3	1973.5	1863.5	1766.1	1678.9	1600.0	1528.2
920	—	4884.6	3841.2	3287.4	2918.9	2646.7	2433.0	2258.2	2111.2	1984.8	1874.3	1776.4	1688.7	1609.5	1537.4
925	—	4908.1	3860.5	3304.4	2934.3	2660.9	2446.3	2270.7	2123.0	1996.0	1885.0	1786.7	1698.6	1619.0	1546.5
930	—	4931.6	3879.9	3321.4	2949.8	2675.2	2459.6	2283.3	2134.9	2007.3	1895.8	1797.0	1708.5	1628.6	1555.7
935	—	4955.1	3899.2	3338.4	2965.2	2689.5	2472.9	2295.8	2146.8	2018.6	1906.6	1807.4	1718.5	1638.1	1565.0
940	—	4978.7	3918.6	3355.5	2980.7	2703.8	2486.3	2308.4	2158.7	2030.0	1917.4	1817.7	1728.4	1647.7	1574.2
945	—	5002.2	3938.0	3372.6	2996.2	2718.2	2499.7	2321.0	2170.6	2041.3	1928.3	1828.1	1738.4	1657.3	1583.5
950	—	5025.8	3957.4	3389.7	3011.8	2732.5	2513.1	2333.7	2182.6	2052.7	1939.2	1838.6	1748.4	1667.0	1592.8
955	—	5049.5	3976.8	3406.8	3027.4	2746.9	2526.6	2346.3	2194.6	2064.2	1950.1	1849.0	1758.5	1676.6	1602.1
960	—	5073.1	3996.3	3424.0	3043.0	2761.3	2540.0	2359.0	2206.6	2075.6	1961.0	1859.5	1768.5	1686.3	1611.4
965	—	5096.8	4015.8	3441.2	3058.6	2775.7	2553.5	2371.7	2218.7	2087.1	1972.0	1870.0	1778.6	1696.0	1620.8

TABLE 6.9 Thermo-mechanical exergy of water vapour at vacuum pressures. Dashed lines indicate that water is not a pure gas at that temperature and pressure (it is a mixture of liquid and gas). (Continued)

e^{tm} (kJ/kg)

T (°C)	0.001	0.072	0.144	0.215	0.286	0.358	0.429	0.500	0.572	0.643	0.715	0.786	0.857	0.929	1.000
								Pressure (bar)							
970	—	5120.5	4035.3	3458.4	3074.2	2790.2	2567.0	2384.5	2230.8	2098.6	1983.0	1880.5	1788.7	1705.8	1630.2
975	—	5144.2	4054.8	3475.6	3089.9	2804.7	2580.6	2397.2	2242.8	2110.1	1994.0	1891.1	1798.9	1715.5	1639.6
980	—	5167.9	4074.4	3492.9	3105.6	2819.2	2594.1	2410.0	2255.0	2121.6	2005.0	1901.6	1809.0	1725.3	1649.1
985	—	5191.7	4094.0	3510.2	3121.3	2833.7	2607.7	2422.8	2267.1	2133.2	2016.0	1912.2	1819.2	1735.1	1658.5
990	—	5215.5	4113.6	3527.5	3137.0	2848.3	2621.3	2435.6	2279.3	2144.8	2027.1	1922.9	1829.4	1745.0	1668.0
995	—	5239.3	4133.2	3544.8	3152.8	2862.9	2635.0	2448.5	2291.5	2156.4	2038.2	1933.5	1839.7	1754.8	1677.5
1000	—	5263.1	4152.8	3562.2	3168.6	2877.5	2648.6	2461.4	2303.7	2168.0	2049.3	1944.2	1849.9	1764.7	1687.1
1020	—	5358.7	4231.7	3631.8	3231.9	2936.1	2703.5	2513.1	2352.8	2214.8	2094.1	1987.1	1891.2	1804.5	1725.5
1040	—	5454.6	4310.9	3701.8	3295.6	2995.1	2758.7	2565.2	2402.2	2261.9	2139.2	2030.4	1932.8	1844.6	1764.2
1060	—	5550.9	4390.4	3772.1	3359.7	3054.5	2814.3	2617.7	2452.0	2309.4	2184.7	2074.0	1974.8	1885.1	1803.4
1080	—	5647.5	4470.3	3842.9	3424.2	3114.2	2870.3	2670.6	2502.2	2357.3	2230.5	2118.0	2017.2	1926.0	1842.8
1100	—	5744.4	4550.6	3913.9	3489.0	3174.3	2926.6	2723.8	2552.8	2405.5	2276.7	2162.4	2059.9	1967.2	1882.7
1120	—	5841.8	4631.2	3985.3	3554.1	3234.7	2983.3	2777.3	2603.7	2454.1	2323.2	2207.1	2103.0	2008.8	1922.9
1140	—	5939.4	4712.1	4057.1	3619.6	3295.5	3040.3	2831.2	2654.9	2503.1	2370.1	2252.2	2146.4	2050.7	1963.4
1160	—	6037.4	4793.4	4129.2	3685.5	3356.6	3097.7	2885.4	2706.5	2552.3	2417.3	2297.6	2190.2	2093.0	2004.3
1180	—	6135.7	4875.0	4201.6	3751.6	3418.1	3155.3	2940.0	2758.4	2601.9	2464.9	2343.3	2234.3	2135.6	2045.5
1200	—	6234.3	4957.0	4274.4	3818.1	3479.8	3213.3	2994.9	2810.6	2651.8	2512.8	2389.4	2278.7	2178.5	2087.1
1220	—	6333.3	5039.2	4347.4	3884.9	3541.9	3271.7	3050.1	2863.2	2702.1	2561.0	2435.8	2323.4	2221.7	2129.0
1240	—	6432.6	5121.8	4420.8	3952.0	3604.3	3330.3	3105.6	2916.0	2752.6	2609.5	2482.5	2368.5	2265.3	2171.1
1260	—	6532.2	5204.7	4494.5	4019.5	3667.0	3389.3	3161.5	2969.2	2803.5	2658.4	2529.5	2413.8	2309.1	2213.6
1280	—	6632.1	5287.9	4568.5	4087.2	3730.1	3448.6	3217.6	3022.7	2854.7	2707.5	2576.8	2459.5	2353.3	2256.4
1300	—	6732.3	5371.4	4642.8	4155.3	3793.4	3508.1	3274.0	3076.5	2906.2	2756.9	2624.4	2505.5	2397.8	2299.5
1320	—	6832.8	5455.2	4717.4	4223.6	3857.0	3568.0	3330.8	3130.6	2958.0	2806.7	2672.4	2551.8	2442.6	2343.0
1340	—	6933.6	5539.2	4792.3	4292.2	3921.0	3628.1	3387.8	3185.0	3010.0	2856.7	2720.6	2598.4	2487.7	2386.6
1360	—	7034.7	5623.6	4867.5	4361.2	3985.2	3688.6	3445.2	3239.6	3062.4	2907.1	2769.1	2645.2	2533.0	2430.6
1380	—	7136.1	5708.3	4943.0	4430.4	4049.7	3749.3	3502.8	3294.6	3115.1	2957.7	2817.9	2692.4	2578.7	2474.9
1400	—	7237.8	5793.2	5018.7	4499.9	4114.4	3810.3	3560.7	3349.8	3168.0	3008.5	2866.9	2739.8	2624.6	2519.4
1420	—	7339.8	5878.5	5094.8	4569.6	4179.5	3871.6	3618.8	3405.3	3221.2	3059.7	2916.3	2787.5	2670.8	2564.3
1440	—	7442.0	5963.9	5171.1	4639.6	4244.8	3933.2	3677.3	3461.1	3274.6	3111.1	2965.9	2835.4	2717.2	2609.4
1460	—	7544.5	6049.7	5247.6	4709.9	4310.4	3995.0	3735.9	3517.1	3328.4	3162.8	3015.7	2883.7	2764.0	2654.7
1480	—	7647.3	6135.7	5324.5	4780.5	4376.2	4057.0	3794.9	3573.4	3382.4	3214.8	3065.9	2932.1	2811.0	2700.3

TABLE 6.9 Thermo-mechanical exergy of water vapour at vacuum pressures. Dashed lines indicate that water is not a pure gas at that temperature and pressure (it is a mixture of liquid and gas). (*Continued*)

161

e^{tm} (kJ/kg)

T (°C)	0.001	0.072	0.144	0.215	0.286	0.358	0.429	0.500	0.572	0.643	0.715	0.786	0.857	0.929	1.000
									Pressure (bar)						
1500	—	7750.3	6222.0	5401.5	4851.3	4442.3	4119.4	3854.1	3630.0	3436.6	3267.0	3116.3	2980.9	2858.2	2746.2
1520	—	7853.6	6308.6	5478.9	4922.4	4508.6	4182.0	3913.6	3686.8	3491.1	3319.4	3166.9	3029.9	2905.7	2792.3
1540	—	7957.2	6395.4	5556.5	4993.7	4575.2	4244.8	3973.3	3743.9	3545.9	3372.2	3217.8	3079.1	2953.4	2838.7
1560	—	8061.0	6482.4	5634.3	5065.3	4642.1	4307.9	4033.2	3801.2	3600.8	3425.1	3268.9	3128.6	3001.4	2885.3
1580	—	8165.0	6569.7	5712.4	5137.1	4709.2	4371.2	4093.4	3858.7	3656.1	3478.3	3320.3	3178.3	3049.6	2932.1
1600	—	8269.3	6657.2	5790.7	5209.1	4776.5	4434.7	4153.9	3916.5	3711.5	3531.7	3371.9	3228.3	3098.1	2979.2
1620	—	8373.9	6745.0	5869.2	5281.4	4844.0	4498.5	4214.5	3974.5	3767.2	3585.4	3423.7	3278.5	3146.8	3026.5
1640	—	8478.7	6833.0	5948.0	5353.9	4911.8	4562.5	4275.4	4032.7	3823.2	3639.3	3475.8	3328.9	3195.7	3074.1
1660	—	8583.7	6921.2	6027.0	5426.6	4979.8	4626.8	4336.5	4091.2	3879.3	3693.4	3528.1	3379.5	3244.9	3121.8
1680	—	8688.9	7009.6	6106.2	5499.6	5048.0	4691.2	4397.9	4149.8	3935.7	3747.7	3580.6	3430.4	3294.2	3169.8
1700	—	8794.4	7098.3	6185.7	5572.7	5116.5	4755.9	4459.4	4208.7	3992.3	3802.2	3633.3	3481.5	3343.8	3218.0
1720	—	—	7187.2	6265.4	5646.1	5185.1	4820.8	4521.2	4267.8	4049.1	3857.0	3686.2	3532.8	3393.6	3266.5
1740	—	—	7276.3	6345.2	5719.7	5254.0	4885.9	4583.2	4327.2	4106.1	3912.0	3739.4	3584.3	3443.6	3315.1
1760	—	—	7365.6	6425.3	5793.5	5323.1	4951.2	4645.4	4386.7	4163.3	3967.2	3792.7	3636.0	3493.8	3363.9
1780	—	—	7455.1	6505.6	5867.5	5392.4	5016.7	4707.7	4446.4	4220.7	4022.5	3846.3	3687.9	3544.2	3412.9
1800	—	—	7544.8	6586.1	5941.7	5461.8	5082.4	4770.3	4506.3	4278.3	4078.1	3900.0	3740.0	3594.8	3462.2
1820	—	—	7634.8	6666.8	6016.1	5531.5	5148.3	4833.1	4566.5	4336.1	4133.9	3954.0	3792.3	3645.6	3511.6
1840	—	—	7724.9	6747.7	6090.7	5601.4	5214.4	4896.1	4626.8	4394.1	4189.8	4008.1	3844.8	3696.6	3561.2
1860	—	—	7815.2	6828.8	6165.5	5671.5	5280.7	4959.3	4687.3	4452.3	4246.0	4062.5	3897.5	3747.8	3611.1
1880	—	—	7905.7	6910.0	6240.5	5741.7	5347.2	5022.6	4748.0	4510.7	4302.4	4117.0	3950.3	3799.2	3661.1
1900	—	—	7996.4	6991.5	6315.7	5812.2	5413.9	5086.2	4808.9	4569.3	4358.9	4171.7	4003.4	3850.8	3711.2
1920	—	—	8087.3	7073.1	6391.0	5882.8	5480.7	5149.9	4869.9	4628.0	4415.6	4226.6	4056.7	3902.5	3761.6
1940	—	—	8178.3	7155.0	6466.6	5953.6	5547.7	5213.8	4931.2	4687.0	4472.5	4281.7	4110.1	3954.4	3812.2
1960	—	—	—	7237.0	6542.3	6024.5	5614.9	5277.9	4992.6	4746.1	4529.5	4336.9	4163.7	4006.5	3862.9
1980	—	—	—	7319.1	6618.1	6095.7	5682.3	5342.1	5054.2	4805.3	4586.8	4392.3	4217.4	4058.8	3913.8
2000	—	—	—	7401.5	6694.2	6167.0	5749.9	5406.5	5115.9	4864.8	4644.2	4447.9	4271.4	4111.2	3964.8

TABLE 6.9 Thermo-mechanical exergy of water vapour at vacuum pressures. Dashed lines indicate that water is not a pure gas at that temperature and pressure (it is a mixture of liquid and gas). (Continued)

162

	e^{tm} (kJ/kg)														
	Pressure (bar)														
T (°C)	250	300	350	400	450	500	550	600	650	700	750	800	850	900	1000
0	29.401	34.338	39.266	44.185	49.094	53.994	58.885	63.766	68.639	73.503	78.357	83.204	88.041	92.871	102.50
10	26.441	31.390	36.330	41.260	46.179	51.089	55.989	60.880	65.761	70.633	75.496	80.350	85.195	90.031	99.678
20	25.010	29.964	34.908	39.842	44.766	49.680	54.584	59.479	64.364	69.239	74.106	78.963	83.811	88.650	98.302
30	25.006	29.960	34.905	39.839	44.763	49.677	54.581	59.475	64.360	69.236	74.102	78.959	83.807	88.646	98.298
40	26.340	31.291	36.232	41.162	46.083	50.994	55.895	60.787	65.669	70.542	75.406	80.261	85.106	89.943	99.590
50	28.930	33.875	38.809	43.734	48.649	53.555	58.451	63.337	68.214	73.082	77.941	82.791	87.632	92.465	102.10
60	32.703	37.639	42.566	47.482	52.389	57.287	62.175	67.054	71.924	76.785	81.637	86.480	91.315	96.141	105.77
70	37.594	42.518	47.434	52.340	57.237	62.124	67.003	71.872	76.733	81.585	86.428	91.263	96.089	100.91	110.52
80	43.541	48.452	53.354	58.247	63.132	68.007	72.874	77.732	82.581	87.422	92.255	97.079	101.90	106.70	116.30
90	50.490	55.386	60.272	65.151	70.020	74.881	79.734	84.578	89.415	94.243	99.063	103.87	108.68	113.48	123.05
100	58.394	63.271	68.140	73.000	77.853	82.697	87.534	92.363	97.183	102.00	106.80	111.60	116.39	121.17	130.72
110	67.208	72.064	76.912	81.753	86.586	91.412	96.230	101.04	105.84	110.64	115.43	120.21	124.98	129.75	139.26
120	76.896	81.727	86.552	91.370	96.180	100.98	105.78	110.57	115.35	120.13	124.90	129.66	134.42	139.17	148.65
130	87.423	92.227	97.024	101.82	106.60	111.38	116.15	120.92	125.68	130.44	135.18	139.93	144.66	149.39	158.83
140	98.760	103.53	108.30	113.06	117.82	122.57	127.32	132.06	136.79	141.52	146.25	150.97	155.68	160.39	169.78
150	110.88	115.62	120.35	125.08	129.81	134.53	139.24	143.95	148.66	153.36	158.06	162.75	167.44	172.12	181.47
160	123.77	128.47	133.16	137.85	142.54	147.22	151.91	156.58	161.26	165.93	170.60	175.26	179.92	184.57	193.87
170	137.41	142.06	146.71	151.36	156.00	160.65	165.29	169.93	174.57	179.20	183.83	188.46	193.09	197.72	206.95
180	151.79	156.38	160.98	165.58	170.17	174.77	179.37	183.97	188.57	193.16	197.76	202.35	206.94	211.53	220.70
190	166.90	171.43	175.96	180.50	185.05	189.59	194.14	198.69	203.24	207.79	212.34	216.90	221.45	226.00	235.10
200	182.73	187.18	191.65	196.13	200.61	205.09	209.58	214.08	218.58	223.08	227.58	232.09	236.60	241.11	250.12
210	199.28	203.66	208.04	212.44	216.85	221.27	225.70	230.13	234.57	239.02	243.47	247.92	252.38	256.84	265.76
220	216.57	220.84	225.14	229.45	233.78	238.12	242.47	246.83	251.21	255.59	259.98	264.37	268.77	273.18	282.00
230	234.59	238.75	242.94	247.15	251.38	255.63	259.90	264.19	268.48	272.79	277.11	281.44	285.78	290.13	298.84
240	253.37	257.40	261.46	265.55	269.67	273.82	277.99	282.19	286.40	290.62	294.87	299.12	303.39	307.67	316.26
250	272.94	276.80	280.71	284.67	288.66	292.69	296.75	300.84	304.95	309.08	313.24	317.41	321.60	325.80	334.25
260	293.32	296.99	300.72	304.51	308.36	312.25	316.18	320.14	324.14	328.17	332.23	336.30	340.40	344.52	352.81
270	314.56	317.99	321.51	325.11	328.78	332.51	336.29	340.12	343.99	347.89	351.83	355.80	359.80	363.83	371.94
280	336.73	339.87	343.13	346.50	349.96	353.50	357.10	360.77	364.49	368.26	372.07	375.92	379.80	383.72	391.63

TABLE 6.10 Thermo-mechanical exergy of water in the supercritical region from 250 to 1000 bar.

163

e^{tm} (kJ/kg)

T (°C)	Pressure (bar)														
	250	300	350	400	450	500	550	600	650	700	750	800	850	900	1000
290	359.90	362.68	365.63	368.72	371.93	375.24	378.64	382.11	385.66	389.27	392.93	396.64	400.40	404.19	411.89
300	384.18	386.51	389.07	391.82	394.72	397.76	400.92	404.18	407.53	410.95	414.44	418.00	421.61	425.26	432.71
310	409.73	411.47	413.54	415.86	418.40	421.12	423.99	426.99	430.11	433.32	436.61	439.99	443.43	446.93	454.10
320	436.78	437.73	439.16	440.96	443.04	445.37	447.89	450.59	453.43	456.39	459.46	462.63	465.88	469.21	476.06
330	465.63	465.50	466.09	467.21	468.73	470.57	472.68	475.01	477.53	480.20	483.01	485.94	488.98	492.10	498.60
340	496.79	495.11	494.55	494.78	495.58	496.82	498.42	500.31	502.44	504.78	507.29	509.94	512.73	515.63	521.73
350	531.14	527.05	524.85	523.87	523.74	524.23	525.20	526.56	528.24	530.17	532.32	534.66	537.16	539.81	545.46
360	570.39	562.11	557.46	554.79	553.42	552.94	553.13	553.84	554.96	556.42	558.16	560.13	562.31	564.65	569.79
370	619.02	601.82	593.09	587.96	584.88	583.13	582.33	582.24	582.69	583.59	584.84	586.38	588.18	590.19	594.75
380	697.85	649.43	633.00	624.00	618.50	615.06	612.98	611.89	611.53	611.74	612.41	613.46	614.82	616.44	620.34
390	949.50	713.49	679.47	663.90	654.80	649.02	645.28	642.92	641.56	640.95	640.94	641.40	642.25	643.43	646.59
400	1050.7	822.93	736.66	709.20	694.52	685.45	679.49	675.51	672.91	671.31	670.48	670.26	670.53	671.20	673.51
410	1111.5	958.83	812.51	761.95	738.64	724.86	715.92	709.87	705.72	702.92	701.12	700.09	699.68	699.77	701.13
420	1158.1	1048.1	907.73	824.82	788.15	767.84	754.93	746.21	740.13	735.87	732.91	730.95	729.75	729.17	729.45
430	1197.3	1111.0	996.88	898.21	843.92	814.80	796.82	784.75	776.30	770.27	765.94	762.88	760.77	759.43	758.49
440	1231.7	1160.8	1068.8	973.29	905.76	865.94	841.70	825.60	814.33	806.21	800.27	795.93	792.79	790.58	788.28
450	1262.9	1203.0	1126.8	1042.1	969.93	920.83	889.50	868.75	854.23	843.70	835.93	830.13	825.81	822.63	818.80
460	1291.7	1240.2	1175.5	1102.0	1032.0	977.56	939.74	913.98	895.88	882.70	872.88	865.47	859.84	855.58	850.07
470	1318.7	1273.8	1218.0	1154.1	1089.4	1033.6	991.31	960.88	939.06	923.06	911.05	901.89	894.84	889.41	882.07
480	1344.3	1304.9	1256.1	1200.1	1141.4	1087.2	1042.7	1008.7	983.41	964.56	950.28	939.29	930.75	924.06	914.78
490	1368.7	1333.9	1290.8	1241.3	1188.5	1137.4	1092.6	1056.5	1028.4	1006.9	990.36	977.52	967.44	959.46	948.13
500	1392.2	1361.3	1323.0	1278.8	1231.3	1183.9	1140.4	1103.3	1073.3	1049.6	1031.0	1016.4	1004.8	995.49	982.08
510	1415.0	1387.5	1353.2	1313.6	1270.7	1227.0	1185.5	1148.7	1117.7	1092.3	1072.0	1055.7	1042.6	1032.0	1016.5
520	1437.1	1412.6	1381.8	1346.1	1307.2	1267.1	1228.1	1192.2	1160.9	1134.6	1112.9	1095.1	1080.7	1068.9	1051.4
530	1458.8	1436.9	1409.1	1376.8	1341.5	1304.6	1268.1	1233.7	1202.8	1176.0	1153.4	1134.5	1118.9	1106.0	1086.5
540	1479.9	1460.4	1435.3	1406.0	1373.8	1339.9	1305.9	1273.2	1243.1	1216.4	1193.2	1173.5	1157.0	1143.1	1121.9
550	1500.7	1483.4	1460.7	1434.0	1404.5	1373.3	1341.7	1310.8	1281.8	1255.5	1232.2	1212.0	1194.8	1180.1	1157.3
560	1521.2	1505.8	1485.3	1460.9	1433.9	1405.1	1375.7	1346.7	1319.0	1293.3	1270.2	1249.8	1232.1	1216.8	1192.6
570	1541.4	1527.8	1509.2	1487.0	1462.1	1435.5	1408.2	1380.9	1354.6	1329.8	1307.2	1286.8	1268.9	1253.2	1227.9
580	1561.4	1549.4	1532.6	1512.3	1489.4	1464.8	1439.4	1413.8	1388.9	1365.1	1343.0	1322.9	1304.9	1289.0	1262.9
590	1581.1	1570.6	1555.5	1536.9	1515.8	1493.1	1469.4	1445.5	1421.9	1399.1	1377.8	1358.1	1340.2	1324.2	1297.5

TABLE 6.10 Thermo-mechanical exergy of water in the supercritical region from 250 to 1000 bar. (Continued)

e^tm (kJ/kg)

Pressure (bar)

T (°C)	250	300	350	400	450	500	550	600	650	700	750	800	850	900	1000
600	1600.7	1591.6	1578.0	1561.0	1541.6	1520.5	1498.5	1476.0	1453.7	1432.1	1411.5	1392.3	1374.7	1358.8	1331.8
610	1620.2	1612.4	1600.1	1584.6	1566.7	1547.2	1526.6	1505.6	1484.5	1464.0	1444.3	1425.7	1408.5	1392.7	1365.6
620	1639.5	1632.9	1622.0	1607.8	1591.3	1573.2	1554.0	1534.3	1514.5	1494.9	1476.1	1458.2	1441.4	1425.9	1398.9
630	1658.8	1653.3	1643.5	1630.6	1615.4	1598.6	1580.7	1562.2	1543.6	1525.1	1507.1	1489.8	1473.6	1458.4	1431.6
640	1677.9	1673.5	1664.9	1653.1	1639.1	1623.5	1606.8	1589.5	1571.9	1554.4	1537.3	1520.7	1505.0	1490.2	1463.9
650	1697.0	1693.6	1686.0	1675.3	1662.5	1648.0	1632.4	1616.2	1599.6	1583.0	1566.7	1550.9	1535.8	1521.4	1495.6
660	1716.0	1713.6	1706.9	1697.3	1685.5	1672.1	1657.6	1642.3	1626.7	1611.0	1595.5	1580.4	1565.9	1552.0	1526.8
670	1735.0	1733.4	1727.7	1719.1	1708.2	1695.8	1682.3	1668.0	1653.3	1638.5	1623.8	1609.3	1595.4	1582.0	1557.5
680	1754.0	1753.2	1748.4	1740.6	1730.7	1719.2	1706.6	1693.2	1679.4	1665.4	1651.4	1637.7	1624.3	1611.4	1587.6
690	1772.9	1772.9	1768.9	1762.0	1753.0	1742.4	1730.6	1718.1	1705.1	1691.9	1678.6	1665.5	1652.7	1640.4	1617.3
700	1791.8	1792.6	1789.4	1783.3	1775.1	1765.3	1754.4	1742.6	1730.4	1717.9	1705.3	1692.9	1680.7	1668.8	1646.5
710	1810.7	1812.2	1809.8	1804.4	1797.0	1788.0	1777.8	1766.9	1755.4	1743.6	1731.7	1719.8	1708.2	1696.8	1675.3
720	1829.6	1831.8	1830.1	1825.4	1818.7	1810.5	1801.0	1790.8	1780.0	1768.9	1757.6	1746.4	1735.2	1724.4	1703.7
730	1848.5	1851.4	1850.3	1846.4	1840.4	1832.8	1824.1	1814.5	1804.4	1793.9	1783.3	1772.6	1762.0	1751.6	1731.7
740	1867.5	1870.9	1870.5	1867.2	1861.9	1855.0	1846.9	1838.0	1828.5	1818.6	1808.6	1798.4	1788.3	1778.4	1759.3
750	1886.4	1890.5	1890.6	1888.0	1883.3	1877.0	1869.6	1861.3	1852.4	1843.1	1833.6	1824.0	1814.4	1804.9	1786.6
760	1905.4	1910.0	1910.8	1908.7	1904.6	1898.9	1892.1	1884.4	1876.1	1867.4	1858.4	1849.3	1840.2	1831.1	1813.6
770	1924.3	1929.5	1930.8	1929.4	1925.8	1920.7	1914.5	1907.3	1899.6	1891.4	1882.9	1874.3	1865.7	1857.0	1840.3
780	1943.4	1949.1	1950.9	1950.0	1947.0	1942.4	1936.7	1930.2	1922.9	1915.3	1907.3	1899.1	1890.9	1882.7	1866.7
790	1962.4	1968.6	1971.0	1970.6	1968.1	1964.1	1958.9	1952.8	1946.1	1938.9	1931.4	1923.7	1916.0	1908.2	1892.9
800	1981.5	1988.2	1991.1	1991.1	1989.2	1985.7	1981.0	1975.4	1969.2	1962.5	1955.4	1948.2	1940.8	1933.4	1918.8
810	2000.6	2007.8	2011.1	2011.7	2010.2	2007.2	2003.0	1997.9	1992.1	1985.9	1979.2	1972.4	1965.4	1958.4	1944.5
820	2019.8	2027.4	2031.2	2032.2	2031.2	2028.6	2024.9	2020.3	2014.9	2009.1	2002.9	1996.5	1989.9	1983.3	1970.1
830	2039.0	2047.1	2051.3	2052.8	2052.2	2050.1	2046.8	2042.6	2037.7	2032.3	2026.5	2020.4	2014.2	2008.0	1995.4
840	2058.2	2066.7	2071.4	2073.3	2073.1	2071.5	2068.6	2064.8	2060.3	2055.3	2049.9	2044.3	2038.4	2032.5	2020.6
850	2077.5	2086.4	2091.5	2093.8	2094.1	2092.8	2090.3	2087.0	2082.9	2078.3	2073.3	2068.0	2062.5	2056.9	2045.6
860	2096.9	2106.2	2111.6	2114.3	2115.0	2114.1	2112.1	2109.1	2105.4	2101.1	2096.5	2091.6	2086.4	2081.1	2070.5
870	2116.3	2126.0	2131.8	2134.9	2136.0	2135.5	2133.8	2131.2	2127.8	2124.0	2119.7	2115.1	2110.2	2105.3	2095.2
880	2135.8	2145.8	2152.0	2155.5	2156.9	2156.8	2155.5	2153.2	2150.2	2146.7	2142.7	2138.5	2134.0	2129.3	2119.8

TABLE 6.10 Thermo-mechanical exergy of water in the supercritical region from 250 to 1000 bar. (Continued)

eᵗᵐ (kJ/kg)

T (°C)	Pressure (bar)														
	250	300	350	400	450	500	550	600	650	700	750	800	850	900	1000
890	2155.3	2165.6	2172.2	2176.0	2177.8	2178.1	2177.1	2175.2	2172.6	2169.4	2165.8	2161.8	2157.6	2153.3	2144.3
900	2174.8	2185.6	2192.5	2196.7	2198.8	2199.4	2198.7	2197.2	2194.9	2192.0	2188.7	2185.1	2181.2	2177.1	2168.7
910	2194.5	2205.5	2212.8	2217.3	2219.8	2220.7	2220.4	2219.1	2217.2	2214.6	2211.6	2208.3	2204.7	2200.9	2193.0
920	2214.1	2225.5	2233.1	2237.9	2240.7	2242.0	2242.0	2241.1	2239.4	2237.2	2234.5	2231.4	2228.1	2224.6	2217.2
930	2233.9	2245.6	2253.5	2258.6	2261.7	2263.3	2263.6	2263.0	2261.7	2259.7	2257.3	2254.5	2251.5	2248.2	2241.3
940	2253.7	2265.7	2273.9	2279.3	2282.8	2284.6	2285.3	2284.9	2283.9	2282.2	2280.1	2277.6	2274.8	2271.8	2265.4
950	2273.6	2285.8	2294.3	2300.1	2303.8	2306.0	2306.9	2306.9	2306.1	2304.7	2302.8	2300.6	2298.1	2295.3	2289.4
960	2293.5	2306.1	2314.8	2320.9	2324.9	2327.3	2328.5	2328.8	2328.3	2327.2	2325.6	2323.6	2321.3	2318.8	2313.4
970	2313.5	2326.3	2335.4	2341.7	2346.0	2348.7	2350.2	2350.7	2350.5	2349.6	2348.3	2346.6	2344.5	2342.3	2337.3
980	2333.5	2346.7	2356.0	2362.6	2367.1	2370.1	2371.9	2372.7	2372.7	2372.1	2371.0	2369.5	2367.7	2365.7	2361.1
990	2353.6	2367.0	2376.6	2383.5	2388.3	2391.5	2393.6	2394.6	2394.9	2394.5	2393.7	2392.4	2390.9	2389.0	2384.9
1000	2373.8	2387.5	2397.3	2404.4	2409.5	2413.0	2415.3	2416.6	2417.1	2417.0	2416.3	2415.3	2414.0	2412.4	2408.7
1010	2394.1	2408.0	2418.1	2425.4	2430.8	2434.5	2437.0	2438.5	2439.3	2439.4	2439.0	2438.2	2437.1	2435.7	2432.4
1020	2414.4	2428.5	2438.9	2446.5	2452.0	2456.0	2458.8	2460.5	2461.5	2461.9	2461.7	2461.1	2460.2	2459.0	2456.1
1030	2434.7	2449.1	2459.7	2467.6	2473.4	2477.6	2480.6	2482.6	2483.8	2484.3	2484.4	2484.0	2483.3	2482.3	2479.8
1040	2455.2	2469.8	2480.6	2488.7	2494.7	2499.2	2502.4	2504.6	2506.0	2506.8	2507.1	2506.9	2506.4	2505.6	2503.5
1050	2475.7	2490.5	2501.6	2509.9	2516.1	2520.8	2524.2	2526.6	2528.3	2529.3	2529.7	2529.8	2529.5	2528.9	2527.1
1060	2496.3	2511.3	2522.6	2531.1	2537.6	2542.4	2546.1	2548.7	2550.6	2551.8	2552.4	2552.7	2552.6	2552.2	2550.7
1070	2516.9	2532.2	2543.6	2552.4	2559.1	2564.1	2568.0	2570.8	2572.9	2574.3	2575.2	2575.6	2575.7	2575.4	2574.4
1080	2537.6	2553.1	2564.8	2573.7	2580.6	2585.9	2589.9	2593.0	2595.2	2596.8	2597.9	2598.5	2598.8	2598.7	2598.0
1090	2558.4	2574.0	2585.9	2595.1	2602.2	2607.7	2611.9	2615.2	2617.6	2619.4	2620.6	2621.4	2621.9	2622.0	2621.6
1100	2579.2	2595.1	2607.2	2616.5	2623.8	2629.5	2633.9	2637.4	2640.0	2642.0	2643.4	2644.4	2645.0	2645.3	2645.2
1110	2600.1	2616.2	2628.4	2638.0	2645.5	2651.4	2656.0	2659.6	2662.4	2664.6	2666.2	2667.3	2668.1	2668.6	2668.8
1120	2621.1	2637.3	2649.8	2659.5	2667.2	2673.3	2678.1	2681.9	2684.9	2687.2	2689.0	2690.3	2691.2	2691.9	2692.4
1130	2642.1	2658.5	2671.2	2681.1	2689.0	2695.2	2700.2	2704.2	2707.4	2709.9	2711.8	2713.3	2714.4	2715.2	2716.0
1140	2663.2	2679.8	2692.7	2702.8	2710.8	2717.2	2722.4	2726.5	2729.9	2732.5	2734.6	2736.3	2737.6	2738.5	2739.7
1150	2684.3	2701.2	2714.2	2724.4	2732.7	2739.3	2744.6	2748.9	2752.4	2755.3	2757.5	2759.3	2760.8	2761.8	2763.3
1160	2705.6	2722.6	2735.7	2746.2	2754.6	2761.3	2766.9	2771.4	2775.0	2778.0	2780.4	2782.4	2784.0	2785.2	2786.9
1170	2726.9	2744.0	2757.4	2768.0	2776.5	2783.5	2789.2	2793.8	2797.6	2800.8	2803.4	2805.5	2807.2	2808.6	2810.6
1180	2748.2	2765.5	2779.1	2789.9	2798.6	2805.7	2811.5	2816.3	2820.3	2823.6	2826.3	2828.6	2830.4	2832.0	2834.3
1190	2769.6	2787.1	2800.8	2811.8	2820.6	2827.9	2833.9	2838.9	2843.0	2846.4	2849.3	2851.7	2853.7	2855.4	2858.0

TABLE 6.10 Thermo-mechanical exergy of water in the supercritical region from 250 to 1000 bar. (Continued)

e^{tm} (kJ/kg)

T (°C)	Pressure (bar) 250	300	350	400	450	500	550	600	650	700	750	800	850	900	1000
1200	2791.1	2808.8	2822.6	2833.7	2842.8	2850.2	2856.3	2861.4	2865.7	2869.3	2872.3	2874.9	2877.0	2878.8	2881.7
1210	2812.7	2830.5	2844.5	2855.7	2864.9	2872.5	2878.8	2884.1	2888.5	2892.2	2895.4	2898.1	2900.4	2902.3	2905.4
1220	2834.3	2852.3	2866.4	2877.8	2887.2	2894.9	2901.3	2906.7	2911.3	2915.2	2918.5	2921.3	2923.7	2925.8	2929.1
1230	2856.0	2874.1	2888.4	2900.0	2909.4	2917.3	2923.9	2929.4	2934.2	2938.2	2941.6	2944.6	2947.1	2949.3	2952.9
1240	2877.7	2896.0	2910.4	2922.1	2931.8	2939.8	2946.5	2952.2	2957.0	2961.2	2964.8	2967.9	2970.5	2972.9	2976.7
1250	2899.5	2917.9	2932.5	2944.4	2954.1	2962.3	2969.2	2975.0	2980.0	2984.3	2988.0	2991.2	2994.0	2996.4	3000.5
1260	2921.4	2940.0	2954.7	2966.7	2976.6	2984.9	2991.9	2997.8	3002.9	3007.4	3011.2	3014.5	3017.5	3020.0	3024.3
1270	2943.3	2962.0	2976.9	2989.0	2999.1	3007.5	3014.6	3020.7	3026.0	3030.5	3034.5	3037.9	3041.0	3043.7	3048.2
1280	2965.3	2984.2	2999.2	3011.4	3021.6	3030.1	3037.4	3043.6	3049.0	3053.7	3057.8	3061.3	3064.5	3067.3	3072.1
1290	2987.4	3006.4	3021.5	3033.9	3044.2	3052.9	3060.3	3066.6	3072.1	3076.9	3081.1	3084.8	3088.1	3091.0	3096.0
1300	3009.5	3028.6	3043.9	3056.4	3066.8	3075.6	3083.2	3089.6	3095.3	3100.2	3104.5	3108.3	3111.7	3114.7	3119.9
1310	3031.7	3050.9	3066.3	3079.0	3089.5	3098.5	3106.1	3112.7	3118.4	3123.5	3127.9	3131.8	3135.3	3138.5	3143.9
1320	3054.0	3073.3	3088.8	3101.6	3112.3	3121.3	3129.1	3135.8	3141.7	3146.8	3151.3	3155.4	3159.0	3162.3	3167.9
1330	3076.3	3095.7	3111.4	3124.3	3135.1	3144.2	3152.1	3159.0	3164.9	3170.2	3174.8	3179.0	3182.7	3186.1	3191.9
1340	3098.7	3118.2	3134.0	3147.0	3157.9	3167.2	3175.2	3182.2	3188.2	3193.6	3198.4	3202.6	3206.5	3209.9	3215.9
1350	3121.1	3140.8	3156.7	3169.8	3180.8	3190.2	3198.3	3205.4	3211.6	3217.1	3221.9	3226.3	3230.2	3233.8	3240.0
1360	3143.6	3163.4	3179.4	3192.6	3203.8	3213.3	3221.5	3228.7	3235.0	3240.6	3245.6	3250.0	3254.1	3257.7	3264.1
1370	3166.1	3186.1	3202.2	3215.5	3226.8	3236.4	3244.8	3252.0	3258.4	3264.1	3269.2	3273.8	3277.9	3281.7	3288.3
1380	3188.8	3208.8	3225.0	3238.5	3249.9	3259.6	3268.0	3275.4	3281.9	3287.7	3292.9	3297.6	3301.8	3305.7	3312.4
1390	3211.4	3231.6	3247.9	3261.5	3273.0	3282.8	3291.4	3298.8	3305.5	3311.4	3316.6	3321.4	3325.7	3329.7	3336.7
1400	3234.2	3254.4	3270.9	3284.5	3296.1	3306.1	3314.7	3322.3	3329.0	3335.0	3340.4	3345.3	3349.7	3353.7	3360.9
1410	3257.0	3277.3	3293.9	3307.7	3319.3	3329.4	3338.1	3345.8	3352.7	3358.7	3364.2	3369.2	3373.7	3377.8	3385.2
1420	3279.8	3300.3	3316.9	3330.8	3342.6	3352.8	3361.6	3369.4	3376.3	3382.5	3388.1	3393.1	3397.7	3402.0	3409.5
1430	3302.7	3323.3	3340.1	3354.0	3365.9	3376.2	3385.1	3393.0	3400.0	3406.3	3412.0	3417.1	3421.8	3426.1	3433.8
1440	3325.7	3346.4	3363.2	3377.3	3389.3	3399.6	3408.7	3416.7	3423.8	3430.2	3435.9	3441.1	3445.9	3450.3	3458.2
1450	3348.7	3369.5	3386.4	3400.6	3412.7	3423.2	3432.3	3440.4	3447.6	3454.0	3459.9	3465.2	3470.1	3474.6	3482.6
1460	3371.8	3392.7	3409.7	3424.0	3436.2	3446.7	3456.0	3464.1	3471.4	3478.0	3483.9	3489.3	3494.3	3498.9	3507.0
1470	3394.9	3415.9	3433.1	3447.4	3459.7	3470.3	3479.7	3487.9	3495.3	3501.9	3508.0	3513.5	3518.5	3523.2	3531.5
1480	3418.1	3439.2	3456.4	3470.9	3483.3	3494.0	3503.4	3511.8	3519.2	3526.0	3532.1	3537.7	3542.8	3547.5	3556.0

TABLE 6.10 Thermo-mechanical exergy of water in the supercritical region from 250 to 1000 bar. (Continued)

e^tm (kJ/kg)

Pressure (bar)

T (°C)	250	300	350	400	450	500	550	600	650	700	750	800	850	900	1000
1490	3441.4	3462.5	3479.9	3494.4	3506.9	3517.7	3527.2	3535.7	3543.2	3550.0	3556.2	3561.9	3567.1	3571.9	3580.6
1500	3464.7	3485.9	3503.4	3518.0	3530.6	3541.5	3551.1	3559.6	3567.2	3574.1	3580.4	3586.1	3591.4	3596.3	3605.1
1510	3488.0	3509.4	3526.9	3541.6	3554.3	3565.3	3575.0	3583.6	3591.3	3598.3	3604.6	3610.5	3615.8	3620.8	3629.8
1520	3511.5	3532.9	3550.5	3565.3	3578.1	3589.1	3598.9	3607.6	3615.4	3622.5	3628.9	3634.8	3640.2	3645.3	3654.4
1530	3534.9	3556.5	3574.1	3589.1	3601.9	3613.0	3622.9	3631.7	3639.5	3646.7	3653.2	3659.2	3664.7	3669.8	3679.1
1540	3558.5	3580.1	3597.8	3612.8	3625.7	3637.0	3646.9	3655.8	3663.7	3671.0	3677.6	3683.6	3689.2	3694.4	3703.8
1550	3582.0	3603.7	3621.6	3636.7	3649.7	3661.0	3671.0	3679.9	3688.0	3695.3	3701.9	3708.1	3713.7	3719.0	3728.6
1560	3605.7	3627.4	3645.4	3660.6	3673.6	3685.0	3695.1	3704.1	3712.3	3719.6	3726.4	3732.6	3738.3	3743.7	3753.4
1570	3629.4	3651.2	3669.2	3684.5	3697.6	3709.1	3719.3	3728.4	3736.6	3744.0	3750.8	3757.1	3762.9	3768.4	3778.2
1580	3653.1	3675.0	3693.2	3708.5	3721.7	3733.3	3743.5	3752.7	3760.9	3768.5	3775.4	3781.7	3787.6	3793.1	3803.0
1590	3676.9	3698.9	3717.1	3732.5	3745.8	3757.4	3767.8	3777.0	3785.3	3792.9	3799.9	3806.3	3812.3	3817.8	3827.9
1600	3700.7	3722.8	3741.1	3756.6	3770.0	3781.7	3792.1	3801.4	3809.8	3817.5	3824.5	3831.0	3837.0	3842.6	3852.9
1610	3724.6	3746.8	3765.2	3780.7	3794.2	3805.9	3816.4	3825.8	3834.3	3842.0	3849.1	3855.7	3861.8	3867.5	3877.8
1620	3748.6	3770.8	3789.3	3804.9	3818.4	3830.3	3840.8	3850.3	3858.8	3866.6	3873.8	3880.4	3886.6	3892.4	3902.8
1630	3772.6	3794.9	3813.4	3829.1	3842.7	3854.6	3865.2	3874.8	3883.4	3891.3	3898.5	3905.2	3911.5	3917.3	3927.9
1640	3796.6	3819.0	3837.6	3853.4	3867.0	3879.0	3889.7	3899.3	3908.0	3916.0	3923.3	3930.0	3936.3	3942.2	3953.0
1650	3820.7	3843.2	3861.8	3877.7	3891.4	3903.5	3914.3	3923.9	3932.7	3940.7	3948.1	3954.9	3961.3	3967.2	3978.1
1660	3844.9	3867.4	3886.1	3902.1	3915.9	3928.0	3938.8	3948.6	3957.4	3965.5	3972.9	3979.8	3986.2	3992.2	4003.2
1670	3869.1	3891.7	3910.5	3926.5	3940.3	3952.6	3963.4	3973.2	3982.1	3990.3	3997.8	4004.7	4011.2	4017.3	4028.4
1680	3893.3	3916.0	3934.9	3950.9	3964.9	3977.1	3988.1	3998.0	4006.9	4015.1	4022.7	4029.7	4036.3	4042.4	4053.6
1690	3917.6	3940.4	3959.3	3975.4	3989.4	4001.8	4012.8	4022.7	4031.8	4040.0	4047.7	4054.7	4061.3	4067.5	4078.9
1700	3942.0	3964.8	3983.8	4000.0	4014.1	4026.5	4037.5	4047.5	4056.6	4065.0	4072.7	4079.8	4086.4	4092.7	4104.1
1710	3966.4	3989.3	4008.3	4024.6	4038.7	4051.2	4062.3	4072.4	4081.5	4089.9	4097.7	4104.9	4111.6	4117.9	4129.5
1720	3990.8	4013.8	4032.9	4049.2	4063.4	4076.0	4087.2	4097.3	4106.5	4114.9	4122.8	4130.0	4136.8	4143.1	4154.8
1730	4015.3	4038.3	4057.5	4073.9	4088.2	4100.8	4112.0	4122.2	4131.5	4140.0	4147.9	4155.2	4162.0	4168.4	4180.2
1740	4039.9	4062.9	4082.2	4098.6	4113.0	4125.6	4136.9	4147.2	4156.5	4165.1	4173.0	4180.4	4187.3	4193.7	4205.6
1750	4064.5	4087.6	4106.9	4123.4	4137.8	4150.5	4161.9	4172.2	4181.6	4190.2	4198.2	4205.6	4212.6	4219.1	4231.1
1760	4089.1	4112.3	4131.7	4148.2	4162.7	4175.5	4186.9	4197.3	4206.7	4215.4	4223.4	4230.9	4237.9	4244.5	4256.6
1770	4113.8	4137.0	4156.5	4173.1	4187.6	4200.4	4211.9	4222.4	4231.9	4240.6	4248.7	4256.2	4263.3	4269.9	4282.1
1780	4138.5	4161.8	4181.3	4198.0	4212.6	4225.5	4237.0	4247.5	4257.1	4265.8	4274.0	4281.6	4288.7	4295.4	4307.7
1790	4163.3	4186.7	4206.2	4222.9	4237.6	4250.5	4262.1	4272.7	4282.3	4291.1	4299.3	4307.0	4314.1	4320.9	4333.3

TABLE 6.10 Thermo-mechanical exergy of water in the supercritical region from 250 to 1000 bar. (Continued)

168

e^{tm} (kJ/kg)

T (°C)	Pressure (bar)														
	250	300	350	400	450	500	550	600	650	700	750	800	850	900	1000
1800	4188.1	4211.5	4231.1	4247.9	4262.6	4275.6	4287.3	4297.9	4307.6	4316.5	4324.7	4332.4	4339.6	4346.4	4358.9
1810	4213.0	4236.4	4256.1	4273.0	4287.7	4300.8	4312.5	4323.1	4332.9	4341.8	4350.1	4357.9	4365.1	4372.0	4384.6
1820	4237.9	4261.4	4281.1	4298.0	4312.8	4326.0	4337.7	4348.4	4358.2	4367.2	4375.6	4383.4	4390.7	4397.6	4410.3
1830	4262.8	4286.4	4306.2	4323.2	4338.0	4351.2	4363.0	4373.8	4383.6	4392.7	4401.1	4408.9	4416.3	4423.2	4436.0
1840	4287.8	4311.5	4331.3	4348.3	4363.2	4376.5	4388.3	4399.1	4409.0	4418.1	4426.6	4434.5	4441.9	4448.9	4461.8
1850	4312.8	4336.6	4356.4	4373.5	4388.5	4401.8	4413.7	4424.6	4434.5	4443.7	4452.2	4460.1	4467.6	4474.6	4487.6
1860	4337.9	4361.7	4381.6	4398.8	4413.8	4427.1	4439.1	4450.0	4460.0	4469.2	4477.8	4485.8	4493.3	4500.3	4513.4
1870	4363.1	4386.9	4406.9	4424.1	4439.1	4452.5	4464.5	4475.5	4485.5	4494.8	4503.4	4511.4	4519.0	4526.1	4539.3
1880	4388.2	4412.1	4432.1	4449.4	4464.5	4477.9	4490.0	4501.0	4511.1	4520.4	4529.1	4537.2	4544.8	4551.9	4565.2
1890	4413.4	4437.4	4457.5	4474.7	4489.9	4503.4	4515.5	4526.6	4536.7	4546.1	4554.8	4562.9	4570.6	4577.8	4591.1
1900	4438.7	4462.7	4482.8	4500.1	4515.4	4528.9	4541.1	4552.2	4562.4	4571.8	4580.5	4588.7	4596.4	4603.7	4617.1
1910	4464.0	4488.0	4508.2	4525.6	4540.9	4554.4	4566.7	4577.8	4588.1	4597.5	4606.3	4614.5	4622.3	4629.6	4643.1
1920	4489.3	4513.4	4533.6	4551.1	4566.4	4580.0	4592.3	4603.5	4613.8	4623.3	4632.1	4640.4	4648.2	4655.5	4669.1
1930	4514.7	4538.8	4559.1	4576.6	4592.0	4605.6	4618.0	4629.2	4639.5	4649.1	4658.0	4666.3	4674.1	4681.5	4695.2
1940	4540.1	4564.3	4584.6	4602.2	4617.6	4631.3	4643.7	4655.0	4665.3	4674.9	4683.9	4692.2	4700.1	4707.5	4721.3
1950	4565.6	4589.8	4610.2	4627.8	4643.2	4657.0	4669.4	4680.8	4691.2	4700.8	4709.8	4718.2	4726.1	4733.6	4747.4
1960	4591.1	4615.4	4635.8	4653.4	4668.9	4682.7	4695.2	4706.6	4717.0	4726.7	4735.7	4744.2	4752.1	4759.6	4773.6
1970	4616.6	4640.9	4661.4	4679.1	4694.6	4708.5	4721.0	4732.4	4742.9	4752.7	4761.7	4770.2	4778.2	4785.8	4799.8
1980	4642.2	4666.6	4687.1	4704.8	4720.4	4734.3	4746.9	4758.3	4768.9	4778.6	4787.7	4796.3	4804.3	4811.9	4826.0
1990	4667.8	4692.2	4712.8	4730.6	4746.2	4760.2	4772.8	4784.3	4794.8	4804.7	4813.8	4822.4	4830.4	4838.1	4852.3
2000	4693.5	4717.9	4738.6	4756.4	4772.0	4786.0	4798.7	4810.2	4820.9	4830.7	4839.9	4848.5	4856.6	4864.3	4878.6

TABLE 6.10 Thermo-mechanical exergy of water in the supercritical region from 250 to 1000 bar. (Continued)

169

References

1. Wagner W, Pruß A. The IAPWS formulation 1995 for the thermodynamic properties of ordinary water substance for general and scientific use. *J Phys Chem Ref Data* 31:387–535 (2002). doi:10.1063/1.1461829.
2. Hill IH, Wronski J, Quoilin S, Lemort V. Pure and pseudo-pure fluid thermophysical property evaluation and the open-source thermophysical property library CoolProp. *Ind Eng Chem Res* 53:2498–2508 (2014). doi:10.1021/ie4033999.
3. Herrig S, Thol M, Harvey AH, Lemmon EW. A reference equation of state for heavy water. *J Phys Chem Ref Data* 47(4):043102 (2018). doi: 10.1063/1.5053993.

CHAPTER **7**

Pure Chemicals

7.1 Thermo-Mechanical Exergy

Pressure-enthalpy diagrams containing thermo-mechanical exergy are provided for a selection of common chemicals.

- acetone   Figure 7.1
- ammonia   Figure 7.2
- benzene   Figure 7.3
- carbon monoxide   Figure 7.4
- dimethyl ether   Figure 7.5
- ethane   Figure 7.6
- ethanol   Figure 7.7
- ethylene   Figure 7.8
- n-hexane   Figure 7.9
- hydrogen   Figure 7.10
- methane   Figure 7.11
- methanol   Figure 7.12
- n-octane   Figure 7.13
- propylene   Figure 7.14
- toluene   Figure 7.15

7.2 Chemical Exergy

The chemical exergy for elemental molecules (such as pure H_2, pure Au, etc.), and their corresponding reference chemicals (such as CO_2, $CaCO_3$, etc.) naturally present in the atmosphere, hydrosphere, and lithosphere are noted in Tables 4.1, 4.3, and 4.4, respectively in Chap. 4. The chemical exergy of other pure chemicals at 25°C and 1.01325 bar (1 atm) using the reference conditions as noted in Chap. 4 are listed in this chapter.

We calculated the chemical exergy of each chemical species by Eq. (7.1)[10]:

$$e_i^{ch} = \Delta g_i^f + \sum_j n_j e_j^{ch} \tag{7.1}$$

where e^{ch} is the chemical exergy, i is the species of interest, j is the elemental species (Ca, O_2, H_2, etc.) that constitute the reference species, and n is the number of moles of elemental species j in one mole of species i. See Chap. 4 for further details.

Tables 7.1 to 7.8 provide the chemical exergy in multiple units. Where available, the enthalpy of formation (h^f), Gibbs free energy of formation (Δg^f), entropy (s), and heat capacity (c_p) used in computing the exergy are also provided, each computed at the reference temperature and pressure as noted in Chap. 4. In the tables, the states include gas (g), liquid (l), aqueous (aq), or crystalline (c). For some aqueous chemicals, the molality m in mol/kg_{H_2O} is given. If no concentration is indicated, the solution is assumed to be dilute.[11]

Although the Gibbs free energies used in this work were determined at the reference temperature and pressure, the reported values in the literature can sometimes have slight variations from source to source due to differences in their means of calculation.[11,12] This will affect chemical exergy values slightly. When differences were found between Refs. 11 and 12, we used the more recent reference.

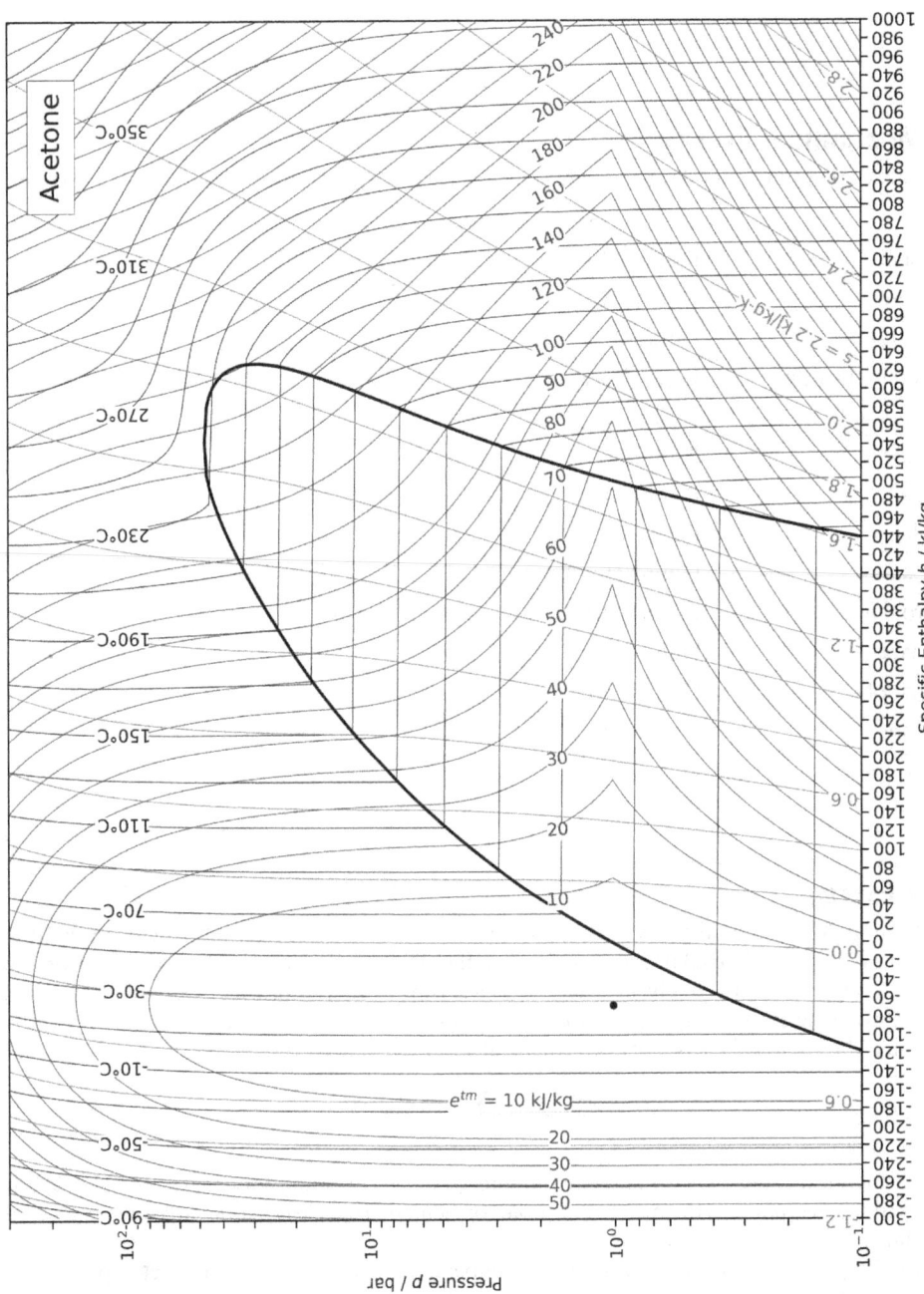

Figure 7.1 Pressure-enthalpy diagram for acetone, including thermo-mechanical exergy e^{tm}. Underlying physical property models based on Ref. 1. For chemical exergy e^{ch}, see Table 7.2.

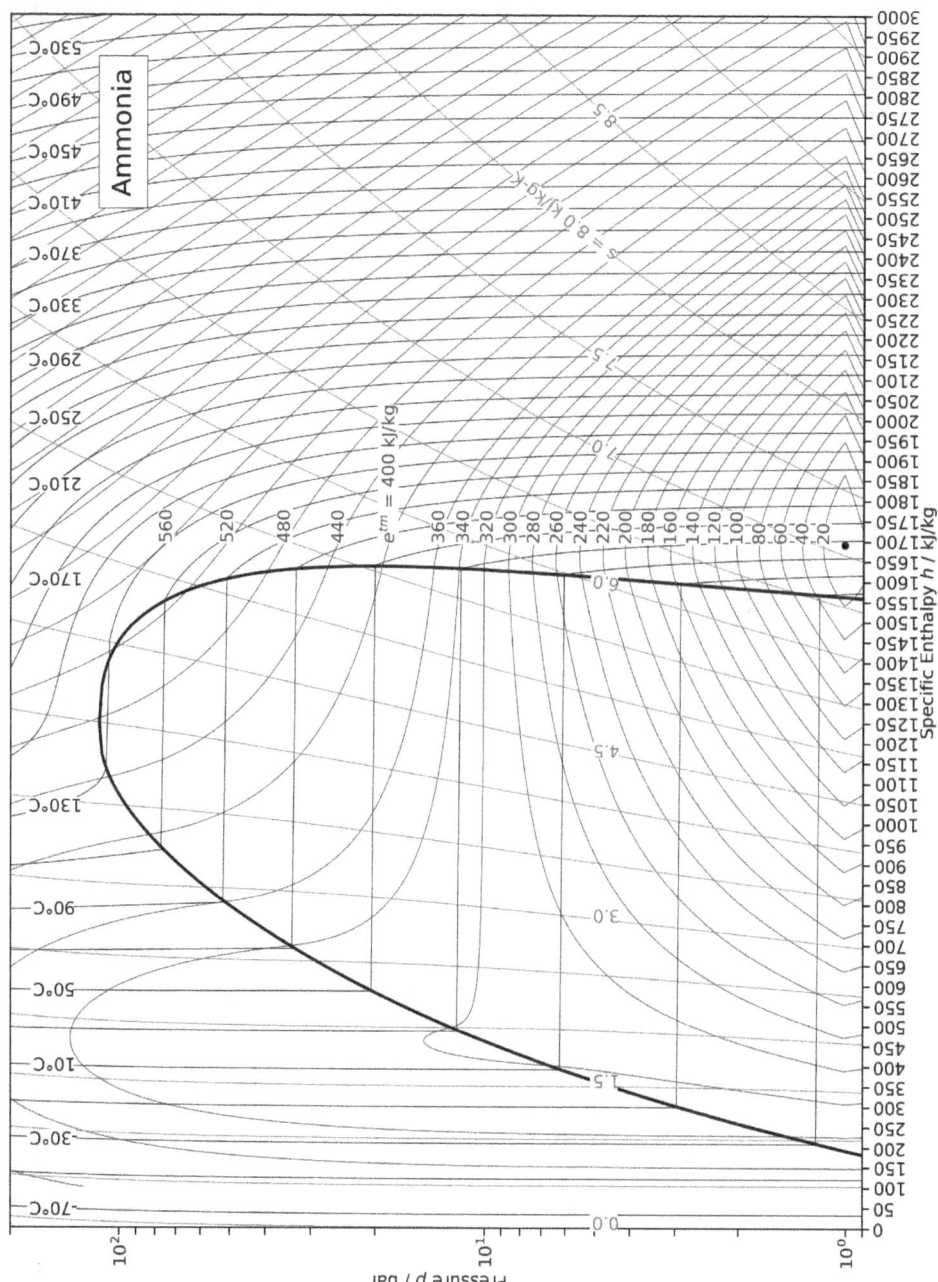

Figure 7.2 Pressure-enthalpy diagram for ammonia, including thermo-mechanical exergy e^{tm}. Underlying physical property models based on CoolProps. For chemical exergy e^{ch}, see Table 7.7. See also Table 7.8.

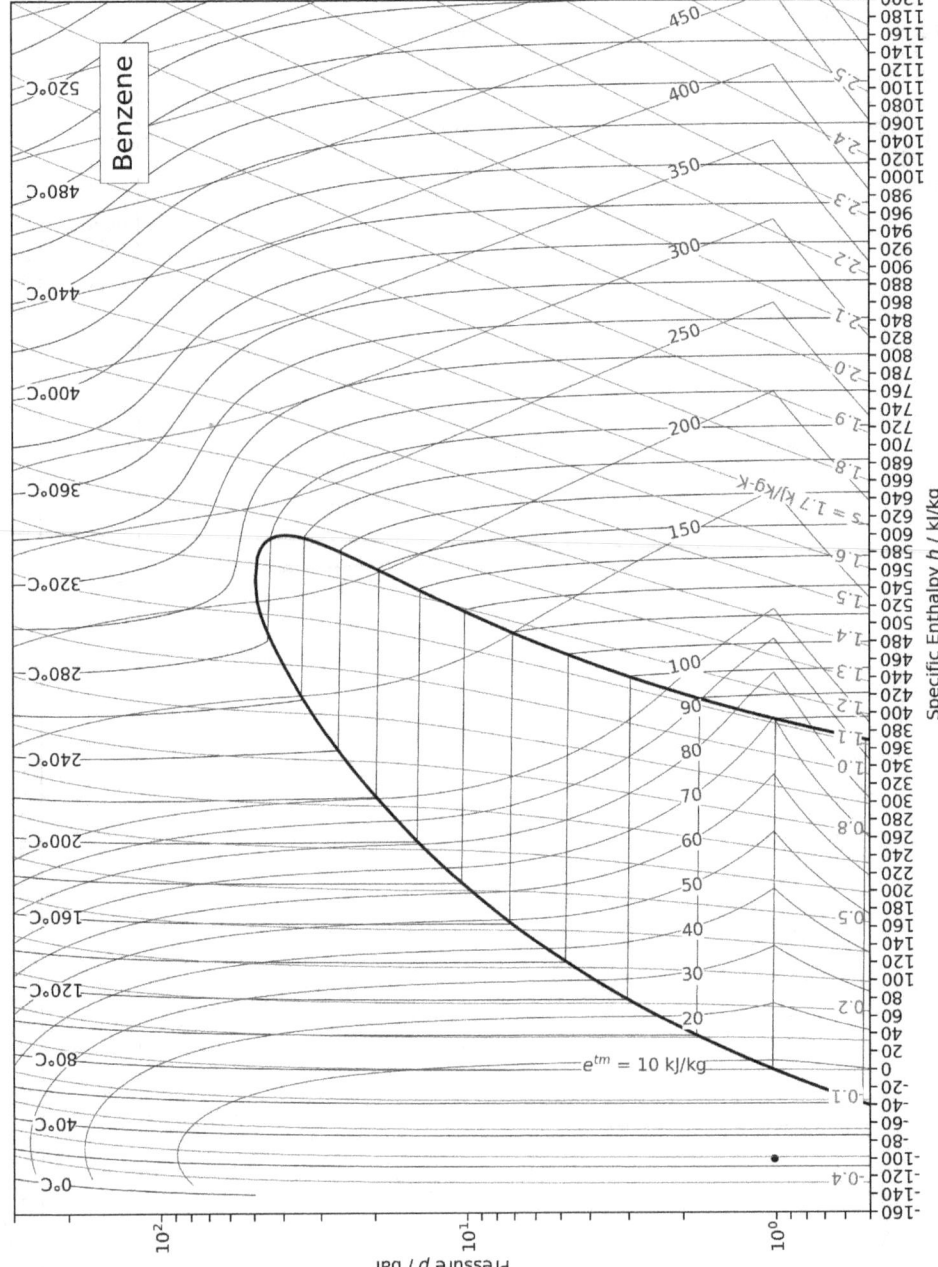

Figure 7.3 Pressure-enthalpy diagram for benzene, including thermo-mechanical exergy e^{tm} in kJ/kg. Underlying physical property models based on Ref. 2. For chemical exergy e^{ch}, see Table 7.1.

174

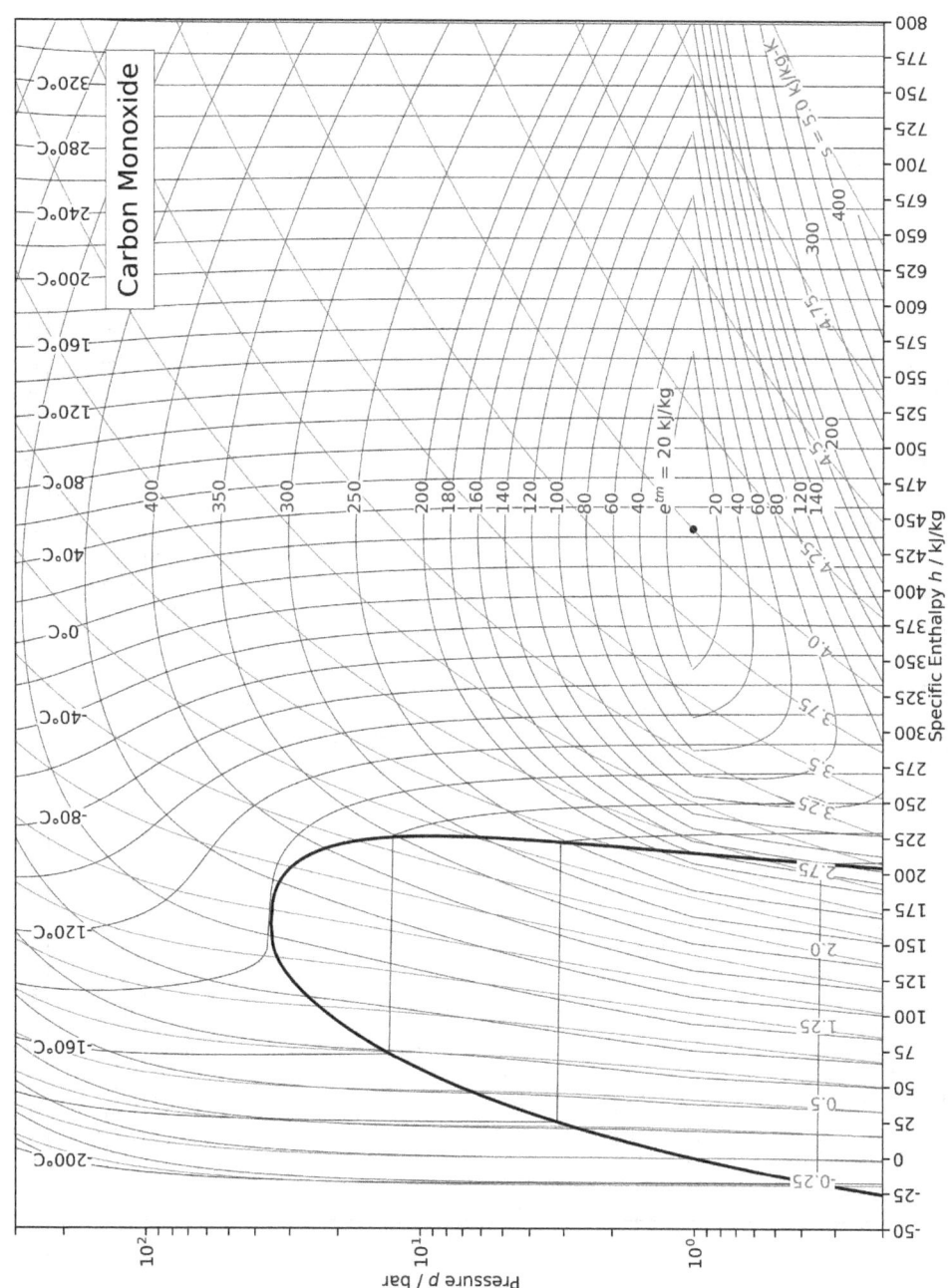

Figure 7.4 Pressure-enthalpy diagram for carbon monoxide, including thermo-mechanical exergy e^{tm}. Underlying physical property models based on Ref. 1. For chemical exergy e^{ch}, see Table 7.7.

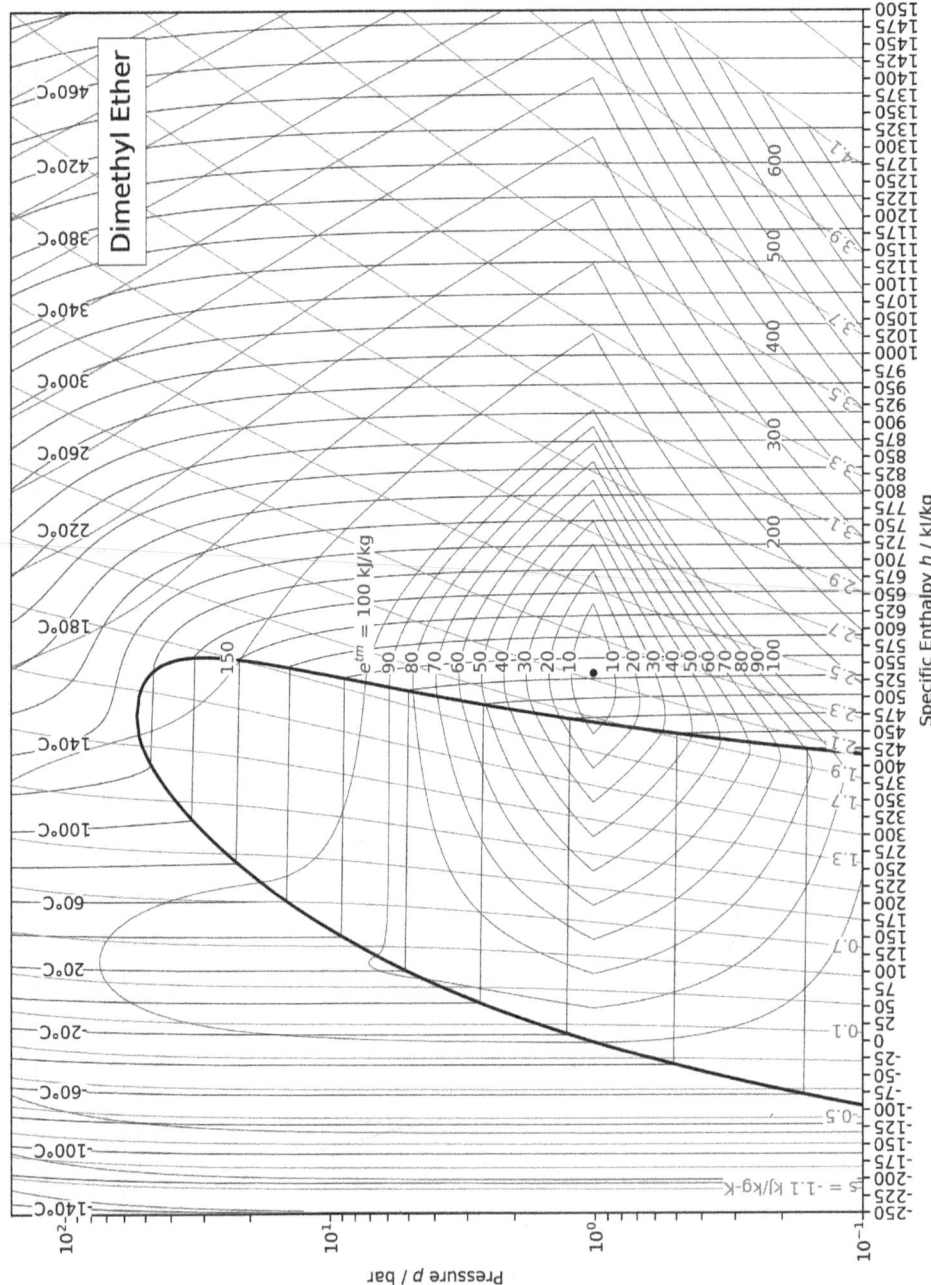

Figure 7.5 Pressure-enthalpy diagram for dimethyl ether, including thermo-mechanical exergy e^{tm}. Underlying physical property models based on Ref. 3. For chemical exergy e^{ch}, see Table 7.2.

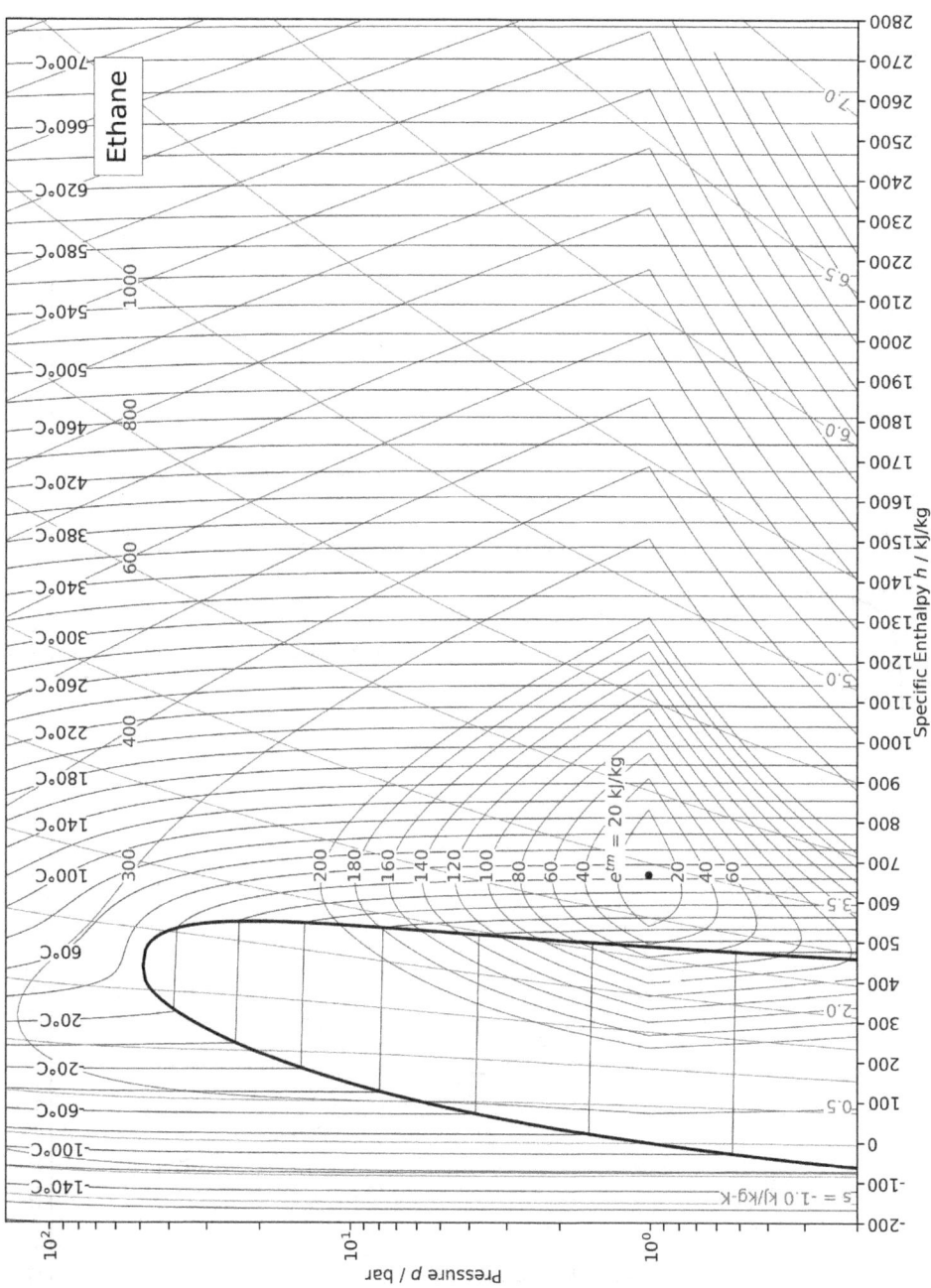

Figure 7.6 Pressure-enthalpy diagram for ethane, including thermo-mechanical exergy e^{tm}. Underlying physical property models based on Ref. 4. For chemical exergy e^{ch}, see Table 7.1.

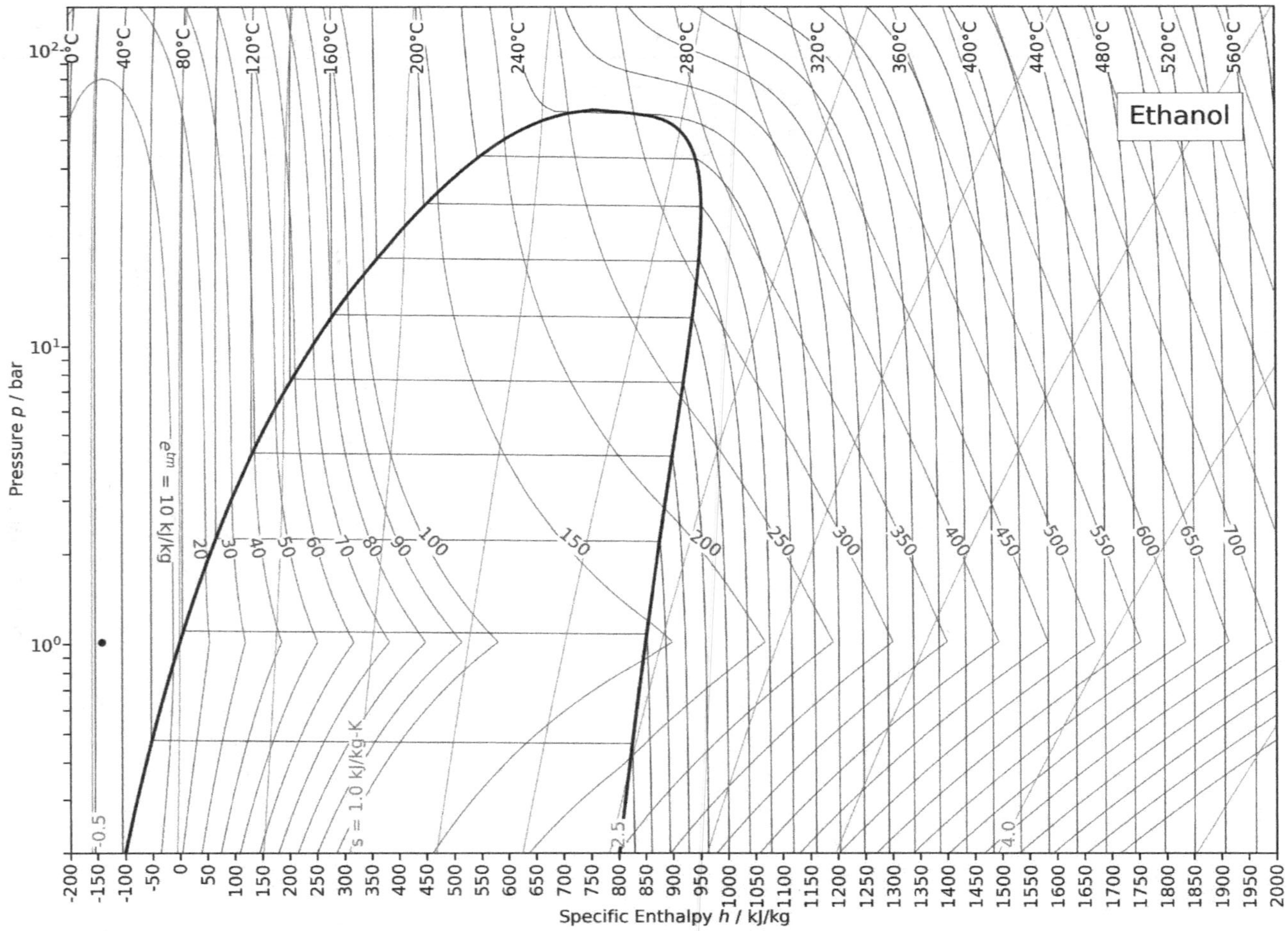

Figure 7.7 Pressure-enthalpy diagram for ethanol, including thermo-mechanical exergy e^{tm}. Underlying physical property models based on Ref. 5. For chemical exergy e^{ch}, see Table 7.2.

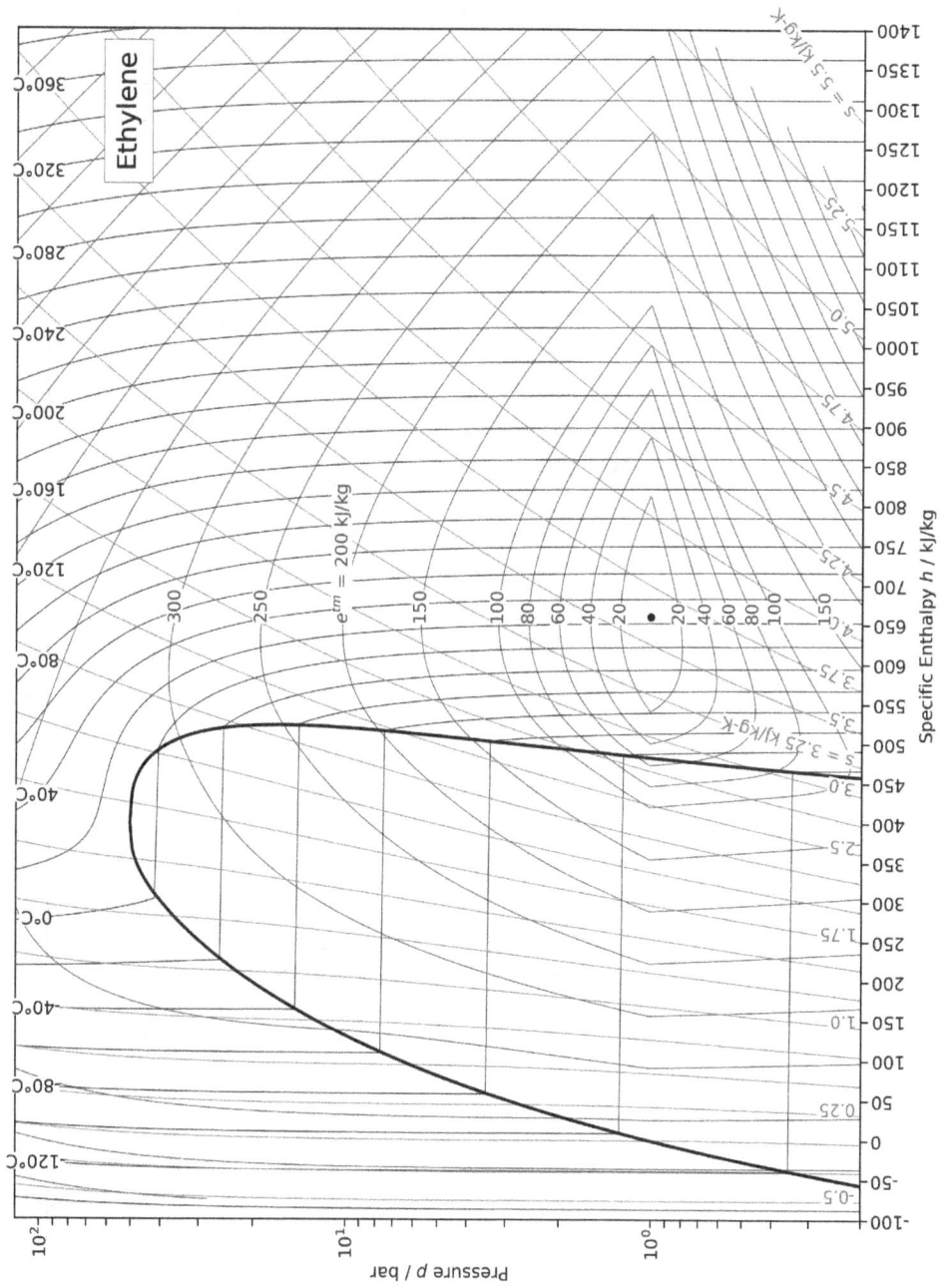

Figure 7.8 Pressure-enthalpy diagram for ethylene, including thermo-mechanical exergy e^{tm}. Underlying physical property models based on Ref. 6. For chemical exergy e^{ch}, see Table 7.1.

179

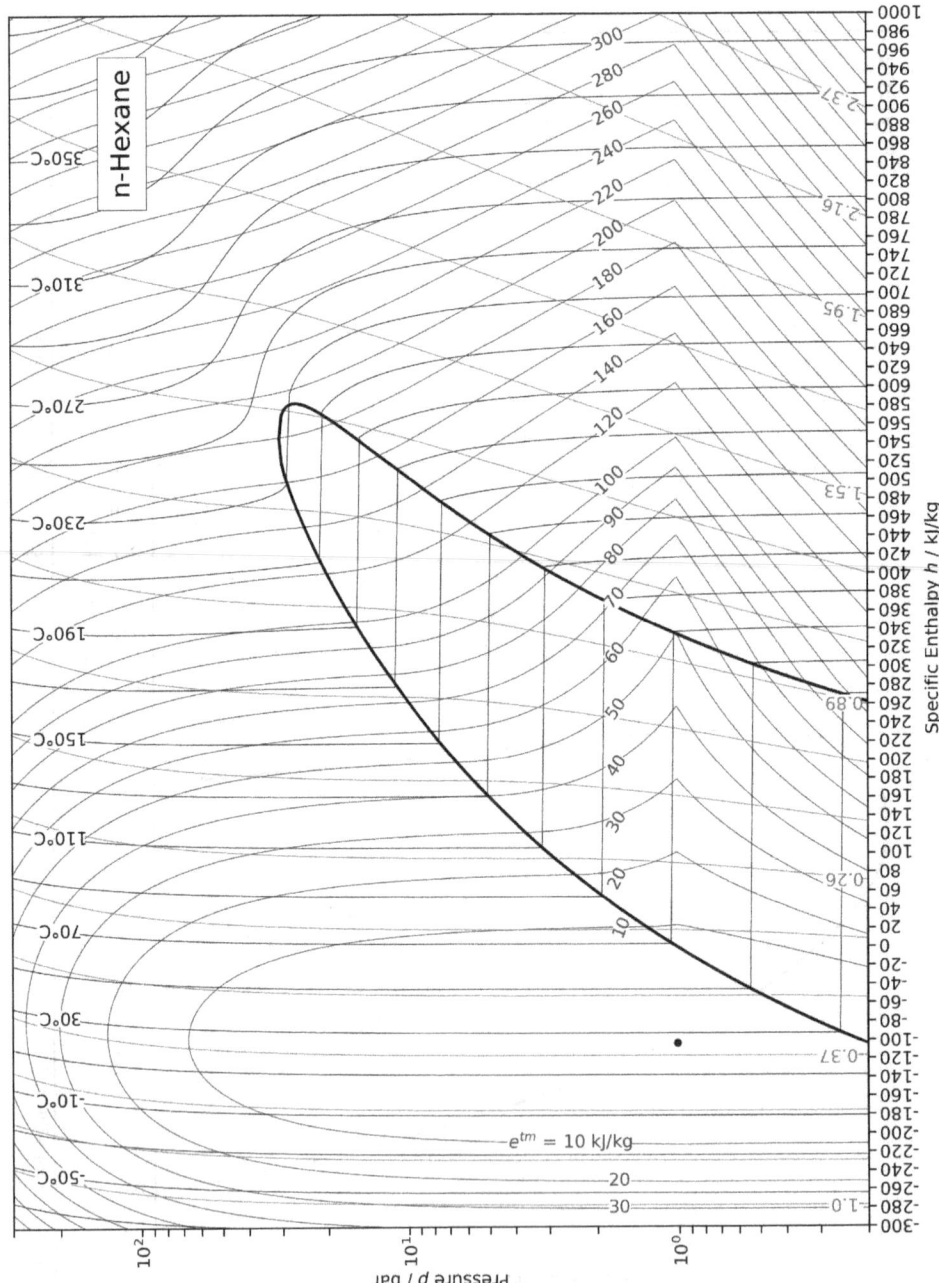

Figure 7.9 Pressure-enthalpy diagram for *n*-hexane, including thermo-mechanical exergy e^{tm}. Underlying physical property models based on CoolProps. For chemical exergy e^{ch}, see Table 7.1.

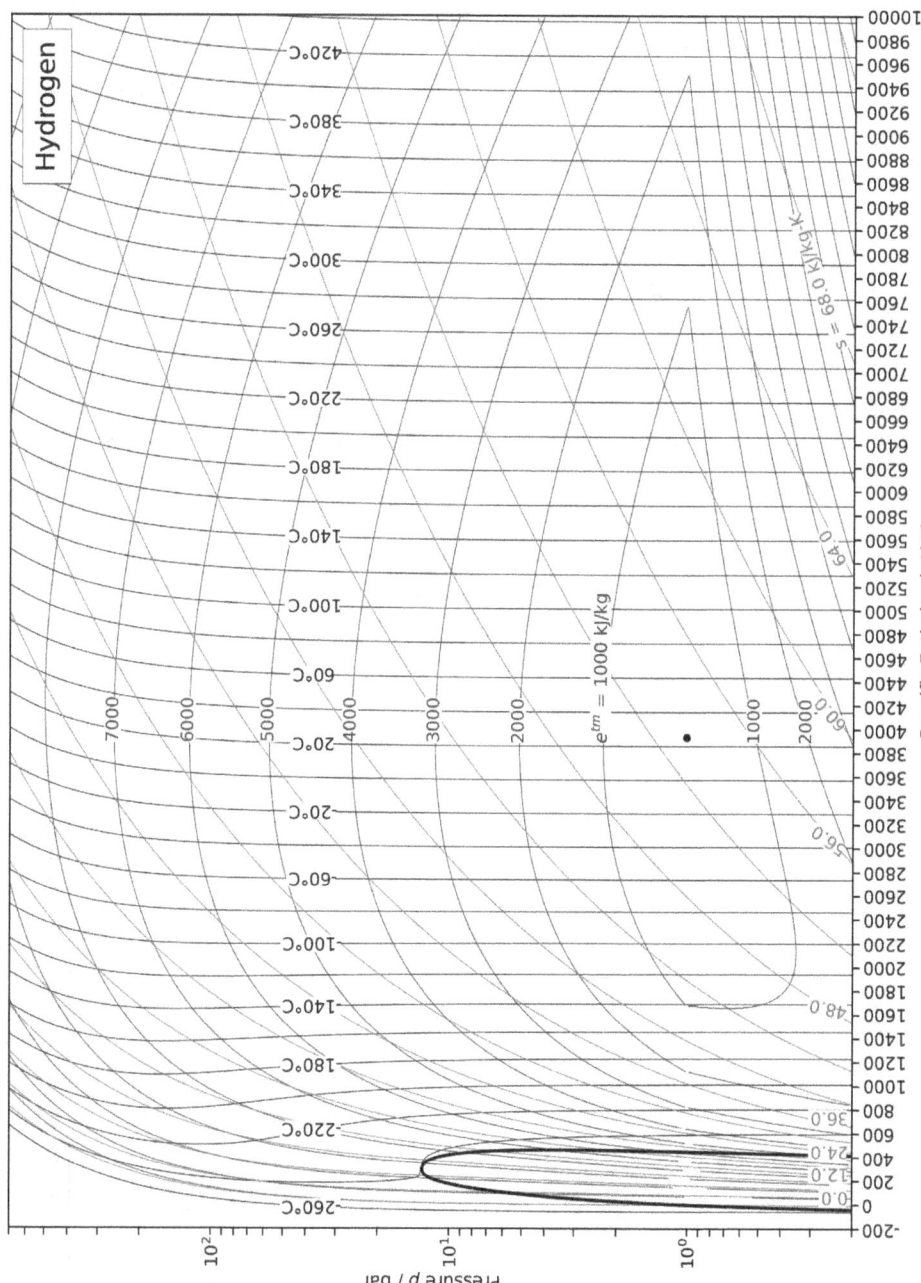

Figure 7.10 Pressure-enthalpy diagram for hydrogen (H₂), including thermo-mechanical exergy e^{tm}. Underlying physical property models based on Ref. 7. For chemical exergy e^{ch}, see Table 7.8.

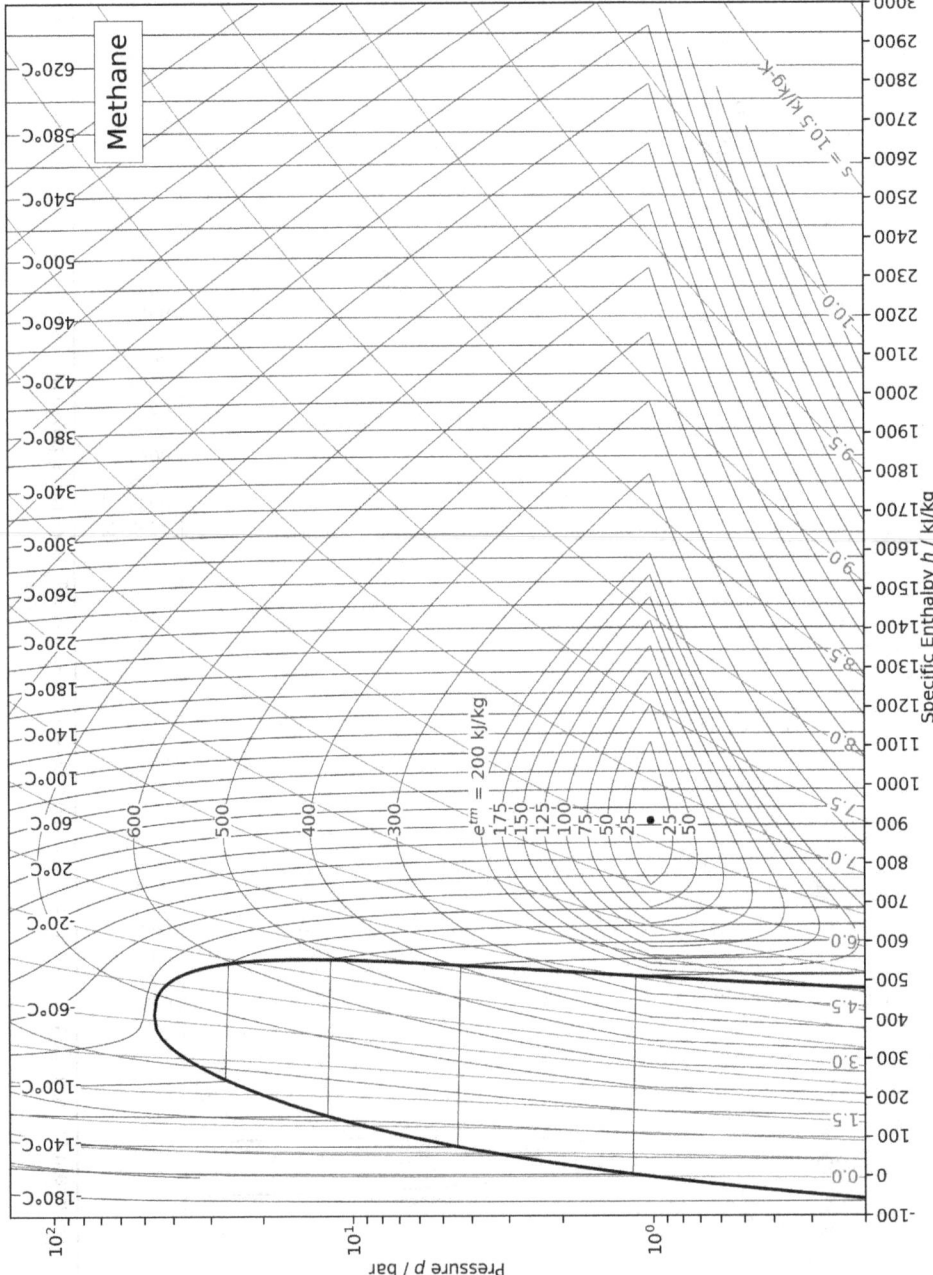

Figure 7.11 Pressure-enthalpy diagram for methane, including thermo-mechanical exergy e^{tm}. Underlying physical property models based on Ref. 8. For chemical exergy e^{ch}, see Table 7.1.

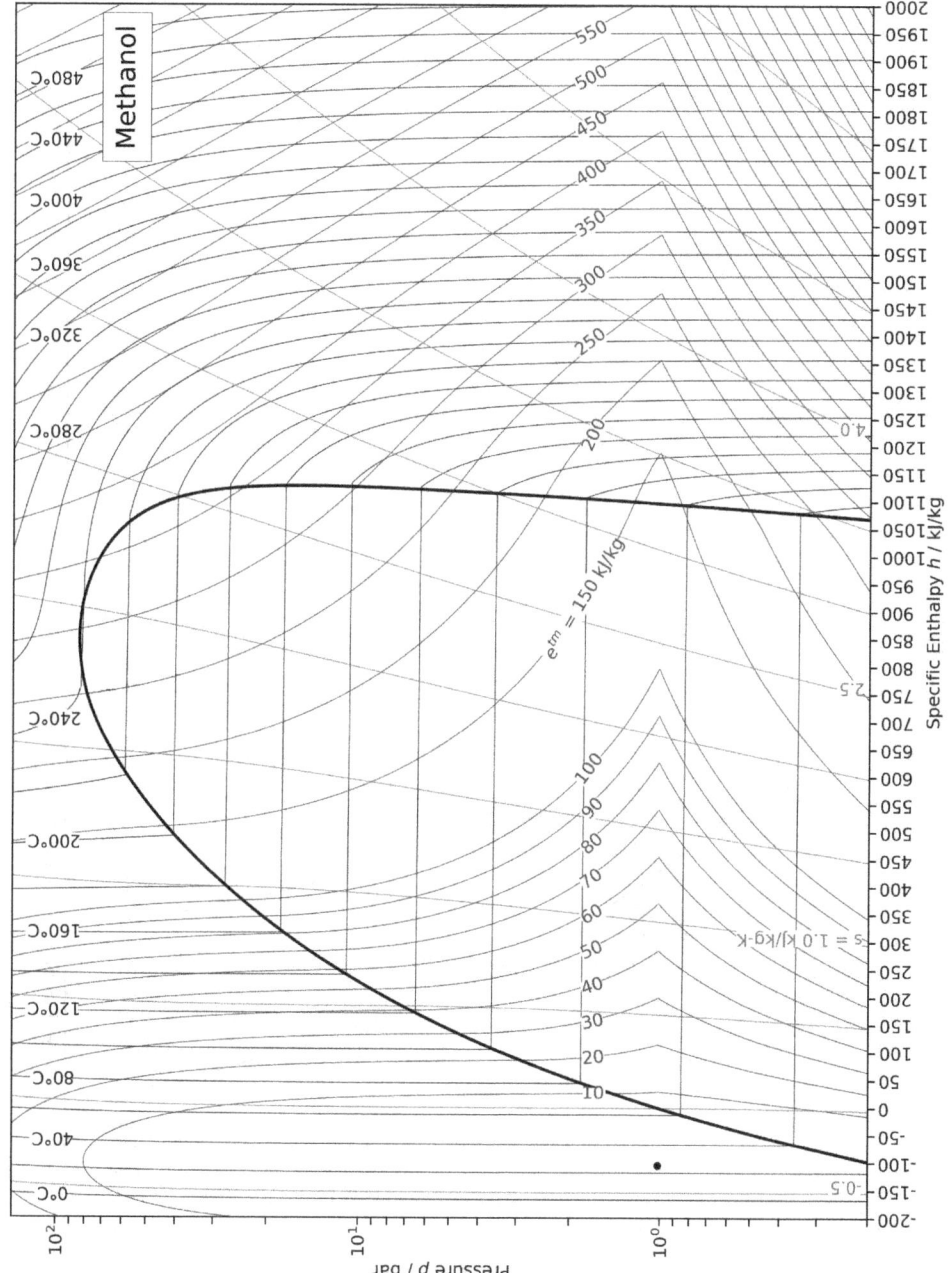

Figure 7.12 Pressure-enthalpy diagram for methanol, including thermo-mechanical exergy e^{tm}. Underlying physical property models based on Ref. 9. For chemical exergy e^{ch}, see Table 7.2.

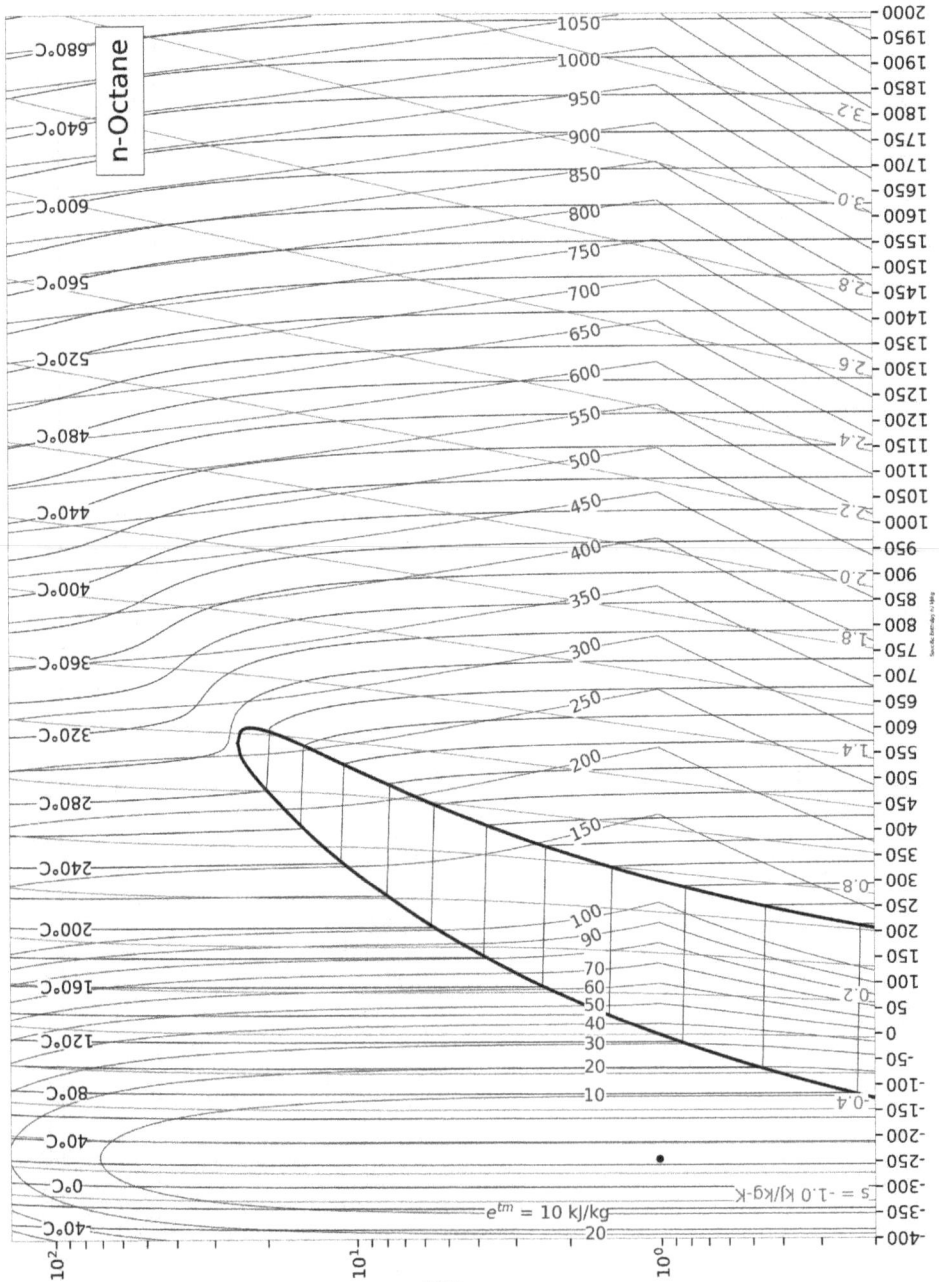

Figure 7.13 Pressure-enthalpy diagram for *n*-octane, including thermo-mechanical exergy e^{tm} (bold lines) in kJ/kg. Underlying physical property models based on CoolProps. For chemical exergy e^{ch}, see Table 7.1.

184

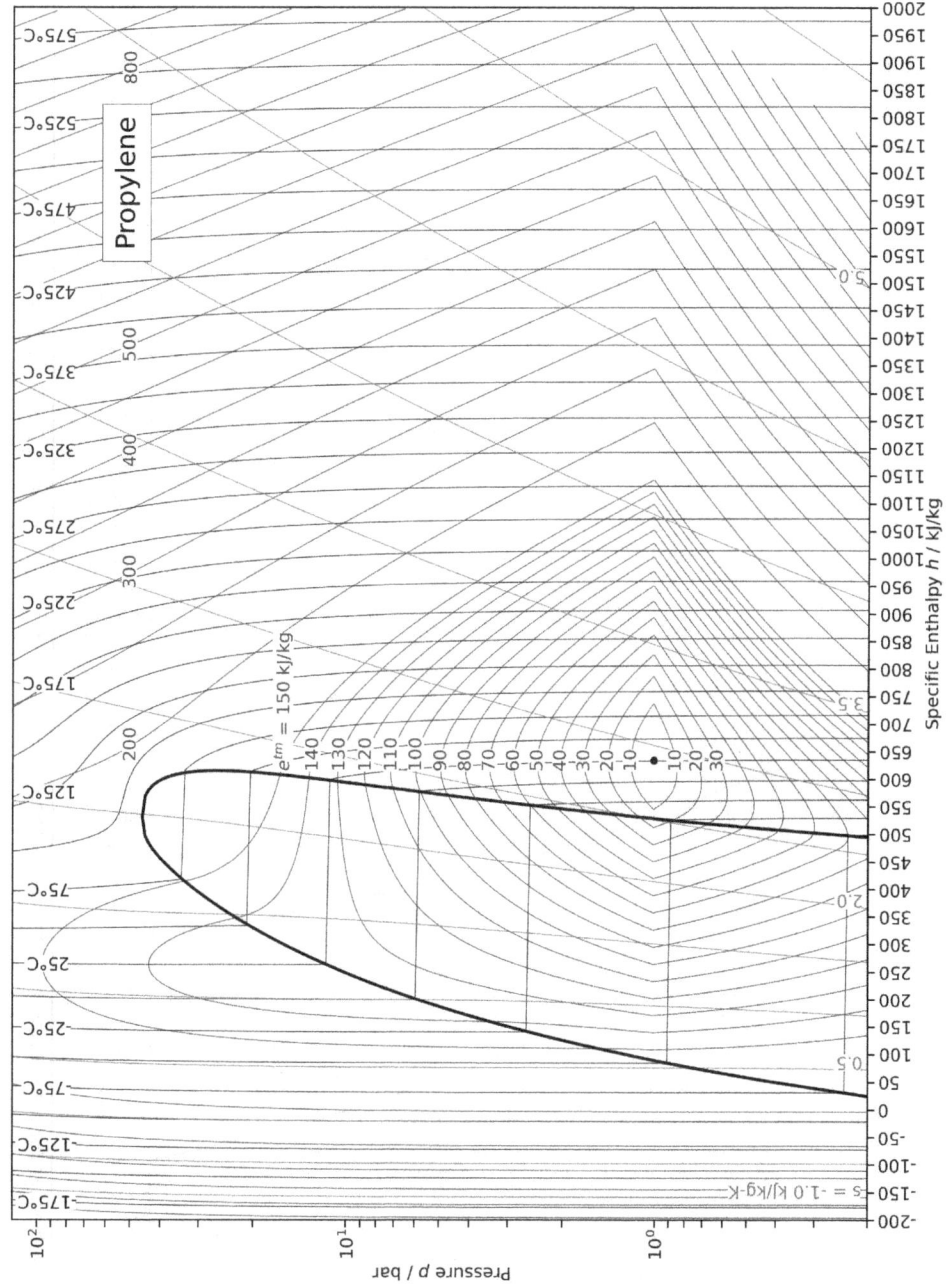

Figure 7.14 Pressure-enthalpy diagram for propylene, including thermo-mechanical exergy e^{tm}. Underlying physical property models based on CoolProps. For chemical exergy e^{ch}, see Table 7.1.

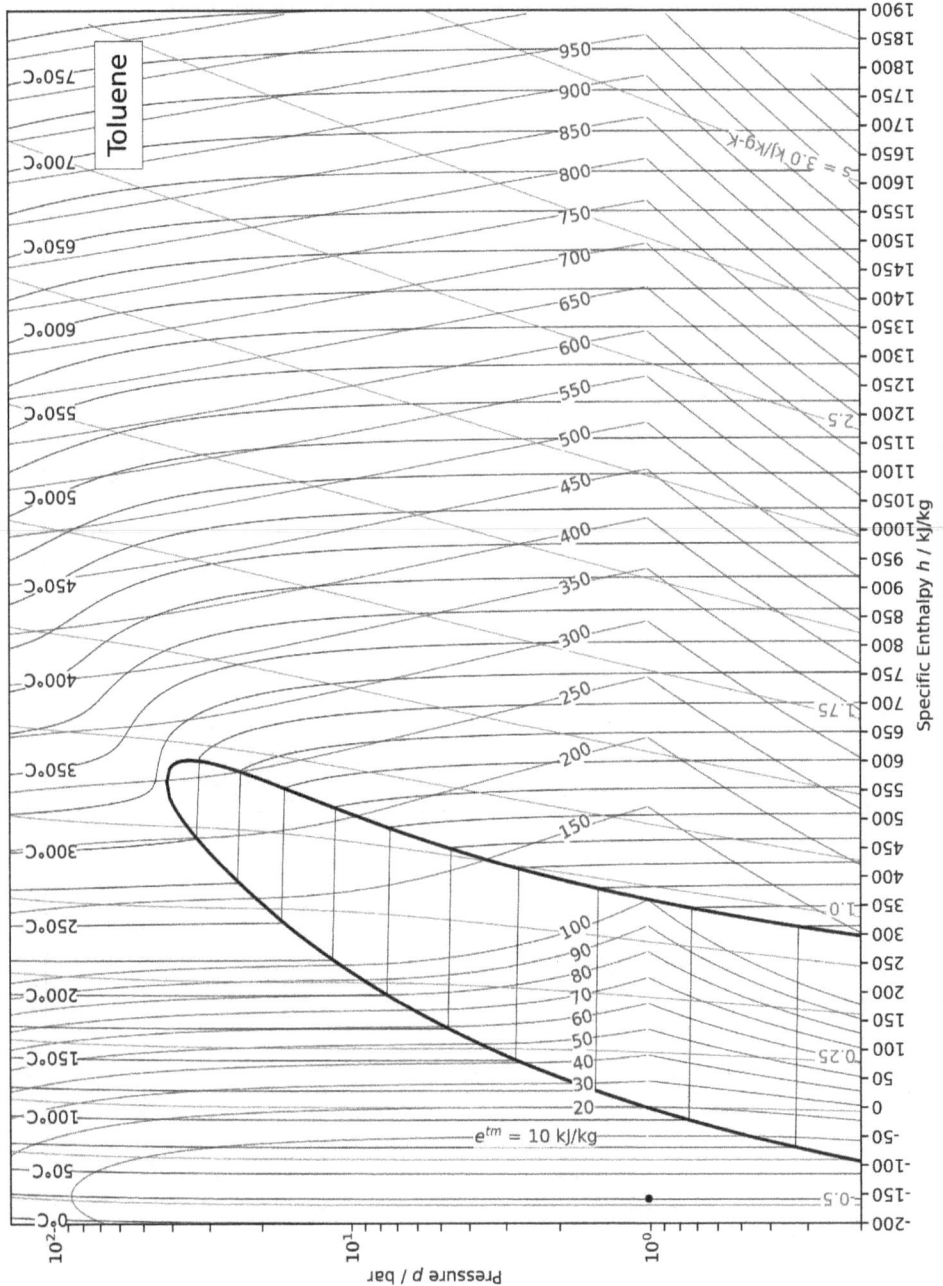

FIGURE 7.15 Pressure-enthalpy diagram for toluene, including thermo-mechanical exergy e^{tm} in kJ/kg. Underlying physical property models based on Ref. 1. For chemical exergy e^{ch}, see Table 7.1.

Tables 7.1 to 7.8 are organized as follows:

- Organic chemicals comprised of C and H (carbon and hydrocarbons)—Table 7.1
- Organic chemicals comprised of C, H, and O—Table 7.2
- Organic chemicals comprised of C, H, and N—Table 7.3
- Organic chemicals comprised of C, H, O, and N—Table 7.4
- Organic chemicals comprised of C, H, and S—Table 7.5
- Other organic chemicals—Table 7.6
- Other inorganic chemicals—Table 7.7
- Hydrogen economy chemicals—Table 7.8

To avoid duplicates, chemicals which have been listed in Chap. 4 (reference chemicals), Chap. 5 (air and atmospheric gases), or Chap. 6 (water) are not listed in these tables.

7.3   Chemicals Relevant to the Hydrogen Economy

The hydrogen economy refers to the concept of molecular hydrogen (H_2) as a fuel and integral part of a global energy system. The motivation is primarily that when H_2 is combusted or otherwise oxidized in a fuel cell to release its energy, it produces no CO_2 unlike most existing carbonaceous transportation fuels. Similarly, although H_2 is primarily produced using fossil fuels today (specifically natural gas reforming), it is possible to produce it with electrolysis techniques using renewable energy (wind, solar, hydroelectric, etc.) or other carbonless energy (nuclear). In theory, this means an entire economy could be developed using a carbon-free pathway. One key challenge in the hydrogen economy is hydrogen transportation and storage since H_2 at room temperatures and pressures has a very low volumetric energy density—only 0.11 MJ/L on a higher heating value basis. Solutions for this usually involve compressing H_2 to very high pressures, liquefying H_2 cryogenically or converting it to other chemicals like methane, ammonia, or methanol and transporting those. Because of the current political and scientific interest, Table 7.8 contains a selection of chemicals at various temperatures and pressures that are relevant to the hydrogen economy, based on the paper by Adams.[13]

Name	Structure	State	h^f kcal/mol	Δg^f kcal/mol	s cal/mol/K	C_p cal/mol/K	e^{ch} kcal/mol	e^{ch} kJ/mol
Acetylene	C_2H_2	g	54.19	50	48	10.5	302.55	1265.85
Acetylene, $m = 1$	C_2H_2	aq	50.54	51.88	29.5		304.43	1273.72
Anthracene	$C_{14}H_{10}$	c	29	68.3	49.58	49.7	1723.26	7210.13
Azulene	$C_{10}H_8$	g	66.9	84.1	80.75	30.69	1290.40	5399.04
2,3-Benzanthracene	$C_{18}H_{12}$	c	38.3	85.79	51.48		2189.41	9160.50
Benzene	C_6H_6	l	11.71	29.72	41.41	19.52	787.36	3294.31
Benzene	C_6H_6	g	19.82	30.99	64.34		788.63	3299.63
Biphenyl	$C_{12}H_{10}$	c	24.02	60.75	49.2	38.8	1519.60	6357.99
Biphenyl	$C_{12}H_{10}$	l	28.5	62.07	59.8		1520.92	6363.52
1,2-Butadiene	C_4H_6	g	38.77	47.43	70.03	19.15	608.95	2547.87
1,3-Butadiene	C_4H_6	g	26.33	36.01	66.62	19.01	597.53	2500.08
Butadiyne	C_4H_2	g	113	106.11	59.76	17.6	554.77	2321.16
Butane	C_4H_{10}	l	−35.29	−3.6	55.2		670.79	2806.58
Butane	C_4H_{10}	g	−30.15	−4.1	74.12	23.29	670.29	2804.48
1-Butene	C_4H_8	g	−0.03	17.04	73.04	20.47	635.00	2656.82
cis-2-Butene	C_4H_8	g	−1.67	15.74	71.9	18.86	633.70	2651.38
trans-2-Butene	C_4H_8	g	−2.67	15.05	70.86	20.99	633.01	2648.50
1-Buten-3-yne	C_4H_4	g	72.8	73.13	66.77	17.49	578.22	2419.28
Butylbenzene	$C_{10}H_{14}$	l	−18.67	27.5			1403.10	5870.55
Butylbenzene	$C_{10}H_{14}$	g	−3.3	34.58	105.04	41.85	1410.18	5900.17
Butylcyclohexane	$C_{10}H_{20}$	g	−50.95	13.49	109.58	49.5	1558.38	6520.26
Butylcyclopentane	C_9H_{18}	g	−40.22	14.67	109.04	42.42	1405.07	5878.82
1-Butyne	C_4H_6	g	39.48	48.3	69.51	19.46	609.82	2551.51
2-Butyne	C_4H_6	g	34.97	44.32	67.71	18.63	605.84	2534.85
Cyclobutane	C_4H_8	g	6.37	26.3	63.43	17.26	644.26	2695.57
Cyclobutene	C_4H_6	g	31	41.76	62.98	16.03	603.28	2524.14
Cycloheptane	C_7H_{14}	l	−37.47	12.92	57.97	29.42	1094.34	4578.73
1,3,5-Cycloheptatriene	C_7H_8	l	34.22	58.09	51.3	38.9	970.22	4059.39
Cyclohexane	C_6H_{12}	l	−37.34	6.37	48.84	37.4	933.30	3904.94
Cyclohexane	C_6H_{12}	g	−29.43	7.59	71.28	25.4	934.52	3910.05
Cyclohexene	C_6H_{10}	l	−9.28	24.28	51.67	34.9	894.78	3743.77
Cyclohexene	C_6H_{10}	g	−1.28	25.54	74.27	25.1	896.04	3749.04
Cyclooctane	C_8H_{16}	l	−40.58	18.6	62.62		1254.51	5248.88
1,3,5,7-Cyclooctatetraene	C_8H_8	l	60.93	85.7	52.65		1095.89	4585.19
Cyclopentadiene	C_5H_6	g	32	42.86	64		702.44	2939.02
Cyclopentane	C_5H_{10}	l	−25.28	8.7	48.82	30.8	781.15	3268.31
Cyclopentane	C_5H_{10}	g	−18.46	9.23	70	19.84	781.68	3270.53
Cyclopentene	C_5H_8	l	1.02	25.93	48.1	29.24	741.94	3104.29
Cyclopentene	C_5H_8	g	7.87	26.48	69.23	17.95	742.49	3106.59
Cyclopropane	C_3H_6	g	12.74	24.95	56.75	13.37	488.42	2043.54
Cyclopropatriene*	C_3	g	196.00	180.31	56.70		474.47	1985.19
Cyclopropene	C_3H_4	g	66	68.42	58.38		475.46	1989.31
cis-Decalin	$C_{10}H_{18}$	l	−52.45	16.47	63.34	55.45	1504.93	6296.62
trans-Decalin	$C_{10}H_{18}$	l	−55.14	13.79	63.32	54.61	1502.25	6285.41
Decane	$C_{10}H_{22}$	l	−71.95	−4.19	101.7	75.16	1597.13	6682.40

TABLE 7.1 Chemical exergy of pure carbon and hydrocarbons (containing only C and H). Underlying state information from Ref. 11 except where marked with * indicating the state information came from Ref. 12.

Name	Structure	State	h^f kcal/mol	Δg^f kcal/mol	s cal/mol/K	C_p cal/mol/K	e^{ch} kcal/mol	e^{ch} kJ/mol
1-Decene	$C_{10}H_{20}$	l	−41.73	25.1	101.58		1569.99	6568.84
1-Decyne	$C_{10}H_{18}$	g	9.85	60.28	125.36	52.51	1548.74	6479.92
Diatomic Carbon*	C_2	g	200.20	186.80	47.63		382.92	1602.12
Dibenzyl	$C_{14}H_{14}$	c	10.53	62.51	64.4	61	1830.34	7658.12
1,2-Diethylbenzene	$C_{10}H_{14}$	g	−4.53	33.72	103.81	43.63	1409.32	5896.58
1,3-Diethylbenzene	$C_{10}H_{14}$	g	−5.22	32.67	104.99	42.27	1408.27	5892.18
1,4-Diethylbenzene	$C_{10}H_{14}$	g	−5.32	32.95	103.73	42.1	1408.55	5893.35
2,2-Dimethylbutane	C_6H_{14}	g	−44.35	−2.2	85.62	33.91	981.17	4105.20
2,3-Dimethylbutane	C_6H_{14}	g	−42.49	−0.98	87.42	33.59	982.39	4110.30
2,3-Dimethyl-1-butene	C_6H_{12}	g	−13.32	18.89	87.39	34.29	945.82	3957.33
2,3-Dimethyl-2-butene	C_6H_{12}	g	−14.15	18.18	87.15	29.54	945.11	3954.36
3,3-Dimethyl-1-butene	C_6H_{12}	g	−10.31	23.46	82.16	30.23	950.39	3976.45
1,1-Dimethylcyclohexane	C_8H_{16}	l	−52.31	6.34	63.87		1242.25	5197.58
1,1-Dimethylcyclohexane	C_8H_{16}	g	−43.26	8.42	87.24	36.9	1244.33	5206.29
cis-1,1-Dimethylcyclohexane	C_8H_{16}	g	−41.15	9.85	89.51	37.4	1245.76	5212.27
trans-1,1-Dimethylcyclohexane	C_8H_{16}	g	−43.02	8.24	88.65	38	1244.15	5205.53
cis-1,2-Dimethylcyclohexane	C_8H_{16}	g	−44.16	7.13	88.54	37.6	1243.04	5200.89
trans-1,2-Dimethylcyclohexane	C_8H_{16}	g	−42.2	8.68	89.91	37.6	1244.59	5207.37
cis-1,4-Dimethylcyclohexane	C_8H_{16}	g	−42.22	9.07	88.54	37.6	1244.98	5209.01
trans-1,4-Dimethylcyclohexane	C_8H_{16}	g	−44.12	7.58	87.19	37.7	1243.49	5202.77
1,1-Dimethylcyclopentane	C_7H_{14}	g	−33.05	9.33	85.87	31.86	1090.75	4563.71
cis-1,2-Dimethylcyclopentane	C_7H_{14}	g	−30.96	10.93	87.51	32.06	1092.35	4570.41
trans-1,2-Dimethylcyclopentane	C_7H_{14}	g	−32.67	9.17	87.67	32.14	1090.59	4563.04
cis-1,3-Dimethylcyclopentane	C_7H_{14}	g	−32.47	9.37	87.67	32.14	1090.79	4563.88
trans-1,3-Dimethylcyclopentane	C_7H_{14}	g	−31.93	9.91	87.67	32.14	1091.33	4566.14
2,2-Dimethylhexane	C_8H_{18}	l	−62.63	0.71	79.33		1293.05	5410.14
2,2-Dimethylhexane	C_8H_{18}	g	−53.71	2.56	103.06		1294.90	5417.88
2,3-Dimethylhexane	C_8H_{18}	l	−60.4	2.17	81.91		1294.51	5416.25
2,3-Dimethylhexane	C_8H_{18}	g	−51.13	4.23	106.11		1296.57	5424.87
2,4-Dimethylhexane	C_8H_{18}	l	−61.47	0.89	82.62		1293.23	5410.89
2,4-Dimethylhexane	C_8H_{18}	g	−52.44	2.8	106.11		1295.14	5418.88
2,5-Dimethylhexane	C_8H_{18}	l	−62.26	0.6	80.96		1292.94	5409.68
2,5-Dimethylhexane	C_8H_{18}	g	−53.21	2.5	104.93		1294.84	5417.63
3,3-Dimethylhexane	C_8H_{18}	l	−61.58	1.23	81.12		1293.57	5412.31
3,3-Dimethylhexane	C_8H_{18}	g	−52.61	3.17	104.7		1295.51	5420.43
3,4-Dimethylhexane	C_8H_{18}	l	−60.23	2.03	82.97		1294.37	5415.66
3,4-Dimethylhexane	C_8H_{18}	g	−50.91	4.14	107.15		1296.48	5424.49
2,2-Dimethylpentane	C_7H_{16}	g	−49.27	0.02	93.9	39.67	1137.87	4760.87
2,3-Dimethylpentane	C_7H_{16}	g	−47.62	0.16	98.96	39.67	1138.01	4761.45

TABLE 7.1   Chemical exergy of pure carbon and hydrocarbons (containing only C and H). Underlying state information from Ref. 11 except where marked with * indicating the state information came from Ref. 12. (*Continued*)

Name	Structure	State	h^f kcal/mol	Δg^f kcal/mol	s cal/mol/K	C_p cal/mol/K	e^{ch} kcal/mol	e^{ch} kJ/mol
2,4-Dimethylpentane	C_7H_{16}	g	−48.28	0.74	94.8	39.67	1138.59	4763.88
3,3-Dimethylpentane	C_7H_{16}	g	−48.17	0.63	95.53	39.67	1138.48	4763.42
2,2-Dimethylpropane	C_5H_{12}	g	−39.67	−0.364	73.23	29.07	828.51	3466.50
1,1-Diphenylethane	$C_{14}H_{14}$	l	11.7	58.58	80.28		1826.41	7641.68
1,2-Diphenylethane	$C_{14}H_{14}$	l	12.31	63.87	64.6		1831.70	7663.81
Diphenylmethane	$C_{13}H_{12}$	l	21.25	66.19	57.2	55.7	1679.53	7027.14
5,6-Dithiadecane	$C_{12}H_{26}$	g	−37.86	12.87	136.91	55.23	1923.17	8046.54
Dodecane	$C_{12}H_{26}$	l	−84.16	6.17	117.26	89.86	1916.47	8018.51
1-Dodecene	$C_{12}H_{24}$	g	−39.52	32.96	147.78	64.43	1886.83	7894.49
1-Dodecyne	$C_{12}H_{22}$	g	−0.01	64.22	143.98	63.44	1861.66	7789.17
Eicosane	$C_{20}H_{42}$	g	−108.93	28.03	223.26	110.73	3174.24	13281.03
1-Eicosene	$C_{20}H_{40}$	g	−78.93	49.03	222.26	108.15	3138.81	13132.78
Ethane	C_2H_6	g	−20.24	−7.84	54.76	12.54	357.57	1496.07
Ethyl radical	C_2H_5	g	26	31	59.2		368.19	1540.52
Ethylbenzene	C_8H_{10}	l	−2.98	28.61	60.99		1095.23	4582.43
Ethylbenzene	C_8H_{10}	g	7.12	31.21	86.15	30.69	1097.83	4593.31
2-Ethyl-1-butene	C_6H_{12}	g	−12.32	19.11	90.01	31.92	946.04	3958.25
Ethylcyclohexane	C_8H_{16}	l	−50.72	6.95	67.14		1242.86	5200.14
Ethylcyclopentane	C_7H_{14}	l	−39.08	8.92	67		1090.34	4562.00
Ethylcyclopentane	C_7H_{14}	g	−30.37	10.65	90.42	31.49	1092.07	4569.23
Ethylene	C_2H_4	g	12.5	16.31	52.39	10.24	325.29	1361.01
3-Ethylhexane	C_8H_{18}	l	−59.88	1.79	84.95		1294.13	5414.66
3-Ethylpentane	C_7H_{16}	g	−45.33	2.63	98.35	39.67	1140.48	4771.79
Ethynylbenzene	C_8H_6	g	78.22	86.46	76.88	27.46	1040.21	4352.26
Fluoranthene	$C_{16}H_{10}$	c	45.75	82.6	55.09		1933.68	8090.50
Heptadecane	$C_{17}H_{36}$	g	−94.15	22.01	195.33	94.33	2704.76	11316.70
1-Heptadecene	$C_{17}H_{34}$	g	−64.15	43	194.33	91.76	2669.31	11168.41
Heptane	C_7H_{16}	l	−53.63	0.42	77.92	53.76	1138.27	4762.54
1-Heptene	C_7H_{14}	l	−23.41	21.22	78.31	50.62	1102.64	4613.46
1-Heptene	C_7H_{14}	g	−14.89	22.9	101.24	37.1	1104.32	4620.49
1-Heptyne	C_7H_{12}	g	24.62	54.18	97.44	36.11	1079.17	4515.25
Hexadecane	$C_{16}H_{34}$	g	−89.23	20	186.02	88.86	2548.26	10661.90
1-Hexadecene	$C_{16}H_{32}$	g	−59.23	40.99	185.02	86.29	2512.81	10513.62
Hexamethylbenzene	$C_{12}H_{18}$	c	−39.19	28.06	71.66	61.5	1712.63	7165.66
Hexane	C_6H_{14}	l	−47.52	−0.91	70.76	45.2	982.46	4110.60
Hexane	C_6H_{14}	g	−39.96	−0.06	92.83	34.2	983.31	4114.15
1-Hexene	C_6H_{12}	l	−17.3	19.93	70.55	43.81	946.86	3961.68
1-Hexene	C_6H_{12}	g	−9.96	20.9	91.93	31.63	947.83	3965.74
cis-2-Hexene	C_6H_{12}	g	−12.51	18.22	92.37	30.04	945.15	3954.53
trans-2-Hexene	C_6H_{12}	g	12.27	18.27	90.97	31.64	945.20	3954.73
cis-3-Hexene	C_6H_{12}	g	−11.38	19.84	90.73	29.55	946.77	3961.30
trans-3-Hexene	C_6H_{12}	g	−13.01	18.55	89.59	31.75	945.48	3955.91

TABLE 7.1 Chemical exergy of pure carbon and hydrocarbons (containing only C and H). Underlying state information from Ref. 11 except where marked with * indicating the state information came from Ref. 12. (*Continued*)

Name	Structure	State	h^f kcal/mol	Δg^f kcal/mol	s cal/mol/K	C_p cal/mol/K	e^{ch} kcal/mol	e^{ch} kJ/mol
1-Hexyne	C_6H_{10}	g	29.55	52.24	88.13	30.65	922.74	3860.75
Indane	C_9H_{10}	l	2.56	36.04	56.01	45.47	1200.71	5023.79
Indene	C_9H_8	l	26.39	52	51.19	44.68	1160.24	4854.46
Isopropylbenzene	C_9H_{12}	l	−9.85	29.7	66.87		1250.81	5233.37
Isopropylbenzene	C_9H_{12}	g	0.94	32.74	92.87	36.26	1253.85	5246.09
Methane	CH_4	g	−17.89	−12.15	44.52	8.54	198.77	831.66
Methyl radical	CH_3	g	34.82	35.35	46.38	9.25	218.05	912.34
2-Methyl-1,3-butadiene	C_5H_8	g	18.1	34.86	75.44	25	750.87	3141.65
3-Methyl-1,2-butadiene	C_5H_8	g	31	47.47	76.4	25.2	763.48	3194.42
2-Methyl-1-butene	C_5H_{10}	g	−8.68	15.68	81.15	26.28	788.13	3297.52
3-Methyl-1-butene	C_5H_{10}	g	−6.92	17.87	79.7	28.35	790.32	3306.68
2-Methyl-2-butene	C_5H_{10}	g	−10.17	14.26	80.92	25.1	786.71	3291.57
3-Methyl-1-butyne	C_5H_8	g	32.6	49.12	76.23	25.02	765.13	3201.32
Methylcyclohexane	C_7H_{14}	l	−45.45	4.86	59.26		1086.28	4545.01
Methylcyclohexane	C_7H_{14}	g	−36.99	6.52	82.06	32.27	1087.94	4551.95
Methylcyclopentane	C_6H_{12}	g	−25.5	8.55	81.24	26.24	935.48	3914.07
1-Methylcyclopentene	C_6H_{10}	g	−1.3	24.41	78	24.1	894.91	3744.31
3-Methylcyclopentene	C_6H_{10}	g	2.07	27.48	79	23.9	897.98	3757.16
4-Methylcyclopentene	C_6H_{10}	g	3.53	29.06	78.6	23.9	899.56	3763.77
Methylene	CH_2	g	92.35	88.25	46.32	8.27	242.74	1015.62
1-Methyl-2-ethylbenzene	C_9H_{12}	g	0.29	31.33	95.42	37.74	1252.44	5240.19
1-Methyl-3-ethylbenzene	C_9H_{12}	g	−0.46	30.22	96.6	36.38	1251.33	5235.55
1-Methyl-4-ethylbenzene	C_9H_{12}	g	−0.79	30.28	95.34	36.22	1251.39	5235.80
2-Methyl-3-ethylpentane	C_8H_{18}	l	−59.69	3.03	81.41		1295.37	5419.84
2-Methyl-3-ethylpentane	C_8H_{18}	g	−50.48	5.08	105.43		1297.42	5428.42
3-Methyl-3-ethylpentane	C_8H_{18}	l	−60.46	2.69	79.97		1295.03	5418.42
3-Methyl-3-ethylpentane	C_8H_{18}	g	−51.28	4.76	103.48		1297.10	5427.08
2-Methylheptane	C_8H_{18}	l	−60.98	0.92	84.16		1293.26	5411.02
2-Methylheptane	C_8H_{18}	g	−51.5	3.05	108.81		1295.39	5419.93
3-Methylheptane	C_8H_{18}	l	−60.34	1.12	85.66		1293.46	5411.85
3-Methylheptane	C_8H_{18}	g	−50.82	3.28	110.32		1295.62	5420.89
4-Methylheptane	C_8H_{18}	l	−60.17	1.87	83.72		1294.21	5414.99
4-Methylheptane	C_8H_{18}	g	−50.69	4	108.35		1296.34	5423.90
2-Methylhexane	C_7H_{16}	l	−54.93	−0.69	77.28	53.28	1137.16	4757.90
2-Methylhexane	C_7H_{16}	g	−46.59	0.77	100.38	39.67	1138.62	4764.01
3-Methylhexane	C_7H_{16}	l	−54.35	−0.39	78.23		1137.46	4759.15
3-Methylhexane	C_7H_{16}	g	−45.96	1.1	101.37	39.67	1138.95	4765.39
Methylidyne CH	C_2H_4	g	142	134.02	43.72	6.97	443.00	1853.50
Methylidyne CH+	C_2H_4	g	388.8	380.1	41	6.98	689.08	2883.10
1-Methylnaphathalene	$C_{11}H_{10}$	l	13.43	46.26	60.9	53.63	1407.05	5887.10
2-Methylnaphthalene	$C_{11}H_{10}$	c	10.72	46.03	52.58	46.84	1406.82	5886.13
2-Methylpentane	C_6H_{14}	g	−41.66	−1.2	90.95	34.46	982.17	4109.38

TABLE 7.1 Chemical exergy of pure carbon and hydrocarbons (containing only C and H). Underlying state information from Ref. 11 except where marked with * indicating the state information came from Ref. 12. (*Continued*)

Name	Structure	State	h^f kcal/mol	Δg^f kcal/mol	s cal/mol/K	C_p cal/mol/K	e^{ch} kcal/mol	e^{ch} kJ/mol
3-Methylpentane	C_6H_{14}	g	−41.02	−0.51	90.77	34.2	982.86	4112.27
2-Methyl-1-pentene	C_6H_{12}	g	−12.49	18.55	91.34	32.41	945.48	3955.91
3-Methyl-1-pentene	C_6H_{12}	g	−10.76	20.66	90.06	34.04	947.59	3964.73
4-Methyl-1-pentene	C_6H_{12}	g	−10.54	21.52	87.89	30.23	948.45	3968.33
cis-2-Methyl-2-pentene	C_6H_{12}	g	−13.8	17.5	90.45	30.26	944.43	3951.51
trans-2-Methyl-2-pentene	C_6H_{12}	g	−14.02	17.04	91.26	30.26	943.97	3949.59
cis-4-Methyl-2-pentene	C_6H_{12}	g	−12.03	19.63	89.23	31.92	946.56	3960.42
trans-4-Methyl-2-pentene	C_6H_{12}	g	−12.99	19.03	88.02	33.8	945.96	3957.91
2-Methylpropane	C_4H_{10}	g	−32.15	−4.99	70.42	23.14	669.40	2800.76
2-Methylpropene	C_4H_8	g	−4.04	13.88	70.17	21.3	631.84	2643.60
α-Methylstyrene	C_9H_{10}	g	27	49.84	91.7	34.7	1214.51	5081.53
cis-β-Methylstyrene	C_9H_{10}	g	29	51.84	91.7	34.7	1216.51	5089.90
trans-β-Methylstyrene	C_9H_{10}	g	28	51.08	90.9	34.9	1215.75	5086.72
Naphthalene	$C_{10}H_8$	c	18	48.05	39.89		1254.35	5248.20
Naphthalene	$C_{10}H_8$	g	35.6	53.44	80.22	31.68	1259.74	5270.75
Nonadecane	$C_{19}H_{40}$	g	−104	26.03	213.95	105.26	3017.75	12626.28
1-Nonadecene	$C_{19}H_{38}$	g	−74	47.02	212.95	102.69	2982.31	12477.99
Nonane	C_9H_{20}	l	−65.84	2.81	94.09		1449.64	6065.31
Nonane	C_9H_{20}	g	−54.74	5.93	120.86	50.6	1452.76	6078.36
1-Nonaethiol	C_9H_{20}	g	−45.61	12.67	136.51	55.61	1459.50	6106.56
1-Nonene	C_9H_{18}	g	−24.74	26.93	119.86	48.03	1417.33	5930.11
Octadecane	$C_{18}H_{38}$	g	−99.08	24.02	204.64	99.8	2861.25	11971.49
1-Octadecene	$C_{18}H_{36}$	g	−69.08	45.01	203.64	97.22	2825.81	11823.20
Octane	C_8H_{18}	l	−59.74	1.77	85.5	45.14	1294.11	5414.57
Octane	C_8H_{18}	g	−49.82	3.92	111.55	45.14	1296.26	5423.57
1-Octene	C_8H_{16}	l	−29.52	22.49	86.15	57.65	1258.40	5265.16
1-Octene	C_8H_{16}	g	−19.82	24.91	110.55	42.56	1260.82	5275.28
1-Octyne	C_8H_{14}	g	19.7	56.26	106.75	41.58	1235.74	5170.34
Pentadecane	$C_{15}H_{32}$	g	−84.31	17.98	176.71	83.4	2391.75	10007.07
1-Pentadecene	$C_{15}H_{30}$	g	−54.31	38.97	175.71	80.82	2356.31	9858.78
1-Pentadecyne	$C_{15}H_{28}$	g	−14.78	70.25	171.91	79.84	2331.15	9753.55
1,2-Pentadiene	C_5H_8	g	34.8	50.29	79.7	25.2	766.30	3206.21
cis-1,3-Pentadiene	C_5H_8	g	18.7	34.84	77.5	22.6	750.85	3141.57
trans-1,3-Pentadiene	C_5H_8	g	18.6	35.07	76.4	24.7	751.08	3142.53
1,4-Pentadiene	C_5H_8	g	25.2	40.69	79.7	25.1	756.70	3166.05
2,3-Pentadiene	C_5H_8	g	33.1	49.21	77.6	24.2	765.22	3201.70
Pentamethylbenzene	$C_{11}H_{16}$	l	−32.33	25.64	70.22	51.74	1555.72	6509.15
Pentane	C_5H_{12}	g	−35	−2	83.4	28.73	826.88	3459.65
cis-1-Pentene	C_5H_{10}	g	−6.71	17.17	82.76	24.32	789.62	3303.75
trans-1-Pentene	C_5H_{10}	g	−7.59	16.71	81.36	25.92	789.16	3301.83
1-Pentyne	C_5H_8	g	34.5	50.25	78.82	25.5	766.26	3206.05
2-Pentyne	C_5H_8	g	30.8	46.41	79.3	23.59	762.42	3189.98
Phenanthrene	$C_{14}H_{10}$	c	27.3	64.12	50.6		1719.08	7192.64

TABLE 7.1   Chemical exergy of pure carbon and hydrocarbons (containing only C and H). Underlying state information from Ref. 11 except where marked with * indicating the state information came from Ref. 12. (*Continued*)

Name	Structure	State	h^f kcal/mol	Δg^f kcal/mol	s cal/mol/K	C_p cal/mol/K	e^{ch} kcal/mol	e^{ch} kJ/mol
Phenylacetylene	C_8H_6	g	78.22	86.46	76.88	27.46	1040.21	4352.26
Propadiene	C_3H_4	g	45.92	48.37	58.3	14.1	455.41	1905.42
Propane	C_3H_8	g	−24.82	−5.63	64.58	17.59	514.27	2151.70
Propylbenzene	C_9H_{12}	g	1.87	32.8	95.76	36.41	1253.91	5246.34
Propylcyclohexane	C_9H_{18}	g	−46.2	11.31	100.27	44.03	1401.71	5864.76
Propylcyclopentane	C_8H_{16}	g	−35.39	12.57	99.73	36.96	1248.48	5223.65
Propylene	C_3H_6	g	4.88	15.02	63.72	15.37	478.49	2001.99
Propyne	C_3H_4	g	44.32	46.47	59.3	14.5	453.51	1897.47
Pyrene	$C_{16}H_{10}$	c	27.44	64.4	53.75	56.4	1915.48	8014.35
Spiropentane	C_5H_8	g	44.27	63.41	67.45	21.06	779.42	3261.11
trans-Stilbene	$C_{14}H_{12}$	c	32.37	75.9	60		1787.29	7478.04
Styrene	C_8H_8	l	24.83	48.37	56.78	43.64	1058.56	4429.00
Styrene	C_8H_8	g	35.22	51.1	82.48	29.18	1061.29	4440.42
Tetradecane	$C_{14}H_{30}$	g	−79.38	15.97	167.4	77.93	2235.25	9352.28
1-Tetradecene	$C_{14}H_{28}$	g	−49.36	36.99	166.4	75.36	2199.84	9204.12
1,2,3,4-Tetramethylbenzene	$C_{10}H_{14}$	l	−23	25.49	69.45		1401.09	5862.14
1,2,3,5-Tetramethylbenzene	$C_{10}H_{14}$	l	−23.54	23.58	99.55	57.5	1399.18	5854.15
1,2,4,5-Tetramethylbenzene	$C_{10}H_{14}$	l	−29.48	24.2	71.83	51.6	1399.80	5856.74
2,2,3,3-Tetramethylbutane	C_8H_{18}	g	−53.99	5.26	93.06		1297.60	5429.17
Toluene	C_7H_8	l	2.87	27.19	52.81	37.58	939.32	3930.11
Toluene	C_7H_8	g	11.95	29.16	76.64	24.77	941.29	3938.35
Tridecane	$C_{13}H_{28}$	g	−74.45	13.97	158.09	72.47	2078.76	8697.53
1-Tridecene	$C_{13}H_{26}$	g	−44.45	34.96	157	69.89	2043.32	8549.24
1,2,3-Trimethylbenzene	C_9H_{12}	l	−14.01	25.68	66.4		1246.79	5216.55
1,2,4-Trimethylbenzene	C_9H_{12}	l	−14.79	24.46	67.93		1245.57	5211.45
1,3,5-Trimethylbenzene	C_9H_{12}	l	−15.18	24.83	65.38		1245.94	5213.00
2,2,3-Trimethylbutane	C_7H_{16}	g	−48.95	1.02	91.61	39.33	1138.87	4765.05
cis-1,3,5-Trimethyl-cyclohexane	C_9H_{18}	g	−51.48	8.1	93.3	42.93	1398.50	5851.33
2,2,3-Trimethylpentane	C_8H_{18}	l	−61.44	2.21	78.3		1294.55	5416.41
2,2,3-Trimethylpentane	C_8H_{18}	g	−52.61	4.09	101.62		1296.43	5424.28
2,2,4-Trimethylpentane	C_8H_{18}	l	−61.97	1.65	78.4		1293.99	5414.07
2,2,4-Trimethylpentane	C_8H_{18}	g	−53.57	3.27	101.15		1295.61	5420.85
2,3,3-Trimethylpentane	C_8H_{18}	l	−60.63	2.54	79.93		1294.88	5417.79
2,3,3-Trimethylpentane	C_8H_{18}	g	−51.73	4.52	103.14		1296.86	5426.08
2,3,4-Trimethylpentane	C_8H_{18}	l	−60.98	2.55	78.71		1294.89	5417.84
2,3,4-Trimethylpentane	C_8H_{18}	g	−51.97	4.52	102.31		1296.86	5426.08
2,4,4-Trimethyl-1-pentene	C_8H_{16}	l	−35.21	20.66	73.2		1256.57	5257.50
2,4,4-Trimethyl-2-pentene	C_8H_{16}	l	−34.44	21.04	74.5		1256.95	5259.09
Triphenylene	$C_{18}H_{12}$	c	33.72	78.68	60.87		2182.30	9130.76
Triphenylethylene	$C_{20}H_{16}$	c	55.8	123			2535.60	10608.96
Triphenylmethane	$C_{19}H_{16}$	c	38.71	98.6	74.6	70.5	2413.14	10096.59
Undecane	$C_{11}H_{24}$	l	−78.05	5.44	109.49		1761.25	7369.07

TABLE 7.1 Chemical exergy of pure carbon and hydrocarbons (containing only C and H). Underlying state information from Ref. 11 except where marked with * indicating the state information came from Ref. 12. (*Continued*)

Name	Structure	State	h^f kcal/mol	Δg^f kcal/mol	s cal/mol/K	C_p cal/mol/K	e^{ch} kcal/mol	e^{ch} kJ/mol
Undecane	$C_{11}H_{24}$	g	−64.6	9.94	139.48	61.53	1765.75	7387.90
1-Undecene	$C_{11}H_{22}$	g	−34.6	30.94	138.48	58.96	1730.32	7239.66
o-Xylene	C_8H_{10}	l	−5.84	26.37	58.91	44.9	1092.99	4573.06
o-Xylene	C_8H_{10}	g	4.54	29.18	84.31	31.85	1095.80	4584.82
m-Xylene	C_8H_{10}	l	−6.08	25.73	60.27	43.8	1092.35	4570.38
m-Xylene	C_8H_{10}	g	4.12	28.41	85.49	30.49	1095.03	4581.59
p-Xylene	C_8H_{10}	l	−5.84	26.31	59.12		1092.93	4572.81
p-Xylene	C_8H_{10}	g	4.29	28.95	84.23	30.32	1095.57	4583.85

TABLE 7.1 Chemical exergy of pure carbon and hydrocarbons (containing only C and H). Underlying state information from Ref. 11 except where marked with * indicating the state information came from Ref. 12. (*Continued*)

Name	Structure	State	h^f kcal/mol	Δg^f kcal/mol	s cal/mol/K	C_p cal/mol/K	e^{ch} kcal/mol	e^{ch} kJ/mol
Acetaldehyde	C_2H_4O	l	−45.96	−30.64	38.30	65.60	278.81	1166.53
Acetaldehyde	C_2H_4O	g	−39.76	−31.86	63.15	13.06	277.59	1161.42
Acetic acid	$C_2H_4O_2$	l	−115.71	−93.20	38.20	29.70	216.72	906.74
Acetic acid	$C_2H_4O_2$	g	−103.93	−90.03	67.52	15.90	219.89	920.00
Acetic acid, ionized; m = 1	$C_2H_4O_2$	aq	−116.16	−88.29	20.70	−1.50	221.63	927.28
Acetic acid, nonionized; m = 1	$C_2H_4O_2$	aq	−116.70	−94.78	42.70		215.14	900.13
Acetic anhydride	$C_4H_6O_3$	l	−149.14	−116.82	64.20		446.11	1866.53
Acetic anhydride	$C_4H_6O_3$	g	−137.60	−113.93	93.20	23.78	449.00	1878.62
Acetone	C_3H_6O	l	−59.18	−37.22	47.90	30.22	426.72	1785.38
Acetone	C_3H_6O	g	−51.78	−36.58	70.49	17.9	427.36	1788.06
Acetophenone	C_8H_8O	l	−34.07	−4.06	59.62		1006.59	4211.59
Acetophenone	C_8H_8O	g	−20.76	0.44	89.12		1011.09	4230.42
Acrolein	C_3H_4O	l	−29.97	−16.17			391.33	1637.34
Acrolein	C_3H_4O	g	−20.50	−15.45			392.05	1640.36
Acrylic acid	$C_3H_4O_2$	g	−80.36	−68.37	75.29	18.59	339.60	1420.90
Adipic acid	$C_6H_{10}O_4$	l	−235.51	−177.17			1675.78	7011.47
Benzaldehyde	C_7H_6O	l	−21.23	2.24			858.41	3591.57
Benzaldehyde	C_7H_6O	g	−9.57	5.85			862.02	3606.67
Benzoic acid	$C_7H_6O_2$	c	−92.03	−58.62	40.05	34.97	798.01	3338.89
Benzophenone	$C_{13}H_{10}O$	c	−8.00	33.50	58.60		1590.87	6656.21
Benzyl alcohol	C_7H_8O	l	−38.49	−6.57	51.80		906.03	3790.82
1-Butanol	$C_4H_{10}O$	l	−78.18	−38.84	54.10	42.31	636.02	2661.09
1-Butanol	$C_4H_{10}O$	g	−65.65	−36.04	86.70	26.29	638.82	2672.81
2-Butanol	$C_4H_{10}O$	l	−81.88	−42.31	53.80	47.50	632.55	2646.58
2-Butanol	$C_4H_{10}O$	g	−69.94	−40.06	85.60	27.08	634.80	2655.99
2-Butanone	C_4H_8O	l	−65.29	−36.18	57.08	37.98	582.25	2436.11
2-Butanone	C_4H_8O	g	−56.26	−34.91	80.81	24.59	583.52	2441.43

TABLE 7.2 Chemical exergy of pure chemicals containing only C, H, and O. Underlying state information from Ref. 11.

Name	Structure	State	h^f kcal/mol	Δg^f kcal/mol	s cal/mol/K	C_p cal/mol/K	e^{ch} kcal/mol	e^{ch} kJ/mol
Butyl ether	$C_8H_{18}O$	g	−87.20	114.96	48.82		1407.77	5890.12
Butyraldehyde	C_4H_8O	g	−49.00	−27.43	82.44	24.52	591.00	2472.72
Butyric acid	$C_4H_8O_2$	l	−127.59	−90.27	54.10	42.10	528.62	2211.76
Citric acid	$C_6H_8O_7$	c	−369.00	−295.50	39.73		521.85	2183.44
Citric acid monohydrate	$C_6H_8O_7 \cdot H_2O$	c	−439.40	−352.00	67.74	1.28	465.35	1947.04
o-Cresol	C_7H_8O	g	−30.74	−8.86	85.47	31.15	903.74	3781.24
m-Cresol	C_7H_8O	g	−31.63	−9.69	85.27	29.27	902.91	3777.76
p-Cresol	C_7H_8O	g	−29.97	−7.38	83.09	29.75	905.22	3787.43
Cyclohexanol	$C_6H_{12}O$	l	−83.22	−31.87	47.70		895.53	3746.91
Cyclohexanone	$C_6H_{10}O$	g	−55.00	−21.69	77.00	26.21	849.28	3553.39
Cyclopentanol	$C_5H_{10}O$	l	−71.74	−30.55	49.20		742.36	3106.05
Decanal	$C_{10}H_{20}O$	g	−79.09	−15.90	138.28	57.29	1529.46	6399.26
1-Decanol	$C_{10}H_{22}O$	l	−114.60	−31.60	102.90		1570.19	6569.68
1-Decanol	$C_{10}H_{22}O$	g	−96.00	−24.90	142.80	59.10	1576.89	6597.71
Dibutyl ether	$C_8H_{18}O$	g	−79.80	−21.16	119.60	48.76	1271.65	5320.60
Diethyl ether	$C_4H_{10}O$	l	−65.30	−27.88	60.50	40.80	646.98	2706.95
Diethyl ether	$C_4H_{10}O$	g	−60.26	−29.24	81.9	26.89	645.62	2701.26
Diisopropyl ether	$C_6H_{14}O$	l	−83.94	−21.10	70.40		963.20	4030.04
Diisopropyl ether	$C_6H_{14}O$	g	−76.20	−29.13	93.27	37.83	955.17	3996.45
Dimethyl ether	C_2H_6O	g	−43.99	−26.99	63.83	15.73	338.89	1417.91
1,4-Dioxane	$C_4H_8O_2$	l	−84.47	−44.96	46.67		573.93	2401.34
1,4-Dioxane	$C_4H_8O_2$	g	−75.30	−43.21	71.65	22.48	575.68	2408.66
1,4-Diphenyl-1,4-butanedione	$C_{16}H_{14}O_2$	c	−61.24	1.87	77.60		1966.75	8228.87
1,4-Diphenyl-2-butene-1,4-dione	$C_{16}H_{12}O_2$	c	−27.55	26.64	76.30		1935.09	8096.40
Diphenyl carbonate	$C_{13}H_{10}O_3$	c	−95.93	−42.05	66.54		1516.26	6344.04
Diphenyl ether	$C_{12}H_{10}O$	l	−3.48	34.47	69.62		1493.79	6250.00
Dipropyl ether	$C_6H_{14}O$	g	−70.00	−25.23	100.98	37.83	958.60	4010.80
meso-Erythritol	$C_4H_{10}O_4$	c	−127.56	−152.12	39.30		524.14	2193.02
Ethanol	C_2H_6O	l	−66.20	−41.63	38.49	26.76	324.25	1356.66
Ethanol	C_2H_6O	g	−56.03	−40.13	67.54	14.64	325.75	1362.93
Ethyl acetate	$C_4H_8O_2$	l	−114.49	−79.52	62.00		539.37	2256.74
Ethyl acetate	$C_4H_8O_2$	g	−105.86	−78.25	86.70	27.16	540.64	2262.06
Ethylene glycol	$C_2H_6O_2$	l	−108.70	−77.25	39.90	35.80	289.10	1209.58
Ethylene glycol	$C_2H_6O_2$	g	−93.05	−72.77	77.33	23.2	293.58	1228.33
Ethylene oxide	C_2H_4O	g	−12.58	−3.13	57.94	11.54	306.32	1281.63
Ethyl methyl ether	C_3H_8O	g	−51.73	−28.12	74.24	21.45	492.25	2059.56
Ethyl propanoate	$C_5H_{10}O_2$	l	−122.16	−79.16			694.22	2904.63
Formaldehyde	CH_2O	g	−27.70	−26.27	52.29	8.46	128.69	538.43
Formaldehyde unhydrolyzed	CH_2O	aq	−35.90	−31.02			123.94	518.56
Formic acid	CH_2O_2	l	−101.51	−86.38	30.82	23.67	69.05	288.89
Formic acid	CH_2O_2	g	−90.49	−83.89	59.45	10.81	71.54	299.31

TABLE 7.2 Chemical exergy of pure chemicals containing only C, H, and O. Underlying state information from Ref. 11. (*Continued*)

Name	Structure	State	h^f kcal/mol	Δg^f kcal/mol	s cal/mol/K	C_p cal/mol/K	e^{ch} kcal/mol	e^{ch} kJ/mol
Formic acid, ionized; $m = 1$	CH_2O_2	aq	−101.71	−83.90	22.00	−21.00	71.53	299.27
Formic acid, nonionized; $m = 1$	CH_2O_2	aq	−101.68	−89.00	39.00		66.43	277.93
Formyl HCO	$C_2H_2O_2$	g	10.40	6.67	53.66	8.27	260.15	1088.49
Formyl HCO+	$C_2H_2O_2$	g	204.00	201.00	48.30	8.62	454.48	1901.56
Fumaric acid	$C_4H_4O_4$	c	−193.84	−156.70	39.70		350.27	1465.53
Furan	C_4H_4O	g	−8.23	0.21	63.86	15.64	505.77	2116.15
Furfuryl alcohol	$C_5H_6O_2$	l	−66.05	36.85	51.50		697.37	2917.80
D-Galactose	$C_6H_{12}O_6$	c	−304.10	−219.60	49.10		710.15	2971.26
α D-Glucose	$C_6H_{12}O_6$	c	−304.26	−217.60	50.70		712.15	2979.63
Glycerol	$C_3H_8O_3$	l	−159.76	−114.01	48.87	35.90	407.30	1704.13
1-Heptanal	$C_7H_{14}O$	g	−63.10	−20.71	110.34	40.89	1061.18	4439.99
Heptanedioic acid	$C_7H_{12}O_4$	g	−44.88	1.91	102.27	39.67	1028.78	4304.41
1-Heptanol	$C_7H_{16}O$	l	−95.80	−34.00	76.50	66.50	1104.32	4620.49
1-Heptanol	$C_7H_{16}O$	g	−79.30	−28.90	114.80	42.70	1109.42	4641.83
Hexadecanoic acid	$C_{16}H_{32}O_2$	c	−213.30	−75.54	108.12		2397.22	10029.98
1-Hexadecanol	$C_{16}H_{34}O$	c, II	−163.40	−23.60	108.00	104.80	2505.13	10481.44
1-Hexadecanol	$C_{16}H_{34}O$	l	−151.86	−23.08	145.00		2505.65	10483.62
Hexanal	$C_6H_{12}O$	g	−59.37	−23.93	101.07	35.43	903.47	3780.13
1-Hexanol	$C_6H_{14}O$	l	−90.70	−36.40	69.20	56.60	947.43	3964.07
1-Hexanol	$C_6H_{14}O$	g	−75.90	−32.40	105.50	37.20	951.43	3980.80
Hydroquinone	$C_6H_6O_2$	c	−87.08	−49.48	33.50	33.90	709.10	2966.86
o-Hydroxybenzoic acid	$C_7H_6O_3$	c	−140.64	−100.70	42.60	38.03	756.40	3164.79
m-Hydroxybenzoic acid	$C_7H_6O_3$	c	−139.80	−99.74	42.30	37.59	757.36	3168.81
p-Hydroxybenzoic acid	$C_7H_6O_3$	c	−139.70	−99.55	42.00	37.08	757.55	3169.60
Ketene	C_2H_2O	g	−14.60	−14.41	57.79	12.37	238.61	998.33
L-Lactic acid	$C_3H_6O_3$	c	−165.89	−124.98	34.00		339.89	1422.12
L-Lactic acid	$C_3H_6O_3$	l	−161.20	−123.84	45.90		341.03	1426.89
β-Lactose	$C_{12}H_{22}O_{11}$	c	−534.10	−374.52	92.30		1428.08	5975.07
Maleic acid	$C_4H_4O_4$	c	−188.94	−149.40	38.10	32.36	357.57	1496.07
L-Malic acid	$C_4H_6O_5$	c	−263.78	−211.45			352.42	1474.52
Maltose	$C_{12}H_{22}O_{11}$	c	−530.80	−412.60			1390.00	5815.74
D-Mannitol	$C_6H_{14}O_6$	c	−139.61	−225.20	57.00		760.98	3183.94
Methanol	CH_4O	l	−57.13	−39.87	30.41	19.40	171.52	717.64
Methanol	CH_4O	g	−48.06	−38.82	57.29	10.49	172.57	722.03
Methyl acrylate	$C_4H_6O_2$	g	−70.10	−56.78			505.68	2115.78
2-Methyl-2-butanol	$C_5H_{12}O$	l	−90.70	−41.90	54.80	59.20	787.45	3294.67
2-Methyl-2-butanol	$C_5H_{12}O$	g	−78.80	−39.50	86.70		789.85	3304.71
Methyl formate	$C_2H_4O_2$	l	−90.60	−71.53	29.00		238.39	997.41
Methyl formate	$C_2H_4O_2$	g	−83.70	−71.03	72.00	15.90	238.89	999.50
Methyl isopropyl ether	$C_4H_{10}O$	g	−60.24	−28.89	80.86	26.55	645.97	2702.72
2-Methyl phenol	C_7H_8O	g	−30.74	−8.86	85.47	31.15	903.74	3781.24
3-Methyl phenol	C_7H_8O	g	−31.63	−9.69	85.27	29.27	902.91	3777.76
4-Methyl phenol	C_7H_8O	g	−29.97	−7.38	83.09	29.75	905.22	3787.43

TABLE 7.2   Chemical exergy of pure chemicals containing only C, H, and O. Underlying state information from Ref. 11. (*Continued*)

Name	Structure	State	h^f kcal/mol	Δg^f kcal/mol	s cal/mol/K	C_p cal/mol/K	e^{ch} kcal/mol	e^{ch} kJ/mol
2-Methyl-1-propanol	$C_4H_{10}O$	g	−67.69	−39.99	85.81	26.60	634.87	2656.28
2-Methyl-2-propanol	$C_4H_{10}O$	l	−85.86	−44.14	46.10	52.61	630.72	2638.92
2-Methyl-2-propanol	$C_4H_{10}O$	g	−74.67	−42.46	77.98	27.10	632.40	2645.95
Methyl propyl ether	$C_4H_{10}O$	g	−56.82	−26.27	83.52	26.89	648.59	2713.69
1-Nonanal	$C_9H_{18}O$	g	−74.16	−17.91	128.97	51.82	1372.96	5744.47
1-Nonanol	$C_9H_{20}O$	l	−109.20	−32.40	91.30	67.50	1414.90	5919.95
1-Octanal	$C_8H_{16}O$	g	−69.23	−19.91	119.66	46.36	1216.47	5089.72
1-Octanol	$C_8H_{18}O$	l	−101.60	−34.20	90.20	77.70	1258.61	5266.04
2-Octanone	$C_8H_{16}O$	l	−91.90	−33.54	89.35	65.31	1202.84	5032.69
Oxacyclobutane	C_3H_6O	g	−19.25	−2.33	65.46		461.61	1931.36
Oxalic acid	$C_2H_2O_4$	c	−197.70	−166.80	28.70		87.62	366.61
Oxalic acid, $m = 1$	$C_2H_2O_4$	aq	−197.20	−161.10	10.90		93.32	390.46
Pentaerythritol	$C_5H_{12}O_4$	c	−220.00	−146.73	47.34	45.51	684.02	2861.95
1-Pentanal	$C_5H_{10}O_2$	g	−54.45	−25.88	91.53	29.96	747.50	3127.55
Pentanoic acid	$C_5H_{10}O_2$	l	−133.71	−89.10	62.10	50.48	684.28	2863.04
1-Pentanol	$C_5H_{12}O$	l	−85.00	−38.30	62.00	49.80	791.05	3309.74
3-Pentanol	$C_5H_{12}O$	l	−88.50	−40.40	57.40	60.00	788.95	3300.95
2-Pentanone	$C_5H_{10}O$	g	−61.82	−32.76	89.91	28.91	740.15	3096.81
Phenol	C_6H_6O	c	−39.44	−12.05	34.42	32.20	746.06	3121.51
Phenol	C_6H_6O	l	−37.80	−11.02			747.09	3125.82
Phenol	C_6H_6O	g	−23.03	−7.86	75.43	24.75	750.25	3139.04
o-Phthalic acid	$C_8H_6O_4$	c	−186.91	−141.39	49.70	45.00	814.24	3406.78
Phthalic anhydride	$C_8H_4O_3$	c	−110.10	−79.12	42.90	38.50	819.61	3429.25
1-Propanol	C_3H_8O	l	−72.66	−40.78	46.50	33.70	479.59	2006.60
1-Propanol	C_3H_8O	g	−61.28	−38.67	77.61	20.82	481.70	2015.42
2-Propanol	C_3H_8O	l	−75.97	−43.09	43.16	36.06	477.28	1996.93
2-Propanol	C_3H_8O	g	−65.11	−41.44	74.07	21.21	478.93	2003.83
2-propen-1-ol	C_3H_6O	g	−31.55	−17.03	73.51	18.17	446.91	1869.86
Propionaldehyde	C_3H_6O	g	−45.90	−31.18	72.83	18.80	432.76	1810.65
Propylene oxide	C_3H_6O	g	−22.17	−6.16	68.53	17.29	457.78	1915.34
Pyruvic acid	$C_3H_4O_3$	l	−139.70	−110.75	42.90		297.69	1245.55
Quinhydrone	$C_{12}H_{10}O_4$	c	−19.79	−77.19	77.90	66.20	1383.53	5788.70
Salicyclic acid	$C_7H_6O_3$	c	−140.90	−99.93	42.60		757.17	3168.01
L-Sorbose	$C_6H_{12}O_6$	c	−303.68	−217.10	52.80		712.65	2981.72
Succinic acid	$C_4H_6O_4$	c	−224.79	−179.64	42.00	35.80	383.76	1605.65
Sucrose	$C_{12}H_{22}O_{11}$	c	2531.90	2369.18	86.10		4171.78	17454.71
Triphenylcarbinol	$C_{19}H_{16}O$	c	−0.80	65.20	78.70		2380.21	9958.81
Valeric acid	$C_5H_{10}O_2$	l	−133.71	−89.10	62.10	50.48	684.28	2863.04

TABLE 7.2 Chemical exergy of pure chemicals containing only C, H, and O. Underlying state information from Ref. 11. (*Continued*)

Name	Structure	State	h^f kcal/mol	Δg^f kcal/mol	s cal/mol/K	C_p cal/mol/K	e^{ch} kcal/mol	e^{ch} kJ/mol
Acetonitrile	C_2H_3N	l	12.8	23.7	35.76	21.86	304.54	1274.20
Acetonitrile	C_2H_3N	g	21	25.24	58.19	12.48	306.08	1280.65
Acetylenedicarbonitrile	C_4N_2	g	127.5	122.1	69.31	20.53	514.49	2152.62
Acrylonitrile	C_3H_3N	g	44.2	46.68	65.47	15.24	425.58	1780.63
Adenine	$C_5H_5N_5$	c	23.21	71.58	36.1		703.35	2942.80
1-Aminobutane	$C_4H_{11}N$	g	−22	11.76	86.76	28.33	714.44	2989.23
2-Aminobutane	$C_4H_{11}N$	g	−24.9	9.71	83.9	27.99	712.39	2980.65
Aniline	$C_6H_5NH_2$	l	7.55	35.63	45.72	45.9	821.57	3437.43
Aniline	$C_6H_5NH_2$	g	20.76	39.84	76.28	25.91	825.78	3455.04
Benzonitrile	C_7H_5N	g	52.3	62.33	76.73	26.07	889.89	3723.30
tert-Butylamine	$C_4H_{11}N$	g	−28.65	6.9	80.76	28.67	709.58	2968.90
Butyronitrile	C_4H_7N	g	8.14	25.97	77.98	23.19	615.79	2576.47
trans-Crotononitrile	C_4H_5N	g	35.77	46.22	71.41	19.62	579.61	2425.08
1-Cyanoguanidine	$C_2H_4N_4$	c	5.4	42.9	30.9	28.4	352.20	1473.60
Diazomethane	CH_2N_2	g	46	52.06	58.02	12.55	206.71	864.87
Diethylamine	$C_4H_{11}N$	g	−17.3	17.23	84.18	27.66	719.91	3012.12
Diethylenediamine	$C_2H_8N_2$	c	−3.2	57.4	20.5		479.40	2005.82
Dimethylamine	C_2H_7N	g	−4.5	16.25	65.24	16.5	409.96	1715.25
Dimethylamine, $m = 1$, $(CH_3)_2NH_2^+$;	C_2H_7N	aq	−16.88	13.85	31.8		407.56	1705.21
Dimethylamine, $m = 1$	C_2H_7N	aq	−28.74	−0.8	41.2		392.91	1643.92
1,1-Dimethylhydrazine	$C_2H_8N_2$	l	11.8	49.4	47.32	39.21	471.40	1972.34
1,2-Dimethylhydrazine	$C_2H_8N_2$	l	13.3	50.8	47.6	40.88	472.80	1978.20
Ethylamine	$C_2H_5NH_2$	g	−11	8.91	68.08	17.36	402.62	1684.54
N-Ethylaniline	$C_8H_{11}N$	l	0.9	45.1	57.2		1140.01	4769.82
Ethyleneimine	C_2H_5N	g	29.5	42.54	59.9	12.55	379.81	1589.14
Hexamethylenetetramine	$C_6H_{12}N_4$	c	30	103.92	39.05		1031.17	4314.43
Hexamethylenetetramine	$C_6H_{12}N_4$	l	−18.7	28.65	73.28		955.90	3999.50
Hydrogen cyanide*	HCN	g	32.30	29.80	48.20		156.15	653.35
Hydrogen cyanide*	HCN	l	26.02	29.86	26.97		156.21	653.60
Isobutyronitrile	C_4H_7N	g	6.07	24.76	74.88	23.04	614.58	2571.40
Melamine	$C_6H_6N_6$	c	−17.3	44.1	35.63		802.22	3356.49
Methylamine	CH_5N	g	−5.5	7.71	57.98	11.97	246.93	1033.14
Methylamine, $m = 1$	CH_5N	aq	−16.77	4.94	29.5		244.16	1021.55
Methylhydrazine	CH_6N_2	l	12.9	43	39.66	32.25	310.51	1299.18
Methylhydrazine	CH_6N_2	g	22.55	44.66	66.61	17	312.17	1306.13
Methyl isocyanide	C_2H_3N	g	35.6	39.6	58.99	12.65	320.44	1340.73
2-Methylpyridine	C_6H_7N	l	13.83	39.8	52.07	37.86	825.74	3454.88
2-Methylpyridine	C_6H_7N	g	24.05	42.32	77.68	23.9	828.26	3465.42
3-Methylpyridine	C_6H_7N	l	15.57	41.16	51.7	37.93	827.10	3460.57
Propionitrile	C_3H_5N	l	3.5	21.31	45.25		456.64	1910.59
Propionitrile	C_3H_5N	g	12.1	22.98	68.5	17.46	458.31	1917.57
1-Propylamine	C_3H_9N	g	−17.3	9.51	77.48	22.89	557.70	2333.44
Pyridine	C_5H_5N	l	23.96	43.34	42.52	31.72	674.79	2823.31
Pyridine	C_5H_5N	g	33.61	45.46	67.59	18.67	676.91	2832.18

TABLE 7.3 Chemical exergy of pure chemicals containing only C, H, and N. Underlying state information from Ref. 11 except where marked with * indicating the state information came from Ref. 12.

Name	Structure	State	h^f kcal/mol	Δg^f kcal/mol	s cal/mol/K	C_p cal/mol/K	e^{ch} kcal/mol	e^{ch} kJ/mol
Pyrrolidine	C_4H_9N	l	−9.84	25.94	48.76		672.19	2812.45
Pyrrolidine	C_4H_9N	g	−0.86	27.41	73.97	19.39	673.66	2818.60
Quinoline	C_9H_7N	l	37.33	65.9	51.9		1146.01	4794.89
p-Quinone	C_9H_7N	c	−44.1	−20	38.9		1060.11	4435.49
2,4,6-Triamino-1,3,5-triazine	$C_3H_6N_6$	c	−17.3	44.1	35.63		508.05	2125.67
2,4,6-Triamino-1,3,5-triazine	$C_3H_3N_3$	g	−17.13	42.33	74.1	20.93	506.28	2118.26
Triethylamine	$C_6H_{15}N$	g	−23.8	26.36	96.8	38.46	1038.02	4343.08
Triethylenediamine	$C_6H_{12}N_2$	c	−3.4	57.28	37.67		984.37	4118.62
Trimethylamine	C_3H_9N	g	−5.7	23.64	69.02	21.93	571.83	2392.56
Trimethylammonium ion, $m = 1$	$C_3H_{10}N$	aq	−26.99	8.9	47		585.31	2448.94

TABLE 7.3 Chemical exergy of pure chemicals containing only C, H, and N. Underlying state information from Ref. 11 except where marked with * indicating the state information came from Ref. 12. (*Continued*)

Name	Structure	State	h^f kcal/mol	Δg^f kcal/mol	s cal/mol/K	C_p cal/mol/K	e^{ch} kcal/mol	e^{ch} kJ/mol
α-Alanine D	$C_3H_7NO_2$	c	−134.03	−88.23	31.6		404.47	1692.31
α-Alanine L	$C_3H_7NO_2$	c	−133.96	−88.49	30.88		404.21	1691.22
α-Alanine DL	$C_3H_7NO_2$	c	−134.55	−88.92	31.6		403.78	1689.42
α-Alanylglycine DL	$C_5H_{10}N_2O_3$	c	−185.64	−117	51		657.01	2748.94
α-Alanylglycine L	$C_5H_{10}N_2O_3$	c	−197.52	−127.3	46.62		646.71	2705.84
Allantoin	$C_4H_6N_4O_3$	c	−171.5	−106.65	46.6		456.60	1910.42
Alloxan monohydrate	$C_4H_2N_2O_4 \cdot H_2O$	c	−239.08	−182.08	44.6		268.62	1123.89
D-Arginine	$C_6H_{14}N_4O_2$	c	−149.05	−57.43	59.9		927.19	3879.38
L-Asparagine	$C_4H_8N_2O_3$	c	−188.5	−126.73	41.7		492.79	2061.85
L-Aspartic acid	$C_4H_7NO_4$	c	−232.47	−174.53	40.66		417.17	1745.42
ε-Caprolactam	$C_6H_{11}NO$	c	−78.54	−22.72	40.3		876.55	3667.47
Creatine	$C_4H_9N_3O_2$	c	−128.16	−63.32	45.3		584.03	2443.58
Creatinine	$C_4H_7N_3O$	c	−56.77	−6.97	40.1		583.48	2441.28
1,2-Dinitrobenzene	$C_6H_4N_2O_4$	c	2.06	50.56	51.7		753.80	3153.91
1,3-Dinitrobenzene	$C_6H_4N_2O_4$	c	−4.04	44.13	52.8		747.37	3127.01
Ethyl nitrate	$C_2H_5NO_3$	g	−36.8	−8.81	83.25	23.27	329.87	1380.18
Formamide	CH_3NO	g	−44.5	−33.71	59.41	10.84	149.54	625.69
D-Glutamic acid	$C_5H_9NO_4$	c	−240.19	−173.87	45.7		572.32	2394.57
L-Glutamic acid	$C_5H_9NO_4$	c	−241.32	−174.78	44.98		571.41	2390.76
Glycine	$C_2H_5NO_2$	c	−126.22	−88.09	24.74	23.71	250.12	1046.51
Glycine Ionized; $m = 1$	$C_2H_5NO_2$	aq	−112.28	−75.28	26.54		262.93	1100.11
Glycine Nonionized; $m = 1$	$C_2H_5NO_2$	aq	−122.85	−88.62	37.84		249.59	1044.29

TABLE 7.4 Chemical exergy of pure chemicals containing C, H, O, and N. Underlying state information from Ref. 11 except where marked with * indicating the state information came from Ref. 12.

Name	Structure	State	h^f kcal/mol	Δg^f kcal/mol	s cal/mol/K	C_p cal/mol/K	e^{ch} kcal/mol	e^{ch} kJ/mol
Glycine $NH_3^+CH_2COOH$; $m = 1$	$C_2H_5NO_2$	aq	–123.78	–91.82	45.46		246.39	1030.90
Glycylglycine	$C_4H_8N_2O_3$	c	–178.51	–117.25	45.4		502.27	2101.51
Guanidine carbonate	$C_3H_{12}N_6O_3$	c	–232.1	–133.23	70.6	61.87	501.42	2097.94
Guanine	$C_5H_5N_5O$	c	–43.72	11.33	38.3		643.57	2692.68
Hippuric acid	$C_9H_9NO_3$	c	–145.63	–88.33	57.2		1049.62	4391.59
Hypoxanthene	$C_5H_4N_4O$	c	–26.47	18.39	34.8		622.33	2603.83
Isocyanic acid*	HNCO	g	–27.90	–25.66	56.91		101.16	423.26
Isopropyl nitrate	$C_3H_7NO_2$	g	–45.65	–9.72	89.2	28.84	482.98	2020.79
D-Leucine	$C_6H_{13}NO_2$	c	–152.36	–82.97	49.71		873.20	3653.46
L-Leucine	$C_6H_{13}NO_2$	c	–154.6	–82.76	50.62	48.03	873.41	3654.34
DL-Leucylglycine	$C_8H_{16}N_2O_3$	c	–205.7	–112.14	67.2		1125.34	4708.42
Methyl nitrate	CH_3NO_3	l	–38	–10.4	51.9	37.6	173.79	727.15
Methyl nitrate	CH_3NO_3	g	–29.8	–9.4	76.1		174.79	731.33
Methyl nitrite	CH_3NO_2	g	–15.3	0.24	67.95	15.11	183.96	769.70
2-Nitroaniline	$C_6H_6N_2O_2$	c	–3.45	42.6	42.1	39.3	801.34	3352.80
3-Nitroaniline	$C_6H_6N_2O_2$	c	–4.46	41.6	42.1	40.2	800.34	3348.61
4-Nitroaniline	$C_6H_6N_2O_2$	c	–9.91	36.1	42.1	40.4	794.84	3325.60
Nitrobenzene	$C_6H_5NO_2$	l	3.8	34.95	53.6	44.4	765.39	3202.40
2-Nitrobenzoic acid	$C_7H_5NO_4$	c	–94.25	–46.95	49.8		782.49	3273.93
3-Nitrobenzoic acid	$C_7H_5NO_4$	c	–100.25	–52.71	49		776.73	3249.83
4-Nitrobenzoic acid	$C_7H_5NO_4$	c	–101.25	–53.07	50.2	43.3.	776.37	3248.32
1-Nitrobutane	$C_4H_9NO_2$	g	–34.4	2.42	94.28	29.85	649.61	2717.97
2-Nitrobutane	$C_4H_9NO_2$	g	–39.1	–1.49	91.62	29.51	645.70	2701.61
Nitroethane	$C_2H_5NO_2$	g	–24.4	–1.17	75.39	18.69	337.04	1410.18
Nitromethane	CH_3NO_2	l	–27.03	–3.47	41.05	25.33	180.25	754.18
Nitromethane	CH_3NO_2	g	–17.86	–1.66	65.73	13.7	182.06	761.75
1-Nitropropane	$C_3H_7NO_2$	g	–30	0.08	85	24.26	492.78	2061.80
2-Nitropropane	$C_3H_7NO_2$	g	–33.21	–3.06	83.1	24.26	489.64	2048.66
Oxamide	$C_2H_4N_2O_2$	c	–123	–81.9	28.2		228.18	954.69
β-Phenyl-q-alanine	$C_9H_{11}NO_2$	c	–11.9	–50.6	51.06	48.52	1143.31	4783.60
α-Piperidione	$C_9H_{15}NO_2$	c	–73.3	–26.79	39.4		1110.22	4645.15
Propyl nitrate	$C_3H_7NO_3$	g	–41.6	–6.53	92.1	28.99	486.64	2036.10
Semicarbazide, $m = 1$	CH_5N_3O	aq	–39.90	–9.70	71.20		230.15	962.93
L-Tryptophan	$C_{11}H_{12}N_2O_2$	c	–99.8	–28.54	60	56.92	1389.78	5814.84
L-Tyrosine	$C_9H_{11}NO_3$	c	–163.4	–92.18	51.15	51.73	1102.20	4611.59
Urea	CH_4N_2O	c	–79.71	–47.19	25	22.26	164.36	687.68
Uric acid	$C_5H_4N_4O_3$	c	–147.73	–85.75	41.4		519.13	2172.03
L and DL- Valine	$C_5H_{11}NO_2$	c	–148.2	–85.8	42.75	40.35	715.88	2995.24
Xanthine	$C_5H_4N_4O_2$	c	–90.49	–39.64	38.5		564.77	2362.99

TABLE 7.4 Chemical exergy of pure chemicals containing C, H, O, and N. Underlying state information from Ref. 11 except where marked with * indicating the state information came from Ref. 12. (*Continued*)

Name	Structure	State	h^f kcal/mol	Δg^f kcal/mol	s cal/mol/K	C_p cal/mol/K	e^{ch} kcal/mol	e^{ch} kJ/mol
Benzenethiol	C_6H_6S	l	15.32	32.02	53.25	41.4	935.29	3913.23
Benzenethiol	C_6H_6S	g	26.66	35.28	80.51	25.07	938.55	3926.87
1-Butanethiol	$C_4H_{10}S$	l	−29.79	0.79	65.96		820.80	3434.24
1-Butanethiol	$C_4H_{10}S$	g	−21.05	2.64	89.68	28.24	822.65	3441.98
2-Butanethiol	$C_4H_{10}S$	l	−31.13	−0.04	64.87		819.97	3430.77
2-Butanethiol	$C_4H_{10}S$	g	−23	1.29	87.65	28.51	821.30	3436.34
Cyclopentanethiol	$C_5H_{10}S$	g	−11.45	13.63	86.38	25.79	931.70	3898.24
1-Decanethiol	$C_{10}H_{22}S$	g	−50.54	14.68	145.82	61.08	1761.63	7370.65
2,4-Dimethyl-3-thiapentane	$C_6H_{14}S$	g	−33.76	6.48	99.3	40.45	1135.47	4750.81
2,3-Dithiabutane	$C_2H_6S_2$	l	−14.82	1.67	56.26	34.92	658.33	2754.46
3,4-Dithiahexane	$C_4H_{10}S_2$	l	−28.69	2.28	72.9		967.92	4049.78
4,5-Dithiaoctane	$C_4H_{10}S_2$	l	−40.95	4.56	89.28		970.20	4059.32
Ethanethiol	C_2H_6S	g	−11.02	−1.12	70.77	17.37	509.92	2133.49
1-Heptanethiol	$C_7H_{16}S$	g	−35.76	8.65	117.89	44.68	1292.13	5406.28
1-Hexanethiol	$C_6H_{14}S$	g	−30.83	6.65	108.58	39.21	1135.64	4751.53
Methanethiol	CH_4S	g	−5.49	−2.37	60.96	12.01	354.18	1481.88
2-Methyl-2-butanethiol	$C_5H_{12}S$	l	−38.9	0.56	69.34		975.06	4079.66
2-Methyl-2-butanethiol	$C_5H_{12}S$	g	−30.36	2.2	92.48	34.3	976.70	4086.53
2-Methyl-1-propanethiol	$C_4H_{10}S$	g	−23.24	1.33	86.73	28.28	821.34	3436.50
2-Methyl-2-propanethiol	$C_4H_{10}S$	g	−26.17	0.17	80.79	28.91	820.18	3431.65
3-Methyl-2-thiabutane	$C_4H_{10}S$	g	−21.61	3.21	85.87	28	823.22	3444.37
2-Methylthiophene	C_5H_6S	l	10.75	27.35	52.22	29.43	832.56	3483.42
3-Methylthiophene	C_5H_6S	l	10.38	27	52.19	29.38	832.21	3481.96
1-Octanethiol	$C_8H_{18}S$	g	−40.68	10.67	127.2	50.14	1448.64	6061.11
1-Pentanethiol	$C_5H_{12}S$	l	−35.72	2.28	74.18		976.78	4086.86
1-Propanethiol	C_3H_8S	g	−16.22	0.52	80.4	22.65	666.04	2786.73
2-Propanethiol	C_3H_8S	g	−18.22	−0.61	77.51	22.94	664.91	2782.00
2-Thiabutane	C_3H_8S	l	−21.89	1.79	57.14		667.31	2792.05
2-Thiabutane	C_3H_8S	g	−14.25	2.73	79.62	22.73	668.25	2795.98
Thiacyclobutane	C_3H_6S	g	14.61	25.69	68.17	16.57	634.78	2655.93
Thiacycloheptane	$C_6H_{12}S$	g	−14.66	20.09	86.5	29.78	1092.65	4571.65
Thiacyclohexane	$C_5H_{10}S$	l	−25.32	9.96	52.16		928.03	3882.88
Thiacyclohexane	$C_5H_{10}S$	g	−15.12	12.68	77.26	25.86	930.75	3894.26
Thiacyclopentane	C_4H_8S	l	−17.39	8.97	49.67		772.55	3232.36
Thiacyclopentane	C_4H_8S	g	−8.08	11	73.94	21.72	774.58	3240.85
Thiacyclopropane	C_2H_4S	l	12.41	22.52	38.84		477.12	1996.29
Thiacyclopropane	C_2H_4S	g	19.65	23.16	61.01	12.83	477.76	1998.97
2-Thiaheptane	$C_6H_{14}S$	g	−29.34	8.39	107.73	39.1	1137.38	4758.81

TABLE 7.5 Chemical exergy of pure chemicals containing only C, H, and S. Underlying state information from Ref. 11.

Name	Structure	State	h^f kcal/mol	Δg^f kcal/mol	s cal/mol/K	C_p cal/mol/K	e^{ch} kcal/mol	e^{ch} kJ/mol
3-Thiaheptane	$C_6H_{14}S$	g	−29.92	7.65	108.27	38.71	1136.64	4755.71
4-Thiaheptane	$C_6H_{14}S$	l	−40.62	5.12	80.85		1134.11	4745.12
4-Thiaheptane	$C_6H_{14}S$	g	−29.96	7.96	107.16	38.53	1136.95	4757.01
2-Thiahexane	$C_5H_{12}S$	l	−34.15	4.08	73.49		978.58	4094.39
2-Thiahexane	$C_5H_{12}S$	g	−24.42	6.37	98.43	33.64	980.87	4103.97
3-Thiahexane	$C_5H_{12}S$	l	−34.58	3.5	73.98		978.00	4091.96
3-Thiahexane	$C_5H_{12}S$	g	−25	5.63	98.97	33.25	980.13	4100.88
5-Thianonane	$C_8H_{18}S$	l	−52.74	7.66	96.82		1445.63	6048.52
5-Thianonane	$C_8H_{18}S$	g	−39.99	11.76	125.76	49.46	1449.73	6065.67
2-Thiapentane	$C_4H_{10}S$	l	−28.21	2.79	65.14		822.80	3442.61
2-Thiapentane	$C_4H_{10}S$	g	−19.54	4.4	88.84	28.05	824.41	3449.35
3-Thiapentane	$C_4H_{10}S$	l	−28.43	2.81	64.36	40.97	822.82	3442.70
3-Thiapentane	$C_4H_{10}S$	g	−19.95	4.25	87.96	27.97	824.26	3448.72
2-Thiapropane	C_2H_6S	g	−8.97	1.66	68.32	17.71	512.70	2145.12
Thiophene	C_4H_4S	l	19.24	28.97	43.3		679.69	2843.82
Thiophene	C_4H_4S	g	27.66	30.3	66.65	17.42	681.02	2849.38

TABLE 7.5 Chemical exergy of pure chemicals containing only C, H, and S. Underlying state information from Ref. 11. (*Continued*)

Name	Structure	State	h^f kcal/mol	Δg^f kcal/mol	s cal/mol/K	C_p cal/mol/K	e^{ch} kcal/mol	e^{ch} kJ/mol
Acetyl chloride	CH_3COCl	l	−65.44	−49.73	48	28	246.28	1030.45
Acetyl chloride	CH_3COCl	g	−58.3	−46.29	70.47	16.21	249.72	1044.84
Acetylenedicarbonitrile	C_4N_2	g	127.5	122.1	69.31	20.53	514.49	2152.62
Allyl chloride	C_3H_5Cl	g	−0.15	10.42	73.29	18.01	460.45	1926.54
2-Aminoethanesulfonic acid	$C_2H_7NO_3S$	c	−187.7	−134.3	36.8	33.6	406.44	1700.54
2-Aminoethanesulfonic acid, ionized, $m = 1$	$C_2H_7NO_3S$	aq	−171.92	−121.76	47.8		418.98	1753.01
2-Aminoethanesulfonic acid, nonionized, $m = 1$	$C_2H_7NO_3S$	aq	−181.92	−134.12	55.7		406.62	1701.29
Bromobenzene	C_6H_5Br	l	14.5	30.12	52	37.17	771.61	3228.43
1-Bromobutane	C_4H_9Br	g	−25.65	−3.08	88.39	26.13	655.16	2741.20
2-Bromobutane	C_4H_9Br	l	−37.2	−4.6			653.64	2734.84
2-Bromobutane	C_4H_9Br	g	−28.7	−6.16	88.5	26.48	652.08	2728.31
Bromochlorodifluoromethane	$CBrClF_2$	g	−112.7	−107.18	76.14		138.62	579.98
Bromochlorofluoromethane	$CHBrClF$	g	−70.5	−66.58	72.88		146.99	615.01
Bromochloromethane	CH_2BrCl	g	−12	−9.39	68.67		171.95	719.44
Bromodichlorofluoromethane	$CHBrClF$	g	−64.4	−58.98	78.87		154.59	646.80
Bromodichloromethane	$CHBrCl_2$	g	−14	−10.16	75.56		157.75	660.02
Bromodifluoromethane	$CHBrF_2$	g	−110.8	−106.9	70.51		152.33	637.36
Bromoethane	C_2H_5Br	l	−21.99	−6.64	47.5	24.1	342.62	1433.54
Bromoethane	C_2H_5Br	g	−15.3	−6.29	68.71	15.45	342.97	1435.00

TABLE 7.6 Chemical exergy of other organic chemicals including acids or organic chemicals containing Br, Cl, F, I, N, Pb, S, and Si. Underlying state information from Ref. 11 except where marked with * indicating the state information came from Ref. 12.

Name	Structure	State	h^f kcal/mol	Δg^f kcal/mol	s cal/mol/K	C_p cal/mol/K	e^{ch} kcal/mol	e^{ch} kJ/mol
Bromoethene	C_2H_3Br	g	18.73	19.3	65.83	13.26	312.13	1305.96
Bromofluoromethane	CH_2BrF	g	−60.4	−57.71	65.97		169.29	708.32
Bromoiodomethane	CH_2BrI	g	12	9.36	73.49		196.92	823.89
Bromomethane	CH_3Br	g	−9.02	−6.75	58.76	10.15	188.02	786.70
2-Bromo-2-methylpropane	C_4H_9Br	g	−32	−6.73	79.34	27.85	651.51	2725.93
1-Bromopentane	$C_5H_{11}Br$	g	−30.87	−1.37	97.7	31.6	811.36	3394.73
1-Bromopropane	C_3H_7Br	g	−21	−5.37	79.08	20.66	498.38	2085.23
2-Bromopropane	C_3H_7Br	g	−23.2	−6.51	75.53	21.37	497.24	2080.46
Bromotrichloromethane	$CBrCl_3$	g	−8.9	−2.96	79.55		151.51	633.94
Bromotrifluoromethane	$CBrF_3$	g	−155.1	−148.8	71.16	16.57	142.66	596.89
Carbonyl dibromide	$COBr_2$	g	−23.00	−26.50	73.85		96.17	402.36
Carbonyl fluoride	COF_2	g	−150.3	−149.3	61.84		70.14	293.45
Chlorobenzene	C_6H_5Cl	l	2.58	21.32	50	35.9	765.53	3202.96
1-Chlorobutane	C_4H_9Cl	g	−35.2	−9.27	85.58	25.71	651.68	2726.65
2-Chlorobutane	C_4H_9Cl	g	−38.6	−12.78	85.94	25.93	648.17	2711.96
2-Chloro-1,1-difluoroethylene	C_2HClF_2	g	−79.2	−72.9	72.28		287.10	1201.24
Chlorodifluoromethane	$CHClF_2$	g	−115.6	−108.1	67.12	13.35	153.84	643.69
Chloroethane	C_2H_5Cl	g	−26.83	−14.46	65.91	14.97	337.52	1412.17
Chloroethylene	C_2H_3Cl	g	8.5	12.4	63.07	12.84	307.94	1288.44
Chloroethyne	C_2HCl	g	51	47	57.81	12.98	286.11	1197.10
Chlorofluoromethane	CH_2ClF	g	−63.2	−57.11	63.16	11.24	172.61	722.18
Chloroform	$CHCl_3$	l	−31.6	−17.17	48.5		153.45	642.04
Chloroform	$CHCl_3$	g	−24.6	−16.76	70.63	15.63	153.86	643.75
Chloroiodomethane	CH_2ClI	g	3	3.69	70.78		193.96	811.52
Chloromethane	CH_3Cl	g	−19.59	−13.97	55.97	9.74	183.52	767.84
1-Chloro-2-methylpropane	C_4H_9Cl	g	−38.1	−11.87	84.56	25.93	649.08	2715.77
2-Chloro-2-methylpropane	C_4H_9Cl	g	−43.8	−15.32	77	27.3	645.63	2701.33
1-Chloropentane	$C_5H_{11}Cl$	g	−41.8	−9.94	94.89	21.18	805.50	3370.23
1-Chloropropane	C_3H_7Cl	g	−31.1	−12.11	76.27	20.23	494.36	2068.38
2-Chloropropane	C_3H_7Cl	g	−35	−14.94	72.7	20.87	491.53	2056.54
2-Chloro-1-propene	C_3H_5Cl	g	−0.15	10.42	73.29	18.01	460.45	1926.54
Chlorotrifluoromethane	$CClF_3$	g	−169.2	−159.38	68.16	15.98	134.79	563.98
Diazocarbene*	CN_2	g	139.0	137.00	55.35		235.22	984.15
1,2-Dibromobutane	$C_4H_8Br_2$	g	−23.7	−3.14	97.7	30.38	638.96	2673.39
Dibromochlorofluoromethane	CBr_2ClF	g	−55.4	−53.4	81.99		119.88	501.60
Dibromochloromethane	$CHBr_2Cl$	g	−5	−4.5	78.31		160.70	672.35
Dibromodichloromethane	CBr_2Cl_2	g	−7	−4.67	83.23		147.09	615.43
Dibromodifluoromethane	CBr_2F_2	g	−102.7	−100.16	77.66		142.93	598.00
1,2-Dibromoethane	$C_2H_4Br_2$	l	−19.4	−5	53.37	32.51	328.12	1372.84
Dibromofluoromethane	$CHBr_2F$	g	−53.4	−52.84	75.7		97.57	408.24
Dibromomethane	CH_2Br_2	g	−3.53	−3.87	70.1	13.04	174.76	731.19
1,2-Dibromopropane	$C_3H_6Br_2$	g	−17.4	−4.22	89.9	24.57	483.39	2022.49
Dichloroacetylene	C_2Cl_2	g	50	47	65	15.67	272.68	1140.89

TABLE 7.6 Chemical exergy of other organic chemicals including acids or organic chemicals containing Br, Cl, F, I, N, Pb, S, and Si. Underlying state information from Ref. 11 except where marked with * indicating the state information came from Ref. 12. (*Continued*)

Name	Structure	State	h^f kcal/mol	Δg^f kcal/mol	s cal/mol/K	c_p cal/mol/K	e^{ch} kcal/mol	e^{ch} kJ/mol
1,2-Dichlorobenzene	$C_6H_4Cl_2$	g	7.16	19.76	81.61	27.12	750.53	3140.23
1,3-Dichlorobenzene	$C_6H_4Cl_2$	g	6.32	18.78	82.09	27.2	749.55	3136.13
1,4-Dichlorobenzene	$C_6H_4Cl_2$	g	5.5	18.44	80.47	27.22	749.21	3134.71
Dichlorodifluoromethane	CCl_2F_2	g	−117.9	−108.51	71.91	17.31	140.00	585.77
1,1-Dichloroethane	$C_2H_4Cl_2$	l	−38.3	−18.1	50.61	30.18	320.44	1340.73
1,1-Dichloroethane	$C_2H_4Cl_2$	g	31.1	−17.52	72.91	18.25	321.02	1343.16
1,2-Dichloroethane	$C_2H_4Cl_2$	l	−39.49	−19.03	49.84	30.9	319.51	1336.84
1,2-Dichloroethane	$C_2H_4Cl_2$	g	−31	−17.65	73.66	18.8	320.89	1342.62
1,1-Dichloroethylene	$C_2H_2Cl_2$	l	−5.8	5.85	48.17	26.6	287.96	1204.83
1,1-Dichloroethylene	$C_2H_2Cl_2$	g	0.3	5.78	68.85	16.02	287.89	1204.54
cis-1,2-Dichloroethylene	$C_2H_2Cl_2$	l	−6.6	5.27	47.42	27	287.38	1202.40
cis-1,2-Dichloroethylene	$C_2H_2Cl_2$	g	0.45	5.82	69.2	15.55	287.93	1204.71
trans-1-2-Dichloroethylene	$C_2H_2Cl_2$	g	1	6.35	69.29	15.93	288.46	1206.92
Dichlorofluoromethane	$CHCl_2F$	g	−68.1	−60.77	70.04	14.58	155.51	650.67
Dichloromethane	CH_2Cl_2	l	−29.7	−16.83	42.7		167.22	699.67
Dichloromethane	CH_2Cl_2	g	−22.8	−16.46	64.61	12.16	167.59	701.21
1,2-Dichloropropane	$C_3H_6Cl_2$	g	−39.6	−19.86	84.8	23.47	473.17	1979.75
1,3-Dichloropropane	$C_3H_6Cl_2$	g	−38.6	−19.74	87.76	23.81	473.29	1980.25
2,2-Dichloropropane	$C_3H_6Cl_2$	g	−42	−20.21	77.92	25.3	472.82	1978.29
1,2-Difluorobenzene	$C_6H_4F_2$	l	−79.04	−59.41	53.2	38.01	762.69	3191.08
1,3-Difluorobenzene	$C_6H_4F_2$	g	−74.09	−61.43	76.57	25.4	760.67	3182.63
1,4-Difluorobenzene	$C_6H_4F_2$	g	−73.43	−60.43	75.43	25.55	761.67	3186.81
2,2-Difluorochloroethylnene	$C_2H_5F_2N$	g	−75.4	−69.1	72.39	17.23	389.06	1627.84
1,1-Difluoroethane	$C_2H_4F_2$	g	−119.7	−105.87	67.5	16.24	324.00	1355.60
1,1-Difluoroethylene	$C_2H_2F_2$	g	−82.5	−76.84	63.38	14.14	296.60	1240.96
Difluoromethane	CH_2F_2	g	−108.24	−101.66	58.94	10.25	173.72	726.84
Difluoromethylene	CF_2	g	−43.50	−45.80	57.53		173.15	724.44
1,2-Diiodoethane	$C_2H_4I_2$	g	15.9	18.76	83.3	19.67	369.73	1546.96
Diiodomethane	CH_2I_2	g	28.3	24.24	73.95	13.8	220.72	923.50
Ethyl chloride	C_2H_5Cl	g	−26.83	−14.46	65.91	14.97	337.52	1412.17
Fluorobenzene	C_6H_5F	g	−27.86	−16.5	72.33	22.57	773.37	3235.77
Fluoroethane	C_2H_5F	g	−62.9	−50.44	63.34	14.21	347.20	1452.68
Fluoromethane	CH_3F	g	−56.8	−51.09	53.25	8.96	192.06	803.58
1-Fluoropropane	C_3H_7F	g	−67.2	−47.87	72.71	19.75	504.26	2109.81
2-Fluoropropane	C_3H_7F	g	−69	−48.81	69.82	19.6	503.32	2105.88
4-Fluorotoluene	$C_7H_{13}F$	l	−44.8	−19.06	56.67		1094.59	4579.77
Formyl fluoride	$CHFO$	g	−90	−88	59	9.66	99.19	415.00
Hexachlorobenzene	C_6Cl_6	c	−31.3	0.25	62.2	48.11	677.29	2833.78
Hexachlorobenzene	C_6Cl_6	g	−8.1	10.56	105.45	41.4	687.60	2876.92
Hexachloroethane	C_2Cl_6	g	−33.2	−13.13	95.3	32.68	271.68	1136.71
Hexadecafluoroheptane	C_7F_{16}	l	−817.6	−739.24	134.28		914.27	3825.33
Hexadecafluoroheptane	C_7F_{16}	g	−808.9	−737.87	158.88		915.64	3831.06
Hexafluorobenzene	C_6F_6	l	−237.27	−211.43	66.9	52.96	739.58	3094.41

TABLE 7.6 Chemical exergy of other organic chemicals including acids or organic chemicals containing Br, Cl, F, I, N, Pb, S, and Si. Underlying state information from Ref. 11 except where marked with * indicating the state information came from Ref. 12. (*Continued*)

Name	Structure	State	h^f kcal/mol	Δg^f kcal/mol	s cal/mol/K	C_p cal/mol/K	e^{ch} kcal/mol	e^{ch} kJ/mol
Hexafluorobenzene	C_6F_6	g	−228.64	−210.18	91.59	37.43	740.83	3099.64
Hexafluoroethane	C_2F_6	g	−320.9	−300.15	79.3	25.43	258.63	1082.12
Hexamethyldisiloxane	$C_6H_{18}OSi_2$	l	−194.7	−129.5	103.69	74.42	967.20	4046.76
Iodobenzene	C_6H_5I	g	38.85	44.88	79.84	24.08	795.30	3327.54
Iodoethane	C_2H_5I	l	−9.6	3.5	50.6	27.5	361.69	1513.31
Iodoethane	C_2H_5I	g	−2	5.1	70.82	15.76	363.29	1520.01
Iodomethane	CH_3I	l	−3.29	3.61	38.9		207.31	867.39
Iodomethane	CH_3I	g	3.29	3.72	60.64	10.54	207.42	867.85
2-Iodo-2-methylpropane	C_3H_9Cl	g	−17.6	5.65	81.79	28.27	568.55	2378.80
1-Iodopropane	C_3H_7I	g	−7.3	6.68	80.32	21.48	519.36	2173.00
2-Iodopropane	C_3H_7I	g	−10	4.8	77.55	21.53	517.48	2165.13
Isothiocyanic acid	CHNS	g	30.5	26.98	59.28	11.09	298.96	1250.85
2-Mercaptopropionic acid	$C_3H_6O_2S$	l	−11.9	−82.19	54.7		527.84	2208.49
L-Methionine	$C_5H_{11}NO_2S$	c	−180.4	−120.88	55.32		826.43	3457.76
Methyl chloride	CH_3Cl	g	−19.59	−13.97	55.97	9.74	183.52	767.84
Methyl isothiocyanate	C_2H_3NS	g	31.3	34.5	69.29	15.65	460.97	1928.69
Octafluorocyclobutane	C_4F_8	g	−365.2	−334.33	95.69	37.32	541.46	2265.45
Pentachloroethane	C_2HCl_5	g	−34.8	−16.79	91.17	28.22	281.45	1177.60
Pentachlorofluoroethane	C_2Cl_5F	g	−75.8	−55.93	93.54		274.54	1148.68
Pentachlorophenol	C_6HCl_5O	c	−70.6	−34.44	60.21	48.27	656.50	2746.80
Pentafluoroethane	C_2HF_5	g	−264	−246	79.76	22.88	280.55	1173.84
Perfluoropiperidine	$C_{12}F_{23}N$	l	−482.9	−422.67	94.02	70.93	2144.32	8971.85
Phosgene	$COCl_2$	g	−52.8	−49.42	67.82	13.79	78.67	329.16
Tetrabromomethane	CBr_4	g	19	15.61	85.53	21.78	161.95	677.58
1,1,1,2-Tetrachlorodifluoroethane	$C_2Cl_4F_2$	g	−117.1	−97.3	91.5	29.5	278.83	1166.64
1,1,1,2-Tetrachloroethane	$C_2H_2Cl_4$	g	−35.7	−19.2	85.05	24.67	292.48	1223.72
1,1,2,2-Tetrachloroethane	$C_2H_2Cl_4$	l	−47	−22.7	59	39.6	288.98	1209.08
1,1,2,2-Tetrachloroethane	$C_2H_2Cl_4$	g	−36.5	−20.45	86.69	24.09	291.23	1218.49
Tetrachloroethylene	C_2Cl_4	g	−3.4	4.9	81.46	22.69	260.14	1088.45
Tetrachloromethane	CCl_4	l	−31.75	−14.97	51.67		142.22	595.04
Tetrachloromethane	CCl_4	g	−22.9	−12.8	74.07	19.94	144.39	604.12
Tetraethyllead	$C_8H_{20}Pb$	l	12.7	80.4	112.92		1462.33	6118.37
1,1,1,2-Tetrafluoroethane	CH_2FCF_3	g	−214.1	−197.46	75.58	20.62	296.86	1242.08
Tetrafluoroethylene	C_2F_4	g	−157.4	−149.07	71.69	19.24	288.82	1208.44
Tetrafluoromethane	CF_4	g	−223	−212.3	62.45	14.59	127.54	533.61
Tetraiodomethane	CI_4	g	62.84	51.89	93.6	22.91	233.93	978.78
Tetramethyllead	$C_4H_{12}Pb$	l	23.5	62.8	76.5		826.77	3459.20
Tetramethyllead	$C_4H_{12}Pb$	g	32.6	64.7	100.5	34.42	828.67	3467.15
Tetramethylsilane	$C_4H_{12}Si$	g	−68.5	−23.92	86.3	31.12	706.90	2957.67
Thioacetic acid	C_2H_4OS	g	−43.49	−36.81	74.86	19.33	418.26	1750.01
Thiourea	CH_4N_2S	c	−21.13	5.2	27.7		361.91	1514.22
Tribromochloromethane	CBr_3Cl	g	3	2.17	85.36		151.28	632.70
Tribromofluoromethane	CBr_3F	g	−45.4	−46.14	82.65		102.97	430.57

TABLE 7.6 Chemical exergy of other organic chemicals including acids or organic chemicals containing Br, Cl, F, I, N, Pb, S, and Si. Underlying state information from Ref. 11 except where marked with * indicating the state information came from Ref. 12. (*Continued*)

Name	Structure	State	h^f kcal/mol	Δg^f kcal/mol	s cal/mol/K	C_p cal/mol/K	e^{ch} kcal/mol	e^{ch} kJ/mol
Tribromomethane	$CHBr_3$	g	4	1.78	7.01	16.96	164.33	687.27
1,1,1-Trichloroethane	$C_2H_3Cl_3$	g	−34.01	−18.21	76.49	22.07	306.86	1284.07
1,1,2-Trichloroethane	$C_2H_3Cl_3$	g	−33.1	−18.52	80.57	21.47	306.55	1282.77
Trichloroethylene	C_2HCl_3	g	−1.4	4.75	77.63	19.17	273.39	1144.02
Trichlorofluoromethane	CCl_3F	g	−68.1	−58.68	74.06	18.66	139.41	603.21
Trichloromethyl	CCl_3	g	19	22	70.9	15.21	164.37	687.87
1,2,3-Trichloropropane	$C_3H_5Cl_3$	g	−44.4	−23.37	91.52	26.82	456.19	1908.86
Trifluoroacetonitrile	C_2F_3N	g	−118.4	−110.4	71.3	18.7	267.13	1117.67
1,1,1-Trifluoroethane	$C_2H_3F_3$	g	−178.2	−162.11	68.67	18.76	299.99	1255.14
Trifluoroethylene	C_2HF_3	g	−118.5	−112.22	69.94	16.54	293.44	1227.77
Trifluoroiodomethane	CF_3I	g	−141	−136.7	73.5		163.69	684.87
Trifluoromethane	CHF_3	g	−165.71	−157.48	62.04	12.22	150.13	628.13
Trifluoromethyl	$CF_3\cdot$	g	−112.4	−109.2	63.3	11.9	170.19	712.08
Trifluoromethyl	$CF_3\cdot$	g	100.6	103.1	60.8	11.87	382.49	1600.34
Trifluoromethylbenzene	$C_7H_5F_3$	l	−152.4	−123.98	64.89		884.83	3702.15
Trifluoromethylbenzene	$C_7H_5F_3$	g	−143.42	−122.2	89.05	31.17	886.61	3709.60
Trifluoromethylhypofluorite	CF_4O	g	−183	−169	77.06	18.97	171.30	716.74
Triiodomethane	CHI_3	g	50.4	42.54	84.97	17.94	231.80	969.86
Vinyl bromide	C_2H_3Br	g	18.7	19.3	65.9	13.27	312.13	1305.96
Vinyl chloride	C_2H_3Cl	g	8.5	12.4	63.07	12.84	307.94	1288.44

TABLE 7.6 Chemical exergy of other organic chemicals including acids or organic chemicals containing Br, Cl, F, I, N, Pb, S, and Si. Underlying state information from Ref. 11 except where marked with * indicating the state information came from Ref. 12. (*Continued*)

Structure	State	h^f (kJ/mol)	s (J mol/K)	Δg^f (kJ/mol)	e^{ch} (kJ/mol)
		Ag—Silver			
Ag_2	(g)	409.99	257.02	358.78	557.38
$Ag_2C_2O_4$	(s)	−673.21	209.20	−584.09	442.91
Ag_2CO_3	(s)	−505.85	167.36	−436.81	177.95
Ag_2CrO_4	(s)	−731.74	217.57	−641.83	149.02
Ag_2MoO_4	(s)	−840.57	213.38	−748.10	189.65
Ag_2O	(s)	−31.05	121.34	−11.21	189.35
Ag_2O_2	(s)	−24.27	117.15	27.61	230.14
Ag_2O_3	(s)	33.89	100.42	121.34	325.82
Ag_2S	(s beta)	−29.41	150.62	−39.46	768.44
Ag_2S	(s alpha orthorhombic)	−32.59	144.01	−40.67	767.23
Ag_2Se	(s)	−37.66	150.71	−44.35	501.75
Ag_2SeO_3	(s)	−365.26	230.12	−304.18	247.81
Ag_2SeO_4	(s)	−420.49	248.53	−334.30	219.65
Ag_2SO_3	(s)	−490.78	158.16	−411.29	402.50

TABLE 7.7 Chemical exergy (e^{ch}) of other chemicals, including inorganic chemicals, organometallic chemicals, oxides of carbon, and sulfides of carbon. Inorganic chemicals that are found in other tables in this book, such as in reference states (Chap. 4), atmosphere gases (Chap. 5), water (Chap. 6), and molecular hydrogen (Table 7.8) are not duplicated here. The enthalpy of formation (h^f), Gibbs free energy of formation (Δg^f), and entropy (s) are referenced from Ref. 12.

Structure	State	h^f (kJ/mol)	s (J mol/K)	Δg^f (kJ/mol)	e^{ch} (kJ/mol)
Ag—Silver					
Ag_2SO_4	(s)	−715.88	200.41	−618.48	197.27
Ag_2Te	(s)	−37.24	154.81	43.10	571.00
$AgBr$	(s)	−100.37	107.11	−96.90	52.90
$AgBrO_3$	(s)	−27.20	152.72	54.39	210.08
$AgClO_2$	(s)	8.79	134.56	75.73	240.81
$AgCN$	(s)	146.02	107.19	156.90	666.81
$AgF \cdot 2H_2O$	(s)	−800.82	174.89	−671.11	157.23
AgI	(s)	−61.84	115.48	−66.19	120.96
$AgIO_3$	(s)	−171.13	149.37	−93.72	99.32
AgN_3	(s)	308.78	104.18	376.14	476.45
$AgNO_2$	(s)	−45.06	128.20	19.08	122.64
$AgNO_3$	(s)	−124.39	140.92	−33.47	72.05
AgO	(s)	−11.42	57.78	14.23	115.49
$AgOCN$	(s)	−95.40	121.34	−58.16	453.71
$AgReO_4$	(s)	−736.38	153.13	−635.55	31.20
$AgSCN$	(s)	87.86	130.96	101.38	1220.59
Al—Aluminum					
$Al(BH_4)_3$	(l)	−16.32	289.11	144.77	4241.43
$Al(BH_4)_3$	(g)	12.55	379.07	146.44	4243.10
$Al(CH_3)_3$	(l)	−136.40	209.41	−10.04	3078.97
$Al(NO_3)_3 \cdot 6H_2O$	(s)	−2850.48	467.77	−2203.88	38.92
$Al(NO_3)_3 \cdot 9H_2O$	(s)	−3757.06	569.02	−2929.64	27.38
$Al_2(CH_3)_6$	(g)	−230.91	524.67	−9.79	6168.23
Al_2Br_6	(g)	−1020.90	547.27	−947.26	947.14
Al_2Cl_6	(g)	−1295.37	475.51	−1220.89	741.61
Al_2F_6	(g)	−2631.74	387.02	−2539.69	569.11
Al_2I_6	(g)	−506.26	584.09	−560.66	1557.84
Al_2O	(g)	−131.38	259.41	−161.08	1432.28
Al_2O_3	(l)	−1581.13	89.58	−1499.25	98.03
Al_2O_3	(s gamma-corundum)	−1656.86	59.83	−1562.72	34.56
Al_2O_3	(s alpha-corundum)	−1675.27	50.92	−1581.97	15.32
$Al_2O_3 \cdot 3H_2O$	(s gibbsite)	−2562.70	140.21	−2287.39	24.11
$Al_2O_3 \cdot H_2O$	(s boehmite)	−1974.85	96.86	−1825.48	9.88
$Al_2Si_2O_7 \cdot 2H_2O$	(s halloysite)	−4079.82	203.34	−3759.32	31.96
$Al_2Si_2O_7 \cdot 2H_2O$	(s kaolinite)	−4098.65	202.92	−3778.15	13.13
Al_4C_3	(s)	−207.28	104.60	−238.45	4175.17
Al_4C_3	(g)	−215.89	89.12	−203.34	4210.27
Al_6BeO_{10}	(l)	−5299.45	314.89	−5034.19	363.93

TABLE 7.7 Chemical exergy (e^{ch}) of other chemicals, including inorganic chemicals, organometallic chemicals, oxides of carbon, and sulfides of carbon. Inorganic chemicals that are found in other tables in this book, such as in reference states (Chap. 4), atmosphere gases (Chap. 5), water (Chap. 6), and molecular hydrogen (Table 7.8) are not duplicated here. The enthalpy of formation (h^f), Gibbs free energy of formation (Δg^f), and entropy (s) are referenced from Ref. 12. (Continued)

Structure	State	h^f (kJ/mol)	s (J mol/K)	Δg^f (kJ/mol)	e^{ch} (kJ/mol)
			Al—Aluminum		
Al_6BeO_{10}	(s)	−5624.13	175.56	−5317.45	80.68
$Al_6Si_2O_{13}$	(s mullite)	−6819.92	274.89	−6443.36	66.35
$AlBO_2$	(g)	−541.41	269.45	−550.61	877.11
$AlBr_3$	(s)	−511.12	180.25	−488.31	458.89
$AlBr_3$	(l)	−501.20	206.48	−486.26	460.94
$AlBr_3$	(g)	−410.87	349.07	−438.48	508.72
AlC	(g)	689.52	223.34	633.04	1839.01
AlCl	(g)	−51.46	227.86	−77.82	779.73
$AlCl_2$	(g)	−288.70	288.28	−299.57	619.83
$AlCl_3$	(g)	−584.50	314.30	−570.07	411.18
$AlCl_3$	(s)	−705.63	109.29	−630.07	351.18
$AlCl_3$	(l)	−674.80	172.92	−618.19	363.06
$AlCl_3 \cdot 6H_2O$	(s)	−2691.57	376.56	−2269.40	140.28
AlF	(g)	−265.27	215.06	−290.79	757.81
AlF_2	(g)	−732.20	263.17	−740.57	560.93
AlF_3	(s)	−1510.42	66.48	−1430.93	123.47
AlF_3	(g)	−1209.18	276.77	−1192.86	361.54
$AlF_3 \cdot 3H_2O$	(s)	−2297.43	209.20	−2051.83	216.78
AlH	(g)	259.24	187.78	231.17	1144.92
AlI_3	(l)	−297.06	219.66	−301.25	758.00
AlI_3	(g)	−205.02	363.17	−251.04	808.21
AlI_3	(s)	−309.62	189.54	−305.43	753.82
AlN	(s)	−317.98	20.17	−287.02	509.01
AlN	(g)	435.14	211.71	410.03	1206.07
AlO	(g)	83.68	218.28	57.74	855.40
AlOCl	(s)	−793.29	54.39	−737.26	122.25
AlOCl	(g)	−348.11	248.82	−350.20	509.31
AlOF	(g)	−586.60	234.26	−587.02	463.55
AlOH	(g)	−179.91	216.31	−184.10	731.62
$AlPO_4$	(s berlinite)	−1692.01	90.79	−1601.22	63.63
AlS	(g)	200.83	230.50	150.21	1555.21
AuH	(g)	294.97	211.05	265.68	434.34
			B—Boron		
B_2	(g)	830.52	201.79	774.04	2030.24
BBr	(g)	238.07	224.89	195.39	873.99
BBr_3	(g)	−205.64	324.13	−232.46	547.14
BBr_3	(l)	−239.74	229.70	−238.49	541.11
$B(CH_3)_3$	(l)	−143.09	238.91	−32.22	2889.19
$B(CH_3)_3$	(g)	−124.26	314.64	−35.98	2885.43

TABLE 7.7   Chemical exergy (e^{ch}) of other chemicals, including inorganic chemicals, organometallic chemicals, oxides of carbon, and sulfides of carbon. Inorganic chemicals that are found in other tables in this book, such as in reference states (Chap. 4), atmosphere gases (Chap. 5), water (Chap. 6), and molecular hydrogen (Table 7.8) are not duplicated here. The enthalpy of formation (h^f), Gibbs free energy of formation (Δg^f), and entropy (s) are referenced from Ref. 12. (*Continued*)

Structure	State	h^f (kJ/mol)	s (J mol/K)	Δg^f (kJ/mol)	e^{ch} (kJ/mol)
B—Boron					
B_4C	(s)	−71.13	27.11	−71.13	2851.54
BCl	(g)	149.49	213.13	120.92	810.87
BCl_3	(g)	−403.76	289.99	−388.74	424.91
BCl_3	(l)	−427.19	206.27	−387.44	426.21
B_2Cl_4	(l)	−523.00	262.34	−464.84	1038.76
$BClF_2$	(g)	−890.36	271.96	−876.13	319.62
BCl_2F	(g)	−645.17	284.51	−631.37	373.33
BF	(g)	−122.17	200.37	−149.79	731.21
BF_3	(g)	−1137.00	254.01	−1120.35	266.45
BH	(g)	449.61	171.75	419.61	1165.77
B_2H_6	(g)	35.56	232.00	86.61	2051.14
B_5H_9	(l)	42.68	184.22	171.67	4374.66
BN	(g)	647.47	212.17	614.50	1242.94
BN	(s)	−254.39	14.81	−228.45	399.99
$B_3N_3H_6$	(l)	−540.99	199.58	−392.79	2200.84
BO	(g)	25.10	203.43	−4.18	625.88
BO_2	(g)	−300.41	229.45	−305.85	326.17
B_2O_2	(g)	−454.80	242.38	−462.33	797.79
B_2O_3	(g)	−843.79	279.70	−831.99	430.10
B_2O_3	(s amorphous)	−1254.53	77.82	−1182.40	79.69
B_2O_3	(s)	−1272.77	53.97	−1193.70	68.39
Ba—Barium					
$BaBr_2$	(s)	−757.30	146.44	−736.80	139.60
$BaBr_2$	(g)	−439.32	330.54	−472.79	403.61
$BaBr_2 \cdot 2H_2O$	(s)	−1366.08	225.94	−1230.51	122.03
$Ba(BrO_3)_2$	(s)	−752.66	242.67	−577.39	310.78
$Ba(BrO_3)_2 \cdot H_2O$	(s)	−1054.79	292.46	−824.62	301.62
$BaCO_3$	(s witherite)	−1216.29	112.13	−1137.63	53.93
$BaCl_2$	(s)	−858.14	123.68	−810.44	88.66
$BaCl_2$	(l)	−832.45	143.51	−790.15	108.95
$BaCl_2$	(g)	−498.73	325.64	−510.70	388.40
$BaCl_2 \cdot 2H_2O$	(s)	−1460.13	202.92	−1296.45	78.79
$Ba(ClO_3)_2$	(s)	−680.32	196.65	−531.37	379.51
$Ba(ClO_4)_2 \cdot 3H_2O$	(s)	−1691.59	393.30	−1270.68	358.33
$BaCrO_4$	(s)	−1445.99	158.57	−1345.28	22.37
BaF_2	(s)	−1208.76	96.40	−1158.55	122.65
BaF_2	(l)	−1171.31	121.25	−1128.38	152.82
BaF_2	(g)	−803.75	301.16	−814.50	466.70

TABLE 7.7 Chemical exergy (e^{ch}) of other chemicals, including inorganic chemicals, organometallic chemicals, oxides of carbon, and sulfides of carbon. Inorganic chemicals that are found in other tables in this book, such as in reference states (Chap. 4), atmosphere gases (Chap. 5), water (Chap. 6), and molecular hydrogen (Table 7.8) are not duplicated here. The enthalpy of formation (h^f), Gibbs free energy of formation (Δg^f), and entropy (s) are referenced from Ref. 12. (*Continued*)

Structure	State	h^f (kJ/mol)	s (J mol/K)	Δg^f (kJ/mol)	e^{ch} (kJ/mol)
Ba—Barium					
BaI_2	(g)	−302.92	348.11	−353.42	597.68
BaI_2	(l)	−585.89	183.68	−587.39	363.71
BaI_2	(s)	−605.42	165.14	−601.41	349.69
$Ba(IO_3)_2$	(s)	−1027.17	249.37	−864.83	98.04
$Ba(IO_3)_2 \cdot H_2O$	(s)	−1322.14	297.06	−1104.16	96.79
$BaMoO_4$	(s)	−1548.08	138.07	−1439.71	74.83
$Ba(N_3)_2 \cdot H_2O$	(s)	−308.36	188.28	−105.02	910.46
BaO	(s)	−548.10	72.09	−520.41	256.96
BaO	(l)	−491.62	96.57	−471.24	306.12
BaO	(g)	−123.85	235.35	−144.81	632.55
$Ba(OH)_2 \cdot 8H_2O$	(s)	−3342.18	426.77	−2793.24	126.77
BaS	(s)	−460.24	78.24	−456.06	928.64
$BaSeO_3$	(s)	−1040.56	167.36	−968.18	160.61
$BaSeO_4$	(s)	−1146.42	175.73	−1044.74	86.00
$BaSiF_6$	(s)	−2952.23	163.18	−2794.08	353.72
$BaSiO_3$	(s)	−1623.60	109.62	−1540.26	96.03
$BaTiO_3$	(s)	−1659.79	107.95	−1572.35	116.14
Ba_2TiO_4	(s)	−2243.04	196.65	−2133.00	332.85
$BaZrO_3$	(s)	−1779.46	124.68	−1694.52	169.77
Be—Beryllium					
$Be(OH)_2$	(s beta)	−905.84	46.02	−816.72	27.62
Be_2C	(s)	−117.15	16.32	−87.86	1531.01
Be_3N_2	(s cubic)	−588.27	34.14	−533.04	1280.53
$BeAl_2O_4$	(s)	−2300.78	66.27	−2178.61	24.94
$BeBr_2$	(s)	−369.87	106.27	−353.13	352.17
BeC_2	(g)	564.84	218.40	506.26	1931.11
$BeCl_2$	(s beta)	−496.22	75.81	−449.53	278.47
BeF_2	(a alpha)	−1026.75	53.35	−979.47	130.63
BeH	(g)	326.77	170.87	298.32	1020.67
BeI_2	(s)	−192.46	120.50	−209.20	570.80
BeO	(s alpha)	−608.35	13.77	−579.07	27.20
BeO	(g)	129.70	197.53	104.18	710.44
$BeSO_4$	(s alpha)	−1205.20	77.99	−1093.86	127.58
$BeSO_4 \cdot 4H_2O$	(s)	−2423.75	232.97	−2080.66	93.08
$BeWO_4$	(s)	−1514.61	88.37	−1405.82	34.83
Br—Bromine					
Br	(g)	111.88	174.91	82.43	132.93
$BrCl$	(g)	14.64	239.99	−0.96	111.39
BrF	(g)	−93.85	228.86	−109.16	194.24

TABLE 7.7 Chemical exergy (e^{ch}) of other chemicals, including inorganic chemicals, organometallic chemicals, oxides of carbon, and sulfides of carbon. Inorganic chemicals that are found in other tables in this book, such as in reference states (Chap. 4), atmosphere gases (Chap. 5), water (Chap. 6), and molecular hydrogen (Table 7.8) are not duplicated here. The enthalpy of formation (h^f), Gibbs free energy of formation (Δg^f), and entropy (s) are referenced from Ref. 12. (Continued)

Structure	State	h^f (kJ/mol)	s (J mol/K)	Δg^f (kJ/mol)	e^{ch} (kJ/mol)
		Br—Bromine			
BrF_3	(l)	−300.83	178.24	−240.58	568.62
BrF_3	(g)	−255.60	292.42	−229.45	579.75
BrF_5	(l)	−458.57	225.10	−351.87	963.13
BrO	(g)	125.77	237.44	108.24	160.70
		C—Carbon (inorganic)			
CNBrV	(g)	181.38	247.15	160.62	621.73
CNCl	(g)	132.21	235.48	125.48	597.94
CNI	(g)	225.10	256.60	196.15	694.60
CNI	(s)	160.25	128.87	169.37	667.83
CO	(g)	−110.54	197.90	−137.28	274.96
C_3O_2	(l)	−117.28	181.08	−105.02	1129.72
C_3O_2	(g)	−93.72	276.35	−109.83	1124.91
COS	(g)	−138.41	231.46	−165.64	855.89
CS_2	(g)	117.07	237.78	66.90	1695.77
CS_2	(l)	89.70	151.34	65.27	1694.14
		Ca—Calcium			
$Ca_3(AsO_4)_2$	(s)	−3298.67	225.94	−3063.11	125.09
$CaBr_2$	(g)	−384.93	314.64	−420.95	409.15
$CaBr_2$	(s)	−683.25	129.70	−664.13	165.97
$CaBr_2$	(l)	−663.00	147.86	−649.31	180.79
$CaBr_2 \cdot 6H_2O$	(s)	−2506.22	410.03	−2153.09	105.45
CaC_2	(s)	−59.83	69.96	−64.85	1484.79
$CaC_2O_4 \cdot H_2O$	(s)	−1674.86	156.48	−1513.98	281.59
$CaCl_2$	(s)	−795.80	104.60	−748.10	104.70
$CaCl_2$	(l)	−774.04	123.85	−732.20	120.60
$CaCl_2$	(g)	−471.54	289.95	−479.07	373.73
$Ca(ClO_4)_2 \cdot 4H_2O$	(s)	−1948.91	433.46	−1476.83	343.96
$CaCrO_4$	(s)	−1379.05	133.89	−1277.38	43.97
CaF_2	(g)	−782.41	273.63	−794.96	439.94
CaF_2	(s)	−1219.64	68.87	−1167.34	67.56
CaF_2	(l)	−1184.07	92.59	−1142.23	92.67
CaH_2	(s)	−186.19	41.84	−147.28	817.93
$CaHPO_4$	(s)	−1814.39	111.38	−1681.26	35.05
$CaHPO_4 \cdot 2H_2O$	(s)	−2403.58	189.45	−2154.76	37.69
$Ca(H_2PO_4)_2 \cdot H_2O$	(s)	−3409.67	259.83	−3058.42	119.27
CaI_2	(l)	−500.16	178.95	−506.52	398.28
CaI_2	(s)	−536.81	145.27	−533.13	371.67
CaI_2	(g)	−258.15	327.44	−308.78	596.02
$Ca(IO_3)_2$	(s)	−1002.49	230.12	−839.31	77.26

TABLE 7.7 Chemical exergy (e^{ch}) of other chemicals, including inorganic chemicals, organometallic chemicals, oxides of carbon, and sulfides of carbon. Inorganic chemicals that are found in other tables in this book, such as in reference states (Chap. 4), atmosphere gases (Chap. 5), water (Chap. 6), and molecular hydrogen (Table 7.8) are not duplicated here. The enthalpy of formation (h^f), Gibbs free energy of formation (Δg^f), and entropy (s) are referenced from Ref. 12. (*Continued*)

Structure	State	h^f (kJ/mol)	s (J mol/K)	Δg^f (kJ/mol)	e^{ch} (kJ/mol)
Ca—Calcium					
$Ca(IO_3)_2 \cdot 6H_2O$	(s)	−2780.69	451.87	−2267.73	77.28
$Ca(OH)_2$	(s)	−986.17	83.39	−898.51	70.62
$Ca[Mg(CO_3)_2]$	(s dolomite)	−2326.30	155.18	−2163.55	24.77
$CaMoO_4$	(s)	−1541.39	122.59	−1434.69	33.56
CaO	(s)	−635.13	38.20	−603.54	127.52
CaO	(l)	−557.35	62.30	−532.96	198.10
$CaO \cdot Al_2O_3$	(s)	−2326.30	114.22	−2208.73	119.62
$CaO \cdot 2Al_2O_3$	(s)	−3977.73	177.82	−3770.62	155.02
$CaO \cdot B_2O_3$	(s)	−2030.96	104.85	−1924.10	69.05
$CaO \cdot 2B_2O_3$	(s)	−3360.25	134.72	−3167.12	88.12
$CaO \cdot Fe_2O_3$	(s)	−1520.34	145.35	−1412.81	72.74
$CaO \cdot V_2O_5$	(s)	−2329.27	179.08	−2169.70	13.78
$Ca_2P_2O_7$	(s beta)	−3338.83	189.24	−3132.14	62.39
Ca_2SiO_4	(s beta)	−2307.48	127.74	−2192.83	128.21
$Ca_3(PO_4)_2$	(s beta)	−4120.82	235.98	−3884.84	40.75
$Ca_3(PO_4)_2$	(s alpha)	−4109.94	240.91	−3875.64	49.96
$Ca_{10}(PO_4)_6(OH)_2$	(s hydroxyapatite)	−13476.66	780.73	−12677.52	68.41
$Ca_{10}(PO_4)_6F_2$	(s fluorapatite)	−13744.44	775.71	−12982.95	28.74
CaS	(s)	−474.88	56.48	−469.86	868.54
$CaSO_3 \cdot H_2O$	(s)	−1752.68	184.10	−1555.19	27.17
$CaSO_4$	(s anhydrite insoluble)	−1434.11	106.69	−1321.85	24.40
$CaSO_4$	(s alpha soluble)	−1425.24	108.37	−1313.48	32.77
$CaSO_4$	(s beta soluble)	−1420.80	108.37	−1309.05	37.20
$CaSO_4 \cdot 2H_2O$	(s)	−2022.63	194.14	−1797.45	24.95
$CaSe$	(s)	−368.19	66.94	−363.17	713.43
$CaSeO_4 \cdot 2H_2O$	(s)	−1706.65	221.75	−1486.99	73.60
$CaSiO_3$	(s pseudowollastonite)	−1628.41	87.36	−1544.73	45.25
$CaSiO_3$	(s wollastonite)	−1634.94	81.92	−1549.71	40.28
Ca_2SiO_4	(s gamma)	−2317.94	120.79	−2201.20	119.85
$CaTiO_3$	(s perovskite)	−1660.63	93.64	−1575.28	66.91
$CaTiSiO_5$	(s sphene)	−2603.28	129.20	−2461.87	39.25
$CaWO_4$	(s)	−1645.15	126.40	−1538.50	26.95
$CaZrO_3$	(s)	−1766.90	100.08	−1681.13	136.86
Cd—Cadmium					
$CdBr_2$	(s)	−316.18	137.24	−296.31	103.09
$CdBr_2 \cdot 4H_2O$	(s)	−1492.56	316.31	−1248.03	103.66
$CdCl_2$	(s)	−391.50	115.27	−343.97	78.13
CdF_2	(s)	−700.40	77.40	−647.68	156.52
CdI_2	(s)	−202.92	161.08	−201.38	272.72

TABLE 7.7 Chemical exergy (e^{ch}) of other chemicals, including inorganic chemicals, organometallic chemicals, oxides of carbon, and sulfides of carbon. Inorganic chemicals that are found in other tables in this book, such as in reference states (Chap. 4), atmosphere gases (Chap. 5), water (Chap. 6), and molecular hydrogen (Table 7.8) are not duplicated here. The enthalpy of formation (h^f), Gibbs free energy of formation (Δg^f), and entropy (s) are referenced from Ref. 12. (*Continued*)

Structure	State	h^f (kJ/mol)	s (J mol/K)	Δg^f (kJ/mol)	e^{ch} (kJ/mol)
Cd—Cadmium					
CdO	(s)	−258.15	54.81	−228.45	71.92
CdS	(s)	−161.92	64.85	−156.48	751.22
CdSb	(s)	−14.39	92.88	−13.01	723.59
$CdSeO_3$	(s)	−575.30	142.26	−497.90	153.89
$CdSeO_4$	(s)	−633.04	164.43	−531.79	121.96
$CdSiO_3$	(s)	−1189.09	97.49	−1105.41	53.87
$CdSO_4$	(s)	−933.28	123.04	−822.78	92.77
$CdSO_4 \cdot 8/3H_2O$	(s)	−1729.37	229.63	−1465.34	85.07
$CdSO_4 \cdot H_2O$	(s)	−1239.55	154.03	−1068.84	84.78
CdTe	(s)	−92.47	100.42	−92.05	535.65
Cl—Chlorine					
Cl	(g)	121.29	165.06	105.31	167.16
ClF	(g)	−54.48	217.78	−55.94	258.81
ClF_3	(g)	−158.99	281.50	−118.83	701.72
$ClF_3 \cdot HF$	(g)	−450.62	359.82	−384.09	807.41
ClF_5	(g)	−238.49	310.62	−146.44	1179.91
Cl_2F_6	(g)	−339.32	489.53	−237.23	1403.87
ClO	(g)	101.21	226.56	97.49	161.30
Cl_2O	(g)	80.33	267.86	97.49	223.15
ClO_2	(g)	102.51	256.77	120.33	186.11
ClO_3F	(g)	−27.15	278.86	44.85	365.49
Co—Cobalt					
$Co(IO_3)_2 \cdot 2H_2O$	(s)	−1081.98	267.78	−795.80	181.22
$Co(OH)_2$	(s pink)	−539.74	79.50	−454.38	99.05
Co_3O_4	(s)	−910.02	114.22	−794.96	153.09
$CoCl_2 \cdot 2H_2O$	(s)	−922.99	188.28	−764.84	148.41
$CoCl_2 \cdot 6H_2O$	(s)	−2115.43	343.09	−1725.48	140.05
$CoCl_3$	(g)	−163.59	334.09	−154.52	344.43
CoF_2	(s)	−692.03	81.96	−647.26	171.94
CoF_3	(s)	−790.78	94.56	−719.65	352.45
CoO	(s)	−237.94	52.97	−214.22	101.14
CoSi	(s)	−100.42	43.10	−98.74	1069.66
$CoSO_4$	(s)	−888.26	117.99	−782.41	148.14
$CoSO_4 \cdot 6H_2O$	(s)	−2683.62	367.61	−2235.72	123.26
$CoSO_4 \cdot 7H_2O$	(s)	−2979.93	406.06	−2473.83	123.22
Cr—Chromium					
$Cr_{23}C_6$	(s)	−364.84	610.03	−373.63	15529.20
Cr_3C_2	(s)	−85.35	85.44	−86.32	2487.43
Cr_7C_3	(s)	−161.92	200.83	−166.94	5154.67

TABLE 7.7 Chemical exergy (e^{ch}) of other chemicals, including inorganic chemicals, organometallic chemicals, oxides of carbon, and sulfides of carbon. Inorganic chemicals that are found in other tables in this book, such as in reference states (Chap. 4), atmosphere gases (Chap. 5), water (Chap. 6), and molecular hydrogen (Table 7.8) are not duplicated here. The enthalpy of formation (h^f), Gibbs free energy of formation (Δg^f), and entropy (s) are referenced from Ref. 12. (*Continued*)

Structure	State	h^f (kJ/mol)	s (J mol/K)	Δg^f (kJ/mol)	e^{ch} (kJ/mol)
Cr—Chromium					
$CrCl_2$	(s)	−395.39	115.31	−356.06	352.04
$CrCl_3$	(s)	−556.47	123.01	−486.18	283.77
CrF_3	(s)	−1158.97	93.89	−1087.84	255.26
Cr_2N	(s)	−125.52	64.85	−102.22	1066.92
CrN	(g)	505.01	230.45	471.91	1056.65
CrN	(s)	−117.15	37.70	−92.80	491.93
CrO	(g)	188.28	239.16	154.60	740.96
CrO_2	(g)	−75.31	269.11	−87.36	500.96
CrO_2Cl_2	(l)	−579.48	221.75	−510.87	201.16
CrO_2Cl_2	(g)	−538.06	329.70	−501.66	210.36
CrO_3	(g)	−292.88	266.06	−273.47	316.82
Cr_2O_3	(s)	−1134.70	81.17	−1053.11	121.57
Cr_2O_3	(l)	−1018.39	125.60	−950.06	224.63
Cs—Caesium					
$CsBr$	(s)	−405.68	113.39	−384.93	70.17
$CsCl$	(s)	−442.83	101.18	−414.22	52.23
$CsCl$	(l)	−434.30	101.71	−406.27	60.18
$CsCl$	(g)	−240.16	255.98	−257.73	208.72
CsF	(s)	−554.80	88.28	−525.51	131.99
CsF	(l)	−543.84	90.08	−515.09	142.41
CsF	(g)	−356.48	243.09	−373.21	284.29
CsH	(g)	121.34	214.43	101.67	624.33
CsI	(s)	−336.81	125.52	−333.72	158.73
Cs_2O	(g)	−92.05	317.98	−104.60	706.56
$CsOH$	(s)	−416.73	98.74	−362.33	162.28
$CsOH$	(g)	−259.41	255.14	−259.83	264.79
$CsOH$	(l)	−406.02	118.45	−365.89	158.73
Cu—Copper					
Cu_2	(g)	484.17	241.46	431.96	697.16
Cu_2O	(s)	−168.62	93.14	−146.02	121.14
Cu_2S	(s alpha)	−79.50	120.92	−86.19	788.31
$CuBr$	(s)	−104.60	96.11	−100.83	82.27
$CuCN$	(s)	94.98	90.00	108.37	651.57
$CuCO_3 \cdot Cu(OH)_2$	(s malachite)	−1051.44	186.19	−893.70	160.29
$CuCl$	(s)	−137.24	86.19	−119.87	74.58
$CuCl_2$	(s)	−205.85	108.07	−161.92	94.38
$CuCl_2 \cdot 2H_2O$	(s)	−821.32	167.36	−656.05	76.39
CuF	(s)	−192.46	64.85	−171.54	213.96
CuF_2	(s)	−548.94	68.62	−499.15	139.25

TABLE 7.7 Chemical exergy (e^{ch}) of other chemicals, including inorganic chemicals, organometallic chemicals, oxides of carbon, and sulfides of carbon. Inorganic chemicals that are found in other tables in this book, such as in reference states (Chap. 4), atmosphere gases (Chap. 5), water (Chap. 6), and molecular hydrogen (Table 7.8) are not duplicated here. The enthalpy of formation (h^f), Gibbs free energy of formation (Δg^f), and entropy (s) are referenced from Ref. 12. (*Continued*)

Structure	State	h^f (kJ/mol)	s (J mol/K)	Δg^f (kJ/mol)	e^{ch} (kJ/mol)
Cu—Copper					
$CuFe_2O_4$	(s)	−965.21	141.00	−858.81	30.24
$CuFeO_2$	(s)	−532.62	88.70	−479.90	30.92
CuI	(s)	−67.78	96.65	−69.45	151.00
$Cu(IO_3)_2 \cdot H_2O$	(s)	−692.03	247.27	−468.61	89.54
CuN_3	(s)	279.07	100.42	344.76	478.37
CuO	(s)	−157.32	42.63	−129.70	4.86
CuS	(s)	−53.14	66.53	−53.56	688.34
$CuSO_4$	(s)	−771.36	108.78	−661.91	87.84
$CuSO_4 \cdot 3H_2O$	(s)	−1684.31	221.33	−1400.18	63.79
$CuSO_4 \cdot 5H_2O$	(s)	−2279.65	300.41	−1880.06	60.06
$CuSO_4 \cdot H_2O$	(s)	−1085.83	146.02	−918.22	69.60
F—Fluorine					
F	(g)	78.99	158.66	61.92	314.82
FNO_3	(g)	10.46	292.88	73.64	332.76
Fe—Iron					
$FeAl_2O_4$	(s)	−1966.48	106.27	−1849.33	124.22
$FeAsS$	(s)	−41.84	121.34	−50.21	1425.99
$FeBr_2$	(s)	−249.78	140.67	−237.23	238.07
Fe_3C	(s alpha-cementite)	25.10	104.60	20.08	1553.26
$FeCO_3$	(s siderite)	−740.57	92.88	−666.72	123.74
$Fe(CO)_5$	(l)	−774.04	338.07	−705.42	1730.05
$Fe(CO)_5$	(g)	−733.87	445.18	−697.26	1738.21
$FeCl_2$	(s)	−341.79	117.95	−302.34	195.66
$FeCl_3$	(s)	−399.49	142.26	−334.05	225.80
$FeCr_2O_4$	(s)	−1444.74	146.02	−1343.90	207.05
FeF_2	(s)	−702.91	86.99	−661.07	219.03
FeF_3	(s)	−1041.82	98.32	−970.69	162.31
FeI_2	(s)	−104.60	167.36	−112.97	437.03
$FeMoO_4$	(s)	−1075.29	129.29	−974.87	138.58
Fe_4N	(s)	−10.46	156.06	3.77	1501.30
FeO	(s)	−271.96	60.75	−251.46	124.80
Fe_3O_4	(s magnetite)	−1118.38	146.44	−1015.46	115.29
$FePO_4 \cdot 2H_2O$	(s strengite)	−1888.24	171.25	−1657.70	61.89
FeS	(s pyrrhotite)	−100.00	60.29	−100.42	883.18
FeS_2	(s pyrite)	−178.24	52.93	−166.94	1425.96
Fe_7S_8	(s pyrrhotite)	−736.38	485.76	−748.52	6745.98
$FeSO_4$	(s)	−928.43	120.92	−825.08	166.36
$FeSO_4 \cdot 7H_2O$	(s)	−3014.57	409.20	−2510.27	147.68
$Fe_2(SO_4)_3$	(s)	−2581.53	307.52	−2263.13	336.92

TABLE 7.7   Chemical exergy (e^{ch}) of other chemicals, including inorganic chemicals, organometallic chemicals, oxides of carbon, and sulfides of carbon. Inorganic chemicals that are found in other tables in this book, such as in reference states (Chap. 4), atmosphere gases (Chap. 5), water (Chap. 6), and molecular hydrogen (Table 7.8) are not duplicated here. The enthalpy of formation (h^f), Gibbs free energy of formation (Δg^f), and entropy (s) are referenced from Ref. 12. (Continued)

Structure	State	h^f (kJ/mol)	s (J mol/K)	Δg^f (kJ/mol)	e^{ch} (kJ/mol)
		Fe—Iron			
FeSi	(s)	−73.64	46.02	−73.64	1155.66
FeSi$_2$	(s beta-lebanite)	−81.17	55.65	−78.24	2006.06
Fe$_3$Si	(s)	−93.72	103.76	−94.56	1883.34
Fe$_2$SiO$_4$	(s fayalite)	−1479.88	145.18	−1379.05	232.40
FeWO$_4$	(s)	−1154.78	131.80	−1054.37	156.28
		H—Hydrogen			
HBO$_2$	(s orthorhombic)	−788.77	50.21	−721.74	28.34
HBO$_2$	(s monoclinic)	−794.25	37.66	−723.41	26.67
H$_3$BO$_3$	(s)	−1094.33	88.83	−969.01	19.14
HBr	(g)	−36.44	198.61	−53.51	115.04
H$_2$CS$_3$	(l)	25.10	223.01	27.82	2502.11
HCl	(g)	−92.30	186.77	−95.31	84.59
HClO	(g)	−92.05	236.61	−79.50	102.37
HI	(g)	26.48	206.48	1.72	207.62
H$_2$MoO$_4$	(g)	−851.03	355.64	−787.43	187.83
HN$_3$	(g)	294.14	238.86	328.03	447.09
HNCS	(g)	127.61	247.69	112.97	1250.93
HNO$_2$	(g cis)	−76.57	249.32	−41.84	80.47
HNO$_2$	(g trans)	−78.66	249.16	−43.93	78.38
HNO$_3$	(l)	−173.22	155.60	−79.91	44.36
HNO$_3$	(g)	−135.06	266.27	−74.77	49.51
H$_2$O$_2$	(g)	−136.11	232.88	−105.48	134.56
H$_2$O$_2$	(l)	−187.78	109.62	−120.42	119.62
HOF	(g)	−98.32	226.65	−85.65	287.27
H$_3$PO$_4$	(l)	−1254.36	150.62	−1111.69	111.63
H$_3$PO$_4$	(s)	−1266.92	110.54	−1112.53	110.79
HReO$_4$	(s)	−762.32	158.16	−664.84	20.67
H$_2$S	(g)	−20.17	205.77	−33.05	812.36
HSO$_3$F	(g)	−753.12	297.06	−690.36	295.78
H$_2$Se	(g)	29.71	218.91	15.90	599.51
H$_2$Se	(g)	29.71	218.91	15.90	599.51
H$_2$SiO$_3$	(s)	−1188.67	133.89	−1092.44	4.55
H$_4$SiO$_4$	(s)	−1481.14	192.46	−1333.02	2.05
H$_2$SO$_4$	(l)	−814.00	156.90	−690.07	163.19
H$_2$SO$_4$	(g)	−740.57	289.11	−656.05	197.21
H$_2$WO$_4$	(s)	−1131.77	146.44	−1004.16	68.30
H$_2$WO$_4$	(g)	−905.42	351.46	−839.73	232.73

TABLE 7.7 Chemical exergy (e^{ch}) of other chemicals, including inorganic chemicals, organometallic chemicals, oxides of carbon, and sulfides of carbon. Inorganic chemicals that are found in other tables in this book, such as in reference states (Chap. 4), atmosphere gases (Chap. 5), water (Chap. 6), and molecular hydrogen (Table 7.8) are not duplicated here. The enthalpy of formation (h^f), Gibbs free energy of formation (Δg^f), and entropy (s) are referenced from Ref. 12. (*Continued*)

Structure	State	h^f (kJ/mol)	s (J mol/K)	Δg^f (kJ/mol)	e^{ch} (kJ/mol)
Hg—Mercury					
$HgBr_2$	(s)	−170.71	170.33	−153.13	55.77
Hg_2Br_2	(s)	−206.90	218.74	−181.08	135.72
$HgCl$	(g)	84.10	259.78	62.76	232.51
Hg_2Cl_2	(s)	−265.22	192.46	−210.78	128.72
$Hg(CH_3)_2$	(l)	59.83	209.20	140.16	1776.94
$Hg(CH_3)_2$	(g)	94.39	305.43	146.02	1782.80
Hg_2CO_3	(s)	−553.54	179.91	−468.19	163.77
HgF_2	(s)	−422.58	116.32	−372.38	241.32
Hg_2F_2	(s)	−485.34	158.99	−426.77	294.83
HgH	(g)	239.99	219.49	216.02	441.97
HgI	(g)	132.38	281.42	88.45	284.20
HgI_2	(g)	−17.15	336.02	−59.83	223.77
HgI_2	(s red)	−105.44	181.17	−101.67	181.93
Hg_2I_2	(s)	−121.34	242.67	−111.00	280.50
$Hg_2(N_3)_2$	(s)	594.13	205.02	746.43	964.23
HgO	(s yellow)	−90.46	71.13	−58.43	51.44
HgO	(s red hexagonal)	−89.54	71.13	−58.24	51.62
HgO	(s red orthorhombic)	−90.83	70.29	−58.56	51.31
HgS	(s red)	−58.16	82.42	−50.63	666.57
HgS	(s black)	−53.56	88.28	−47.70	669.50
Hg_2SO_4	(s)	−743.12	200.66	−625.88	207.07
$HgSe$	(g)	75.73	267.02	31.38	486.78
$HgSe$	(s)	−46.02	94.14	−38.07	417.33
$HgTe$	(s)	−33.89	106.69	−28.03	409.17
I—Iodine					
I	(g)	106.84	180.68	70.28	158.13
IBr	(g)	40.84	258.66	3.72	142.07
ICl	(l)	−23.89	135.14	−13.60	136.10
ICl	(g)	17.78	247.44	−5.44	144.26
ICl_3	(s)	−89.54	167.36	−22.34	251.06
IF	(g)	−95.65	236.06	−118.49	222.26
IF_5	(g)	−840.15	334.72	−771.53	580.82
IF_7	(g)	−943.91	346.44	−818.39	1039.76
IO	(g)	175.06	245.39	149.79	239.60
K—Potassium					
$KAlCl_4$	(s)	−1196.62	196.65	−1096.21	313.59
K_3AlCl_6	(s)	−2092.00	376.56	−1938.45	328.45

TABLE 7.7 Chemical exergy (e^{ch}) of other chemicals, including inorganic chemicals, organometallic chemicals, oxides of carbon, and sulfides of carbon. Inorganic chemicals that are found in other tables in this book, such as in reference states (Chap. 4), atmosphere gases (Chap. 5), water (Chap. 6), and molecular hydrogen (Table 7.8) are not duplicated here. The enthalpy of formation (h^f), Gibbs free energy of formation (Δg^f), and entropy (s) are referenced from Ref. 12. (*Continued*)

Structure	State	h^f (kJ/mol)	s (J mol/K)	Δg^f (kJ/mol)	e^{ch} (kJ/mol)
		K—Potassium			
$KAl(SO_4)_2$	(s)	−2465.38	204.60	−2235.47	161.23
KBF_4	(s)	−1886.98	133.89	−1784.89	221.51
KBH_4	(s)	−226.77	106.61	−159.83	1307.19
KBO_2	(s)	−994.96	80.00	−978.64	20.09
$K_2B_4O_7$	(s)	−3334.23	208.36	−3136.74	122.79
KBr	(s)	−392.17	96.44	−379.20	38.00
$KBrO_3$	(s)	−332.21	149.16	−243.51	179.58
K_2CO_3	(s)	−1150.18	155.52	−1064.41	85.15
KCl	(s)	−435.89	82.68	−408.32	20.23
KCl	(g)	−215.89	239.49	−235.14	193.41
$KClO_3$	(s)	−391.20	142.97	−289.91	144.53
$KClO_4$	(s)	−430.12	151.04	−300.41	135.99
KCN	(s)	−113.47	127.78	−102.05	675.26
KF	(s)	−568.61	66.57	−538.90	80.70
$KF \cdot 2H_2O$	(s)	−1158.97	150.62	−1015.46	80.29
KH	(s)	−57.82	50.21	−34.06	450.70
KH_2AsO_4	(s)	−1135.96	155.14	−991.61	111.65
KHF_2	(s)	−931.36	104.27	−863.16	127.40
KI	(s)	−327.65	106.40	−322.29	132.26
KIO_3	(s)	−508.36	151.46	−425.51	34.92
$KMnO_4$	(s)	−813.37	171.71	−713.79	148.46
KO_2	(s)	−284.51	122.59	−240.58	130.04
K_2O	(s)	−363.17	94.14	−322.17	413.19
K_2O_2	(s)	−495.80	112.97	−429.70	307.63
KOH	(s)	−425.85	78.87	−379.07	107.65
K_2SO_4	(s)	−1433.69	175.73	−1316.37	34.18
K_2SiO_3	(s)	−1548.08	146.15	−1455.61	138.67
		Li—Lithium			
$LiAlF_4$	(g)	−1853.51	326.35	−1811.67	388.33
Li_3AlF_6	(s)	−3383.60	187.86	−3223.77	267.43
$LiAlH_4$	(s)	−117.15	87.86	−48.53	1612.09
$LiAlO_2$	(s)	−1189.51	53.35	−1127.17	65.16
$LiBH_4$	(s)	−190.46	75.81	−124.77	1368.25
$LiBO_2$	(s)	−1019.22	51.71	−963.16	61.57
$Li_2B_4O_7$	(s)	−3363.94	155.64	−3171.47	140.06
$LiBeF_3$	(s)	−1651.84	89.12	−1576.11	179.59
Li_2BeF_4	(s)	−2273.59	130.54	−2171.50	229.80
$LiBr$	(s)	−350.91	74.06	−341.62	101.58

TABLE 7.7 Chemical exergy (e^{ch}) of other chemicals, including inorganic chemicals, organometallic chemicals, oxides of carbon, and sulfides of carbon. Inorganic chemicals that are found in other tables in this book, such as in reference states (Chap. 4), atmosphere gases (Chap. 5), water (Chap. 6), and molecular hydrogen (Table 7.8) are not duplicated here. The enthalpy of formation (h^f), Gibbs free energy of formation (Δg^f), and entropy (s) are referenced from Ref. 12. (*Continued*)

Structure	State	h^f (kJ/mol)	s (J mol/K)	Δg^f (kJ/mol)	e^{ch} (kJ/mol)
Li—Lithium					
Li_2CO_3	(s)	−1216.04	90.17	−1132.19	69.37
LiCl	(s)	−408.27	59.29	−384.05	70.50
$LiCl \cdot H_2O$	(s)	−712.58	103.76	−632.62	60.00
$LiClO_4$	(s)	−380.74	125.52	−253.97	208.43
LiF	(s)	−616.93	35.65	−588.69	56.91
LiH	(s)	−90.63	20.04	−68.45	442.30
LiI	(s)	−270.08	85.77	−269.66	210.89
Li_3N	(s)	−197.48	37.66	−153.97	1024.46
LiO	(g)	83.68	210.87	60.46	455.12
Li_2O	(s)	−598.73	37.91	−561.91	225.45
Li_2O_2	(s)	−632.62	56.48	−571.12	218.21
LiOH	(s)	−484.93	42.80	−438.90	73.82
$LiOH \cdot H_2O$	(s)	−789.81	92.05	−689.52	61.27
$Li_2Si_2O_5$	(s)	−2561.03	125.52	−2417.10	88.11
Li_2SiO_3	(s)	−1649.33	80.33	−1558.96	87.33
Li_2TiO_3	(s)	−1670.67	91.76	−1579.88	118.61
Mg—Magnesium					
$MgAl_2O_4$	(s)	−2312.92	88.70	−2190.32	35.83
$Mg_2Al_4Si_5O_{19}$	(s cordierite)	−9108.57	407.10	−8598.12	150.76
$MgBr_2$	(s)	−524.26	117.15	−503.75	224.15
$MgBr_2 \cdot 6H_2O$	(s)	−2409.98	397.48	−2056.02	100.32
$MgCO_3$	(s)	−1095.79	65.69	−1012.11	30.95
$MgCl_2$	(s)	−641.62	89.62	−592.12	158.48
$MgCl_2 \cdot 6H_2O$	(s)	−2499.02	366.10	−2114.97	64.06
$MgCl_2 \cdot 6H_2O$	(s)	−1279.72	179.91	−1118.13	1060.90
$Mg(ClO_4)_2 \cdot 6H_2O$	(s)	−2445.55	520.91	−1863.14	331.60
$MgCr_2O_4$	(s)	−1783.64	106.02	−1669.00	134.55
MgF_2	(s)	−1124.24	57.24	−1071.10	61.60
$MgFe_2O_4$	(s)	−1428.42	123.85	−1317.12	66.23
Mg_2Ge	(s)	−108.78	86.48	−105.86	1705.64
MgH_2	(s)	−75.31	31.09	−35.98	827.03
MgI_2	(s)	−364.01	129.70	−358.15	444.45
$MgMoO_4$	(s)	−1400.85	118.83	−1295.74	70.31
Mg_3N_2	(s)	−460.66	87.86	−400.83	1480.54
$Mg(NO_3)_2$	(s)	−790.65	164.01	−589.53	49.82
MgO	(s microcrystal)	−597.98	27.91	−565.97	62.89
MgO	(s periclase)	−601.66	26.94	−569.02	59.84

TABLE 7.7 Chemical exergy (e^{ch}) of other chemicals, including inorganic chemicals, organometallic chemicals, oxides of carbon, and sulfides of carbon. Inorganic chemicals that are found in other tables in this book, such as in reference states (Chap. 4), atmosphere gases (Chap. 5), water (Chap. 6), and molecular hydrogen (Table 7.8) are not duplicated here. The enthalpy of formation (h^f), Gibbs free energy of formation (Δg^f), and entropy (s) are referenced from Ref. 12. (*Continued*)

Structure	State	h^f (kJ/mol)	s (J mol/K)	Δg^f (kJ/mol)	e^{ch} (kJ/mol)
		Mg—Magnesium			
$Mg(OH)_2$	(s)	−924.66	63.18	−833.87	33.06
MgS	(s)	−346.02	50.33	−341.83	894.37
$MgSO_4$	(s)	−1284.91	91.63	−1170.68	73.37
$MgSO_4 \cdot 6H_2O$	(s)	−3086.96	348.11	−2632.15	40.33
$MgSO_4 \cdot 7H_2O$	(s)	−3388.71	372.38	−2871.90	38.66
$MgSO_4 \cdot H_2O$	(s)	−1602.05	126.36	−1428.84	53.29
Mg_2Si	(s)	−77.82	66.94	−75.31	2033.49
$MgSiO_3$	(s clinoenstatite)	−1549.00	67.78	−1462.14	25.65
Mg_2SiO_4	(s forsterite)	−2174.01	95.14	−2055.18	61.47
$Mg_3Si_2O_5(OH)_4$	(s chrysotile)	−4365.59	221.33	−4037.98	42.60
$Mg_3(PO_4)_2$	(s)	−3780.66	189.20	−3538.83	80.17
$MgTi_2O_5$	(s)	−2509.56	127.28	−2366.89	84.22
$MgTiO_3$	(s)	−1572.77	74.56	−1484.06	55.92
Mg_2TiO_4	(s)	−2164.38	109.33	−2047.65	121.20
$MgWO_4$	(s)	−1532.60	101.17	−1420.89	42.36
$Mg(VO_3)_2$	(s)	−2201.58	160.67	−2039.41	41.87
$Mg_2V_2O_7$	(s)	−2835.92	200.41	−2645.29	64.84
		Mn—Manganese			
Mn_3C	(s)	4.60	98.74	5.44	1878.81
$MnCO_3$	(s natural)	−894.12	85.77	−816.72	87.14
$MnCO_3$	(s precipitated)	−882.82	112.97	−811.70	92.16
$MnC_2O_4 \cdot 2H_2O$	(s)	−1628.41	200.83	−1415.03	377.21
$MnCl_2$	(s)	−481.29	118.24	−440.53	170.87
$MnCl_2 \cdot 2H_2O$	(s)	−1092.02	218.82	−942.24	145.31
$MnCl_2 \cdot 4H_2O$	(s)	−1687.41	303.34	−1423.82	139.87
$MnCl_2 \cdot H_2O$	(s)	−789.94	174.05	−696.22	153.25
MnF_2	(s)	−790.78	−748.94	92.26	1085.76
$Mn(IO_3)_2$	(s)	−669.44	263.59	−520.49	154.68
MnO	(s)	−385.22	59.71	−362.92	126.74
Mn_2O_3	(s)	−958.97	110.46	−881.15	100.14
Mn_3O_4	(s)	−1387.83	155.64	−1283.23	187.72
$Mn(OH)_2$	(s precipitated/amorphous)	−695.38	99.16	−615.05	112.69
MnS	(s green)	−214.22	78.24	−218.40	878.60
$MnSO_4$	(s)	−1065.25	112.13	−957.42	147.42
$MnSe$	(s)	−106.69	90.79	−111.71	723.49
$MnSiO_3$	(s)	−1320.89	89.12	−1240.56	108.03
Mn_2SiO_4	(s)	−1730.50	163.18	−1632.18	206.07

TABLE 7.7 Chemical exergy (e^{ch}) of other chemicals, including inorganic chemicals, organometallic chemicals, oxides of carbon, and sulfides of carbon. Inorganic chemicals that are found in other tables in this book, such as in reference states (Chap. 4), atmosphere gases (Chap. 5), water (Chap. 6), and molecular hydrogen (Table 7.8) are not duplicated here. The enthalpy of formation (h^f), Gibbs free energy of formation (Δg^f), and entropy (s) are referenced from Ref. 12. (*Continued*)

Structure	State	h^f (kJ/mol)	s (J mol/K)	Δg^f (kJ/mol)	e^{ch} (kJ/mol)
		N—Nitrogen			
N	(g)	472.70	153.19	455.58	455.91
NF_3	(g)	−131.38	260.66	−89.96	669.08
N_2F_2	(g trans)	81.17	262.55	120.50	626.97
N_2F_2	(g cis)	66.94	259.83	108.78	615.25
NH	(g imidogen)	377.23	181.13	371.25	489.64
NH_2	(g amidogen)	190.37	194.60	199.83	436.27
NH_3	(g)	−46.11	192.34	−16.48	338.01
NH_4	(s carbamate)	−645.05	133.47	−448.06	24.49
N_2H_2	(g trans)	182.42	220.12	212.97	449.75
N_2H_2	(g cis-diimide)	213.38	218.40	243.09	479.87
N_2H_4	(g)	95.40	238.36	159.28	632.17
N_2H_4	(l)	50.63	121.21	149.24	622.13
$NH_4Al(SO_4)_2$	(s)	−2352.24	216.31	−2038.44	464.11
NH_4Br	(s)	−270.83	112.97	−175.31	347.75
NH_4Cl	(s)	−314.43	94.56	−202.97	331.44
NH_4ClO_4	(s)	−295.31	184.18	−88.91	453.34
NH_4F	(s)	−463.96	71.96	−348.78	376.68
$NH_4H_2AsO_4$	(s)	−2189.49	172.05	−833.03	376.08
$NH_4H_2PO_4$	(s)	−1445.07	151.96	−1210.56	367.26
NH_4HCO_3	(s)	−849.35	120.92	−666.09	340.68
NH_4HF_2	(s)	−802.91	115.52	−651.03	445.38
NH_4HS	(s)	−156.90	97.49	−50.63	1149.28
NH_4HSe	(s)	−133.05	96.65	−23.43	914.68
NH_4I	(s)	−201.42	117.15	−112.55	447.86
NH_4N_3	(s)	115.48	112.55	274.05	747.61
NH_4NO_3	(s)	−365.56	151.08	−184.01	294.76
$(NH_4)_2O$	(l)	−430.70	267.52	−267.11	679.97
NH_4OH	(l)	−361.20	165.56	−254.14	338.44
NH_4ReO_4	(s)	−945.58	232.63	−774.88	265.13
$(NH_4)_2SiF_6$	(s hexagonal)	−2681.69	280.24	−2365.55	951.96
$(NH_4)_2SO_4$	(s)	−1180.85	220.08	−901.90	660.36
NH_4VO_3	(s)	−1053.11	140.58	−888.26	311.48
NO	(g)	90.25	210.65	86.57	88.86
NO_2	(g)	33.18	239.95	51.30	55.56
NO_3	(g)	70.92	252.55	114.47	120.70
N_2O	(g)	82.05	219.74	104.18	106.81
N_2O_2	(g)	170.37	287.52	202.88	207.48

TABLE 7.7 Chemical exergy (e^{ch}) of other chemicals, including inorganic chemicals, organometallic chemicals, oxides of carbon, and sulfides of carbon. Inorganic chemicals that are found in other tables in this book, such as in reference states (Chap. 4), atmosphere gases (Chap. 5), water (Chap. 6), and molecular hydrogen (Table 7.8) are not duplicated here. The enthalpy of formation (h^f), Gibbs free energy of formation (Δg^f), and entropy (s) are referenced from Ref. 12. *(Continued)*

Structure	State	h^f (kJ/mol)	s (J mol/K)	Δg^f (kJ/mol)	e^{ch} (kJ/mol)
N—Nitrogen					
N_2O_3	(g)	83.72	312.17	139.41	145.97
N_2O_4	(l)	−19.58	209.24	97.40	105.92
N_2O_4	(g)	9.16	304.18	97.82	106.34
N_2O_4	(s)	−35.02	150.29	99.54	108.06
N_2O_5	(s)	11.30	347.19	117.70	128.18
NOBr	(g)	82.17	273.55	82.42	135.22
NOCl	(g)	51.71	261.58	66.07	130.21
NO_2Cl	(g)	12.55	272.04	54.39	120.50
NOF	(g)	−65.69	247.99	−50.29	204.91
NO_2F	(g)	−79.50	260.24	−37.24	219.92
NOF_3	(g)	−163.18	278.40	−96.23	664.77
Na—Sodium					
$NaAlCl_4$	(s)	−1142.23	188.28	−1041.82	337.98
Na_3AlCl_6	(s)	−1979.03	347.27	−1828.41	348.49
Na_3AlF_6	(s cryolite)	−3309.54	238.49	−3142.18	181.02
Na_3AlF_6	(l)	−3238.42	286.60	−3088.21	234.99
$NaAlO_2$	(s)	−1133.03	70.42	−1069.43	66.89
$NaBH_4$	(s)	−191.84	101.38	−127.11	1309.91
$NaBO_2$	(s)	−975.71	73.51	−919.22	49.50
$Na_2B_4O_7$	(s)	−3276.07	189.54	−3083.61	115.93
NaBr	(s)	−361.41	86.82	−349.28	37.92
Na_2CO_3	(s)	−1130.94	135.98	−1047.67	41.89
NaCl	(s)	−410.99	72.38	−384.05	14.50
$NaClO_4$	(s)	−382.75	142.26	−254.35	152.05
NaF	(s)	−575.30	51.21	−545.18	44.42
NaH	(s)	−56.44	40.00	−33.56	421.20
NaH	(g)	125.02	187.99	103.68	558.43
$NaHCO_3$	(s)	−947.68	102.09	−851.86	19.05
NaI	(s)	−288.03	98.32	−284.51	140.04
Na_2O	(s)	−415.89	72.80	−376.56	298.80
Na_2O_2	(s)	−513.38	94.81	−449.78	227.54
NaOH	(l)	−416.89	75.86	−374.13	82.58
NaOH	(s)	−426.73	64.43	−379.07	77.65
$NaOH \cdot H_2O$	(s)	−732.91	84.52	−623.42	71.37
Na_2S	(s)	−373.21	97.91	−359.82	922.88
Na_2SO_3	(s)	−1090.35	146.02	−1002.07	286.52
Na_2SO_4	(s)	−1384.49	149.49	−1266.83	23.72
Na_2SiO_3	(s)	−1518.79	113.80	−1426.74	107.54
Na_2WO_4	(s)	−1543.90	160.25	−1430.93	78.82

TABLE 7.7 Chemical exergy (e^{ch}) of other chemicals, including inorganic chemicals, organometallic chemicals, oxides of carbon, and sulfides of carbon. Inorganic chemicals that are found in other tables in this book, such as in reference states (Chap. 4), atmosphere gases (Chap. 5), water (Chap. 6), and molecular hydrogen (Table 7.8) are not duplicated here. The enthalpy of formation (h^f), Gibbs free energy of formation (Δg^f), and entropy (s) are referenced from Ref. 12. (*Continued*)

Structure	State	h^f (kJ/mol)	s (J mol/K)	Δg^f (kJ/mol)	e^{ch} (kJ/mol)
Ni—Nickel					
$Ni(CO)_4$	(g)	−602.91	410.45	−587.27	1304.27
$Ni(CO)_4$	(l)	−633.04	313.38	−588.27	1303.27
$NiCl_2$	(s)	−305.33	97.65	−259.06	107.24
$NiCl_2 \cdot 2H_2O$	(s)	−922.15	175.73	−760.23	82.21
$NiCl_2 \cdot 4H_2O$	(s)	−1516.70	242.67	−1235.12	83.47
$NiCl_2 \cdot 6H_2O$	(s)	−2103.17	344.34	−1713.52	81.22
NiF_2	(s)	−651.45	73.60	−604.17	144.23
$Ni(IO_3)_2$	(s)	−489.11	213.38	−326.35	103.72
NiS	(s)	−82.01	52.97	−79.50	772.40
Ni_3S_2	(s)	−202.92	133.89	−197.07	1749.33
$NiSO_4$	(s)	−872.91	97.07	−759.81	99.93
$NiSO_4 \cdot 6H_2O$	(s tetrahedral)	−2682.82	332.17	−2224.97	63.22
$NiSO_4 \cdot 7H_2O$	(s)	−2976.33	378.94	−2462.24	64.01
O—Oxygen					
O	(g)	249.17	160.95	231.75	233.71
O_3	(g)	142.67	238.82	163.18	169.06
OF	(g)	124.26	217.74	120.12	374.98
OF_2	(g)	24.52	247.32	41.76	549.52
O_2F_2	(g)	19.79	268.11	61.42	571.15
P—Phosphorous					
P_2	(g)	146.19	218.03	103.76	1826.36
P_4	(g)	128.87	128.87	72.38	3517.58
PBr_3	(g)	−139.33	347.98	−162.76	850.04
PBr_3	(l)	−184.51	240.16	−175.73	837.07
PCl_3	(l)	−319.66	217.15	−272.38	774.47
PCl_3	(g)	−287.02	311.67	−267.78	779.07
PCl_5	(g)	−342.67	364.47	−278.24	892.31
PF_3	(g)	−918.81	273.13	−897.47	722.53
PF_5	(g)	−1576.95	300.83	−1508.75	617.05
PH	(g)	255.22	196.23	221.75	1201.11
PH_3	(g)	23.01	210.20	25.52	1240.99
P_2H_4	(l)	−5.02	167.36	66.94	2261.76
PH_4Br	(s)	−127.61	110.04	−47.70	1336.32
PH_4I	(s)	−69.87	123.01	0.84	1422.21
PN	(g)	32.47	211.08	10.33	871.97
PO	(g)	−12.13	222.68	−41.17	822.09
P_4O_{10}	(s hexagonal)	−2940.10	228.86	−2675.25	789.57
$POBr_3$	(g)	−389.11	359.70	−390.91	623.85

TABLE 7.7 Chemical exergy (e^{ch}) of other chemicals, including inorganic chemicals, organometallic chemicals, oxides of carbon, and sulfides of carbon. Inorganic chemicals that are found in other tables in this book, such as in reference states (Chap. 4), atmosphere gases (Chap. 5), water (Chap. 6), and molecular hydrogen (Table 7.8) are not duplicated here. The enthalpy of formation (h^f), Gibbs free energy of formation (Δg^f), and entropy (s) are referenced from Ref. 12. (*Continued*)

Structure	State	h^f (kJ/mol)	s (J mol/K)	Δg^f (kJ/mol)	e^{ch} (kJ/mol)
P—Phosphorous					
$POCl_2F$	(g)	−748.94	320.29	−711.28	528.58
$POCl_3$	(g)	−542.25	325.35	−502.50	546.31
$POCl_3$	(l)	−597.06	222.46	−520.91	527.90
$POClF_2$	(g)	−953.95	301.58	−912.11	518.80
POF_3	(g)	−1236.79	285.31	−1193.70	428.27
P_4S_3	(g)	−81.17	319.16	−120.50	5152.60
P_4S_3	(s)	−154.81	200.83	−158.99	5114.11
P_4S_3	(l)	−151.04	207.11	−156.90	5116.20
$PSBr_3$	(g)	−263.59	372.71	−288.70	1333.40
$PSCl_3$	(g)	−363.17	337.23	−347.69	1308.46
PSF_3	(g)	−991.61	298.03	−973.62	1255.68
Pb—Lead					
PbB_2O_4	(s)	−1556.45	130.54	−1450.17	63.07
PbB_4O_7	(s)	−2857.67	166.94	−2667.30	108.04
$PbBr$	(g)	71.13	272.38	31.80	331.50
$PbBr_2$	(g)	−104.39	339.28	−140.83	209.37
$PbBr_2$	(s)	−277.40	161.13	−260.75	89.45
$PbBr_2$	(l)	−267.40	173.89	−254.55	95.65
PbC_2O_4	(s)	−851.44	146.02	−750.19	327.40
$PbCl$	(g)	15.06	259.49	−9.62	301.43
$PbCl_2$	(s)	−359.41	135.98	−314.18	58.72
$PbCl_2$	(l)	−344.26	153.18	−304.22	68.68
$PbCl_2$	(g)	−174.05	317.11	−182.80	190.10
$PbClF$	(s)	−534.72	121.75	−488.27	75.68
PbF	(g)	−80.33	249.83	−105.02	397.08
PbF_2	(s alpha)	−676.97	112.97	−630.95	124.05
PbF_2	(s beta)	−676.13	114.43	−631.16	123.84
PbF_4	(g)	−1133.45	333.51	−1092.69	168.11
PbI_2	(l)	−157.69	198.91	−161.88	263.02
PbI_2	(s)	−175.39	174.85	−173.59	251.31
PbI_4	(g)	−224.47	466.14	−274.89	325.71
$Pb(IO_3)_2$	(s)	−495.39	312.96	−351.46	85.22
$PbMoO_4$	(s)	−1051.86	166.10	−951.44	36.91
$Pb(N_3)_2$	(s monoclinic)	478.23	148.11	624.67	875.88
$Pb(N_3)_2$	(s orthorhombic)	476.14	149.37	622.16	873.37
PbO	(s red)	−218.99	66.53	−189.24	61.92
PbO	(s yellow)	−218.07	68.70	−188.66	62.51
$PbO \cdot PbCO_3$	(s)	−918.39	204.18	−816.72	99.80

TABLE 7.7   Chemical exergy (e^{ch}) of other chemicals, including inorganic chemicals, organometallic chemicals, oxides of carbon, and sulfides of carbon. Inorganic chemicals that are found in other tables in this book, such as in reference states (Chap. 4), atmosphere gases (Chap. 5), water (Chap. 6), and molecular hydrogen (Table 7.8) are not duplicated here. The enthalpy of formation (h^f), Gibbs free energy of formation (Δg^f), and entropy (s) are referenced from Ref. 12. (*Continued*)

Structure	State	h^f (kJ/mol)	s (J mol/K)	Δg^f (kJ/mol)	e^{ch} (kJ/mol)
Pb—Lead					
PbO_2	(s)	−274.47	71.80	−215.48	37.65
Pb_3O_4	(s)	−718.81	212.13	−601.66	153.79
$Pb_3(PO_4)_2$	(s)	−2595.34	353.13	−2432.58	53.32
PbS	(s)	−98.32	91.34	−96.73	761.77
$PbSO_4$	(s)	−919.94	148.57	−813.20	53.15
$PbSe$	(s)	−102.93	102.51	−101.67	495.03
$PbSeO_4$	(s)	−609.19	167.78	−505.01	99.54
$PbSiO_3$	(s)	−1145.16	110.04	−1061.06	49.02
Pb_2SiO_4	(s)	−1376.54	187.02	−1267.75	93.50
$PbTe$	(s)	−70.71	110.04	−69.45	509.05
Ra–Radium					
$RaCl_2 \cdot 2H_2O$	(s)	−1464.40	213.38	−1302.90	121.21
$Ra(IO_3)_2$	(s)	−1026.75	271.96	−868.60	143.14
$Ra(NO_3)_2$	(s)	−991.61	221.75	−796.22	40.49
Rb—Rubidium					
$RbBr$	(s)	−389.24	104.93	−378.15	61.05
$RbCl$	(s)	−435.05	94.56	−412.04	38.51
$RbClO_3$	(s)	−392.46	151.88	−292.04	164.39
$RbClO_4$	(s)	−434.59	160.67	−306.23	152.17
RbI	(s)	−328.44	118.03	−325.52	151.03
S—Sulfur					
S_2	(g)	129.03	228.07	80.08	1298.68
S_8	(g)	101.25	430.20	49.16	4923.56
S_2Cl_2	(g)	−19.50	319.45	−29.25	1313.05
SF_2Cl	(g)	−1048.09	319.07	−949.35	227.60
SF_4	(g)	−728.43	291.12	−684.84	936.06
SF_6	(g)	−1220.89	291.71	−1115.87	1010.83
SO	(g)	4.88	221.84	−21.17	590.09
SO_2	(g)	−296.83	248.11	−300.19	313.03
SO_3	(g)	−395.72	256.65	−371.08	244.11
SO_3	(l)	−441.04	95.60	−368.36	246.83
SO_3	(s beta)	−454.51	52.30	−368.99	246.20
$SOCl_2$	(g)	−212.55	309.66	−198.32	536.64
SO_2Cl_2	(g)	−354.80	311.83	−310.45	426.47
SO_2F_2	(g)	−758.56	283.93	−712.12	406.91

TABLE 7.7 Chemical exergy (e^{ch}) of other chemicals, including inorganic chemicals, organometallic chemicals, oxides of carbon, and sulfides of carbon. Inorganic chemicals that are found in other tables in this book, such as in reference states (Chap. 4), atmosphere gases (Chap. 5), water (Chap. 6), and molecular hydrogen (Table 7.8) are not duplicated here. The enthalpy of formation (h^f), Gibbs free energy of formation (Δg^f), and entropy (s) are referenced from Ref. 12. (*Continued*)

Structure	State	h^f (kJ/mol)	s (J mol/K)	Δg^f (kJ/mol)	e^{ch} (kJ/mol)
Si—Silicon					
Si_2	(g)	594.13	229.79	535.55	2245.55
$SiBr_4$	(l)	−457.31	277.82	−443.92	613.08
$SiBr_4$	(g)	−415.47	377.77	−431.79	625.21
SiC	(s beta cubic)	−73.22	16.61	−70.71	1194.56
SiC	(s alpha hexagonal)	−71.55	16.48	−69.04	1196.24
SiC	(g)	615.05	236.61	552.29	1817.56
$SiCl_2$	(g)	−165.64	280.33	−177.19	801.51
$SiCl_4$	(g)	−662.75	330.62	−622.58	479.82
$SiCl_4$	(l)	−687.01	239.74	619.90	1722.30
SiF	(g)	7.11	225.68	−24.27	1083.63
SiF_2	(g)	−587.85	256.81	−598.31	762.49
SiF_4	(g)	−1614.94	282.38	−1572.68	293.92
SiH_4	(g)	30.54	204.51	56.90	1384.12
Si_2H_6	(g)	80.33	272.55	127.19	2545.52
SiH_3Cl	(g)	−200.83	250.54	−179.91	1091.10
$SiHCl_3$	(l)	−539.32	227.61	−482.58	676.02
SiH_3F	(g)	−439.32	238.28	−418.40	1043.66
SiN	(g)	486.52	216.65	456.10	1311.43
Si_3N_4	(s)	−743.50	112.97	−581.58	1984.76
$SiOF_2$	(g)	−966.50	271.17	−949.77	412.99
SiS	(g)	112.47	223.55	60.92	1525.22
SiS_2	(s)	−213.38	80.33	−212.55	1861.05
Sn—Tin					
$SnBr_4$	(g)	−314.64	411.83	−331.37	422.43
$SnBr_4$	(s)	−377.40	264.43	−350.20	403.60
$SnCl_4$	(l)	−511.28	258.57	−440.16	359.04
SnH_4	(g)	162.76	227.57	188.28	1212.30
SnO	(s)	−285.77	56.48	−256.90	296.86
SnS	(s)	−100.42	76.99	−98.32	1062.78
Sr—Strontium					
$Sr_3(AsO_4)_2$	(s)	−3317.08	255.22	−3080.26	170.04
$SrBr_2$	(s)	−717.97	143.43	−699.77	151.03
$SrBr_2 \cdot 6H_2O$	(s)	−2531.32	405.85	−2174.42	104.81
$Sr(BrO_3)_2 \cdot H_2O$	(s)	−1104.58	280.33	−791.19	309.45
$SrCl_2$	(s)	−828.85	114.85	−781.03	92.47
$SrCl_2 \cdot 2H_2O$	(s)	−1438.04	217.57	−1281.98	67.67
$SrCl_2 \cdot 6H_2O$	(s)	−2623.79	390.79	−2241.24	60.69

TABLE 7.7 Chemical exergy (e^{ch}) of other chemicals, including inorganic chemicals, organometallic chemicals, oxides of carbon, and sulfides of carbon. Inorganic chemicals that are found in other tables in this book, such as in reference states (Chap. 4), atmosphere gases (Chap. 5), water (Chap. 6), and molecular hydrogen (Table 7.8) are not duplicated here. The enthalpy of formation (h^f), Gibbs free energy of formation (Δg^f), and entropy (s) are referenced from Ref. 12. (*Continued*)

Structure	State	h^f (kJ/mol)	s (J mol/K)	Δg^f (kJ/mol)	e^{ch} (kJ/mol)
		Sr—Strontium			
SrF_2	(s)	−1217.13	82.13	−1165.58	90.02
$SrHPO_4$	(s)	−1821.71	121.34	−1688.66	48.34
SrI_2	(s)	−561.49	159.12	−558.73	366.77
$Sr(IO_3)_2$	(s)	−1019.22	234.30	−855.21	82.06
$Sr(IO_3)_2 \cdot 6H_2O$	(s)	−2789.89	456.06	−2274.84	90.87
SrO	(s)	−592.04	55.52	−562.41	189.35
SrS	(s)	−453.13	68.20	−448.52	910.58
$SrSO_4$	(s)	−1453.10	117.57	−1340.97	25.98
$SrSiO_3$	(s)	−1633.85	96.65	−1549.75	60.93
Sr_2SiO_4	(s)	−2304.55	153.13	−2191.16	171.29
$SrTiO_3$	(s)	−1672.39	108.78	−1588.41	74.47
Sr_2TiO_4	(s)	−2287.39	158.99	−2178.61	236.04
$SrWO_4$	(s)	−1639.71	138.07	−1531.34	54.81
$SrZrO_3$	(s)	−1767.32	115.06	−1682.80	155.88
		Ti—Titanium			
TiB	(s)	−160.25	34.73	−159.83	1375.47
TiB_2	(s)	−280.33	28.45	−271.96	1891.44
$TiBr_2$	(s)	−405.85	108.37	−383.25	624.95
$TiBr_3$	(s)	−550.20	176.44	−525.51	533.19
$TiBr_4$	(s)	−617.98	243.63	−590.78	518.42
$TiBr_4$	(g)	−550.20	398.94	−569.02	540.18
$TiBr_4$	(l)	−605.42	284.18	−589.94	519.26
TiC	(s)	−184.10	24.23	−180.33	1137.14
$TiCl_2$	(s)	−513.80	87.45	−464.42	566.48
$TiCl_3$	(s)	−720.90	139.75	−653.54	439.21
$TiCl_4$	(g)	−763.16	354.80	−726.76	427.84
$TiCl_4$	(l)	−804.16	252.34	−737.22	417.38
$TiCl_4$	(s)	−815.04	208.78	−735.13	419.47
TiF_2	(g)	−686.18	255.22	−694.54	718.46
TiF_3	(s)	−1435.11	87.86	−1362.31	303.59
TiF_4	(g)	−1551.43	314.64	−1515.44	403.36
TiF_4	(s)	−1649.33	133.97	−1559.38	359.42
TiH_2	(s)	−144.35	29.71	−105.02	1038.29
TiI_2	(s)	−267.78	121.34	−259.41	823.49
TiI_3	(s)	−322.17	192.46	−317.98	852.77
TiI_4	(l)	−348.32	311.83	−362.92	895.68

TABLE 7.7 Chemical exergy (e^{ch}) of other chemicals, including inorganic chemicals, organometallic chemicals, oxides of carbon, and sulfides of carbon. Inorganic chemicals that are found in other tables in this book, such as in reference states (Chap. 4), atmosphere gases (Chap. 5), water (Chap. 6), and molecular hydrogen (Table 7.8) are not duplicated here. The enthalpy of formation (h^f), Gibbs free energy of formation (Δg^f), and entropy (s) are referenced from Ref. 12. (*Continued*)

Structure	State	h^f (kJ/mol)	s (J mol/K)	Δg^f (kJ/mol)	e^{ch} (kJ/mol)
Ti—Titanium					
TiI_4	(s)	−375.72	246.02	−370.70	887.90
TiN	(s)	−338.07	30.25	−309.62	597.92
TiO	(s alpha)	−542.66	34.77	−513.38	395.79
Ti_2O_3	(s)	−1520.88	78.78	−1434.28	386.01
Ti_3O_5	(s alpha)	−2459.36	129.29	−2317.52	413.89
Ti_4O_7	(s)	−3404.52	198.74	−3213.31	429.22
$TiOCl_2$	(g)	−545.59	320.91	−535.13	497.73
$TiOF_2$	(g)	−924.66	284.60	−907.93	507.03
U—Uranium					
UBr_3	(s)	−711.70	205.02	−689.10	659.00
UBr_4	(s)	−822.57	234.30	−788.68	609.92
UCl_3	(s)	−891.19	158.95	−823.83	558.32
UCl_4	(s)	−1051.02	198.32	−962.32	481.68
UCl_5	(s)	−1096.63	242.67	−993.28	512.57
UCl_6	(s)	−1139.72	285.77	−1010.44	557.26
UF_3	(s)	−1493.69	117.15	−1418.38	536.92
UF_4	(s)	−1853.51	151.04	−1761.46	446.74
UF_5	(s)	−2041.79	197.90	−1928.82	532.28
UF_6	(s)	−2112.92	227.61	−2029.24	684.76
UH_3	(s)	−127.19	63.89	−72.59	1478.17
UI_3	(s)	−479.90	234.30	−482.42	977.73
UI_4	(s)	−531.37	280.33	−527.60	1020.40
$UICl_3$	(s)	−920.06	225.94	−855.21	614.79
UN	(s)	−334.72	62.34	−313.80	883.13
U_2N_3	(s)	−891.19	121.34	−811.70	1582.51
UO_2	(s)	−1129.68	77.82	−1075.29	125.24
UO_3	(s)	−1263.57	98.62	−1184.07	18.41
$UO_2(NO_3)_2$	(s)	−1377.37	276.14	−1142.65	70.32
$UO_2(NO_3)_2 \cdot H_2O$	(s)	−1693.68	317.98	−1402.90	48.15
$UO_2(NO_3)_2 \cdot 2H_2O$	(s)	−2008.32	355.64	−1659.37	29.74
$UO_2(NO_3)_2 \cdot 3H_2O$	(s)	−2310.40	393.30	−1902.46	24.72
$UO_2(NO_3)_2 \cdot 6H_2O$	(s)	−3197.83	505.64	−2615.00	26.40
$UO_2SO_4 \cdot 3H_2O$	(s)	−2789.89	263.59	−2451.82	80.07
W—Tungsten					
WBr_5	(s)	−311.71	271.96	−269.45	811.55
WBr_6	(s)	−343.09	313.80	−288.70	842.80
WCl_2	(s)	−257.32	130.54	−220.08	732.12

TABLE 7.7   Chemical exergy (e^{ch}) of other chemicals, including inorganic chemicals, organometallic chemicals, oxides of carbon, and sulfides of carbon. Inorganic chemicals that are found in other tables in this book, such as in reference states (Chap. 4), atmosphere gases (Chap. 5), water (Chap. 6), and molecular hydrogen (Table 7.8) are not duplicated here. The enthalpy of formation (h^f), Gibbs free energy of formation (Δg^f), and entropy (s) are referenced from Ref. 12. (*Continued*)

Structure	State	h^f (kJ/mol)	s (J mol/K)	Δg^f (kJ/mol)	e^{ch} (kJ/mol)
W—Tungsten					
WCl_4	(s)	−443.50	198.32	−359.82	716.08
WCl_5	(s)	−514.63	217.57	−401.66	736.09
WCl_6	(s)	−594.13	238.49	−456.06	743.54
WF_6	(l)	−1748.49	249.37	−1631.76	714.14
WF_6	(g)	−961.06	347.69	−835.96	1509.94
WO_2	(s)	−589.69	50.54	−533.88	298.55
WO_3	(s)	−842.91	75.90	−764.12	70.26
W_3O_8	(g)	−1711.26	493.71	−1581.55	919.65
WO_2Cl_2	(s)	−780.32	200.83	−702.91	253.21
$WOCl_4$	(s)	−671.11	172.80	−549.36	528.50
$WOCl_4$	(g)	−573.21	376.98	−510.45	567.41
WOF_4	(g)	−1334.70	334.72	−1276.12	565.94
Zn—Zinc					
$ZnBr_2$	(s)	−328.65	138.49	−312.13	133.57
$ZnBr_2$	(s)	−405.85	117.15	−380.74	64.96
$ZnBr_2 \cdot 2H_2O$	(s)	−937.22	198.74	−799.56	122.28
$ZnC_2O_4 \cdot 2H_2O$	(s)	−1564.82	195.39	−1345.99	303.25
$ZnCl_2$	(s)	−415.05	108.37	−369.43	98.97
$Zn(ClO_4)_2 \cdot 6H_2O$	(s)	−2133.38	545.59	−1555.61	356.92
ZnF_2	(s)	−764.42	73.68	−713.37	137.13
ZnI_2	(s)	−208.03	161.08	−208.95	311.45
$Zn(NO_3)_2 \cdot 6H_2O$	(s)	−2306.64	456.89	−1773.14	12.44
ZnO	(s)	−348.28	43.64	−318.32	28.34
$Zn(OH)_2$	(s beta)	−641.91	81.17	−553.17	31.57
$Zn(OH)_2$	(s epsilon)	−639.06	81.59	−555.13	29.60
ZnS	(s sphalerite)	−205.98	57.74	−201.29	752.71
$ZnSO_4$	(s)	−982.82	128.03	−874.46	87.39
$ZnSO_4 \cdot 6H_2O$	(s)	−2777.46	363.59	−2324.80	65.48
$ZnSO_4 \cdot 7H_2O$	(s)	−3077.75	388.69	−2563.08	65.28
$ZnSO_4 \cdot H_2O$	(s)	−1304.49	138.49	−1132.02	67.90
$ZnSe$	(s)	163.18	83.68	163.18	855.38
$ZnSeO_3 \cdot H_2O$	(s)	−930.94	163.18	−792.87	143.29
Zn_2SiO_4	(s)		131.38	−1523.23	29.02

TABLE 7.7 Chemical exergy (e^{ch}) of other chemicals, including inorganic chemicals, organometallic chemicals, oxides of carbon, and sulfides of carbon. Inorganic chemicals that are found in other tables in this book, such as in reference states (Chap. 4), atmosphere gases (Chap. 5), water (Chap. 6), and molecular hydrogen (Table 7.8) are not duplicated here. The enthalpy of formation (h^f), Gibbs free energy of formation (Δg^f), and entropy (s) are referenced from Ref. 12. (*Continued*)

Chemical	Temp. °C	Pres. (bar)	State	e^{ch} (MJ/kg)	e^{tm} (MJ/kg)	e^{conv} (MJ/kg)	e^{ch} (kJ/mol)	e^{tm} (kJ/mol)	e^{conv} (kJ/mol)
H_2	25	1.01325	g	117.12	0	117.12	236.11	0.00	236.11
H_2 (slightly compressed)	25	1.5	g	117.12	0.483	117.60	236.11	0.97	237.08
H_2 (slightly compressed)	25	3	g	117.12	1.336	118.45	236.11	2.69	238.80
H_2 (compressed)	25	30	sc	117.12	4.187	121.31	236.11	8.44	244.55
H_2 (compressed)	25	300	sc	117.12	7.221	124.34	236.11	14.56	250.67
H_2 (compressed)	25	350	sc	117.12	7.450	124.57	236.11	15.02	251.13
H_2 (compressed)	25	700	sc	117.12	8.577	125.70	236.11	17.29	253.40
H_2 (cryo-compressed)	−231	300	sc	117.12	11.497	128.62	236.11	23.18	259.29
H_2 (cryo-liquid)	−253	1.01325	l	117.12	12.011	129.13	236.11	24.21	260.32
Ammonia (liquid)	25	11	l	19.848	0.319	20.167	338.01	5.44	343.45
Ethanol	25	1.01325	l	29.449	0	29.449	1356.66	0.00	1356.66
Formic acid (pure)	25	1.01325	l	6.277	0	6.277	288.89	0.00	288.89
Methane (compressed)	25	250	sc	51.849	0.802	52.651	831.66	12.86	844.52
Methane (cryo-liquid)	−162	1.01325	l	51.849	1.083	52.932	831.66	17.37	849.03
Methanol	25	1.01325	l	22.397	0	22.397	717.64	0.00	717.64
n-Octane	25	1.01325	l	47.401	0	47.401	5414.57	0.00	5414.57

TABLE 7.8 Selected chemicals relevant to the hydrogen economy. The chemical, thermo-mechanical, and conventional exergies ($e^{ch} + e^{tm} = e^{conv}$) are provided. Abbreviations: g, gas; l, liquid; sc, supercritical fluid

References

1. Lemmon EW, Span R. Short fundamental equations of state for 20 industrial fluids. *J Chem Eng Data* 51:785–850 (2006). doi:10.1021/je050186n.
2. Thol M, Lemmon EW, Span R. Equation of state for benzene for temperatures from the melting line up to 725 K with pressures up to 500 MPa. *High Temperatures-High Pressures* 41:81–97 (2012).
3. Wu J, Zhou Y, Lemmon LW. An equation of state for the thermodynamic properties of dimethyl ether. *J Phys Chem Ref Data* 40(2):023104 (2011). doi:10.1063/1.3582533.
4. Buecker D, Wagner W. A reference equation of state for the thermodynamic properties of ethane for temperatures from the melting line to 675 K and pressures up to 900 MPa. *J Phys Chem Ref Data* 35(1):205–266 (2006). doi:10.1063/1.1859286.
5. Schroeder JA, Penoncello SG, Schroeder JS. A fundamental equation of state for ethanol. *J Phys Chem Ref Data* 43(4):043102 (2014). doi:10.1063/1.4895394.
6. Smukala J, Span R, Wagner W. New equation of state for ethylene covering the fluid region for temperatures from the melting line to 450 K at pressures up to 300 MPa. *J Phys Chem Ref Data* 29(5):1053–1121 (2000). doi:10.1063/1.1329318.
7. Leachman JW, Jacobsen RT, Penoncello SG, Lemmon EW. Fundamental equations of state for parahydrogen, normal hydrogen, and orthohydrogen. *J Phys Chem Ref Data* 38(3):721–748 (2009). doi:10.1063/1.3160306.
8. Setzmann U, Wagner W. A new equation of state and tables of thermodynamic properties for methane covering the range from the melting line to 625 K at pressures up to 1000 MPa. *J Phys Chem Ref Data* 20(6):1061–1151 (1991). doi:10.1063/1.555898.
9. Piazza L, Span R. An equation of state for methanol including the association term of SAFT. *Fluid Phase Equilib* 349:12–24 (2013). doi:10.1016/j.fluid.2013.03.024.
10. Szargut J. *Exergy Method: Technical and Ecological Applications*. Vol. 18. WIT Press: Massachusetts (2005).
11. Dean JA, Gokel GW. *Dean's Handbook of Organic Chemistry*. McGraw-Hill Professional Pub: New York. (2003).
12. Dean JA. *Lange's Handbook of Chemistry*. 12th ed. McGraw-Hill: New York (1979); pp. 9-4–9.94.
13. Adams TA II. How Canada can supply Europe with critical energy by creating a Trans-Atlantic energy bridge. *Canadian J Chem Eng*, 101(4):1729–1742 (2023). doi: 10.1002/cjce.24787.

CHAPTER 8

Gaseous Fuel Mixtures

This chapter contains the composition and chemical exergy of a selection of gaseous fuel mixtures encountered in energy-intensive chemical processes. Table 8.1 contains a variety of syngas blends produced from various processes such as gas reforming and coal gasification, steel refinery off-gases, conventional average pipeline natural gas, and hydrogen product gases from various production methods. Table 8.2 contains a list of raw natural gases from various wells around the world, and Table 8.3 contains a similar list of shale gases. Note that chemical composition might not add up to exactly 100 for each gas mixture due to rounding.

We computed each chemical exergy at the environmental reference conditions (25°C and 1.01325 bar, see Chap. 4) using the following equation[1]:

$$e_{\text{mix}}^{\text{ch}} = \sum_i x_i e_i^{\text{ch}} + RT_0 \sum_i x_i \ln x_i \qquad (8.1)$$

where i is the species in the mixture, x is the mole fraction of that species in the mixture at reference conditions (25°C and 1.01325 bar), e_i^{ch} is the chemical exergy of that species at the reference state if it were a pure chemical. The equation used to calculate the chemical exergy of species i is described in Chap. 4. Note that thermo-mechanical exergy is not computed, since the temperature and pressure of the gas will vary for each specific application. References for the underlying compositions of the mixtures are provided where applicable.

Gas	Mole percent (%)									
	Ar	O_2	CO_2	H_2O	N_2	CH_4	CO	C_2H_2	C_2H_4	H_2
U.S. pipeline natural gas			1.00		0.8	93.9				
Coke oven gas		0.72	1.59		9.75	23.63	4.81	1.85	2.00	53.36
Blast furnace gas			22.3		49.59		22.55			5.56
Basic oxygen furnace gas		0.63	20.82		30.88		45.74			1.93
Live oak pyrolysis (at 500 °C) gas			25.6			15.1	46.6			12.7
Syngas from steam reforming of U.S. natural gas	0.01*	trac	4.80	30.93	0.10	0.20	11.81			52.15
Wet, shifted syngas from steam reforming of natural gas	0.01*	trac	16.02	19.72	0.10	0.20	0.60			63.36
Syngas from partial oxidation of U.S. natural gas	0.59	0.18	3.63	17.71	0.56	0.02	25.08			52.23
Syngas from autothermal reforming of U.S. natural gas	0.30	trac	6.87	30.16	0.33	0.52	13.69			48.13
Syngas from wood gasification	4.33*		8.86	33.08	16.71	0.5*	26.57			14.79
Syngas for Fischer-Tropsch production	0.30		2.70	0.20	1.30	2.60	30.83			62.06
Syngas for dimethyl ether production	0.00		14.30	0.40	0.30	1.20	27.80			56.00
Syngas from petcoke gasification	0.10	trac	10.08	37.73	0.30	30.9*	30.55			20.17
Sweet syngas from petcoke gasification	1.00	trac	4.30	0.60	4.50	45*	29.30			60.30
Syngas from gasified petcoke/reformed natural gas blend	0.70	trac	2.60	0.80	3.40	1.50	29.67			61.34
Coal syngas from oven gas process			2.29		6.24	35.14	7.17			49.17
Coal syngas from producer gas process			4.53		51.21	3.02	27.26			13.98
Coal syngas from water gas process			3.02		3.32	0.50	43.02			50.15
Coal syngas from carbureted water gas process			4.32		8.76	11.53	37.03			38.36
Coal syngas from synthetic coal gas process			31.74			11.05	16.02			41.18
Typical coal syngas from Lurgi dry ash gasifiers			4.12		2.06	6.19	58.76			28.87
Typical coal syngas from Koppers-Totzek gasifier			2.03		2.03	0.51	65.99			29.44
Typical coal syngas from Texaco (a.k.a. GE) coal slurry-fed gasifier			10.62		1.06	0.21	52.02			36.09
Steam reforming of acetaldehyde over Ni-Rh/CeO$_2$ catalyst			12.60			9.50	20.70		1.80	52.00
Oxidative steam reforming of acetaldehyde over Ni-Rh/CeO$_2$ catalyst			35.00			9.60	3.30			52.00
Gas composition with lignite as feedstock in GSP gasification process			9.00		4.70	0.05	54.99			30.99
Gas composition with waste residue as feedstock in GSP gasification process			6.00		4.90	0.05	42.98			45.98
Gas composition with petrol coke as feedstock in GSP gasification process			7.70		13.32	0.05	46.87			30.79
Hydrogen production from conventional steam methane reforming reactor			15.90			15.70	1.20			67.20
Hydrogen from sorption-enhanced reaction process			0.00			5.60	0.3*			94.40
Hydrogen from decomposed formic acid			50							50
Carbon monoxide from decomposed formic acid				50			50			
Humidified hydrogen 90%				10						90
Humidified hydrogen 80%				20						80
Humidified hydrogen 70%				30						70
Humidified hydrogen 60%				40						60
Humidified hydrogen 50%				50						50
Humidified hydrogen 40%				60						40
Humidified hydrogen 30%				70						30
Humidified hydrogen 20%				80						20
Humidified hydrogen 10%				90						10

TABLE 8.1 The chemical exergy of selected process gases at environmental reference conditions. Symbol * means ppm. *Trac* means it contains less than 10^{-9} mol%

												Mole percent (%)	
CS$_2$	C$_4$H$_4$S	H$_2$S	C$_2$H$_6$	C$_3$H$_6$O	C$_3$H$_8$	C$_4$H$_{10}$	C$_3$H$_6$O	C$_5$H$_{12}$	NH$_3$	HCl	COS	e_{mix}^{ch} (kJ/mol)	Ref.
			3.2		0.7	0.4						854.51	
0.0087	0.003	0.378	1.9									415.05	[2]
												76.96	
												131.82	
												285.67	[3]
			0.05*	0.01*	0.01*						trac	158.39	[4]
			0.05*	0.01*	0.01*						trac	155.61	
			8.53*	1.63*	1.38*							192.07	
			6.22*	1.19*	1.01*							156.86	
		0.12*						0.07*	4.81*			109.21	[5]
		1.06*										251.27	
		0.01*							trac			218.91	
		1.00						trac	120*		0.07	142.57	[6]
												221.61	
			2.44*	trac	trac							237.16	
												425.62	[7]
												131.30	
												239.12	
												285.88	
												236.37	
												279.39	
												253.57	
												229.62	
						3.40						343.17	
						0.10						217.81	
		0.27										226.14	
		0.10										226.58	
		1.28										210.90	
												293.45	
		trac										268.93	
												126.25	
												140.50	
												212.64	
												189.54	
												166.61	
												143.79	
												121.08	
												98.46	
												75.96	
												53.57	
												31.34	

Table 8.2 The chemical exergy of selected natural gases at environmental reference conditions

Country	Basin	Well	Mole percent (%)								e^{ch}_{mix} (kJ/mol)	Ref.
			CO_2	N_2	CH_4	H_2S	C_2H_6	C_3H_8	C_4H_{10}	C_5H_{12}		
Canada		Natural gas from Alberta region	3.18		96.91		2.13	0.96			805.07	[8]
Iraq		Natural gas			59.00	7.73	23.20	6.89			1046.56	
Libya		Natural gas	1.24		70.06		16.27	12.43			1091.62	
New Zealand		Natural gas	44.20		44.20		11.60				547.49	
United Kingdom		Natural gas from Hewett region			95.37		3.71	0.93			868.00	
United States		Natural gas from California region	0.6		88.70		7.00	1.90			898.64	
United States	San Juan Basin	Nye no.11	0.78	2.52	80.27		9.72	4.45	1.69	0.57	903.97	
United States	San Juan Basin	Nye no. 14	4.40	2.94	92.55		0.11				919.95	
United States	San Juan Basin	Hubbell no.19	2.12	3.43	93.44		0.91	0.09	0.01		824.15	
United States	San Juan Basin	Schultz com. no. F11	1.39	1.42	87.33		6.51	2.24	0.81	0.30	801.79	
United States	San Juan Basin	Bullack no. 1F	0.60	2.52	74.95		10.43	7.61	2.93	0.95	946.99	
United States	San Juan Basin	Blanco 30-12 no. A3	0.64	1.69	76.03		12.65	6.79	1.27	0.92	940.73	
United States	San Juan Basin	Harmon no. 1	0.70	3.81	78.80		10.41	3.96	1.59	0.73	784.15	
United States	San Juan Basin	Lindrith no. 22	0.33	5.85	75.64		11.35	4.35	1.78	0.69	920.62	
United States	San Juan Basin	Lindrith no. 87	0.40	3.31	75.39		11.77	5.63	2.54	0.96	919.73	[9]
United States	San Juan Basin	Lindrith no. 14	0.74	2.15	73.48		13.80	7.96	1.88		982.05	
United States	San Juan Basin	Jicarilla no. 1P	0.96	2.44	88.56		6.08	1.39	0.56		958.15	
United States	San Juan Basin	Canyon largo no.1	0.38	1.74	88.74		5.87	2.14	1.13		926.04	
United States	San Juan Basin	Canyon largo no. 30	1.12	2.96	82.46		9.66	2.72	1.08		935.91	
United States	San Juan Basin	Largo no.1	1.10	1.70	94.90		1.84	0.29	0.17		896.68	
United States	San Juan Basin	San Juan 27-5 no.62	1.45	3.24	83.54		7.76	2.89	1.12		947.53	
United States	San Juan Basin	San Jan 28-6 no.77	2.03	4.09	93.25		0.57	0.05	0.01		907.05	
United States	San Juan Basin	Burrought comm no.1	1.48	1.46	88.18		6.53	1.69	0.65		926.08	
United States	San Juan Basin	Day no. A13	1.58	3.53	84.49		6.85	2.53	1.02		897.65	
United States	San Juan Basin	Day no. A14	1.63	1.72	86.77		7.11	2.01	0.75		925.56	
United States	San Juan Basin	Roelofs no. E5	4.44	3.40	92.05		0.11				892.22	
United States	San Juan Basin	Grambling no.9	3.03	3.14	93.36		0.43	0.03			888.32	
United States	San Juan Basin	Hubbell no. 10	4.44	3.42	92.03		0.11				838.07	
United States	San Juan Basin	San Juan 30-4 no.14	0.39	2.68	90.12		4.62	1.61	0.58		827.99	
United States	San Juan Basin	Turner no. 4	0.76	25.54	67.08		4.88	1.30	0.45		957.38	

Mole percent (%)

Country	Basin	Well	CO$_2$	N$_2$	CH$_4$	H$_2$S	C$_2$H$_6$	C$_3$H$_8$	C$_4$H$_{10}$	C$_5$H$_{12}$	e$_{mix}^{ch}$ (kJ/mol)	Ref.
United States	San Juan Basin	Schumacher no.2	5.13	3.26	91.56		0.06				920.55	
United States	San Juan Basin	Bruington no. 5	5.69	2.50	91.74		0.07				964.19	
United States	San Juan Basin	Neil no. A3	0.78	2.52	80.27		9.72	4.45	1.69	0.57	1082.48	
United States	San Juan Basin	Scott no. 20	4.40	2.94	92.55		0.11				886.64	
United States	San Juan Basin	San Juan 27-5 no. 59	2.12	3.43	93.44		0.91	0.09	0.01		977.19	
United States	San Juan Basin	San Juan 28-6 no.77	1.39	1.42	87.33		6.51	2.24	0.81	0.30	971.65	
United States	San Juan Basin	Schnerdtfeger no. A15	0.60	2.52	74.95		10.43	7.61	2.93	0.95	966.85	
United States	San Juan Basin	San Juan 28-6 no. 62	0.64	1.69	76.03		12.65	6.79	1.27	0.92	964.93	
United States	San Juan Basin	Day no. A1	0.70	3.81	78.80		10.41	3.96	1.59	0.73	984.73	
United States	San Juan Basin	San Juan no. A20	0.33	5.85	75.64		11.35	4.35	1.78	0.69	1009.33	
United States	San Juan Basin	Nye no. 14 Well	0.40	3.31	75.39		11.77	5.63	2.54	0.96	935.20	
United States	San Juan Basin	Hubbell no. 10	0.74	2.15	73.48		13.80	7.96	1.88		933.93	
United States	San Juan Basin	San Juan 30-5 no.7	0.96	2.44	88.56		6.08	1.39	0.56		822.63	
United States	San Juan Basin	San Juan 30-6 no. 39	0.38	1.74	88.74		5.87	2.14	1.13		806.34	
United States	San Juan Basin	San Juan 30-6 no. 31	1.12	2.96	82.46		9.66	2.72	1.08		926.89	
United States	San Juan Basin	San Juan 30-6 no. A8	1.10	1.70	94.90		1.84	0.29	0.17		1050.71	
United States	San Juan Basin	Howell no. 2A	1.45	3.24	83.54		7.76	2.89	1.12		918.79	
United States	San Juan Basin	Florance no. 2	2.03	4.09	93.25		0.57	0.05	0.01		949.59	
United States	San Juan Basin	Kerneghan no. 4	1.48	1.46	88.18		6.53	1.69	0.65		907.52	
United States	San Juan Basin	Sheets no. 1	1.58	3.53	84.49		6.85	2.53	1.02		927.21	
United States	San Juan Basin	Mudge no. 10	1.63	1.72	86.77		7.11	2.01	0.75		968.10	
United States	San Juan Basin	Neil no. A3	4.44	3.40	92.05		0.11				974.24	
United States	San Juan Basin	Allison no. 25	3.03	3.14	93.36		0.43	0.03			771.48	
United States	San Juan Basin	Allison no. 12	4.44	3.42	92.03		0.11				792.61	
United States	San Juan Basin	Fields no.1	0.39	2.68	90.12		4.62	1.61	0.58		903.90	
United States	San Juan Basin	Nageezi no. 1	0.76	25.54	67.08		4.88	1.30	0.45		1056.18	
United States	San Juan Basin	Huerfano no. 255	5.13	3.26	91.56		0.06				1033.24	
United States	San Juan Basin	Shumacher no. 13	5.69	2.50	91.74		0.07				964.25	
United States	San Juan Basin	Jicarilla-Otero 11-28 no.2	0.78	2.52	80.27		9.72	4.45	1.69	0.57	964.41	
United States	San Juan Basin	Canyon Largo no. 249	4.40	2.94	92.55		0.11				1026.47	
United States	San Juan Basin	Nageezi no. 3	2.12	3.43	93.44		0.91	0.09	0.01		1039.34	

TABLE 8.2 The chemical exergy of selected natural gases at environmental reference conditions (*Continued*)

Country	Basin	Well	Mole percent (%)									e_{mix}^{ch} (kJ/mol)	Ref.
			CO_2	N_2	CH_4	H_2S	C_2H_6	C_3H_8	C_4H_{10}	C_5H_{12}		
United States	San Juan Basin	San Juan 27-5 no. 59	1.39	1.42	87.33		6.51	2.24	0.81	0.30	872.14	
United States	San Juan Basin	San Juan 28-6 no. 169	0.60	2.52	74.95		10.43	7.61	2.93	0.95	902.49	
United States	San Juan Basin	Fillan no. 6	0.64	1.69	76.03		12.65	6.79	1.27	0.92	917.52	
United States	San Juan Basin	San Juan 28-6 no. 128	0.70	3.81	78.80		10.41	3.96	1.59	0.73	827.34	
United States	San Juan Basin	Farmington no. A1	0.33	5.85	75.64		11.35	4.35	1.78	0.69	903.16	
United States	San Juan Basin	San Juan 30-6 no. 31	0.40	3.31	75.39		11.77	5.63	2.54	0.96	785.14	
United States	San Juan Basin	Schumacher no. 12	0.74	2.15	73.48		13.80	7.96	1.88		884.75	
United States	San Juan Basin	Bruington POD no. 4	0.96	2.44	88.56		6.08	1.39	0.56		887.02	
United States	San Juan Basin	Mudge no. 21	0.38	1.74	88.74		5.87	2.14	1.13		891.21	
United States	San Juan Basin	Allison no. 25	1.12	2.96	82.46		9.66	2.72	1.08		767.23	
United States	San Juan Basin	Allison no. 32	1.10	1.70	94.90		1.84	0.29	0.17		783.52	
United States	San Juan Basin	Allison no. 12	1.45	3.24	83.54		7.76	2.89	1.12		767.06	
United States	San Juan Basin	Barker no. 2	2.03	4.09	93.25		0.57	0.05	0.01		868.59	
United States	San Juan Basin	Ignacio 33-7 no.1	1.48	1.46	88.18		6.53	1.69	0.65		669.56	
United States	San Juan Basin	Gearhart no. 1-5	1.58	3.53	84.49		6.85	2.53	1.02		762.45	
United States	San Juan Basin	Burkett no. 1-3	1.63	1.72	86.77		7.11	2.01	0.75		764.29	

TABLE 8.2 The chemical exergy of selected natural gases at environmental reference conditions (*Continued*)

Country	Basin	Well	Mole percent (%)							e_{mix}^{ch} (kJ/mol)	Ref.
			CO_2	N_2	CH_4	C_2H_6	C_3H_8	C_4H_{10}	C_5H_{12}		
Argentina	NqSa 1148				64.00	8.50	8.80	9.60	9.10	1429.98	
Argentina	NqSa 1148				66.20	7.30	9.20	9.90	7.40	1388.64	
Argentina	NqSa 1148				72.63	5.99	6.89	8.09	6.39	1287.75	
Argentina	NqSa 1148				74.60	7.60	6.10	6.20	5.50	1227.26	
Argentina	NqSa 1148				47.40	11.40	14.40	17.30	9.50	1684.96	
Argentina	NqSa 1148				73.55	6.49	6.59	7.19	6.19	1263.74	
Argentina	ADC 1016				74.13	11.99	7.79	4.20	1.90	1144.64	
Argentina	ADC 1016				72.70	12.80	8.30	4.40	1.80	1158.11	
Argentina	ADC 1016				69.83	13.49	9.39	5.09	2.20	1201.07	
Argentina	ADC 1016				71.67	12.71	8.41	4.90	2.30	1182.05	
Argentina	ADC 1016				79.32	12.19	5.99	2.10	0.40	1041.88	[10]
Argentina	ADC 1016				58.30	21.50	12.90	7.30		1286.08	
Argentina	ADC 1016				96.50	2.70	0.50	0.30		861.71	
Argentina	ADC 1016				53.35	18.52	17.02	11.11		1395.57	
Argentina	ADC 1016				97.20	2.20	0.30	0.30		855.79	
Argentina	ADC 1016				96.20	3.00	0.50	0.30		863.65	
Argentina	ADC 1016				97.50	2.00	0.30	0.20		852.50	
Argentina	ADC 1016				73.43	23.58	2.00	1.00		1032.67	
Argentina	ADC 1016				97.80	1.90	0.20	0.10		848.62	
Argentina	ADC 1016				98.10	1.60	0.20	0.10		846.64	
Argentina	ADC 1016				98.70	1.20	0.10			840.78	
China	Sichuan	NH1	0.04		99.45	0.50	0.01			834.71	
China	Sichuan	N211	0.91		98.74	0.32	0.03			826.59	
China	Sichuan	Z104	0.07		99.40	0.52	0.01			834.58	
China	Sichuan	YH1-1	0.01		99.51	0.47	0.01			834.75	
China	Sichuan	NH2-1			99.48	0.42	0.10			835.70	
China	Sichuan	NH2-2			99.52	0.47	0.01			834.84	
China	Sichuan	NH2-3	0.59		98.98	0.42	0.01			829.62	
China	Sichuan	NH2-4			99.55	0.44	0.01			834.65	
China	Sichuan	NH2-5	0.16	1.37	97.98	0.48	0.01			822.01	
China	Sichuan	NH2-6	0.39	0.21	98.92	0.47	0.01			829.82	[11]
China	Sichuan	NH2-7	0.48	0.35	98.69	0.47	0.01			827.90	
China	Sichuan	NH3-1		1.14	98.32	0.53	0.01			825.60	
China	Sichuan	NH3-2		0.24	99.30	0.44	0.02			832.73	
China	Sichuan	NH3-3		1.10	98.81	0.09				822.94	
China	Sichuan	NH3-4	0.46	1.79	97.17	0.57	0.01			816.57	
China	Sichuan	NH3-5	0.72	0.51	98.22	0.54	0.01			825.02	
China	Sichuan	NH3-6	0.68	0.32	98.47	0.52	0.01			826.83	
China	Sichuan	YH1-3		1.41	97.99	0.58	0.02			823.78	
China	Sichuan	YH1-5	0.26	0.15	99.09	0.49	0.01			831.53	
China	Weiyuan	Wei201	0.36	0.81	98.36	0.46	0.01			824.94	
China	Weiyuan	Wei201	0.42		99.10	0.48				831.29	
China	Weiyuan	Wei201-H1	1.07	2.95	95.64	0.32	0.01			800.14	
China	Weiyuan	Wei201-H1	1.06		98.57	0.37				825.30	
China	Weiyuan	Wei201-H2	1.07	2.95	95.64	0.32	0.01			800.14	
China	Weiyuan	Wei202	0.02	0.01	99.27	0.68	0.02			836.08	

TABLE 8.3 The chemical exergy of selected shale gases at environmental reference conditions

Country	Basin	Well	Mole percent (%)							e^{ch}_{mix} (kJ/mol)	Ref.
			CO_2	N_2	CH_4	C_2H_6	C_3H_8	C_4H_{10}	C_5H_{12}		
China	Weiyuan	Wei203	1.05	0.08	98.30	0.57				826.01	
China	Changning-Zhaotong	Ning201-H1	0.04	0.30	99.15	0.50	0.01			832.15	
China	Changning-Zhaotong	Ning201-H1	0.40		99.06	0.54				831.85	
China	Changning-Zhaotong	Ning211	0.91	0.17	98.57	0.32	0.03			825.16	
China	Changning-Zhaotong	NingH2-1		0.40	99.08	0.42	0.10			832.29	
China	Changning-Zhaotong	NingH2-2		0.23	99.29	0.47	0.01			832.88	
China	Changning-Zhaotong	NingH2-3	0.59	0.37	98.61	0.42	0.01			826.50	
China	Changning-Zhaotong	NingH2-4		0.40	99.15	0.44	0.01			831.25	
China	Changning-Zhaotong	Zhao104	0.07	0.15	99.25	0.52	0.01			833.30	
China	Changning-Zhaotong	YSL1-1H	0.01	0.03	99.48	0.47	0.01			834.50	
China	Jiaoshiba	JY1	0.32	0.43	98.53	0.67	0.05			830.36	
China	Jiaoshiba	JY1-2	0.13	0.34	98.81	0.70	0.02			832.50	
China	Jiaoshiba	JY1-3	0.17	0.41	98.67	0.72	0.03			831.84	
China	Jiaoshiba	JY6-2	0.02	0.39	98.94	0.63	0.02			832.54	
China	Jiaoshiba	JY7-2	0.14	0.32	98.84	0.67	0.03			832.52	
China	Jiaoshiba	JY8-2	0.21	0.32	98.75	0.70	0.02			832.01	
China	Jiaoshiba	JY9-2	0.20	0.52	98.57	0.69	0.02			830.34	
China	Jiaoshiba	JY10-2	0.26	0.36	98.66	0.70	0.02			831.26	
China	Jiaoshiba	JY11-2	0.23	0.42	98.64	0.69	0.02			830.93	
China	Jiaoshiba	JY12-2	0.09	0.43	98.70	0.74	0.04			832.59	
China	Jiaoshiba	JY13-2	0.03	0.42	98.88	0.65	0.02			832.33	
China	Jiaoshiba	JY12-1	1.16	0.47	97.67	0.68	0.02			822.78	
China	Jiaoshiba	JY12-3		0.44	98.87	0.67	0.02			832.54	
China	Jiaoshiba	JY12-4		0.57	98.75	0.66	0.02			831.38	
China	Jiaoshiba	JY13-1	0.39	0.64	98.35	0.60	0.02			827.17	
China	Jiaoshiba	JY13-3	0.25	0.51	98.56	0.66	0.02			829.81	
China	Jiaoshiba	JY20-2		0.89	98.38	0.71	0.02			829.01	
China	Jiaoshiba	JY42-1	0.38	0.38	98.54	0.68	0.02			829.97	
China	Jiaoshiba	JY42-2		0.39	98.90	0.69	0.02			833.09	
China	Jiaoshiba	JY4-1		1.07	98.28	0.62	0.02			826.88	
China	Jiaoshiba	JY4-2		1.36	98.06	0.57	0.01			824.01	
China	Fushun-Yongchuan	Hai201-H		4.05	95.33	0.60	0.02			801.73	
China	Fushun-Yongchuan	Dong202-H1	1.74	0.67	97.15	0.41	0.03			814.70	
China	Fushun-Yongchuan	Yang101	1.60	0.53	97.59	0.27	0.01			815.86	
China	Fushun-Yongchuan	Yang201-H2	0.06	0.01	99.59	0.33	0.01			833.34	
China	Fushun-Yongchuan	Lai101	1.48	0.61	97.67	0.23	0.01			815.90	

TABLE 8.3   The chemical exergy of selected shale gases at environmental reference conditions (*Continued*)

Country	Basin	Well	Mole percent (%)							e^{ch}_{mix} (kJ/mol)	Ref.
			CO_2	N_2	CH_4	C_2H_6	C_3H_8	C_4H_{10}	C_5H_{12}		
China	Pengshui	PY1	0.35	0.15	98.78	0.71	0.01			832.23	
China	Pengshui	PY3	0.93	0.05	98.46	0.55	0.01			827.25	
China	Yanye13		0.23	0.27	89.11	6.96	2.79	0.64		922.09	
China	Yanye5		0.02	0.32	82.06	9.97	5.74	1.90		1006.59	
China	Yanye11		0.03	0.85	90.18	5.63	2.32	0.99		910.86	
China	YanyeH1		0.01	0.33	87.13	8.34	3.36	0.83		943.73	
China	Yanye22		11.27	0.85	82.21	4.25	0.95	0.46		781.37	
China	Liuping179		0.20	0.50	78.76	10.96	6.66	2.92		1042.40	
China	Liuping177		0.49	0.10	91.68	5.55	1.77	0.41		894.27	
China	Xin59		0.38	0.49	82.53	8.68	5.31	2.61		1002.05	
China	Liuping177		0.32	2.16	89.49	5.35	1.95	0.71		885.34	
China	Liuping179		1.30	1.77	85.58	6.97	3.16	1.22		917.03	
China	Xin59		0.89	3.20	77.56	8.86	6.17	3.33		1001.76	
China	Xinye2		0.63	0.89	94.13	3.48	0.85	0.02		853.17	
United States	Barnett	Barnett Shale Gas Well 1	1.40	7.90	80.30	8.10	2.30			837.02	
United States	Barnett	Barnett Shale Gas Well 2	0.30	1.50	81.20	11.80	5.20			962.17	
United States	Barnett	Barnett Shale Gas Well 3	2.30	1.10	91.80	4.40	0.40			837.43	
United States	Barnett	Barnett Shale Gas Well 4	2.70	1.00	93.70	2.60				817.96	
United States	Marcellus	Marcellus Shale Gas Well 1	0.10	0.40	79.40	16.10	4.00			985.72	
United States	Marcellus	Marcellus Shale Gas Well 2	0.10	0.30	82.10	14.00	3.50			966.14	
United States	Marcellus	Marcellus Shale Gas Well 3	0.90	0.30	83.80	12.00	3.00			939.78	
United States	Marcellus	Marcellus Shale Gas Well 4	0.30	0.20	95.50	3.00	1.00			860.13	
United States	Fayetteville	Fayetteville Shale Gas	1.00	0.70	97.30	1.00				823.98	[13]
United States	New Albany	New Albany Shale Gas Well 1	8.10		87.70	1.70	2.50			809.00	
United States	New Albany	New Albany Shale Gas Well 2	10.40		88.00	0.80	0.80			762.05	
United States	New Albany	New Albany Shale Gas Well 3	7.40		91.00	1.00	0.60			785.26	
United States	New Albany	New Albany Shale Gas Well 4	5.60		92.80	1.00	0.60			800.00	
United States	Artrim	Artrim Shale Gas Well 1	3.00	65.00	27.50	3.50	1.00			301.37	
United States	Artrim	Artrim Shale Gas Well 2		35.90	57.30	4.90	1.90			588.71	
United States	Artrim	Artrim Shale Gas Well 3	3.30	14.30	77.50	4.00	0.90			722.61	
United States	Artrim	Artrim Shale Gas Well 4	9.00	0.70	85.60	4.30	0.40			785.28	
United States	Haynesville	Haynesville Shale Gas	4.80	0.10	95.00	0.10				792.01	

TABLE 8.3 The chemical exergy of selected shale gases at environmental reference conditions (*Continued*)

References

1. Moran MJ, Shapiro HN, Boettner DD, Bailey MB. *Fundamentals of Engineering Thermodynamics*. Wiley: New Jersey, 2014.

2. Deng L, Adams TA II. Optimization of coke oven gas desulfurization and combined cycle power plant electricity generation. *Industrial & Engineering Chemistry Research* 57(38):12816–12828 (2018).

3. Amini E, Safdari MS, DeYoung JT, Weise DR, Fletcher TH. Characterization of pyrolysis products from slow pyrolysis of live and dead vegetation native to the southern United States. *Fuel* 235:1475–1491 (2019).

4. Adams TA II, Barton PI. High-efficiency power production from natural gas with carbon capture. *Journal of Power Sources* 195(7):1971–1983 (2010).

5. Scott JA, Adams TA II. Biomass-gas-and-nuclear-to-liquids (BGNTL) processes part I: model development and simulation. *The Canadian Journal of Chemical Engineering* 96(9):1853–1871 (2018).

6. Okeke IJ, Adams TA II. Combining petroleum coke and natural gas for efficient liquid fuels production. *Energy* 163:426–442 (2018).

7. Liu K, Chi Z, Fletcher TH. Coal gasification. In: *Hydrogen and Syngas Production and Purification Technologies* Editors Lie K, Song C, Subramani V. John Wiley & Sons: New Jersey (2010).

8. Subramani V, Sharma P, Zhang L, Liu K. Catalytic steam reforming technology for the production of hydrogen and syngas. In: *Hydrogen and Syngas Production and Purification Technologies*. Editors: Lie K, Song C, Subramani V. John Wiley & Sons: New Jersey (2010).

9. Rice DD. Relation of natural gas composition to thermal maturity and source rock type in San Juan Basin, northwestern New Mexico and southwestern Colorado. *AAPG Bulletin* 67(8):1199–1218 (1983).

10. Ostera H, García R, Malizia D, Kokot P, Wainstein L, Ricciutti M. Shale gas plays, Neuquén Basin, Argentina: chemostratigraphy and mud gas carbon isotopes insights. *Brazilian Journal of Geology* 46:181–196 (2016).

11. Feng Z, Liu D, Huang S, Wu W, Dong D, Peng W, Han E. Carbon isotopic composition of shale gas in the Silurian Longmaxi formation of the changning area, Sichuan Basin. *Petroleum Exploration and Development* 43(5):769–777 (2016).

12. Dai J, Zou C, Dong D, Ni Y, Wu W, Gong D, Wang Y, Huang S, Huang J, Fang C, Liu D. Geochemical characteristics of marine and terrestrial shale gas in China. *Marine and Petroleum Geology* 76:444–463 (2016).

13. Bullin KA, Krouskop PE. Composition variety complicates processing plans for US shale gas. *Oil and Gas Journal* 107(10):50–55 (2009).

CHAPTER 9

Liquid Fuel Mixtures

This chapter contains the chemical exergy of liquid fuel blends involving gasoline, diesel, alcohols, and various biofuels. In this chapter, only mixtures of chemicals are considered—for pure chemicals, see Chap. 7.

The fuels in this chapter are often complex mixtures of compounds that can be hard or impossible to characterize or identify exactly, particularly for biofuels. Without knowing chemical compositions exactly, it is not possible to determine chemical exergies as a function of the chemical exergies of its components (for example, a weighted sum). However, it is possible using standard chemistry laboratory techniques to determine the heating value and atomic composition of a fuel, which can be useful in estimating its chemical exergy.

The strong correlation between the heating value of a fuel and its chemical exergy is evident, because chemical exergy is strongly influenced by chemical bond energy for fuels as discussed in Chaps. 2 and 3. Therefore, it is possible to use known heating value and exergy data for pure chemicals and create an empirical correlation between the heating value of those pure chemicals and chemical exergy. Then, the correlation could be applied to liquid fuel mixtures to estimate the exergy of those mixtures.

Although heating value can be measured experimentally, it is difficult to find sets of data that would be consistent from one fuel to another, or even for a single fuel. For example, the composition of diesel and gasoline blends changes regularly depending on the crude oil from which they were derived, the season, and a patchwork of government regulations from market to market that regulate certain additives and qualities. For this book, it is important for chemical exergies to be consistent from chemical to chemical, so that any combination of chemicals and fuels can be used together in the same exergy analysis.

We used the method of Ghazikhani et al.[1] in which we derived an empirical correlation that predicts the HHV of a chemical from its atomic composition. We first calculated the HHV of 523 pure chemicals in Tables 7.1 to 7.4 that contain C, H, O, and N using the following equation:

$$HHV = \sum_p x_p h_p^f - h^f \tag{9.1}$$

where HHV is higher heating value and h^f is the enthalpy of formation of the pure chemical of interest at the standard reference conditions of 25°C and 1 atm (1.01325 bar). The summation term considers the products if the chemical is combusted completely and adiabatically with a stoichiometric amount of pure oxygen into a set of product chemicals p (all C becomes CO_2, all H becomes H_2O, etc.). h_p^f is the heat of formation of each product chemical p which has mole fraction x_p in the combustion mixture, also at 25°C and 1 atm. For water, $h_{H_2O}^f$ uses the liquid state heat of formation. Thus, this represents all thermal heat available by combusting the chemical and recovering the heat until the product mixture returns to 25°C, including the latent heat available from condensing water.*

Next, we derived the correlation between HHV and chemical compound C, H, N, and O for that data set. We get the following correlation with an R^2 of more than 0.998.

$$HHV_{mol} = 425.696\ C + 118.06\ H - 180.544\ O + 14.329\ N \tag{9.2}$$

where C, H, O, and N represent the number of moles of C, H, O, and N in 1 mole of the chemical of interest, and HHV is in kJ/mol. We were able to verify the accuracy of the correlation using experimentally determined HHVs for some chemicals found in the literature. For example, the ethanol HHV calculated using Eq. (9.2) is 29.94 MJ/kg, while the experimental HHV is about 29.7 MJ/kg.[2]

---

*This is equivalent to the heat of combustion. We note that there is inconsistency in the literature about whether HHV is the heat of combustion, or whether it is the lower heating value plus the latent heat of vaporization of water in product set p. Experimental methods often use the latter in practice depending on the standards and methods used. However, they are not the same as the latter does not include the specific heat of mixture p between 25°C and 150°C. This means that HHV estimated with the heat of combustion approach will be slightly higher than those estimated with the LHV + heat of vaporization approach. This missing specific heat is usually quite small compared to the rest of the terms; using an experimentally determined HHV that uses the LHV + latent heat approach versus a complete heat of combustion approach will only result in a small difference in the estimated chemical exergy.

Third, we found a simple correlation between chemical exergy and HHV for the 523 pure chemicals using the same procedure. Two correlations were determined, one on a mass basis and the other on a mole basis.

$$e_{\text{mass}}^{\text{ch}} = 1.0088\ \text{HHV}_{\text{mass}} + 1.9007 \tag{9.3}$$

$$e_{\text{mol}}^{\text{ch}} = 0.9883\ \text{HHV}_{\text{mol}} + 46.12 \tag{9.4}$$

where $e_{\text{mass}}^{\text{ch}}$ and $e_{\text{mol}}^{\text{ch}}$ represent the chemical exergy of the fuel in MJ/kg and kJ/mol, respectively; HHV_{mass} and HHV_{mol} represent the higher heating value in MJ/kg, and kJ/mol, respectively. The R^2 for both models was above 0.947.

These correlations are valid for the range of pure chemicals in Tables 7.1 to 7.4 containing C, H, O, and N. This correlation is not valid for pure chemicals containing other atoms.

We applied those correlations developed for pure chemicals to the complex chemical liquid fuel mixtures listed in this chapter. This is an extrapolation. However, liquid fuel mixtures in this chapter are all similar on an atomic level, containing C, H, O, and N elements only, and with commonly occurring bond types (e.g., C—H, C—C, C—OH) also found in the pure chemical dataset on which it was based. Furthermore, we were able to verify the HHV correlation for using experimental data that was available for some of the blends. For example, we estimated the HHV of unleaded gasoline using Eq. (9.3) to be 46.5 MJ/kg from the composition data in Ref. 1, while some literature reported values were 46.4 MJ/kg[2] and 46.52 MJ/kg.[3] The latter value, from the Pacific Northwest National Laboratory, was computed with the same HHV definition we used.

In this chapter, we consider liquid fuel blends. For liquid fuels with known chemical formula, we calculated the blended fuel's chemical formula as the weighted arithmetic mean. For example, the exergy of unleaded gasoline-ethanol blends was computed in Table 9.1. We know the approximate chemical composition of unleaded gasoline ($C_{6.97}H_{14.02}$) from the literature and we also know the chemical composition of ethanol (C_2H_6O). For ethanol blends with 5 mol%, we calculate the formula of the blend by taking the weighted arithmetic mean:

$$0.95 C_{6.97}H_{14.02} + 0.05 C_2H_6O = C_{6.7214}H_{13.619}O_{0.05} \tag{9.5}$$

With this chemical formula of the blended fuel, we then estimate its HHV with Eq. (9.2) and e^{ch} using Eq. (9.4).

In this chapter, we use fuel names like E5 or E10 to mean the fuel has 5 mol% or 10 mol% of ethanol, and 95 mol%, 90 mol% of the other major fuel blended together. Similarly, M5 means 5 mol% of methanol and 95 mol% of the other major fuel blended together, etc. Note that these blends may not quite match strict definitions of various government-regulated labels like "E10" used in some jurisdictions for the commercial sale of everyday fuels (for example, additives are neglected) but it will be close. Tables 9.6 to 9.10 have fuel blended in volume percent, names like VO5, DMS5 which mean 5 vol% of vegetable oil, and 5 vol% of methyl soyate blended with 95 vol% of diesel, etc.

For liquid fuels with unknown chemical formula that are listed in our tables (Tables 9.6 to 9.10), we use the HHV_{mass} data available from the reference literature in Eq. (9.3). Usually, the referenced literature provides HHV_{mass} with blends at various blend ratios, but not for the whole range from 0 to 100 vol% at intervals of 5 vol%. Hence for those cases, we use a similar linear regression of the experimental data to fill in the gaps for HHV_{mass} at some intermediate data points.

Notice that the data for diesel and gasoline may vary from table to table, because different samples and sources were used in determining their properties. This reflects the inherent variability in these fuel blends.

Fuel	Formula	HHV (MJ/kg)	HHV (kJ/mol)	e^{ch} (kJ/mol)	e^{ch} (MJ/kg)
Gasoline[1]	$C_{6.97}H_{14.02}$	46.52	5126.60	5112.74	48.83
E5	$C_{6.72}H_{13.62}O_{0.05}$	46.17	4937.92	4926.26	48.48
E10	$C_{6.47}H_{13.22}O_{0.1}$	45.79	4753.49	4744.00	48.09
E15	$C_{6.2245}H_{12.817}O_{0.15}$	45.39	4564.81	4557.52	47.69
E20	$C_{5.976}H_{12.416}O_{0.2}$	44.95	4377.12	4372.03	47.25
E25	$C_{5.7275}H_{12.015}O_{0.25}$	44.50	4189.75	4186.85	46.79
E30	$C_{5.479}H_{11.614}O_{0.3}$	44.00	4002.38	4001.67	46.29
E35	$C_{5.2305}H_{11.213}O_{0.35}$	43.48	3815.01	3816.50	45.76
E40	$C_{4.982}H_{10.812}O_{0.4}$	42.91	3627.64	3631.32	45.19
E45	$C_{4.7335}H_{10.411}O_{0.45}$	42.30	3440.27	3446.14	44.57
E50	$C_{4.485}H_{10.01}O_{0.5}$	41.63	3252.90	3260.96	43.90
E55	$C_{4.2365}H_{9.609}O_{0.55}$	40.91	3065.53	3075.79	43.18
E60	$C_{3.988}H_{9.208}O_{0.6}$	40.13	2878.16	2890.61	42.39
E65	$C_{3.7395}H_{8.807}O_{0.65}$	39.27	2690.79	2705.43	41.52
E70	$C_{3.491}H_{8.406}O_{0.7}$	38.33	2503.43	2520.26	40.57
E75	$C_{3.2425}H_{8.005}O_{0.75}$	37.30	2316.06	2335.08	39.52
E80	$C_{2.994}H_{7.604}O_{0.8}$	36.14	2128.69	2149.90	38.36
E85	$C_{2.7455}H_{7.203}O_{0.85}$	34.86	1941.32	1964.72	37.07
E90	$C_{2.497}H_{6.802}O_{0.9}$	33.42	1753.95	1779.55	35.62
E95	$C_{2.2485}H_{6.401}O_{0.95}$	28.83	1566.58	1594.37	33.97
E100	C_2H_6O	29.94	1379.21	1356.66	29.45

TABLE 9.1 Chemical exergy of gasoline-ethanol blends

Fuel	Formula	HHV (MJ/kg)	HHV (kJ/mol)	e^{ch} (kJ/mol)	e^{ch} (MJ/kg)
Unleaded gasoline[4]	$C_{7.93}H_{14.83}$	47.24	4622.30	4614.34	49.56
E5	$C_{7.63}H_{14.39}O_{0.05}$	46.82	4460.15	4454.08	49.13
E10	$C_{7.34}H_{13.95}O_{0.1}$	46.38	4297.99	4293.83	48.69
E15	$C_{7.04}H_{13.51}O_{0.15}$	45.91	4135.84	4133.57	48.22
E20	$C_{6.744}H_{13.064}O_{0.2}$	45.42	3973.68	3973.31	47.72
E25	$C_{6.4475}H_{12.6225}O_{0.25}$	44.89	3811.53	3813.05	47.19
E30	$C_{6.151}H_{12.181}O_{0.3}$	44.34	3649.37	3652.80	46.63
E35	$C_{5.8545}H_{11.7395}O_{0.35}$	43.74	3487.22	3492.54	46.03
E40	$C_{5.558}H_{11.298}O_{0.4}$	43.11	3325.06	3332.28	45.39
E45	$C_{5.2615}H_{10.8565}O_{0.45}$	42.43	3162.91	3172.02	44.70
E50	$C_{4.965}H_{10.415}O_{0.5}$	41.70	3000.76	3011.77	43.97
E55	$C_{4.6685}H_{9.9735}O_{0.55}$	40.92	2838.60	2851.51	43.18
E60	$C_{4.372}H_{9.532}O_{0.6}$	40.08	2676.45	2691.25	42.33
E65	$C_{4.0755}H_{9.0905}O_{0.65}$	39.17	2514.29	2530.99	41.41
E70	$C_{3.779}H_{8.649}O_{0.7}$	38.18	2352.14	2370.74	40.42
E75	$C_{3.4825}H_{8.2075}O_{0.75}$	37.11	2189.98	2210.48	39.34
E80	$C_{3.186}H_{7.766}O_{0.8}$	35.94	2027.83	2050.22	38.16
E85	$C_{2.8895}H_{7.3245}O_{0.85}$	34.66	1865.67	1889.96	36.86
E90	$C_{2.593}H_{6.883}O_{0.9}$	33.24	1703.52	1729.71	35.44
E95	$C_{2.2965}H_{6.4415}O_{0.95}$	31.68	1541.36	1569.45	33.86
E100	C_2H_6O	29.94	1379.21	1356.66	28.82

TABLE 9.2 Chemical exergy of unleaded gasoline-ethanol blends (alternative sample to Table 9.1)

Fuel	Formula	HHV (MJ/kg)	HHV (kJ/mol)	e^{ch} (kJ/mol)	e^{ch} (MJ/kg)
Unleaded gasoline[4]	$C_{6.97}H_{14.02}$	47.24	4622.30	4614.34	49.56
M5	$C_{6.6715}H_{13.519}O_{0.05}$	46.82	4427.06	4421.38	49.13
M10	$C_{6.373}H_{13.018}O_{0.1}$	46.37	4231.81	4228.42	48.68
M15	$C_{6.0745}H_{12.517}O_{0.15}$	45.88	4036.57	4035.46	48.19
M20	$C_{5.776}H_{12.016}O_{0.2}$	45.36	3841.32	3842.50	47.66
M25	$C_{5.4775}H_{11.515}O_{0.25}$	44.79	3646.07	3649.54	47.09
M30	$C_{5.179}H_{11.014}O_{0.3}$	44.18	3450.83	3456.57	46.47
M35	$C_{4.8805}H_{10.513}O_{0.35}$	43.52	3255.58	3263.61	45.80
M40	$C_{4.582}H_{10.012}O_{0.4}$	42.79	3060.34	3070.65	45.06
M45	$C_{4.2835}H_{9.511}O_{0.45}$	41.99	2865.09	2877.69	44.26
M50	$C_{3.985}H_{9.01}O_{0.5}$	41.11	2669.85	2684.73	43.37
M55	$C_{3.6865}H_{8.509}O_{0.55}$	40.14	2474.60	2491.77	42.39
M60	$C_{3.388}H_{8.008}O_{0.6}$	39.05	2279.36	2298.81	41.30
M65	$C_{3.0895}H_{7.507}O_{0.65}$	37.84	2084.11	2105.85	40.08
M70	$C_{2.791}H_{7.006}O_{0.7}$	36.48	1888.87	1912.89	38.70
M75	$C_{2.4925}H_{6.505}O_{0.75}$	34.93	1693.62	1719.92	37.13
M80	$C_{2.194}H_{6.004}O_{0.8}$	33.15	1498.37	1526.96	35.34
M85	$C_{1.8955}H_{5.503}O_{0.85}$	31.09	1303.13	1334.00	33.27
M90	$C_{1.597}H_{5.002}O_{0.9}$	28.69	1107.88	1141.04	30.84
M95	$C_{1.2985}H_{4.501}O_{0.95}$	25.83	912.64	948.08	27.96
M100	CH_4O	22.39	717.39	717.64	24.49

TABLE 9.3   Chemical exergy of unleaded gasoline-methanol blends

Fuel	Formula	HHV (MJ/kg)	HHV (kJ/mol)	e^{ch} (kJ/mol)	e^{ch} (MJ/kg)
Diesel[5]	$C_{13}H_{23}$	46.00	8249.43	8199.03	48.31
M5	$C_{12.4}H_{22.05}O_{0.05}$	45.78	7872.83	7826.83	48.09
M10	$C_{11.8}H_{21.1}O_{0.1}$	45.54	7496.22	7454.64	47.84
M15	$C_{11.2}H_{20.15}O_{0.15}$	45.28	7119.62	7082.44	47.58
M20	$C_{10.6}H_{19.2}O_{0.2}$	44.99	6743.02	6710.25	47.29
M25	$C_{10}H_{18.25}O_{0.25}$	44.68	6366.42	6338.05	46.97
M30	$C_{9.4}H_{17.3}O_{0.3}$	44.32	5989.82	5965.86	46.61
M35	$C_{8.8}H_{16.35}O_{0.35}$	43.93	5613.22	5593.66	46.22
M40	$C_{8.2}H_{15.4}O_{0.4}$	43.49	5236.61	5221.47	45.77
M45	$C_{7.6}H_{14.45}O_{0.45}$	42.99	4860.01	4849.27	45.27
M50	$C_7H_{13.5}O_{0.5}$	42.42	4483.41	4477.07	44.70
M55	$C_{6.4}H_{12.55}O_{0.55}$	41.77	4106.81	4104.88	44.04
M60	$C_{5.8}H_{11.6}O_{0.6}$	41.01	3730.21	3732.68	43.27
M65	$C_{5.2}H_{10.65}O_{0.65}$	40.12	3353.60	3360.49	42.37
M70	$C_{4.6}H_{9.7}O_{0.7}$	39.05	2977.00	2988.29	41.30
M75	$C_4H_{8.75}O_{0.75}$	37.76	2600.40	2616.10	40.00
M80	$C_{3.4}H_{7.8}O_{0.8}$	36.16	2223.80	2243.90	38.38
M85	$C_{2.8}H_{6.85}O_{0.85}$	34.12	1847.20	1871.71	36.32
M90	$C_{2.2}H_{5.9}O_{0.9}$	31.44	1470.60	1499.51	33.62
M95	$C_{1.6}H_{4.95}O_{0.95}$	27.76	1093.99	1127.31	29.91
M100	CH_4O	22.39	717.39	717.64	24.49

TABLE 9.4   Chemical exergy of diesel-methanol blends

Fuel	Formula	HHV (MJ/kg)	HHV (kJ/mol)	e^{ch} (kJ/mol)	e^{ch} (MJ/kg)
Diesel[5]	$C_{13}H_{23}$	46.00	8249.43	8199.03	48.31
E5	$C_{12.45}H_{22.15}O_{0.05}$	45.79	7905.92	7859.54	48.09
E10	$C_{11.9}H_{21.3}O_{0.1}$	45.56	7562.41	7520.05	47.86
E15	$C_{11.35}H_{20.45}O_{0.15}$	45.31	7218.90	7180.55	47.61
E20	$C_{10.8}H_{19.6}O_{0.2}$	45.03	6875.38	6841.06	47.33
E25	$C_{10.25}H_{18.75}O_{0.25}$	44.74	6531.87	6501.57	47.03
E30	$C_{9.7}H_{17.9}O_{0.3}$	44.41	6188.36	6162.08	46.70
E35	$C_{9.15}H_{17.05}O_{0.35}$	44.05	5844.85	5822.59	46.34
E40	$C_{8.6}H_{16.2}O_{0.4}$	43.65	5501.34	5483.09	45.94
E45	$C_{8.05}H_{15.35}O_{0.45}$	43.21	5157.83	5143.60	45.49
E50	$C_{7.5}H_{14.5}O_{0.5}$	42.72	4814.32	4804.11	45.00
E55	$C_{6.95}H_{13.65}O_{0.55}$	42.16	4470.81	4464.62	44.44
E60	$C_{6.4}H_{12.8}O_{0.6}$	41.53	4127.30	4125.13	43.80
E65	$C_{5.85}H_{11.95}O_{0.65}$	40.81	3783.79	3785.63	43.07
E70	$C_{5.3}H_{11.1}O_{0.7}$	39.98	3440.27	3446.14	42.23
E75	$C_{4.75}H_{10.25}O_{0.75}$	39.01	3096.76	3106.65	41.25
E80	$C_{4.2}H_{9.4}O_{0.8}$	37.86	2753.25	2767.16	40.10
E85	$C_{3.65}H_{8.55}O_{0.85}$	36.48	2409.74	2427.67	38.70
E90	$C_{3.1}H_{7.7}O_{0.9}$	34.79	2066.23	2088.18	37.00
E95	$C_{2.55}H_{6.85}O_{0.95}$	46.00	8249.43	7859.54	48.31
E100	C_2H_6O	29.94	1379.21	1356.66	32.10

TABLE 9.5 Chemical exergy of diesel-ethanol blends

Fuel	Methyl soyate blend vol. %	HHV (MJ/kg)[6]	e^{ch} (MJ/kg)
Diesel	0.00	45.17	47.47
MS5	5.00	44.90	47.19
MS10	10.00	44.62	46.91
MS15	15.00	44.35	46.64
MS20	20.00	44.07	46.36
MS25	25.00	43.80	46.08
MS30	30.00	43.52	45.80
MS35	35.00	43.25	45.53
MS40	40.00	42.97	45.25
MS45	45.00	42.70	44.97
MS50	50.00	42.42	44.69
MS55	55.00	42.15	44.42
MS60	60.00	41.87	44.14
MS65	65.00	41.60	43.86
MS70	70.00	41.32	43.58
MS75	75.00	41.05	43.31
MS80	80.00	40.77	43.03
MS85	85.00	40.50	42.75
MS90	90.00	40.22	42.47
MS95	95.00	39.95	42.20
Methyl soyate	100	39.67	41.92

TABLE 9.6 Chemical exergy of diesel-methyl soyate

Fuel	Vegetable oil blend vol. %	HHV (MJ/kg)[7]	e^{ch} (MJ/kg)
Diesel	0.00	45.09	47.39
VO5	5.00	44.80	47.09
VO10	10.00	44.51	46.80
VO15	15.00	44.22	46.51
VO20	20.00	43.93	46.22
VO25	25.00	43.64	45.93
VO30	30.00	43.36	45.64
VO35	35.00	43.07	45.35
VO40	40.00	42.78	45.05
VO45	45.00	42.49	44.76
VO50	50.00	42.20	44.47
VO55	55.00	41.91	44.18
VO60	60.00	41.62	43.89
VO65	65.00	41.33	43.60
VO70	70.00	41.04	43.30
VO75	75.00	40.75	43.01
VO80	80.00	40.47	42.72
VO85	85.00	40.18	42.43
VO90	90.00	39.89	42.14
VO95	95.00	39.60	41.85
Vegetable oil	100.00	39.31	41.56

TABLE 9.7 Chemical exergy of diesel-vegetable oil blends

Fuel	Rapeseed oil blend vol. %	HHV (MJ/kg)[8]	e^{ch} (MJ/kg)
Diesel	0.00	45.09	47.39
RO5	5.00	44.80	47.09
RO10	10.00	44.51	46.80
RO15	15.00	44.22	46.51
RO20	20.00	43.93	46.22
RO25	25.00	43.65	45.93
RO30	30.00	43.36	45.64
RO35	35.00	43.07	45.35
RO40	40.00	42.78	45.06
RO45	45.00	42.49	44.77
RO50	50.00	42.20	44.48
RO55	55.00	41.91	44.18
RO60	60.00	41.63	43.89
RO65	65.00	41.34	43.60
RO70	70.00	41.05	43.31
RO75	75.00	40.76	43.02
RO80	80.00	40.47	42.73
RO85	85.00	40.18	42.44
RO90	90.00	39.90	42.15
RO95	95.00	39.61	41.86
Rapeseed oil	100.00	39.32	41.56

TABLE 9.8 Chemical exergy of diesel-rapeseed oil blends

Fuel	Rapeseed methyl ester blend vol. %	HHV (MJ/kg)[9]	e^{ch} (MJ/kg)
Diesel	0.00	45.09	47.39
RME5	5.00	45.05	47.35
RME10	10.00	44.79	47.08
RME15	15.00	44.53	46.82
RME20	20.00	44.26	46.55
RME25	25.00	44.00	46.29
RME30	30.00	43.74	46.03
RME35	35.00	43.48	45.76
RME40	40.00	43.22	45.50
RME45	45.00	42.95	45.23
RME50	50.00	42.69	44.97
RME55	55.00	42.43	44.70
RME60	60.00	42.17	44.44
RME65	65.00	41.91	44.18
RME70	70.00	41.64	43.91
RME75	75.00	41.38	43.65
RME80	80.00	41.12	43.38
RME85	85.00	40.86	43.12
RME90	90.00	40.60	42.85
RME95	95.00	40.33	42.59
Rapeseed methyl ester	100.00	40.07	42.33

TABLE 9.9 Chemical exergy of diesel-rapeseed methyl ester blends

Fuel	Butanol blend vol. %	HHV (MJ/kg)[10]	e^{ch} (MJ/kg)
Diesel	0.00	45.63	47.93
B5	5.00	45.14	47.43
B10	10.00	44.65	46.94
B15	15.00	44.16	46.45
B20	20.00	43.67	45.96
B25	25.00	43.19	45.47
B30	30.00	42.70	44.97
B35	35.00	42.21	44.48
B40	40.00	41.72	43.99
B45	45.00	41.23	43.50
B50	50.00	40.75	43.00
B55	55.00	40.26	42.51
B60	60.00	39.77	42.02
B65	65.00	39.28	41.53
B70	70.00	38.79	41.04
B75	75.00	38.31	40.54
B80	80.00	37.82	40.05
B85	85.00	37.33	39.56
B90	90.00	36.84	39.07
B95	95.00	36.35	38.57
Butanol	100.00	35.87	36.06

TABLE 9.10 Chemical exergy of diesel-butanol blends

Fuel	Pentanol blend vol. %	HHV (MJ/kg)[10]	e^{ch} (MJ/kg)
Diesel	0.00	45.88	48.19
P5	5.00	45.49	47.79
P10	10.00	45.09	47.39
P15	15.00	44.70	46.99
P20	20.00	44.30	46.59
P25	25.00	43.91	46.20
P30	30.00	43.51	45.80
P35	35.00	43.12	45.40
P40	40.00	42.73	45.00
P45	45.00	42.33	44.60
P50	50.00	41.94	44.21
P55	55.00	41.54	43.81
P60	60.00	41.15	43.41
P65	65.00	40.75	43.01
P70	70.00	40.36	42.61
P75	75.00	39.96	42.22
P80	80.00	39.57	41.82
P85	85.00	39.17	41.42
P90	90.00	38.78	41.02
P95	95.00	38.39	40.62
Pentanol	100.00	37.99	37.55

TABLE 9.11 Chemical exergy of diesel-pentanol blends

Name	Formula	LHV (MJ/kg)	HHV (MJ/kg)	e^{ch} (MJ/kg)	Ref.
Liquid natural gas		48.63	55.21	57.62	[11]
Jet fuel		43.31	46.12	48.65	[12]
Diesel oil		37.03	39.50	42.12	[12]
#1 Diesel		40.34	42.99	45.57	[13]
#2 Diesel		40.08	42.72	45.29	[13]
FT naphtha		43.89	46.73	49.25	[14]
Methyl T-butyl ether	$C_5H_{12}O$	32.67	34.90	37.58	[15]
Residual oil		39.31	41.90	44.49	[16]
Rice bran oil biodiesel		43.50	46.32	48.85	[17]
Bioethanol		36.30	38.73	41.36	[17]
Rubber seed oil		24.98	26.80	29.59	[18]
Jatropha oil		35.13	37.50	40.15	[19]
Jatropha oil methyl ester		39.77	41.10	43.70	[19]
Mahua oil		38.45	41.60	44.19	[20]
Mahua biodiesel		33.71	36.00	38.67	[20]
Raw sunflower oil	$C_{57}H_{103}O_6$	34.66	37.00	39.65	
Sunflower methyl ester	$C_{55}H_{105}O_6$	37.05	39.53	42.15	
Raw cottonseed oil	$C_{55}H_{102}O_6$	38.05	40.58	43.19	
Cottonseed methyl ester	$C_{54}H_{101}O_6$	37.17	39.65	42.27	
Raw soybean oil	$C_{56}H_{102}O_6$	38.06	40.58	43.19	[21]
Soybean methyl ester	$C_{53}H_{101}O_6$	37.15	39.62	42.24	
Corn oil	$C_{56}H_{103}O_6$	37.28	39.76	42.38	
Opium poppy oil	$C_{57}H_{103}O_6$	35.44	37.83	40.47	
Rapeseed oil	$C_{57}H_{105}O_6$	36.48	38.92	41.55	
Bio-oil		20.10	21.66	24.51	[22]
Glycerol		17.48	18.90	21.79	[22]

TABLE 9.12 Chemical exergy of common liquid fuels

References

1. Ghazikhani M, Hatami M, Safari B. The effect of alcoholic fuel additives on exergy parameters and emissions in a two stroke gasoline engine. *Arabian Journal for Science and Engineering* 39(3):2117–2125 (2014). doi: 10.1007/s13369-013-0738-3.
2. "Fuels—higher and lower calorific values." https://www.engineeringtoolbox.com/fuels-higher-calorific-values-d_169.html (accessed December 23, 2022).
3. "Lower and higher heating values of fuels | hydrogen tools." https://h2tools.org/hyarc/calculator-tools/lower-and-higher-heating-values-fuels (accessed December 23, 2022).
4. Canakci M, Ozsezen AN, Alptekin E, Eyidogan M. Impact of alcohol–gasoline fuel blends on the exhaust emission of an SI engine. *Renewable Energy* 52:111–117 (2013). doi: 10.1016/j.renene.2012.09.062.
5. Jamrozik A. The effect of the alcohol content in the fuel mixture on the performance and emissions of a direct injection diesel engine fueled with diesel-methanol and diesel-ethanol blends. *Energy Conversion and Management* 148:461–476 (2017). doi: 10.1016/j.enconman.2017.06.030.
6. Ali Y, Hanna MA, Leviticus LI. Emissions and power characteristics of diesel engines on methyl soyate and diesel fuel blends. *Bioresource Technology* 52(2):185–195 (1995). doi: 10.1016/0960-8524(95)00024-9.
7. Wang YD, Al-Shemmeri A, Eames P, McMullan J, Hewitt N, Huang Y, Rezvani S. An experimental investigation of the performance and gaseous exhaust emissions of a diesel engine using blends of a vegetable oil. *Applied Thermal Engineering* 26(14):1684–1691 (2006). doi: 10.1016/j.applthermaleng.2005.11.013.
8. Nwafor OMI, Rice G. Performance of rapeseed oil blends in a diesel engine. *Applied Energy* 54(4):345–354 (1996). doi: 10.1016/0306-2619(96)00004-9.
9. Nwafor OMI. Emission characteristics of diesel engine operating on rapeseed methyl ester. *Renewable Energy* 29(1):119–129 (2004). doi: 10.1016/S0960-1481(03)00133-2.
10. Campos-Fernández J, Arnal JM, Gómez J, Dorado MP. A comparison of performance of higher alcohols/diesel fuel blends in a diesel engine. *Applied Energy* 95:267–275 (2012). doi: 10.1016/j.apenergy.2012.02.051.
11. Heating values of hydrogen and fuels. ESSOM. [Online]. Available: https://chemeng.queensu.ca/courses/CHEE332/files/ethanol_heating-values.pdf.
12. Jaber JO, Al-Ghandoor A, Sawalha SA. Energy analysis and exergy utilization in the transportation sector of Jordan. *Energy Policy* 36(8):2995–3000 (2008). doi: 10.1016/j.enpol.2008.04.004.
13. Tat ME, Van Gerpen JH. The specific gravity of biodiesel and its blends with diesel fuel. *Journal of the American Oil Chemists' Society* 77(2):115–119 (2000). doi: 10.1007/s11746-000-0019-3.
14. Dancuart LP. Process for producing synthetic naphtha fuel and synthetic naphtha fuel produced by that process. US Patent 6,656,343 B2, December 2, 2003.
15. Yao C, Li J, Li Q, Huang C, Wei L, Wang J, Tian Z, Li Y, Qi F. Study on combustion of gasoline/MTBE in laminar flame with synchrotron radiation. *Chemosphere* 67(10):2065–2071 (2007). doi: 10.1016/j.chemosphere.2006.10.063.
16. Bertran CA, Marques CST. Study of the particulate matter emitted from residual oil combustion and natural gas reburning. *Journal of the Brazilian Chemical Society* 15:548–555 (2004). doi: 10.1590/S0103-50532004000400017.
17. Venkata Subbaiah G, Raja Gopal K. An experimental investigation on the performance and emission characteristics of a diesel engine fuelled with rice bran biodiesel and ethanol blends. *International Journal of Green Energy* 8(2):197–208 (2011). doi: 10.1080/15435075.2010.548539.
18. Geo VE, Nagarajan G, Nagalingam B. Experimental investigations to improve the performance of rubber seed oil–fueled diesel engine by dual fueling with hydrogen. *International Journal of Green Energy* 6(4):343–358 (2009). doi: 10.1080/15435070903106991.
19. Lakshmi Narayana Rao G, Durga Prasad B, Sampath S, Rajagopal K. Combustion analysis of diesel engine fueled with jatropha oil methyl ester—diesel blends. *International Journal of Green Energy* 4(6):645–658 (2007). doi: 10.1080/15435070701665446.
20. Ghadge SV, Raheman H. Biodiesel production from mahua (Madhuca indica) oil having high free fatty acids. *Biomass and Bioenergy* 28(6):601–605 (2005). doi: 10.1016/j.biombioe.2004.11.009.
21. Altın R, Çetinkaya S, Yücesu HS. The potential of using vegetable oil fuels as fuel for diesel engines. *Energy Conversion and Management* 42(5):529–538 (2001). doi: 10.1016/S0196-8904(00)00080-7.
22. Zhang M, Wu H. Phase behavior and fuel properties of bio-oil/glycerol/methanol blends. *Energy and Fuels* 28(7):4650–4656 (2014). doi: 10.1021/ef501176z.

This chapter contains the chemical exergy of solid fuels. Solid fuels are often heterogenous mixtures of complex chemical compounds that are difficult or impossible to identify and classify exactly. Coal, for example, varies from mine to mine and chunk to chunk. This makes it difficult to determine chemical exergy purely based on theory, since some underlying thermodynamic properties such as the entropy of the reaction cannot be determined with a reasonable degree of accuracy.[1] As a result, chemical exergies of heterogeneous solid fuels cannot be determined directly from Gibbs free energies and must be estimated through some other technique.

One approach is to create an empirical equation that computes chemical exergy as a function of the composition of constituent atoms using data from pure chemicals. Then, it can be applied directly to the solid fuel. For this work, we use the following empirical model from Ref. 2:

$$e^{ch} = 362.008\ C + 1101.841\ H - 86.218\ O + 2.418\ N + 196.701\ S - 21.1\ A \qquad (10.1)$$

where e^{ch} is the chemical exergy of the solid fuel (kJ/kg), and C, H, O, N, S, and A are the mass percentage of atomic C, H, O, N, S, and ash in the solid fuel on a dry basis (DB) (e.g., use C = 80 if the solid is 80% carbon by mass).

The model is accurate for most useful solids. However, as an empirical model, it is limited to the range of data from which it was made. As such, it cannot be used for very weak, high-ash coals ($\geq$71.4 wt.% ash).[2] If the model is used outside of this range, it can result in meaningless results (such as negative chemical exergies). We cross-checked this model against other competing models in Refs. 34 and 38 and found them to be sufficiently similar.

The tables are broken down by solid fuel type as follows:

- Anthracite coals   Table 10.1   p252
- Bituminous coals   Table 10.2   p257
- Subbituminous coals   Table 10.3   p269
- Lignite coals   Table 10.4   p275
- Other coals   Table 10.5   p280
- Biomasses   Table 10.6   p286
- Wastes from Human Activities   Table 10.7   p292

The literature sources characterize the fuels in different ways, either DB, as received (AR), or dry ash free (DAF). We have presented the proximate analysis and ultimate analysis data strictly as reported in the data source. Table 10.1 to 10.4 are AR. Tables 10.5 to 10.7 include some DAF or DB cases. Conversions can be made between AR, DB, and DAF using:

$$DB = (DAF)(100 - Ash)/100 \qquad (10.2)$$

$$DB = (AR)(100)/(100 - moisture) \qquad (10.3)$$

In Table 10.4 to 10.6, unless otherwise specified, the ash and element weight percentage are reported using DB.

Continent	Country	Primary field ID	Coal basin or field	Moisture	Ash	V.M.	F.C.	Wt.% H	C	N	S	O	e^{ch} (MJ/kg)
Asia	China	CN-35-20	Jincheng	4.11	8.75	9.72	77.42	2.48	80.85	1.06	0.30	2.45	33.02
Asia	China	CN-35-21	Jincheng	5.13	9.79	8.00	77.08	2.26	77.76	0.86	1.36	2.84	32.10
Asia	China	CN-35-22	Yicheng-Qinshui	1.58	18.67	8.84	70.91	2.50	73.20	1.20	0.56	2.29	29.24
Asia	China	CN-35-23	Gaoping	1.59	14.70	10.93	72.78	2.96	76.38	1.34	0.26	2.77	30.91
Asia	China	CN-35-24	Yicheng-Qinshui	1.71	18.08	7.10	73.11	1.60	74.19	0.69	3.10	0.63	29.30
Asia	China	CN-35-25	Yicheng-Qinshui	1.14	12.73	9.19	76.94	3.18	78.55	1.16	0.46	2.78	31.89
Asia	China	CN-35-26	Yicheng-Qinshui	2.68	14.36	7.55	75.41	2.31	76.62	1.05	0.30	2.68	30.63
Asia	China	CN-35-27	Gaoping	1.26	18.18	11.30	69.26	2.92	70.18	0.95	4.28	2.23	29.26
Asia	China	CN-35-28	Gaoping	3.75	20.23	12.90	63.12	2.65	63.73	0.84	3.97	4.83	26.94
Asia	China	CN-35-39	Yangquan-Yuxian	1.87	8.05	9.19	80.89	3.30	82.88	1.11	0.36	2.43	33.97
Asia	China	CN-35-40	Yangxian-Yuxian	2.05	18.58	9.27	70.10	2.94	70.29	1.04	1.75	3.35	28.94
Asia	China	CN-35-41	Yangquan-Yuxian	2.40	33.79	10.98	52.83	2.53	54.79	0.79	0.93	4.77	22.22
Asia	China	CN-35-42	Yangxian-Yuxian	2.81	13.13	12.78	71.28	2.74	75.05	1.16	2.67	2.44	31.10
Asia	China	CN-35-43	Pingxi	2.83	14.78	8.78	73.61	2.63	74.66	1.13	1.10	2.87	30.45
Asia	China	CN-35-46	Yangquan-Yuxian	3.22	13.27	8.46	75.05	3.02	73.91	1.12	2.06	3.40	30.91
Asia	China	CN-01-01	Mentougou	4.65	14.63	5.52	75.20	0.53	76.13	0.26	0.17	3.63	28.90
Asia	China	CN-20-01	Meixiang	4.79	12.99	3.88	78.34	1.22	77.91	0.58	0.77	1.74	30.75
Asia	China	CN-20-02	Quren	2.71	12.66	12.08	72.55	3.15	76.65	1.25	0.48	3.10	31.64
Asia	China	CN-27-02		4.79	25.37	8.56	61.28	1.77	60.39	0.52	5.49	1.67	25.44
Asia	China	CN-27-03		2.89	33.41	10.81	52.89	2.20	53.02	1.13	4.70	2.65	22.25
Asia	China	CN-28-06	Furong	1.69	23.74	19.51	55.06	2.33	66.03	0.72	3.54	1.95	26.96
Asia	China	CN-28-13	Songzao	2.93	25.53	9.99	61.55	2.69	62.54	1.06	2.39	2.86	26.06
Asia	China	CN-28-14	Songzao	5.85	15.36	10.11	68.68	2.85	69.66	1.19	3.07	2.02	30.23
Asia	China	CN-31-05	Fengfeng	4.91	16.40	4.73	73.96	1.57	72.98	1.00	0.51	2.63	29.11
Asia	China	CN-95-03	Rujigou	1.42	6.90	9.80	81.88	2.95	84.90	0.74	0.30	2.79	34.15
Asia	China	CN-37-01	Xinmi	1.73	11.06	11.42	75.79	3.42	79.88	1.41	0.27	2.23	32.88
Asia	China	CN-37-03	Xinmi	2.13	16.71	15.70	65.46	3.24	71.44	1.50	0.74	4.24	29.49
Asia	China	CN-37-04	Yinggong	7.52	16.74	7.38	68.36	1.74	68.67	0.59	3.15	1.59	29.09

TABLE 10.1 Chemical exergy of anthracite coals. Coal compositions are as-received as listed in Ref. 3.

Continent	Country	Primary field ID	Coal basin or field	Wt.%									e^{ch} (MJ/kg)
				Moisture	Ash	V.M.	F.C.	H	C	N	S	O	
Asia	China	CN-37-21	Yanlong	4.56	29.69	9.02	56.73	2.27	58.51	1.03	0.51	3.43	23.95
Asia	China	CN-37-22	Jiaozuo	5.28	25.28	7.17	62.27	2.11	63.16	0.95	0.22	3.00	25.80
Asia	China	CN-37-24	Yongxia	1.36	16.10	8.93	73.61	2.93	75.53	1.08	0.40	2.60	30.50
Asia	China	CN-59-01	Yong'an	5.05	20.05	5.32	69.58	0.76	69.89	0.47	0.94	2.84	27.03
Asia	China	CN-59-02	Tianhushan	7.87	14.49	2.85	74.79	0.43	74.27	0.44	0.09	2.41	29.16
Asia	China	CN-59-03	Longyong	5.61	22.13	5.22	67.04	1.38	67.04	0.62	0.56	2.66	26.71
Asia	China	CN-73-02	Huanfengqiao-Lanchun	4.52	24.00	3.57	67.91	0.62	67.27	0.44	0.73	2.42	25.63
Asia	China	CN-73-03	Baisha	6.01	16.39	3.97	73.63	1.41	72.65	1.08	0.81	1.65	29.28
Asia	China	CN-73-04	Baisha	3.94	11.29	6.13	78.64	2.49	78.35	1.27	0.62	2.04	32.08
Asia	China	CN-73-05	Jiedong	6.78	6.41	8.32	78.49	2.51	80.42	1.23	0.67	1.98	34.01
Asia	China	CN-73-08	Lengshuijiang	7.62	9.20	6.21	76.97	2.30	77.09	0.71	0.50	2.58	32.61
Asia	China	CN-77-05	Hongmao	1.24	12.80	9.07	76.89	3.13	77.58	0.77	3.07	1.41	32.15
Asia	China	CN-79-04	Pingxiang	6.14	20.83	8.37	64.66	2.74	65.61	1.17	0.74	2.77	27.96
Asia	China	CN-79-07		2.47	30.76	9.83	56.94	2.51	56.83	0.75	3.15	3.53	23.59
Asia	China	CN-85-01		3.72	18.90	11.44	65.94	2.81	67.54	1.09	2.83	3.11	28.50
Asia	China	CN-85-02		1.56	22.81	8.80	66.83	2.79	63.74	0.69	5.70	2.71	26.97
Asia	China	CN-85-03		4.97	23.82	6.72	64.49	2.39	64.72	0.90	0.31	2.89	26.71
Asia	China	CN-85-04		7.71	25.45	7.96	58.88	2.58	59.93	0.85	0.93	2.55	25.96
Asia	China	CN-85-13	Anshun	11.09	19.42	10.69	58.80	2.34	59.79	0.93	2.48	3.95	26.95
Asia	China	CN-85-14		3.61	21.44	9.77	65.18	2.27	67.11	0.86	2.69	2.02	27.70
Asia	DPR Korea	NK-1	Pyongyang	2.86	16.73	7.90	72.51	2.17	73.85	0.81	0.83	2.75	29.55
Asia	DPR Korea	NK-2	Pyongyang	4.43	8.56	2.51	84.50	0.98	83.17	0.56	0.96	1.34	32.52
Asia	DPR Korea	NK-3	Pyongyang	2.57	12.71	3.38	81.34	0.87	80.76	0.37	0.51	2.21	30.62
Asia	DPR Korea	NK-4	Kangso	3.79	8.00	5.74	82.47	2.01	82.44	0.93	1.05	1.78	33.20
Asia	DPR Korea	NK-5	Kangso	4.39	15.95	4.59	75.07	1.49	75.06	0.82	0.75	1.54	29.80
Asia	DPR Korea	NK-6	Kangso	4.84	19.28	2.99	72.89	0.73	72.12	0.62	0.20	2.21	27.70
Asia	DPR Korea	NK-7	Sunchon	3.37	5.28	5.67	85.68	0.76	85.65	0.22	0.12	4.60	32.45

TABLE 10.1 Chemical exergy of anthracite coals. Coal compositions are as-received as listed in Ref. 3. (Continued)

Continent	Country	Primary field ID	Coal basin or field	Moisture	Wt.% Ash	V.M.	F.C.	H	C	N	S	O	e^{ch} (MJ/kg)
Asia	DPR Korea	NK-8	Sunchon	1.52	5.17	7.25	86.06	0.79	85.98	0.30	0.12	6.12	31.87
Asia	DPR Korea	NK-9	Sunchon	1.32	12.42	8.31	77.95	0.80	78.89	0.29	0.09	6.19	29.05
Asia	DPR Korea	NK-10	Sunchon	2.49	9.87	6.47	81.17	0.79	80.13	0.21	0.14	6.37	29.89
Asia	DPR Korea	NK-11	Kaechon	3.67	14.26	6.34	75.73	0.77	74.41	0.20	0.15	6.54	27.98
Asia	DPR Korea	NK-12	Kaechon	6.08	27.74	4.96	61.22	0.53	60.65	0.22	0.18	4.60	22.99
Asia	DPR Korea	NK-13	Kaechon	6.83	18.42	4.88	69.87	0.44	69.58	0.23	0.20	4.30	26.78
Asia	DPR Korea	NK-14	Pukchang	1.70	6.64	7.54	84.12	0.91	84.07	0.19	0.14	6.35	31.31
Asia	DPR Korea	NK-15	Pukchang	6.86	11.98	3.13	78.03	0.38	77.60	0.26	0.72	2.20	30.29
Asia	DPR Korea	NK-16	Pukchang	2.68	13.64	5.44	78.24	0.65	78.35	0.27	0.22	4.19	29.26
Asia	DPR Korea	NK-17	Tokchon	3.46	12.69	4.18	79.67	0.62	77.53	0.31	0.23	5.16	29.09
Asia	DPR Korea	NK-18	Tokchon	3.52	18.96	4.00	73.52	0.55	73.46	0.23	0.25	3.03	27.56
Asia	DPR Korea	NK-19	Tokchon	3.37	11.51	4.26	80.86	0.64	79.43	0.25	0.25	4.55	29.88
Asia	DPR Korea	NK-20	Tokchon	3.11	13.69	4.35	78.85	0.62	78.83	0.33	0.34	3.08	29.66
Asia	DPR Korea	NK-21	Tokchon	4.10	14.40	5.55	75.95	0.71	75.67	0.36	0.19	4.57	28.69
Asia	DPR Korea	NK-22	Tokchon	3.36	12.82	5.31	78.51	0.62	78.49	0.32	0.20	4.19	29.50
Asia	DPR Korea	NK-23	Kujang	6.81	10.36	4.21	78.62	0.48	77.46	0.24	0.14	4.51	30.04
Asia	DPR Korea	NK-24	Kujang	3.05	14.53	5.84	76.58	0.65	76.10	0.27	0.40	5.00	28.48
Asia	DPR Korea	NK-25	Kowon-Munchon	1.16	8.75	5.09	85.00	0.74	83.92	0.38	0.38	4.67	31.04
Asia	DPR Korea	NK-26	Kowon-Munchon	2.58	6.73	4.47	86.22	0.71	85.40	0.34	0.36	3.88	32.12
Asia	DPR Korea	NK-27	Kowon-Munchon	3.50	7.17	4.60	84.73	0.63	84.20	0.28	0.32	3.90	31.87
Asia	DPR Korea	NK-28	Kowon-Munchon	4.84	15.13	3.63	76.40	0.49	75.52	0.29	0.37	3.36	28.73
Asia	DPR Korea	NK-29	Jonchon	1.22	54.36	7.13	37.29	1.51	38.38	0.43	0.61	3.49	14.41
Asia	DPR Korea	NK-30	Jonchon	3.38	47.69	8.07	40.86	1.63	41.88	0.36	0.54	4.52	16.22
Asia	DPR Korea	NK-31	Kangso	5.89	20.62	5.39	68.10	1.15	68.12	0.65	0.34	3.23	26.86
Asia	DPR Korea	NK-32	Anju	12.82	32.18	30.02	24.98	3.17	37.26	0.55	0.14	13.88	17.36
Asia	DPR Korea	NK-33	Anju	9.49	50.51	22.23	17.77	2.27	25.95	0.50	0.45	10.83	11.03
Asia	DPR Korea	NK-34	Anju	12.88	29.43	29.92	27.77	3.05	38.63	0.78	2.54	12.69	18.52
Asia	DPR Korea	NK-35	Anju	15.75	12.63	37.62	34.00	4.05	50.59	0.81	0.22	15.95	25.14

TABLE 10.1 Chemical exergy of anthracite coals. Coal compositions are as-received as listed in Ref. 3. (Continued)

Continent	Country	Primary field ID	Coal basin or field	Moisture	Ash	V.M.	F.C.	Wt.% H	C	N	S	O	e^{ch} (MJ/kg)
Asia	DPR Korea	NK-36	Sariwon	18.72	9.69	34.63	36.96	3.56	49.17	1.14	3.13	14.59	25.69
Asia	Republic of Korea	Jang Seong Mine	Samcheog	7.64	10.00	2.93	79.43	0.76	78.04	0.38	0.97	2.22	31.27
Asia	Republic of Korea	Doe Gae Mine	Samcheog	6.68	37.50	4.27	51.55	0.72	51.43	0.42	0.32	2.93	19.75
Asia	Republic of Korea	Kyung Dong Mine	Samcheog	3.49	26.64	3.64	66.23	0.72	64.39	0.45	0.19	4.12	24.06
Asia	Republic of Korea	Han Bo Mine	Samcheog	7.41	35.18	3.49	53.92	0.67	52.74	0.41	0.25	3.34	20.36
Asia	Republic of Korea	Tae Back Mine	Samcheog	3.46	14.92	3.17	78.45	0.87	77.56	0.27	0.14	2.78	29.53
Asia	Republic of Korea	Sam Tan Mine	Samcheog	7.17	16.48	2.82	73.56	0.69	72.48	0.50	0.19	2.49	28.52
Asia	Republic of Korea	Dong Won Mine	Samcheog	4.73	9.88	3.28	82.11	0.67	80.50	0.26	0.22	3.74	30.85
Asia	Republic of Korea	Young Wol Mine	Danyang (Youngwol)	5.50	7.72	4.50	82.28	1.82	81.78	0.72	0.36	2.11	33.16
Asia	Republic of Korea	Ma Ro Mine	Boeun	2.81	17.10	2.34	77.75	0.20	77.89	0.07	0.03	1.90	28.71
Asia	Republic of Korea	Tae Meag Mine	Munkyung	4.71	24.45	2.82	68.02	0.23	64.54	0.05	0.39	5.63	23.81
Asia	Republic of Korea	Hwa Sun Mine	Honam	7.25	15.78	2.82	74.15	0.48	73.40	0.37	0.14	2.58	28.65
Asia	Vietnam	3	Cam Pha	2.08	2.68	6.77	88.47	3.28	89.33	0.98	0.46	1.19	36.65
Asia	Vietnam	6	Cam Pha	2.48	5.26	6.55	85.71	3.22	85.56	1.00	0.59	1.89	35.24
Asia	Vietnam	9	Cam Pha	2.83	10.72	7.88	78.57	3.07	80.15	1.06	0.62	1.55	33.10
Asia	Vietnam	8A	Cam Pha	2.30	18.18	7.17	72.35	2.98	72.76	1.19	0.51	2.08	29.85
Asia	Vietnam	10A	Cam Pha	2.01	25.91	7.14	64.94	2.74	65.39	1.00	0.43	2.53	26.55
Asia	Vietnam	10C	Vang danh	3.52	15.30	4.03	77.15	1.52	75.75	0.60	0.79	2.52	29.76
Asia	Malaysia	SI	Silantek	2.09	5.74	23.62	68.55	4.86	81.16	2.37	0.84	2.94	35.27

TABLE 10.1 Chemical exergy of anthracite coals. Coal compositions are as-received as listed in Ref. 3. (Continued)

Continent	Country	Primary field ID	Coal basin or field	Moisture	Ash	V.M.	F.C.	Wt.% H	C	N	S	O	e^{ch} (MJ/kg)
South America	Peru	Peru BP2400	Alto Chicama	5.07	5.86	2.72	86.35	1.08	86.00	0.56	0.27	1.16	33.87
South America	Peru	Peru BP2500	Alto Chicama	5.77	2.17	1.65	90.41	1.03	88.18	0.54	0.28	2.03	34.91
South America	Peru	Peru BP2700	Alto Chicama		8.73	4.12	83.93	1.15	82.90	0.32	0.44	6.46	30.62
South America	Peru	Peru BP2900	Alto Chicama	3.22	9.02	4.26	86.72	0.46	85.66	0.33	0.45	0.86	32.38
South America	Peru	Peru BP3100	Alto Chicama	4.41	3.57	1.67	90.35	1.00	88.13	0.65	0.53	1.71	34.41
South America	Peru	Peru BP3200	Alto Chicama	2.12	61.57	12.37	23.94	1.97	24.99	0.73	0.67	7.95	9.57
South America	Peru	BP0101	Alto Chicama	4.28	30.95	5.85	58.92	0.75	59.22	0.60	0.49	3.71	22.35
South America	Peru	BP0102	Alto Chicama	7.02	28.87	7.87	56.24	0.48	55.96	0.45	0.67	6.55	21.24
South America	Peru	BP0103	Alto Chicama	5.95	22.28	6.80	64.97	0.41	66.37	0.47	0.38	4.14	25.23
South America	Peru	BP0104	Alto Chicama	4.37	10.07	3.38	82.18	0.83	81.45	0.71	0.34	2.23	31.44
South America	Peru	BP0105	Alto Chicama	2.60	11.39	4.48	81.53	2.04	80.70	0.79	0.92	1.56	32.10
South America	Peru	BP0106	Alto Chicama	5.09	25.77	3.40	65.74	0.91	63.93	0.67	0.44	3.19	24.67
South America	Peru	BP0107	Alto Chicama	5.59	16.22	4.38	73.81	1.43	72.67	0.73	0.61	2.75	29.05
South America	Peru	BP0108	Alto Chicama	15.78	26.72	12.55	44.95	0.31	46.84	0.62	1.50	8.23	19.38
South America	Peru	BP0501	Alto Chicama	4.72	3.22	1.78	90.28	1.15	87.95	0.55	0.36	2.05	34.56

TABLE 10.1 Chemical exergy of anthracite coals. Coal compositions are as-received as listed in Ref. 3. (Continued)

Continent	Country	Primary field ID	Coal basin or field	Moisture	Ash	V.M.	F.C.	Wt.%					e^{ch} (MJ/kg)
								H	C	N	S	O	
Africa	Egypt	EG-1 C2 MAIN	Maghara	3.83	14.05	47.23	34.89	5.11	62.74	1.04	2.59	10.64	28.74
Africa	Egypt	EG-2 C2 TAIL	Maghara	3	5.04	55.39	36.57	5.99	72.35	1.11	2.76	9.75	33.39
Africa	Egypt	EG-3 C3 MAIN	Maghara	4.11	21.31	45.51	29.07	4.33	53.1	0.82	5.48	10.85	24.70
Africa	Egypt	EG-4 C3 TAIL	Maghara	3.03	9.1	51.59	36.28	5.4	68.6	1.17	3.83	8.87	31.54
Africa	Egypt	EG-5 C4 MAIN	Maghara	3.17	27.4	42.68	26.75	4.54	50.86	0.73	4.13	9.18	23.61
Africa	Egypt	EG-6 C4 TAIL	Maghara	4.16	12.5	46.22	37.12	5.16	63.71	1.14	3.52	9.81	29.56
Africa	Egypt	EG-7 WASHED	Maghara	2.99	6.21	53.22	37.58	6.02	71.96	1.2	2.71	8.92	33.32
Africa	Egypt	EG-8 FILTER	Maghara	3.76	44	32.23	20.01	3.08	35.24	0.53	2.27	11.12	15.29
Africa	Egypt	EG-9 B4/12	Maghara	2.51	2.17	55.19	40.13	6.36	75.75	1.14	1.89	10.18	34.75
Africa	Egypt	EG-10 B4/6	Maghara	3.4	4.3	48.59	43.71	5.6	72.41	1.37	2.8	10.12	33.10
Africa	Mozambique	WOCQI MZ-1	Moatize	0.96	18.63	15.55	64.86	3.39	72.11	1.8	0.75	2.36	29.68
Africa	Nigeria	Lafia Obi—no. 14	Lafia Obi coal field, middle Benue trough	6.4	14.26	25.65	53.69	3.7	63.28	1.55	0.76	10.05	27.75
Africa	Nigeria	Lafia Obi—no. 15	Lafia Obi coal field, middle Benue trough	10.77	14.66	24.19	50.38	3.38	59.25	1.4	0.7	9.84	27.07
Africa	Nigeria	Lafia Obi—no. 16	Lafia Obi coal field, middle Benue trough	17.47	39.25	15.82	27.46	1.7	31.82	0.81	0.51	8.44	14.47
Asia	Afghanistan	DSF-DTR-3/06-1.3	Darrah-i-Suf	3.99	22.93	25.13	47.95	3.96	59.76	0.81	0.62	7.93	25.99
Asia	Afghanistan	DSF-DTR-3/06-1.1	Darrah-i-Suf	4.43	25.55	22.94	47.08	3.73	57.22	0.74	0.48	7.85	24.80
Asia	Afghanistan	DSF-DTR-3/06-2.3	Darrah-i-Suf	5.24	9.91	28.36	56.49	4.44	71.71	0.76	0.57	7.37	31.79
Asia	Afghanistan	DSF-DTR-3/06-2.1	Darrah-i-Suf	4.99	10.51	27.6	56.9	4.34	71.73	0.73	0.52	7.18	31.59
Asia	Afghanistan	DSF-DTR-3/06-3.3	Darrah-i-Suf	3.61	22.85	28.4	45.14	4.26	59.23	1.22	0.72	8.11	26.04
Asia	Afghanistan	DSF-DTR-3/06-3.1	Darrah-i-Suf	4.56	10.48	29.46	55.5	4.51	71.63	0.73	0.57	7.52	31.58
Asia	Afghanistan	DSF-DDS-3/06-3/1	Darrah-i-Suf	6.68	8.32	26.22	58.78	4.26	71.86	0.67	0.49	7.72	32.11
Asia	Afghanistan	DSF-DTR-6/06-A4.1	Darrah-i-Suf	4.03	24.31	24.77	46.89	3.81	58.67	0.74	0.54	7.9	25.37
Asia	Afghanistan	DSF-DTR-6/06-A4.3	Darrah-i-Suf	3.79	28.24	24.78	43.19	3.66	54.07	0.69	0.51	9.04	23.21
Asia	Afghanistan	DSF-DTR-6/06-A4.5	Darrah-i-Suf	4.05	15.26	28.05	52.64	4.3	65.55	0.91	0.48	9.45	28.59
Asia	Afghanistan	DSF-DTR-6/06-ROM1	Darrah-i-Suf	2.55	17.12	27.76	52.57	4.28	65.61	0.8	0.55	9.09	28.15
Asia	Afghanistan	DSF-SSK-3/06-A1	Darrah-i-Suf	4.06	33.84	26.09	36.01	3.7	47.6	0.81	0.59	9.4	20.74

TABLE 10.2 Chemical exergy of bituminous coals. Coal compositions are as-received as listed in Ref. 3.

Continent	Country	Primary field ID	Coal basin or field	Moisture	Ash	V.M.	F.C.	Wt.%					e^{ch} (MJ/kg)
								H	C	N	S	O	
Asia	Afghanistan	DSF-SSK-3/06-A2	Darrah-i-Suf	17.52	11.37	28.63	42.48	3.06	51.61	1.06	0.45	14.93	25.00
Asia	Afghanistan	DSF-SSK-3/06-3.1	Darrah-i-Suf	3.46	5.53	36.1	54.91	5.1	74.98	1.66	0.72	8.55	33.20
Asia	Afghanistan	DSF-SSK-6/06-ROM1	Darrah-i-Suf	5.56	8.46	34.01	51.97	4.56	68.76	1.14	0.51	11.01	30.59
Asia	Afghanistan	DSF-SSK-6/06-ROM2	Darrah-i-Suf	6.69	16.01	32.42	44.88	4.26	61.32	1.06	0.49	10.17	27.62
Asia	Afghanistan	DSF-SSK-6/06-1	Darrah-i-Suf	4.36	17.02	31.36	47.26	4.45	62.93	1.19	0.57	9.48	27.84
Asia	Afghanistan	TKR-GAZ-5/05-1.1	Farkar	3.05	6.51	24.82	65.62	4.23	80.98	0.77	0.54	3.92	34.66
Asia	Afghanistan	TKR-GAZ-5/05-1.2	Farkar	2.3	10.84	20.61	66.25	3.72	77.63	0.7	0.69	4.12	32.51
Asia	Afghanistan	TKR-GAZ-5/05-2.1	Farkar	15.28	16.76	22.26	45.7	1.74	52.3	0.5	0.27	13.15	22.92
Asia	Afghanistan	TKR-GAZ-5/05-2.2	Farkar	15.8	13.87	22.49	47.84	1.68	54.57	0.46	0.24	13.38	24.00
Asia	Afghanistan	FRK-GAZ-6/06-1.1	Farkar	2.18	27.69	19.4	50.73	3.75	60.53	0.66	0.69	4.5	25.77
Asia	Afghanistan	FRK-GAZ-6/06-A2-ROM	Farkar	10.49	41.92	15.3	32.29	2.03	36.07	0.52	0.31	8.66	15.33
Asia	Afghanistan	FRK-GAZ-6/06-A3-ROM	Farkar	4.14	15.05	20.62	60.19	4.12	70.77	0.73	0.87	4.32	30.92
Asia	Afghanistan	KTW-KLGN-MMI-04-I	Katawaz	7.96	53.78	16.01	22.25	1.72	29.02	0.34	1.46	5.72	12.02
Asia	Afghanistan	KTW-PKRI-MMI-04-II	Katawaz	2.2	57.06	17.1	23.64	1.97	31.23	0.36	1.58	5.6	12.38
Asia	Afghanistan	PK-AND-3/04-1.1	Pol-e-Khomri	8.21	41.2	27.01	23.58	2.62	37.24	0.54	0.55	9.64	16.10
Asia	Afghanistan	PK-AND-3/04-1.2	Pol-e-Khomri	10.84	21.01	30.16	37.99	3.66	53.81	0.83	0.67	9.18	25.14
Asia	Afghanistan	PK-AND-3/04-1.3	Pol-e-Khomri	8.5	49.08	22.37	20.05	2.27	31.03	0.39	0.61	8.12	13.25
Asia	Afghanistan	PK-DKS-3/04-1	Pol-e-Khomri	14.97	26.44	26.66	31.93	3.11	45.99	0.58	2.04	6.88	22.73
Asia	Afghanistan	PK-KKR-3/04-1.1	Pol-e-Khomri	10.27	5.06	35.4	49.27	4.51	69.48	1.01	2.37	7.3	33.27
Asia	Afghanistan	PK-KKR-3/04-1.2	Pol-e-Khomri	8.52	6.53	38.4	46.55	4.71	70.11	0.98	0.87	8.28	32.68
Asia	Afghanistan	PK-NRN-6/06-A1-ROM	Pol-e-Khomri	8.22	1.5	30.2	60.08	4.73	75.75	0.99	0.57	8.24	34.87
Asia	Afghanistan	PK-NRN-6/06-A2-ROM	Pol-e-Khomri	7.59	6.6	25.39	60.42	4.17	72.55	0.84	0.48	7.77	32.62
Asia	Afghanistan	PK-NRN-6/06-A3-ROM	Pol-e-Khomri	6.75	20.82	21.95	50.48	3.7	60.01	0.77	0.44	7.51	26.60
Asia	Afghanistan	PK-AND-12/04-1.2	Pol-e-Khomri	8.87	25.58	29.92	35.63	3.76	51.27	0.8	0.61	9.11	23.59
Asia	Afghanistan	PK-AND-12/04-1.4	Pol-e-Khomri	7.67	31.69	29.18	31.46	3.41	46.54	0.7	0.62	9.37	20.85
Asia	Afghanistan	PK-KRD-12/04-1.1	Pol-e-Khomri	9.3	33.2	26.99	30.51	2.91	44.6	0.67	0.66	8.66	19.89
Asia	Afghanistan	PK-KRD-12/04-1.3	Pol-e-Khomri	11.19	20.76	29.19	38.86	3.68	53.69	0.87	0.76	9.05	25.25
Asia	Afghanistan	PK-KRD-12/04-1.4	Pol-e-Khomri	8.77	42.18	23.39	25.66	2.56	36.93	0.53	0.69	8.34	16.13
Asia	Afghanistan	PK-KRD-12/04-1.6	Pol-e-Khomri	8.73	42.62	24.02	24.63	2.73	36.3	0.55	0.47	8.6	16.00
Asia	Afghanistan	PK-DKS-12/04-1.1	Pol-e-Khomri	7.43	20.03	33.81	38.73	3.95	57.37	0.83	2.21	8.18	26.39

TABLE 10.2 Chemical exergy of bituminous coals. Coal compositions are as-received as listed in Ref. 3. (Continued)

Continent	Country	Primary field ID	Coal basin or field	Wt.%									e^{ch} (MJ/kg)
				Moisture	Ash	V.M.	F.C.	H	C	N	S	O	
Asia	Afghanistan	PK-DKS-12/04-1.2	Pol-e-Khomri	7.56	60.06	18.38	14	1.78	23.01	0.3	0.83	6.46	9.34
Asia	Afghanistan	PK-DKS-12/04-1.3	Pol-e-Khomri	7.41	10.38	38.61	43.6	4.89	66.36	1.01	1.54	8.41	31.07
Asia	Afghanistan	PK-DKS-12/04-2	Pol-e-Khomri	7.45	14.01	34.41	44.13	4.31	63.25	0.81	2.28	7.89	29.30
Asia	Afghanistan	PK-KKR-12/04-1.1	Pol-e-Khomri	3.49	50.06	30.95	15.5	3.15	33.31	0.54	2.63	6.82	14.92
Asia	Afghanistan	PK-KKR-12/04-1.2	Pol-e-Khomri	3.85	43.04	32.86	20.25	3.58	38.5	0.72	4.5	5.81	18.05
Asia	Afghanistan	PK-KKR-12/04-1.3	Pol-e-Khomri	7.23	10.51	39.9	42.36	4.69	64.65	1.14	4.57	7.21	30.86
Asia	Afghanistan	PK-KKR-12/04-1.4	Pol-e-Khomri	8.05	9.24	37.43	45.28	4.62	66.19	1.2	2.92	7.78	31.28
Asia	Afghanistan	PK-KKR-12/04-1g	Pol-e-Khomri	9.62	5.91	31.37	53.1	3.96	69.37	0.78	3.51	6.85	32.59
Asia	Afghanistan	PK-KKR-12/04-2.1	Pol-e-Khomri	12.07	9.19	36.14	42.6	3.99	63.89	0.83	1.11	8.92	30.46
Asia	Afghanistan	PK-KKR-12/04-2.2	Pol-e-Khomri	14.47	4.8	34.08	46.65	4.32	66.54	0.89	0.65	8.33	32.92
Asia	Afghanistan	BOX-MMI-SBZK-11/05-I	Sabzak	2.14	14.55	29.26	54.05	4.14	67.78	0.87	2.97	7.55	29.36
Asia	Afghanistan	BOX-MMI-SBZK-11/05-II	Sabzak	1.42	45.56	25.19	27.83	2.81	39.43	0.56	2.15	8.07	16.37
Asia	China	CN-28-02	Leshan	1.84	32.91	20.62	44.63	3.42	55.27	1.02	0.54	5	23.19
Asia	China	CN-47-07	Xilinhaote	4.09	12.51	24.92	58.48	3.7	69.8	0.89	0.58	8.43	29.69
Asia	China	CN-85-15		1.32	9.99	36.29	52.4	5.04	71.6	1.61	5.57	4.87	32.37
Asia	India	AML-1	Sohagpur	9.71	12.88	32.77	44.64	4	63.35	1.4	0.57	8.09	29.34
Asia	India	AML-2	Sohagpur	11.91	17.15	29.22	41.72	3.57	57.7	1.26	0.81	7.6	27.21
Asia	India	AML-4	Sohagpur	9.11	23.31	23.66	43.92	3.07	55.15	1.2	0.48	7.68	24.52
Asia	India	AML-5	Sohagpur	7.76	22.91	23.65	45.68	3.07	57.08	1.19	0.43	7.56	24.93
Asia	India	AML-6	Sohagpur	11.93	14.08	28.45	45.54	3.59	60.3	1.17	0.96	7.98	28.38
Asia	India	AML-8	Sohagpur	7.78	12.37	35.04	44.81	4.44	63.97	1.46	0.96	9.02	29.50
Asia	India	AML-9	Sohagpur	10.76	13.64	27.6	48	3.57	61.73	1.29	0.76	8.25	28.50
Asia	India	AML-10	Sohagpur	8.63	16.23	26.31	48.83	3.45	61.52	1.23	0.92	8.02	27.60
Asia	India	AML-11	Sohagpur	8.81	19.38	27.7	44.11	3.61	58.43	1.21	0.63	7.93	26.50
Asia	India	AML-13	Sohagpur	13.93	13.71	24.57	47.79	3.23	59.79	1.32	0.74	7.28	28.39
Asia	India	JAM-1	Sohagpur	11.86	20.34	27.66	40.14	3.61	56.17	1.17	0.52	6.33	26.60
Asia	India	JAM-2	Sohagpur	7.19	17.08	28.37	47.36	3.68	63.29	1.3	0.44	7.03	28.11
Asia	India	JAM-3	Sohagpur	7.79	12.36	30.61	49.24	4.16	66.86	1.37	0.42	7.04	30.37
Asia	India	JAM-4	Sohagpur	6.27	18.95	26.61	48.17	3.25	62.87	1.3	0.36	7	27.11

TABLE 10.2 Chemical exergy of bituminous coals. Coal compositions are as-received as listed in Ref. 3. (Continued)

Continent	Country	Primary field ID	Coal basin or field	Moisture	Ash	V.M.	F.C.	Wt.% H	Wt.% C	Wt.% N	Wt.% S	Wt.% O	e^{ch} (MJ/kg)
Asia	India	JAM-5	Sohagpur	8.42	8.53	30.82	52.23	4.19	68.99	1.46	0.42	7.99	31.46
Asia	India	JAM-6	Sohagpur	28.64	22.78	21.72	26.86	1.39	33.9	0.92	0.1	12.27	17.22
Asia	India	SBT-11-2	Sohagpur	2.69	14.42	32.37	50.52	4.08	70.22	1.46	0.48	6.65	29.94
Asia	India	SBT-11-3	Sohagpur	1.83	31.66	26.98	39.53	3.8	55.57	1.15	0.33	5.67	23.65
Asia	India	SBT-11-6	Sohagpur	2.12	21.4	27.01	49.47	3.98	64.54	1.36	0.45	6.15	27.44
Asia	India	SBT-11-7	Sohagpur	2.62	16.32	30.16	50.9	4.39	68.43	1.41	0.46	6.37	29.58
Asia	India	SBT-11-8	Sohagpur	2.31	25.35	25.88	46.46	3.79	60.95	1.32	0.38	5.9	25.87
Asia	India	SBT-16-1	Sohagpur	1.57	20.37	30.05	48.01	3.97	66.51	1.3	0.98	5.3	28.20
Asia	India	SBT-16-3	Sohagpur	1.46	38.65	23.73	36.16	3.24	49.86	0.99	0.48	5.32	20.75
Asia	India	SBT-16-4	Sohagpur	1.95	16.85	31.13	50.07	4.21	68.76	1.43	0.54	6.26	29.32
Asia	India	SBT-16-5	Sohagpur	1.65	33.48	21.69	43.18	2.9	55.3	1.05	0.27	5.36	22.47
Asia	India	SBT-16-6	Sohagpur	1.84	12.16	32.17	53.83	4.14	73.93	1.48	0.5	5.95	31.23
Asia	India	SBT-16-7	Sohagpur	1.75	26.8	25.53	45.92	3.4	61.17	1.17	0.75	4.96	25.49
Asia	India	SBT-16-9	Sohagpur	1.53	23.9	30.42	44.15	3.96	62.21	1.26	2.34	4.8	26.84
Asia	India	SBT-19D	Raniganj	1.88	39.13	26.8	32.19	3.32	47.14	1.06	1.28	6.2	19.99
Asia	India	SBT-19-R1	Sohagpur	5.05	34.9	25.21	34.84	3.09	48.18	1.12	0.71	6.96	20.70
Asia	India	SBT-19-R10	Sohagpur	1.86	37.8	27.78	32.56	3.69	48.6	1.08	0.69	6.28	20.85
Asia	India	SBT-19-R2	Sohagpur	4.81	25.21	30.5	39.48	3.53	57.03	1.43	0.73	7.26	24.71
Asia	India	SBT-19-R3	Sohagpur	5.28	27.84	29.63	37.25	3.32	54.48	1.44	0.77	6.87	23.60
Asia	India	SBT-19-R4	Sohagpur	2.97	35.25	26.82	34.96	3.56	50.08	1.11	0.47	6.56	21.48
Asia	India	SBT-19-R5A	Sohagpur	2.66	36.91	26.36	34.07	3.18	49.13	0.93	0.48	6.71	20.58
Asia	India	SBT-19-R5B	Sohagpur	2.81	35.29	26.12	35.78	3.56	50.13	1.07	0.58	6.56	21.48
Asia	India	SBT-19-R6	Sohagpur	2.11	31.42	29.62	36.85	3.45	54.11	1.29	1.94	5.68	23.11
Asia	India	SBT-19-R7	Sohagpur	1.81	41.43	25.96	30.8	3.28	45.06	0.99	1.08	6.35	19.06
Asia	India	SBT-19-R8	Sohagpur	1.66	49.93	22.82	25.59	2.74	38.39	0.84	1.04	5.4	15.87
Asia	India	SBT-19-R9	Sohagpur	1.94	47.17	23.35	27.54	2.73	40.5	0.84	1.27	5.55	16.77
Asia	India	SBT-23-1	Sohagpur	2.8	23	7.44	66.76	2.17	68.65	1.37	0.5	1.51	27.50
Asia	India	SBT-23-2	Sohagpur	2.7	21.57	8.98	66.75	2.64	69.17	1.48	0.52	1.92	28.20
Asia	India	SBT-23-3	Sohagpur	2.04	19.65	25.65	52.66	3.55	66.96	1.33	2.27	4.2	28.40
Asia	India	SBT-23-4	Sohagpur	2.65	37.43	26.01	33.91	3.3	48.58	1.15	0.69	6.2	20.58

TABLE 10.2 Chemical exergy of bituminous coals. Coal compositions are as-received as listed in Ref. 3. (Continued)

Continent	Country	Coal basin or field	Primary field ID	Wt.%									e^{ch} (MJ/kg)
				Moisture	Ash	V.M.	F.C.	H	C	N	S	O	
Asia	India	Sohagpur	SBT-23-5	2.95	29.26	28.42	39.37	3.79	55.93	1.16	0.98	5.93	24.20
Asia	India	Sohagpur	SBT-9-3	0.57	35.46	17.02	46.95	3.05	56.55	0.87	0.36	3.14	23.02
Asia	India	Sohagpur	SBT-9-5	0.78	45.71	14.45	39.06	2.65	46.61	0.69	0.26	3.3	18.74
Asia	India	Sohagpur	SBT-9-6	0.56	28.46	17.66	53.32	3.54	63.53	0.97	0.38	2.56	26.30
Asia	India	Sohagpur	SBT-9-8	0.66	34.34	15.73	49.27	3.2	58.78	0.86	0.32	1.84	24.15
Asia	India	Sohagpur	SBT-9-9	0.7	15.84	21.69	61.77	4.34	74.75	1.28	0.47	2.62	31.60
Asia	India	Sohagpur	SDA-1	5.27	26.68	28.27	39.78	3.69	56.06	1.13	0.42	6.75	24.60
Asia	India	Sohagpur	SDA-2	7.36	14.41	31.35	46.88	4.32	64.58	1.41	0.59	7.33	29.49
Asia	India	Sohagpur	SDA-3	5.57	25.45	22.95	46.03	3.18	58.13	1.19	0.41	6.07	24.96
Asia	India	Sohagpur	SDA-4	9.53	16.4	29.7	44.37	4.2	60.53	1.35	0.5	7.49	28.35
Asia	India	Sohagpur	SKM-5-4	10.08	25.81	25.94	38.17	3.1	49.71	1.15	0.49	9.66	22.39
Asia	India	Sohagpur	SKM-5-5	10.86	18.17	26.4	44.57	3.43	57.43	1.24	0.44	8.44	26.42
Asia	India	Sohagpur	SKM-5-8	19.11	12.6	28.31	39.98	3.45	54.13	1.14	0.82	8.75	27.87
Asia	India	Sohagpur	SKM-5-9	20.69	11.19	27.76	40.36	3.4	53.71	1.21	0.76	9.05	28.15
Asia	India	Sohagpur	SKM-6-1	3.12	49.94	18.02	28.92	2.43	38.07	0.82	0.35	5.27	15.51
Asia	India	Sohagpur	SKM-6-2	1.63	36.17	22.56	39.64	3.3	51.77	1.12	0.62	5.39	21.63
Asia	India	Sohagpur	SKM-6-3	2.16	33.08	24.05	40.71	3.55	52.46	1.25	0.76	6.74	22.26
Asia	India	Sohagpur	SKM-6-4	1.79	40.17	22.43	35.61	3.08	47.06	1.01	0.71	6.18	19.54
Asia	India	Sohagpur	SKM-6-5	2.28	46.04	22.98	28.7	2.89	39.72	0.91	1.46	6.71	16.68
Asia	India	Sohagpur	SSL-1-1	8.69	21.07	31.92	38.32	3.94	59	1.15	0.47	5.68	27.23
Asia	India	Sohagpur	SSL-1-10	11.17	14.52	32.24	42.07	4.28	61.08	1.31	0.86	6.78	29.39
Asia	India	Sohagpur	SSL-1-3	7.28	28.5	26.12	38.1	3.54	53.64	1.16	0.3	5.59	24.05
Asia	India	Sohagpur	SSL-1-4	4.33	44.9	19.62	31.15	2.6	42.61	0.82	0.21	4.54	17.76
Asia	India	Sohagpur	SSL-1-5A	5.29	41.29	20.77	32.65	2.83	44.15	0.92	0.26	5.26	18.83
Asia	India	Sohagpur	SSL-1-6	6.57	12.99	32.61	47.83	4.36	67.87	1.44	0.35	6.43	30.63
Asia	India	Sohagpur	SSL-1-7	6.56	13.36	31.25	48.83	4.29	67.5	1.37	0.51	6.41	30.43
Asia	India	Sohagpur	SSL-2-1	12.29	17.12	32.56	38.03	3.85	58.99	1.1	0.41	6.25	28.25
Asia	India	Sohagpur	SSL-2-2	6.7	35.1	22.59	35.61	2.64	48.56	0.86	0.29	5.85	20.69
Asia	India	Sohagpur	SSL-2-3	6.12	40.19	17.1	36.59	2.38	44.51	0.75	0.22	5.84	18.57
Asia	India	Sohagpur	SSL-2-4	10.53	22.78	26.16	40.53	3.23	55.88	1.1	0.35	6.13	25.54

TABLE 10.2 Chemical exergy of bituminous coals. Coal compositions are as-received as listed in Ref. 3. (Continued)

Continent	Country	Primary field ID	Coal basin or field	Moisture	Ash	V.M.	F.C.	Wt.%					e^{ch} (MJ/kg)
								H	C	N	S	O	
Asia	India	SSL-2-5	Sohagpur	4.67	53.66	17.43	24.24	1.77	32.54	0.54	0.9	5.92	12.87
Asia	India	SSL-2-6	Sohagpur	10.01	19.7	29.25	41.04	3.22	57.91	1.06	1.72	6.38	26.54
Asia	India	SSL-2-7	Sohagpur	5.73	12.46	30.15	51.66	3.6	70.63	1.29	0.33	5.96	30.58
Asia	India	SSL-2-8	Sohagpur	5.91	20.65	31.31	42.13	3.67	61.49	1.14	0.8	6.34	27.08
Asia	India	SSL-2-9	Sohagpur	5.82	15.3	31.69	47.19	3.5	66.95	1.35	0.76	6.32	29.07
Asia	Japan	Japan Ikesima	Sakito-Matsushima	2.35	24.43	35.24	37.98	4.5	58.93	1.06	0.62	8.11	25.81
Asia	Kazakhstan	AIPET-1-Sbrk	Karaganda	9.78	1.95	40.49	47.78	4.75	67.37	1.35	0.36	14.44	31.49
Asia	Kazakhstan	AIPET-2-Kich		1.33	22.37	18.03	58.27	3.41	66.91	1.06	0.6	4.32	27.62
Asia	Kazakhstan	AIPET-3-Mlds		1.49	26.6	22.58	49.33	3.71	61.64	0.83	0.4	5.33	25.85
Asia	Kazakhstan	AIPET-4-Shtn	Karaganda	1.46	40.99	20.05	37.5	3.29	46.14	0.94	0.26	6.92	19.20
Asia	Kazakhstan	AIPET-5-Dlns	Karaganda	1.13	18.53	21.48	58.86	3.47	68.91	1.11	0.37	6.48	28.21
Asia	Kazakhstan	AIPET-6-Krvs	Karaganda	2.11	22.83	20.44	54.62	3.57	64.57	1.01	0.47	5.44	27.02
Asia	Kazakhstan	AIPET-7-Kar22	Karaganda	1.46	21.09	21.87	55.58	4.01	66.39	1.41	0.35	5.29	28.03
Asia	Kazakhstan	AIPET-8-Shn12	Karaganda	1.44	39.11	20.54	38.91	3.3	47.49	0.94	0.4	7.32	19.74
Asia	Kazakhstan	AIPET-9-Tntk	Karaganda	6.5	9.03	37.05	47.42	4.5	65.93	1.13	0.19	12.72	29.50
Asia	Kazakhstan	AIPET-10-Stpn	Karaganda	1.96	9.86	26.74	61.44	4.54	75.56	1.74	0.41	5.93	32.36
Asia	Kazakhstan	AIPET-11-Kstu	Karaganda	2	19.98	22.99	55.03	3.77	66.7	1.01	0.32	6.22	27.97
Asia	Kazakhstan	AIPET-12-Grba	Karaganda	6.15	9.09	24.52	60.24	3.67	68.18	1.13	0.53	11.25	29.48
Asia	Kazakhstan	AIPET-13-Kdns	Karaganda	1.76	27.26	25.32	45.66	3.34	57.91	1.05	0.35	8.33	23.84
Asia	Kazakhstan	AIPET-15-Bgtr	Ekibastuz	3.38	26.75	21.7	48.17	3.4	57.25	1.03	0.33	7.86	24.11
Asia	Kazakhstan	AIPET-17-Svrn	Ekibastuz	1.8	23.68	22.19	52.33	3.67	63.23	0.97	0.69	5.96	26.54
Asia	Mongolia	Shar-Tgd-1-9/02	Alagtogoo	12.52	4.11	37.95	45.42	4.56	63.67	0.83	0.63	13.68	30.79
Asia	Mongolia	Shar-Tgd-2-9/02	Alagtogoo	11.72	6.24	37.67	44.37	4.51	62.69	0.88	0.91	13.05	30.12
Asia	Mongolia	Khus-A-1-10/02	Altay	6.16	8.27	16.8	68.77	3.43	73.08	1.77	0.4	6.89	31.49
Asia	Mongolia	Khus-B-1-10/02	Altay	4.59	10.42	15.52	69.47	3.87	74.6	1.84	0.41	4.27	32.25
Asia	Mongolia	Zeegt-I-1-11/02	Atlay-Chandmani	10.4	14.2	28.57	46.83	3.24	58.04	2	0.36	11.76	26.05
Asia	Mongolia	Khar-I-1-11/02	Kharkhiraa	5.58	14.08	22.04	58.3	3.42	67.18	1.4	0.34	8	28.78
Asia	Mongolia	Khar-I-2-11/02	Kharkhiraa	5.69	15.24	20.51	58.56	3.16	66.38	1.52	0.39	7.62	28.22
Asia	Mongolia	Byne-28-1-8/02	Ongiyngol	4.31	15.12	40.16	40.41	4.83	62.5	1.39	1.33	10.52	28.20

TABLE 10.2 Chemical exergy of bituminous coals. Coal compositions are as-received as listed in Ref. 3. (Continued)

Continent	Country	Coal basin or field	Primary field ID	Moisture	Ash	V.M.	F.C.	Wt.%					e^{ch} (MJ/kg)
								H	C	N	S	O	
Asia	Mongolia	Ongiyngol	Byne-39-1-8/02	5.61	11.02	38.75	44.62	4.67	66.72	1.26	0.74	9.98	30.04
Asia	Mongolia	Orkhon-Selenge	Jlch-I-1-11/02	9.09	2.38	40.81	47.72	5.2	70.31	1.13	0.68	11.21	33.33
Asia	Mongolia	Orkhon-Selenge	Jlch-2-11/02	9.19	2.82	41.36	46.63	5.07	69.02	1.02	0.89	11.99	32.66
Asia	Mongolia	Orkhon-Selenge	Mogn-1-A-10/02	2.23	10.8	29.33	57.64	4.82	74.81	1.86	1.06	4.42	32.73
Asia	Mongolia	Orkhon-Selenge	Mogn-1-B-10/02	3.36	9.11	28.4	59.13	4.43	76.44	1.95	0.94	3.77	33.35
Asia	Mongolia	Orkhon-Selenge	Nurn-1-1-11/02	6.25	6.97	22.58	64.2	4.09	73.79	2.22	0.4	6.28	32.66
Asia	Mongolia	Orkhon-Selenge	Nurn-2-1-11/02	6.13	11.55	21.23	61.09	3.86	69.92	2.08	0.4	6.06	30.77
Asia	Mongolia	Orkhon-Selenge	Saio-6A-1-8/02	6.83	8.7	38.29	46.18	4.4	67.07	1.13	0.73	11.14	30.19
Asia	Mongolia	Orkhon-Selenge	Saio-6A-2-8-02	5.61	11.02	38.75	44.62	4.67	66.72	1.26	0.74	9.98	30.04
Asia	Mongolia	South Govi	Nar-63-1-9/02	2.37	15.2	31.1	51.33	4.33	69.09	0.84	1.09	7.09	29.77
Asia	Mongolia	South Govi	Nar-71-1-9/02	2.25	9.42	21.98	66.35	4.03	76.09	1.08	0.52	6.61	32.04
Asia	Mongolia	South Govi	Tav-4-1-6/02	3.04	10.67	26.07	60.22	4.43	74.8	2.23	0.71	4.12	32.51
Asia	Mongolia	South Govi	Tav-8-1-6/02	2.03	10.02	31.44	56.51	4.6	74.81	2.02	1.02	5.5	32.33
Asia	Taiwan	Sanhsia-Tachi	Taiwan-1	2.69	13.97	36.03	47.31	4.82	68.74	1.7	0.74	7.34	30.23
Asia	Taiwan	Sanhsia-Tachi	Taiwan-2	2.72	4.07	42.09	51.12	5.51	77.02	1.65	0.74	8.29	34.23
Asia	Taiwan	Sanhsia-Tachi	Taiwan-3	2.75	5.74	39.32	52.19	5.12	76.03	1.83	0.81	7.72	33.46
Asia	Taiwan	Shihting	Taiwan-4	4.61	9.59	42.96	42.84	4.97	67.03	1.64	2.68	9.48	30.67
Asia-Europe	Georgia	Tkibuli-Shaori	WOCQI GA-1	6.78	23.45	29.62	40.15	3.74	53.49	0.88	1.07	10.59	23.91
Europe	Hungary	Mecsek	Zobak 6	3.07	13.37	30.26	53.3	4.46	69.12	1.4	2.9	5.68	30.68
Europe	Hungary	Mecsek	Zobak 4	3.53	23.78	26.94	45.75	3.87	57.61	1.14	6.15	3.93	26.42
Europe	Hungary	Mecsek	Zobak 2	3.09	18.17	27.84	50.9	4.19	64.84	1.31	4.57	3.83	29.18
Europe	Hungary	Mecsek	Zobak 1	3.42	51.6	18.19	26.79	2.46	34.86	0.81	1.79	5.06	14.66
Europe	Romania	Petrosani, Valea Jiului de Vest	Romania-1 Valea de Brazi	2.41	29.17	34.79	33.63	4.2	52.57	1.23	2.64	7.78	23.46
Europe	Romania	Petrosani, Valea Jiului de Vest	Romania-2 Uricani	1.66	18.94	34.17	45.23	4.35	64.51	1.1	0.81	8.63	27.62
Europe	Romania	Petrosani, Valea Jiului de Vest	Romania-3 Barbateni	1.8	18.62	35.53	44.05	4.46	64.33	1.07	1.73	7.99	27.97
Europe	Romania	Petrosani, Valea Jiului de Vest	Romania-4 Barbateni	1.62	12.62	38.42	47.34	4.74	69.94	1.25	1.47	8.36	30.34

TABLE 10.2 Chemical exergy of bituminous coals. Coal compositions are as-received as listed in Ref. 3. (*Continued*)

Continent	Country	Primary field ID	Coal basin or field	Wt.%									e^{ch} (MJ/kg)
				Moisture	Ash	V.M.	F.C.	H	C	N	S	O	
Europe	Romania	Romania-5 Lupeni	Petrosani, Valea Jiului de Vest	2.49	12	37.42	48.09	4.99	69.09	0.8	1.47	9.16	30.52
Europe	Romania	Romania-6 Paroseni	Petrosani, Valea Jiului de Vest	2.1	6.32	42.7	48.88	5.43	75.16	1.35	1.31	8.34	33.30
Europe	Romania	Romania-7 Vulcan	Petrosani, Valea Jiului de Vest	2.57	17.92	38.56	40.95	4.68	63.32	0.92	1.72	8.87	28.00
Europe	Romania	Romania-8 Aninoasa	Petrosani, Valea Jiului de Vest	2.79	10.49	40.49	46.23	4.74	68.28	1.23	4.16	8.31	30.68
Europe	Romania	Romania-9 Livezeni	Petrosani, Valea Jiului de Vest	6.77	18.19	35.93	39.11	4.27	56.99	0.86	1.41	11.51	26.00
Europe	Romania	Romania-10 Lonea	Petrosani, Valea Jiului de Vest	3.28	3.96	42.55	50.21	5.18	74.99	0.98	1.31	10.3	33.23
Europe	Romania	Romania-11 Petrila	Petrosani, Valea Jiului de Vest	5.32	2.25	40.03	52.4	4.99	73.94	0.82	0.77	11.92	33.10
Europe	Ukraine	KMS-1	Donbas	0.99	37.14	19.82	42.05	2.52	50.24	1.03	5.38	2.7	21.22
Europe	Ukraine	KMS-2	Donbas	0.55	8.94	28.72	61.79	4.33	77.89	1.59	5.84	0.86	34.04
Europe	Ukraine	KMS-3	Donbas	6.88	27.57	22.22	43.33	3.03	53.49	1.21	2.78	5.04	23.88
Europe	Ukraine	KMS-4	Donbas	1.37	30.21	22.45	45.97	3.26	57.06	1.13	2.8	4.17	24.14
Europe	Ukraine	GG-1	Donbas	1.06	36.12	16.73	46.09	2.33	54.27	1	1.53	3.69	21.67
Europe	Ukraine	Ukraine 1	Donetsk-Makeyevka	1.04	36.45	18	44.51	3.06	52.26	1.04	2.91	3.24	22.05
Europe	Ukraine	Ukraine 2	Donetsk-Makeyevka	0.87	2.1	23.5	73.53	4.66	86.7	1.65	1.69	2.33	36.93
Europe	Ukraine	Ukraine 3	Central district	1.08	1.63	22.46	74.83	4.54	86.54	1.58	0.98	3.65	36.57
Europe	Ukraine	Ukraine 4	Western district	2.03	1.51	44.94	51.52	5.5	80.42	1.32	1.68	7.54	35.55
Europe	Ukraine	Ukraine 5	Central district	1.5	4.53	24.89	69.08	4.39	80.98	1.72	3.99	2.89	35.12
Europe	Ukraine	Ukraine 6	Krasnoarmeysk	2.38	3.68	39.54	54.4	4.99	76.72	1.52	3.54	7.17	34.09
Europe	Ukraine	D-1	Donetsk-Makeyevka	1.13	3.82	13.67	81.38	3.83	86.2	1.58	3.28	0.16	36.39
Europe	Ukraine	D-4	Donetsk-Makeyevka	5.32	8.41	17.06	69.21	3.45	73.92	1.61	3.97	3.33	32.62
Europe	Ukraine	D-5	Donetsk-Makeyevka	3.33	4.58	13.07	79.02	3.78	81.43	1.63	1.24	4.01	34.60

TABLE 10.2 Chemical exergy of bituminous coals. Coal compositions are as-received as listed in Ref. 3. (Continued)

Continent	Country	Primary field ID	Coal basin or field	Wt.%									e^{ch} (MJ/kg)
				Moisture	Ash	V.M.	F.C.	H	C	N	S	O	
Europe	Ukraine	D-7	Donetsk-Makeyevka	1.37	7.03	14.31	77.29	3.92	82	1.82	3.61	0.25	35.03
Europe	Ukraine	N-1		4.27	7.2	21.84	66.69	4.1	76.5	1.32	4.04	2.57	34.09
Europe	Ukraine	N-4A		4.77	6.64	19.14	69.45	4.09	77.94	1.58	3.08	1.9	34.68
Europe	Ukraine	N-4B		1.42	10.26	19.05	69.27	4.12	77.06	1.54	4.7	0.9	33.55
Europe	Ukraine	N-5		8.05	5.45	14.19	72.31	3.65	76.7	1.41	1.89	2.85	34.59
Europe	Ukraine	G-1	Donetsk-Makeyevka	1.14	1.22	21.32	76.32	4.54	87.55	1.96	0.63	2.96	36.97
Europe	Ukraine	G-2	Donetsk-Makeyevka	1.14	9.64	20.39	68.83	4.28	78.93	1.72	0.55	3.74	33.25
Europe	Ukraine	G-3	Donetsk-Makeyevka	1	0.66	20.63	77.71	4.66	89.04	1.85	0.67	2.12	37.68
Europe	Ukraine	DZ-1	Central district	1.2	38.59	18.05	42.16	3.13	51.12	0.87	0.7	4.39	21.16
Europe	Ukraine	DZ-2	Central district	0.86	11.7	28.61	58.83	4.57	74.86	1.44	4.7	1.87	32.94
Europe	Ukraine	DZ-3	Central district	1.37	32.07	23.82	42.74	3.49	55.78	1.19	2.02	4.08	23.74
Europe	Ukraine	DZ-4	Central district	1.33	1.59	30.03	67.05	5.01	84.47	2.04	0.74	4.82	36.28
Europe	United Kingdom	UK-1	Yorkshire-Nottinghamshire	2.27	8.57	35.13	54.03	4.66	72.92	1.65	1.78	8.15	31.72
Europe	United Kingdom	UK-2	Yorkshire-Nottinghamshire	1.48	5.25	34.24	59.03	4.81	79.4	1.77	1.75	5.54	34.31
Europe	United Kingdom	UK-3	Yorkshire-Nottinghamshire	1.48	5.53	33.8	59.19	4.72	77.49	1.69	1.92	7.17	33.39
Europe	United Kingdom	UK-4	Yorkshire-Nottinghamshire	2.59	8.76	33.38	55.27	4.53	74.5	1.63	1.89	6.1	32.47
Europe	United Kingdom	UK-5	Yorkshire-Nottinghamshire	2.19	28.95	29.18	39.68	3.63	56.06	1.24	1.41	6.53	23.92
Europe	United Kingdom	FMD6EBedsA&B	Northumberland	2.27	8.43	32.38	56.92	4.62	73.33	1.53	2.59	7.23	32.08
Europe	United Kingdom	FMD6EBedsF-J	Northumberland	2.28	4.78	32.42	60.52	4.85	77.4	1.72	1.47	7.51	33.68
Europe	United Kingdom	KS310BedA	Northumberland	3.43	10.87	33.57	52.13	4.55	68.4	1.54	4.05	7.16	30.78

TABLE 10.2 Chemical exergy of bituminous coals. Coal compositions are as-received as listed in Ref. 3. (Continued)

Continent	Country	Primary field ID	Coal basin or field	Wt.%									e^{ch} (MJ/kg)
				Moisture	Ash	V.M.	F.C.	H	C	N	S	O	
Europe	United Kingdom	KS310BedC	Northumberland	3.72	10.35	33.31	52.62	4.62	69.72	1.79	3.01	6.79	31.29
Europe	United Kingdom	KS310BedE-G	Northumberland	3.52	8.36	34.07	54.05	4.81	72.36	1.74	1.13	8.08	31.97
Europe	United Kingdom	K311MainBedA	Northumberland	4.03	13.2	32.36	50.41	4.01	64.16	1.34	6.37	6.89	29.21
Europe	United Kingdom	K311MainBedC	Northumberland	4.08	10.61	34.41	50.9	4.62	68.84	1.67	1.58	8.6	30.61
Europe	United Kingdom	K311MainBedE	Northumberland	3.24	28.92	28.19	39.65	3.69	53.04	1.26	1.21	8.64	22.89
Europe	United Kingdom	K311MainBedG-I	Northumberland	3.89	8.05	33.86	54.2	4.73	72.21	1.74	1.19	8.2	31.96
Europe	United Kingdom	K311TailBedA&B	Northumberland	2.29	36.35	27.86	33.5	3.39	46.4	1.03	2.94	7.6	20.15
Europe	United Kingdom	K311TailBedF	Northumberland	2.21	57.33	20.23	20.23	2.23	28.27	0.61	0.38	8.97	11.03
Europe	United Kingdom	K311TailBedI-K	Northumberland	3.39	18.36	32.77	45.48	4.14	63.13	1.41	0.86	8.71	27.38
Europe	United Kingdom	K311TailBedM	Northumberland	2.22	65.48	17.01	15.29	1.79	21.05	0.43	0.5	8.53	7.75
Europe	United Kingdom	K311TailBedP	Northumberland	3.02	7.02	32.21	57.75	4.66	73.95	1.73	1.45	8.17	32.32
Europe	United Kingdom	HarworthBdsA-H	Yorkshire-Nottinghamshire	1.83	6.51	35.96	55.7	5.16	75.8	1.82	3.06	5.83	33.71
Europe	United Kingdom	KellingleyBedA	Yorkshire-Nottinghamshire	2.35	20.57	32.22	44.86	3.82	57.12	1.3	11.2	3.64	26.98
Europe	United Kingdom	KellingleyBedD	Yorkshire-Nottinghamshire	3.22	7.31	33.39	56.08	4.71	73.05	1.68	2.11	7.92	32.25
Europe	United Kingdom	KellinglyBdF-L	Yorkshire-Nottinghamshire	3.17	5.32	35.47	56.04	5.07	75.42	1.78	1.77	7.48	33.55
Europe	United Kingdom	Maltby18BdsA-F	Yorkshire-Nottinghamshire	1.58	2.93	34.01	61.48	5.05	81.54	1.72	1.67	5.51	35.44

TABLE 10.2 Chemical exergy of bituminous coals. Coal compositions are as-received as listed in Ref. 3. (Continued)

Continent	Country	Coal basin or field	Primary field ID	Wt.%									e^{ch} (MJ/kg)
				Moisture	Ash	V.M.	F.C.	H	C	N	S	O	
Europe	United Kingdom	Yorkshire-Nottinghamshire	Maltby105BedA	1.05	25.62	28.6	44.73	3.8	57.97	1.22	7.17	3.17	26.05
Europe	United Kingdom	Yorkshire-Nottinghamshire	Maltby105BdC-H	1.3	3.98	31.99	62.73	4.89	81.13	1.65	1.9	5.16	35.06
Europe	United Kingdom	Yorkshire-Nottinghamshire	RiccallBedsA-E	4.3	3.42	34.91	57.37	4.82	76.49	1.56	1.32	8.09	33.95
Europe	United Kingdom	Yorkshire-Nottinghamshire	RiccallBedsG-I	3.68	11.2	31.57	53.55	4.21	69.04	1.31	2.63	7.93	30.35
Europe	United Kingdom	Yorkshire-Nottinghamshire	RiccallBedsK&L	3.88	7.06	32.32	56.74	4.61	73.6	1.49	1.67	7.69	32.50
Europe	United Kingdom	Yorkshire-Nottinghamshire	RossingtnBdA-C	6.11	5.17	35.84	52.88	4.81	70.6	1.53	2.29	9.49	32.36
Europe	United Kingdom	Yorkshire-Nottinghamshire	RossingtnBdE-M	6.39	4.35	34.97	54.29	4.76	72.21	1.6	1.12	9.58	32.79
Europe	United Kingdom	Yorkshire-Nottinghamshire	StillingfltBdA	2.12	13.15	34.14	50.59	4.67	69.32	1.52	1.47	7.75	30.23
Europe	United Kingdom	Yorkshire-Nottinghamshire	StillingfltC-J	2.6	5.84	34.3	57.26	4.94	76.32	1.72	1.14	7.44	33.40
Europe	United Kingdom	Yorkshire-Nottinghamshire	WelbeckBedsA-H	2.63	8.02	33.55	55.8	4.67	73.7	1.49	2.19	7.3	32.31
Europe	United Kingdom	Yorkshire-Nottinghamshire	WelbeckBedsJ-K	3.03	5.68	33.13	58.16	4.86	75.63	1.63	1.76	7.41	33.34
Europe	United Kingdom	Yorkshire-Nottinghamshire	WelbeckBedM	2.9	11.26	33.29	52.55	4.8	70.78	1.54	1.56	7.17	31.27
Europe	United Kingdom	Yorkshire-Nottinghamshire	WistowBedsA-K	5.33	10.53	32.47	51.67	4.46	67.34	1.48	2.87	7.99	30.58
South America	Argentina	Rio Turbio	Argentina-BA0101	10.77	25.2	30.34	33.69	3.97	49.62	0.89	0.45	9.11	23.66
South America	Argentina	Rio Turbio	Argentina-BA0201	10.11	16.67	36.49	36.73	4.53	57.83	0.98	0.43	9.45	27.64
South America	Argentina	Rio Turbio	Argentina-BA0301	7.65	12.05	42.37	37.93	5.14	63.25	1.06	1.6	9.25	30.13

TABLE 10.2 Chemical exergy of bituminous coals. Coal compositions are as-received as listed in Ref. 3. (Continued)

Continent	Country	Primary field ID	Coal basin or field	Wt.%									e^{ch} (MJ/kg)
				Moisture	Ash	V.M.	F.C.	H	C	N	S	O	
South America	Argentina	Argentina-BA0501	Pico Quemado	8.18	40.32	25.16	26.34	3.26	38.92	1.1	0.29	7.94	17.65
South America	Brazil	99-163	Santa Terezinha	4.89	36.31	23.19	35.61	2.71	44.23	0.79	0.43	10.64	18.30
South America	Brazil	99-175	Santa Terezinha	0.76	54.1	30.03	15.11	1.95	33.17	0.62	0.35	9.05	12.40
South America	Brazil	99-176	Santa Terezinha	1.23	59.37	18.37	21.03	1.96	29.85	0.62	1.15	5.82	11.58
South America	Brazil	99-182	Santa Terezinha	1.39	36.46	20.28	41.87	2.22	49.67	1.02	2.21	7.03	19.76
South America	Chile	Chile-3	Arauco Basin	3.68	6.56	45.5	44.26	5.54	71.7	1.09	3.57	7.86	33.17
South America	Chile	Lebu #1	Arauco	3.94	1.92	43.82	50.32	5.38	76.55	1.94	0.92	9.35	34.33
South America	Chile	Lebu #2	Arauco	2.08	7.9	45.6	44.42	5.32	72.36	1.54	3.91	6.88	32.75
South America	Colombia	71C/Manto 70	Cerrejon Norte	3.09	1.39	36.79	58.73	5.18	78.72	1.52	0.39	9.71	34.48
South America	Colombia	73C/Manto 75	Cerrejon Norte	3.66	1.01	46.18	49.15	5.33	77.7	1.73	0.47	10.1	34.47
South America	Colombia	67C/Manto 115	Cerrejon Norte	3.89	0.53	39.82	55.76	5.21	75.91	1.7	0.63	12.14	33.60
South America	Colombia	74C/Manto 130	Cerrejon Norte	3.99	4.27	38.72	53.02	5.13	73.82	1.15	0.33	11.31	32.68

TABLE 10.2 Chemical exergy of bituminous coals. Coal compositions are as-received as listed in Ref. 3. (Continued)

Continent	Country	Primary field ID	Coal basin or field	Moisture	Ash	V.M.	F.C.	H	C	N	S	O	e^{ch} (MJ/kg)
								Wt.%					
Africa	Nigeria	Opara—no. 1	Enugu coal field, Anambra basin	6.75	4.51	37.81	50.93	5.12	72.09	1.85	0.65	9.03	33.24
Africa	Nigeria	Opara—no. 2	Enugu coal field, Anambra basin	9.05	4.35	38.77	47.83	4.98	69.59	1.96	0.56	9.51	32.86
Africa	Nigeria	Opara—no. 3	Enugu coal field, Anambra basin	8.55	4.89	38.65	47.91	4.86	68.35	1.85	0.73	10.77	31.95
Africa	Nigeria	Opara—no. 4	Enugu coal field, Anambra basin	7.82	5.16	38.17	48.85	4.80	69.08	1.85	0.64	10.65	31.89
Africa	Nigeria	Onyema—no. 5	Enugu coal field, Anambra basin	7.48	9.50	39.05	43.97	4.95	66.00	1.79	0.53	9.75	30.71
Africa	Nigeria	Onyema—no. 6	Enugu coal field, Anambra basin	9.53	6.56	37.48	46.43	4.84	67.30	1.87	0.53	9.37	31.90
Africa	Nigeria	Onyema—no. 7	Enugu coal field, Anambra basin	8.21	23.42	29.34	39.03	3.67	52.93	1.35	0.30	10.12	23.86
Africa	Nigeria	Okaba-Odagbo—no. 8	Okaba Odagbo coal field, Anambra Basin	12.96	9.62	35.21	42.21	4.18	59.01	1.44	0.66	12.13	28.55
Africa	Nigeria	Okaba-Odagbo—no. 9	Okaba Odagbo coal field, Anambra Basin	24.54	5.67	33.54	36.25	3.64	52.18	1.30	0.66	12.01	28.99
Africa	Nigeria	Okaba-Odagbo—no. 10	Okaba Odagbo coal field, Anambra Basin	20.01	9.90	31.28	38.81	3.31	52.11	1.17	0.58	12.92	26.64
Africa	Nigeria	Okaba-Odagbo—no. 11	Okaba Odagbo coal field, Anambra Basin	23.85	5.69	34.48	35.98	3.93	53.31	1.44	0.61	11.17	29.77
Africa	Nigeria	Okaba-Odagbo—no. 12	Okaba Odagbo coal field, Anambra Basin	24.61	7.12	32.22	36.05	3.49	52.28	1.21	0.57	10.72	28.93

TABLE 10.3 Chemical exergy of subbituminous coals. Coal compositions are as-received as listed in Ref. 3.

269

| Continent | Country | Primary field ID | Coal basin or field | Moisture | Ash | V.M. | F.C. | Wt.% | | | | | e^{ch} (MJ/kg) |
								H	C	N	S	O	
Africa	Nigeria	Orukpa - no. 13	Orukpa coal field, Benue Trough	13.48	4.06	36.20	46.26	4.09	61.68	1.59	0.45	14.65	29.56
Africa	Tanzania	Tanzania KCM-TZ01	Songwe-Kiwira	3.00	35.22	22.90	38.88	2.86	49.01	1.19	0.70	8.02	20.21
Africa	Tanzania	Tanzania KCM-TZ02	Songwe-Kiwira	2.61	29.98	24.40	43.01	3.31	53.70	1.27	1.13	8.00	22.58
Africa	Tanzania	Tanzania KCM-TZ03	Songwe-Kiwira	2.50	38.61	23.24	35.65	3.14	46.38	1.11	1.18	7.08	19.55
Africa	Tanzania	Tanzania KCM-TZ04	Songwe-Kiwira	2.29	52.58	16.21	28.92	2.18	34.91	0.85	0.47	6.72	13.77
Africa	Tanzania	Tanzania KCM-TZ05	Songwe-Kiwira	2.75	20.91	30.97	45.37	4.20	61.90	1.48	0.69	8.07	26.77
Antarctica		95/015		4.00	24.26	33.45	38.29	4.04	55.31	1.21	0.57	10.60	24.13
Antarctica		95/021		3.33	47.76	24.17	24.74	2.75	36.11	0.80	0.45	8.80	14.92
Antarctica		95/024		4.00	28.24	32.85	34.91	3.89	51.28	1.28	0.59	10.72	22.34
Antarctica		95/027		4.08	34.21	29.05	32.66	3.39	46.80	1.13	0.59	9.80	20.05
Antarctica		95/035		3.97	26.95	34.44	34.64	4.09	53.42	1.24	0.53	9.81	23.47
Antarctica		95/042		4.42	30.88	31.27	33.43	3.53	48.26	1.18	0.52	11.21	20.76
Antarctica		95/046		4.05	24.07	34.09	37.79	3.69	52.92	1.28	0.63	13.36	22.61
Antarctica		95/052		5.03	29.76	31.14	34.07	3.10	47.17	1.24	0.44	13.26	19.81
Antarctica		95/054		5.18	24.35	31.80	38.67	3.51	52.08	1.42	0.55	12.91	22.36
Antarctica		95/066		3.84	11.41	39.30	45.45	4.60	65.74	1.77	0.70	11.94	28.85
Antarctica		95/067		5.34	16.48	34.52	43.66	3.72	58.12	1.37	0.44	14.53	24.96
Antarctica		95/123		2.58	20.00	33.64	43.78	4.45	63.05	1.61	0.79	7.52	27.53
Antarctica		95/124		2.46	16.44	33.32	47.78	4.54	66.82	1.61	0.69	7.43	29.06
Antarctica		95/133		6.52	16.90	34.58	42.00	3.51	55.82	1.56	0.63	15.06	24.12
Antarctica		95/160		2.13	18.05	12.98	66.84	2.13	70.33	1.73	0.32	5.31	27.62
Asia	Afghanistan	ARC-SDS-1-8/03	Darrah-i-Suf	14.59	3.90	38.74	42.77	4.57	62.71	0.91	2.22	11.10	31.77
Asia	Afghanistan	ESH-BFK-6/06-A1-ROM	Eshposhteh	19.78	6.80	31.25	42.17	3.82	55.73	0.82	2.09	10.96	29.55
Asia	Afghanistan	ESH-BFK-6/06-A3-ROM	Eshposhteh	18.77	9.86	31.14	40.23	3.80	53.88	0.81	0.88	12.00	27.85
Asia	Afghanistan	ESH-BFK-6/06-A4-ROM	Eshposhteh	16.32	18.37	32.94	32.37	3.61	49.10	0.74	0.39	11.47	24.44
Asia	Afghanistan	ESH-BFK-6/06-A5-ROM	Eshposhteh	19.63	8.34	32.16	39.87	3.86	54.88	0.86	0.65	11.78	28.69
Asia	Afghanistan	ESH-BFK-6/06-A6-ROM	Eshposhteh	19.38	7.17	31.71	41.74	3.87	56.19	0.88	0.52	11.99	29.18
Asia	Afghanistan	ESH-BFK-3/06-1.1	Eshposhteh	18.35	5.36	33.33	42.96	3.98	58.73	0.72	0.59	12.27	30.12
Asia	Afghanistan	ESH-BFK-3/06-1.2p	Eshposhteh	7.07	57.39	26.44	9.10	2.17	22.89	0.33	0.51	9.64	9.40

TABLE 10.3 Chemical exergy of subbituminous coals. Coal compositions are as-received as listed in Ref. 3. (Continued)

Continent	Country	Coal basin or field	Primary field ID	Wt.%									e^{ch}
				Moisture	Ash	V.M.	F.C.	H	C	N	S	O	(MJ/kg)
Asia	Afghanistan	Eshposhteh	ESH-BFK-3/06-1.3	18.73	2.74	32.93	45.60	4.06	60.85	0.74	0.44	12.44	31.33
Asia	Afghanistan	Eshposhteh	ESH-BFK-3/06-2.1	17.31	11.24	33.60	37.85	4.01	53.33	0.81	0.81	12.49	27.30
Asia	Afghanistan	Eshposhteh	ESH-BFK-3/06-2.2	17.14	14.09	31.55	37.22	3.75	51.75	0.71	0.67	11.89	26.16
Asia	Afghanistan	Eshposhteh	ESH-BFK-3/06-2.3	19.25	6.64	31.86	42.25	3.94	55.60	0.87	0.70	13.00	28.91
Asia	Afghanistan	Eshposhteh	ESH-BFK-3/06-2.4	19.39	4.30	32.11	44.20	3.98	58.34	0.74	0.94	12.31	30.44
Asia	Afghanistan	Eshposhteh	ESH-BFK-3/06-A2-ROM	16.94	13.55	30.22	39.29	3.88	52.49	0.79	0.48	11.87	26.56
Asia	Afghanistan	Eshposhteh	ESH-GLN-3/06-1.1	23.56	4.57	30.13	41.74	3.55	54.24	0.72	1.07	12.29	29.57
Asia	Afghanistan	Eshposhteh	ESH-GLN-3/06-1.2	23.33	3.81	28.87	43.99	3.58	55.25	0.68	0.92	12.43	29.97
Asia	Afghanistan	Eshposhteh	ESH-GLN-3/06-1.3	19.96	6.12	28.70	45.22	3.36	56.30	0.63	1.11	12.52	28.85
Asia	Afghanistan	Eshposhteh	ISH-GSN-3/04-1	25.02	4.37	26.54	44.07	3.22	55.23	0.65	0.65	10.86	30.20
Asia	Afghanistan	Eshposhteh	ISH-EMB-1-02	9.82	10.73	37.17	42.28	3.78	56.54	0.85	3.54	14.74	26.43
Asia	Afghanistan	Eshposhteh	ESH-KLJ-6/06-ROM1	19.09	5.35	31.95	43.61	3.89	57.66	0.85	1.60	11.56	30.12
Asia	Afghanistan	Eshposhteh	ESH-KLJ-6/06-ROM2	20.74	4.73	30.41	44.12	3.72	57.71	0.78	0.55	11.77	30.26
Asia	Afghanistan	Eshposhteh	ESH-KMK-6/06-ROM1	9.76	8.72	36.20	45.32	4.40	62.48	0.91	0.55	13.18	29.10
Asia	Afghanistan	Eshposhteh	ESH-KLJ-3/06-ROM1	16.46	5.11	37.74	40.69	4.43	59.17	0.83	1.26	12.74	30.34
Asia	Afghanistan	Eshposhteh	ESH-KLJ-3/06-ROM2	14.73	6.03	41.83	37.41	4.82	59.56	0.92	1.50	12.44	30.46
Asia	Afghanistan	Eshposhteh	ESH-KLJ-3/06-ROM3	16.21	8.44	36.11	39.24	4.27	56.71	0.76	1.25	12.36	28.93
Asia	Afghanistan	Eshposhteh	ARC-BK1-L3-8/03	15.22	6.47	38.07	40.24	4.50	60.04	0.91	0.73	12.13	30.26
Asia	Afghanistan	Eshposhteh	ARC-BK1-R1-8/03	18.32	5.32	34.33	42.03	3.87	57.46	0.82	2.82	11.39	30.03
Asia	Afghanistan	Eshposhteh	ARC-BK1-R3-8/03	17.25	2.54	35.49	44.72	4.35	61.72	0.90	0.88	12.36	31.65
Asia	Afghanistan	Pol-e-Khomri	PK-AND-12/04-1.1	9.03	46.35	24.81	19.81	2.30	31.75	0.46	0.75	9.36	13.62
Asia	Japan	Kushiro	Japan Taiheiyo	6.90	11.28	42.88	38.94	5.09	62.85	1.00	0.24	12.64	29.11
Asia	Kazakhstan		AIPET-14-Ermn	13.76	2.93	34.47	48.84	4.03	63.21	0.75	0.45	14.87	30.23
Asia	Kazakhstan	Ekibastuz	AIPET-16-Vstc	6.63	20.36	28.11	44.90	3.34	57.17	0.86	0.26	11.38	24.65
Asia	Kazakhstan		AIPET-18-Mkbn	15.43	5.91	33.11	45.55	3.73	59.04	0.69	0.64	14.56	28.65
Asia	Kyrgyzstan	South Fergana	K-1	28.92	4.69	31.58	34.81	3.47	50.73	0.92	2.53	8.74	30.73
Asia	Kyrgyzstan	Kavak	K-2	23.68	11.18	24.28	40.86	3.01	52.22	0.66	0.94	8.31	28.11
Asia	Kyrgyzstan	South Fergana	K-3	22.83	27.17	25.50	24.50	2.67	35.54	0.39	2.30	9.10	19.31
Asia	Kyrgyzstan	Issyl-Kul	K-4	8.33	6.61	35.27	49.79	4.13	68.62	0.86	0.76	10.69	31.07
Asia	Kyrgyzstan	Issyk-Kul	K-5	9.02	8.92	29.58	52.48	3.65	66.78	0.83	0.33	10.47	29.87

TABLE 10.3 Chemical exergy of subbituminous coals. Coal compositions are as-received as listed in Ref. 3. (Continued)

271

Continent	Country	Primary field ID	Coal basin or field	Wt.%									e^{ch} (MJ/kg)
				Moisture	Ash	V.M.	F.C.	H	C	N	S	O	
Asia	Kyrgyzstan	K-8	North Fergana	23.07	14.17	29.27	33.49	3.09	47.81	0.72	2.22	8.92	26.10
Asia	Kyrgyzstan	K-9	North Fergana	20.96	13.34	29.83	35.87	3.29	50.33	0.79	1.16	10.13	26.47
Asia	Kyrgyzstan	K-14	North Fergana	16.69	6.81	31.88	44.62	3.89	60.52	0.87	0.85	10.37	30.40
Asia	Kyrgyzstan	K-15	North Fergana	14.46	13.13	31.52	40.89	3.68	56.19	0.90	1.51	10.13	27.53
Asia	Malaysia	M1	Merit-Pila	16.08	2.54	42.06	39.32	4.85	60.08	1.33	0.24	14.88	30.75
Asia	Mongolia	Zeegt-II-1-11/02	Atlay-Chandmani	6.37	7.32	31.36	54.95	4.41	69.72	2.05	0.42	9.71	31.18
Asia	Mongolia	Nal-5-A-1-7/02	Choir-Niarga	25.73	16.26	28.75	29.26	2.81	42.48	0.62	0.62	11.48	23.25
Asia	Mongolia	Nal-5-B-1-7/02	Choir-Niarga	26.39	11.46	29.00	33.15	2.94	45.76	0.70	0.49	12.26	25.27
Asia	Mongolia	Shar-Vln-1-9/02	Orkhon-Selenge	18.40	12.30	33.35	35.95	3.69	52.84	0.81	0.67	11.29	27.08
Asia	Mongolia	Shar-Vls-1-9/02	Orkhon-Selenge	17.89	5.10	30.50	46.51	3.97	60.31	0.65	0.43	11.65	30.67
Asia	Mongolia	Bagn-2A-1-9/02	Tavan suveet	31.29	7.52	28.41	32.78	3.02	46.13	0.58	0.27	11.19	27.59
Asia	Mongolia	Bagn-2A-2-9/02	Tavan suveet	31.08	6.03	28.96	33.93	3.19	47.26	0.57	0.28	11.59	28.37
Asia	Philippines	Pan1Channel		27.90	12.44	30.84	28.82	3.23	42.90	0.81	0.41	12.31	24.75
Asia	Philippines	Pan1Stockpile		26.60	13.01	30.07	30.32	3.22	43.37	0.83	0.58	12.39	24.55
Asia	Philippines	LCP01	Southern Cebu	5.03	4.40	42.69	47.88	5.44	73.45	1.70	0.60	9.38	33.48
Asia	Philippines	LCP02	Southern Cebu	6.69	11.98	39.30	42.03	4.96	64.96	1.40	1.63	8.38	30.36
Asia	Philippines	ILB-LC-1	Zamboanga	5.20	8.81	30.60	55.39	4.98	72.90	1.78	0.39	5.94	32.98
Asia	Philippines	ILB-LC-2	Zamboanga	5.24	10.34	30.71	53.71	4.80	71.34	1.75	0.40	6.12	32.14
Australia		LC1		25.26	21.71	22.40	30.63	2.45	38.24	0.98	0.22	11.14	20.30
Europe	Hungary	Balinka 2	Bakony	22.56	17.50	33.93	26.01	3.25	41.17	0.52	4.16	10.84	23.24
Europe	Hungary	Balinka 3	Bakony	24.43	9.89	39.25	26.43	3.56	47.04	0.78	3.62	10.68	27.17
Europe	Hungary	Balinka 4	Bakony	24.93	9.85	36.54	28.68	3.34	46.06	0.80	4.16	10.86	26.68
Europe	Hungary	Balinka 5	Bakony	25.15	10.10	35.88	28.87	3.26	45.93	0.80	4.15	10.61	26.60
Europe	Hungary	Balinka 6	Bakony	21.97	20.39	32.52	25.12	3.02	39.28	0.75	3.78	10.81	21.70
Europe	Hungary	Armin 1	Bakony	25.57	14.31	32.65	27.47	2.88	43.37	0.72	3.54	9.61	24.78
Europe	Hungary	Armin 4	Bakony	26.79	12.80	32.35	28.06	2.98	43.70	0.88	3.51	9.34	25.57
Europe	Hungary	Armin 6	Bakony	26.14	13.91	33.31	26.64	2.84	42.65	0.70	3.31	10.46	24.41
Europe	Hungary	Armin 7	Bakony	28.22	8.61	32.86	30.31	3.21	46.26	0.99	3.51	9.20	27.86

TABLE 10.3 Chemical exergy of subbituminous coals. Coal compositions are as-received as listed in Ref. 3. (*Continued*)

Continent	Country	Primary field ID	Coal basin or field	Moisture	Ash	V.M.	F.C.	Wt.% H	C	N	S	O	e^{ch} (MJ/kg)
Europe	Hungary	Armin 8	Bakony	23.54	25.11	36.36	14.99	2.10	33.22	0.60	1.93	13.50	17.04
Europe	Hungary	Armin 9	Bakony	20.52	32.54	34.92	12.02	1.94	30.83	0.55	2.79	10.83	15.39
Europe	Hungary	Armin 10	Bakony	24.89	16.95	32.41	25.75	2.94	42.68	0.87	3.79	7.89	24.50
Europe	Hungary	Armin 11	Bakony	22.09	26.87	35.58	15.46	2.03	33.23	0.54	2.32	12.92	16.74
Europe	Hungary	Armin 12	Bakony	21.68	26.73	35.83	15.76	2.17	34.21	0.62	2.28	12.31	17.36
Europe	Hungary	Putnok 6	Borsod	25.81	18.31	29.18	26.70	2.89	37.95	0.64	3.26	11.14	21.86
Europe	Hungary	Putnok 4	Borsod	27.76	13.11	30.49	28.64	2.90	40.85	0.84	3.21	11.33	24.04
Europe	Hungary	Putnok 2	Borsod	27.55	15.00	30.10	27.35	2.97	39.64	0.79	2.25	11.80	23.10
South America	Brazil	99-036	Candiota	10.57	46.69	19.36	23.38	2.21	30.99	0.38	1.20	7.96	13.66
South America	Brazil	99-035	Candiota	11.48	45.44	19.61	23.47	2.22	32.05	0.56	0.80	7.46	14.24
South America	Brazil	99-217	Leao-Butia	11.00	44.41	19.73	24.86	2.22	32.64	0.56	0.31	8.86	14.18
South America	Brazil	99-218	Leao-Butia	10.38	55.86	16.91	16.85	1.64	23.79	0.95	4.93	2.45	11.16
South America	Brazil	99-219	Leao-Butia	11.31	41.67	20.49	26.53	2.34	33.98	0.55	1.94	8.21	15.42
South America	Brazil	99-220	Leao-Butia	12.35	47.14	18.73	21.78	1.83	27.36	0.50	5.32	5.50	13.12
South America	Brazil	99-221	Leao-Butia	10.65	46.74	19.12	23.49	1.92	29.55	0.52	5.29	5.33	13.89
South America	Brazil	99-222	Leao-Butia	10.67	42.84	19.45	27.04	2.27	34.81	0.63	0.34	8.44	15.16
South America	Brazil	99-223	Leao-Butia	9.14	45.16	20.57	25.13	2.33	33.64	0.57	0.28	8.88	14.40
South America	Brazil	99-224	Leao-Butia	11.60	27.11	28.78	32.51	3.03	45.86	0.76	3.07	8.57	21.76
South America	Brazil	99-225	Leao-Butia	14.29	20.46	27.71	37.54	3.22	51.20	0.95	1.30	8.58	24.70
South America	Brazil	99-226	Leao-Butia	9.33	50.46	18.92	21.29	1.53	24.78	0.44	11.46	2.00	12.88
South America	Brazil	99-284	Candiota	5.51	49.18	18.85	26.46	2.06	33.44	0.50	0.60	8.71	13.45
South America	Brazil	99-296	Candiota	7.41	45.91	20.49	26.19	2.18	33.06	0.55	1.65	9.24	13.97
South America	Brazil	99-308	Candiota	7.28	46.14	19.49	27.09	2.21	34.05	0.71	0.82	8.79	14.23
South America	Brazil	99-326	Candiota	5.34	57.15	17.93	19.58	1.88	26.63	0.52	0.85	7.63	10.58
South America	Brazil	292-02		1.53	64.20	14.03	20.24	1.94	21.80	0.43	7.73	2.37	10.15
South America	Brazil	295-02		2.01	56.09	11.67	30.23	2.15	32.62	0.60	5.44	1.10	14.26
South America	Brazil	297-02		2.03	54.94	20.01	23.02	2.70	30.91	0.57	3.75	5.10	13.58
South America	Brazil	299-02		1.41	56.38	20.05	22.16	2.50	30.13	0.55	6.32	2.71	13.68
South America	Brazil	301-02		1.17	50.18	20.41	28.24	2.98	37.12	0.65	3.35	4.55	16.12
South America	Brazil	303-02		2.07	62.28	18.11	17.54	2.33	23.24	0.42	5.73	3.93	10.68

TABLE 10.3 Chemical exergy of subbituminous coals. Coal compositions are as-received as listed in Ref. 3. (Continued)

Continent	Country	Primary field ID	Coal basin or field	Wt.%									e^{ch}
				Moisture	Ash	V.M.	F.C.	H	C	N	S	O	(MJ/kg)
South America	Brazil	305-02		4.32	61.43	15.20	19.05	2.03	24.61	0.43	4.41	2.77	10.95
South America	Brazil	00-23		9.13	43.06	20.18	27.63	2.76	35.45	0.55	0.48	8.57	15.76
South America	Brazil	02-296		2.10	47.90	12.68	37.32	2.57	42.09	0.75	2.53	2.07	17.75
South America	Brazil	02-298		1.52	45.30	23.55	29.63	3.20	42.64	0.80	2.17	4.37	18.34
South America	Brazil	02-300		1.22	27.78	28.95	42.05	3.78	55.99	1.02	9.22	0.99	25.89
South America	Brazil	02-302		1.48	24.46	27.62	46.44	4.19	60.99	1.10	2.18	5.60	26.52
South America	Brazil	02-304		1.53	40.48	25.18	32.81	3.61	45.19	0.89	2.19	6.11	19.69
South America	Brazil	02-419		1.96	60.35	11.35	26.34	2.18	26.16	0.53	5.82	3.00	11.72
South America	Brazil	02-420		1.28	39.27	13.44	46.01	3.11	49.77	1.02	1.61	3.94	20.86
South America	Brazil	02-421		1.36	53.60	21.66	23.38	2.79	32.21	0.60	6.35	3.09	14.79
South America	Brazil	02-422		1.31	50.52	29.24	18.93	2.59	33.81	0.57	10.94	0.26	16.37
South America	Brazil	02-429		4.13	25.72	27.95	42.20	3.81	54.49	1.07	5.17	5.61	24.95
South America	Chile	Chile-1	Valdivia Basin	20.12	5.80	33.74	40.34	4.02	55.57	1.06	0.28	13.15	29.23
South America	Chile	Chile-2	Magallanes Basin	24.02	12.18	37.35	26.45	3.53	46.31	0.63	0.24	13.09	25.42
South America	Chile	Pecket #1	Magallanes	25.54	21.10	28.41	24.95	3.05	38.76	0.45	0.25	10.84	21.57
South America	Chile	Pecket #2	Magallanes	24.71	20.30	29.87	25.12	3.02	39.42	0.46	0.27	11.81	21.52
South America	Colombia	IGM 1238	San Jorge	12.81	2.92	41.05	43.22	4.54	61.64	1.41	0.58	16.10	29.80
South America	Colombia	Interlab 201	San Jorge	13.52	3.37	39.64	43.47	4.27	60.07	1.43	0.47	16.87	28.93

TABLE 10.3 Chemical exergy of subbituminous coals. Coal compositions are as-received as listed in Ref. 3. (Continued)

274

Table 10.4 Chemical exergy of lignite coals. Coal compositions are as-received as listed in Ref. 3.

| Continent | Country | Primary field ID | Coal basin or field | Wt.% | | | | | | | | | e^{ch} |
				Moisture	Ash	V.M.	F.C.	H	C	N	S	O	(MJ/kg)
Asia	Afghanistan	GRBN-PRS-11/04-1	Ghowr Band	5.81	32.66	33.54	27.99	3.34	44.78	1.15	3.61	8.65	20.35
Asia	Afghanistan	KBL-CHLW-5/05-1.1	Kabul	30.03	36.58	17.76	15.63	1.22	23.44	0.50	2.02	6.21	12.75
Asia	Afghanistan	KBL-CHLW-5/05-1.3	Kabul	24.08	50.87	16.31	8.74	0.71	16.53	0.47	2.68	4.66	7.66
Asia	Afghanistan	KBL-CHLW-5/05-1.4	Kabul	30.13	36.37	20.22	13.28	1.19	23.28	0.47	2.50	6.06	12.80
Asia	Afghanistan	KBL-CHLW-3/06-A2-ROM	Kabul	24.82	44.46	19.51	11.21	1.82	20.42	0.54	2.52	5.42	11.29
Asia	Afghanistan	KBL-CHLW-3/06-L2.1	Kabul	26.13	33.23	23.39	17.25	2.28	27.83	0.61	2.71	7.21	15.97
Asia	Afghanistan	KBL-CHLW-3/06-L2.2p	Kabul	21.11	47.66	24.48	6.75	1.61	19.59	0.61	2.20	7.22	9.72
Asia	Afghanistan	KBL-CHLW-3/06-L2.3	Kabul	21.28	51.28	16.19	11.25	1.56	17.56	0.62	2.89	4.81	9.08
Asia	Afghanistan	KBL-CHLW-3/06-L2.5	Kabul	19.42	69.89	9.52	1.17	0.87	4.53	0.49	2.56	2.24	1.78
Asia	Afghanistan	KBL-CHLW-3/06-L2.7	Kabul	28.89	38.98	19.28	12.85	1.76	21.97	0.61	2.13	5.66	12.66
Asia	Afghanistan	KBL-CHLW-3/06-U3.1	Kabul	21.30	65.59	13.09	0.02	0.92	5.30	0.35	1.73	4.81	1.87
Asia	Afghanistan	KBL-CHLW-3/06-U3.3	Kabul	28.59	50.95	18.13	2.33	1.03	11.35	0.51	1.89	5.68	5.67
Asia	Afghanistan	KBL-CHLW-3/06-U3.5	Kabul	30.17	52.96	16.31	0.56	1.03	9.46	0.38	2.02	3.98	5.01
Asia	Afghanistan	KBL-CHLW-3/06-U3.7	Kabul	38.93	31.43	23.34	6.30	1.26	18.75	0.60	2.23	6.80	12.06
Asia	China	CN-24-01	Shenyang	18.08	36.00	23.44	22.48	2.46	30.88	0.73	0.32	11.53	14.89
Asia	China	CN-28-05	Yanyuan	3.66	29.25	16.13	50.96	3.08	57.35	0.64	0.61	5.41	24.07
Asia	China	CN-43-01	Shulan	17.04	45.14	22.85	14.97	2.26	24.16	0.71	0.17	10.52	11.35
Asia	China	CN-47-01	Zhalainuo'er	29.79	7.28	30.20	32.73	3.13	46.29	0.84	0.25	12.42	27.10
Asia	China	CN-47-02	Dayan	30.95	9.93	27.71	31.41	3.01	43.49	0.91	0.24	11.47	25.93
Asia	China	CN-47-03	Yimin	33.04	10.65	26.04	30.27	2.07	40.04	0.45	0.07	13.68	22.98
Asia	China	CN-47-04	Baorixile	23.05	11.15	30.73	35.07	2.78	47.60	0.56	0.15	14.71	24.46
Asia	China	CN-47-05	Huolinhe	32.59	17.28	24.56	25.57	2.37	34.66	0.49	0.39	12.22	20.50
Asia	China	CN-47-06	Yuanbaoshan	22.98	18.48	26.97	31.57	2.73	41.30	0.51	1.63	12.37	21.84
Asia	China	CN-47-08	Xilinhaote	20.71	22.30	30.30	26.69	2.91	38.42	0.54	1.21	13.91	19.78
Asia	China	CN-47-09	Wunite	24.74	7.01	31.47	36.78	2.95	48.58	0.57	0.27	15.88	25.75
Asia	China	CN-77-01	Youjiang	13.86	24.11	30.63	31.40	2.96	42.88	1.21	1.49	13.49	20.21
Asia	China	CN-77-02	Yining	25.91	10.53	32.18	31.38	3.18	45.23	1.37	0.92	12.86	25.28
Asia	China	CN-87-03	Fengmingchun	32.98	14.77	31.33	20.92	2.85	33.41	0.95	1.90	13.14	21.14

Continent	Country	Primary field ID	Coal basin or field	Moisture	Ash	V.M.	F.C.	Wt.%					e^{ch} (MJ/kg)
								H	C	N	S	O	
Asia	China	CN-87-05	Longtan	33.52	7.73	32.09	26.66	2.88	39.23	1.07	0.88	14.69	24.25
Asia	DPR Korea	NK-37		7.44	7.15	40.69	44.72	5.05	65.96	1.95	0.56	11.89	30.66
Asia	DPR Korea	NK-38		2.70	18.83	6.80	71.67	2.07	71.76	1.92	0.58	2.14	28.57
Asia	DPR Korea	NK-39		15.06	20.91	30.49	33.54	3.52	44.79	1.54	1.43	12.76	22.18
Asia	DPR Korea	NK-40		7.17	42.29	29.09	21.45	3.25	35.56	0.82	0.15	10.76	15.80
Asia	DPR Korea	NK-41		18.17	3.51	38.09	40.23	4.63	60.38	1.89	0.57	10.85	31.85
Asia	DPR Korea	NK-42		14.72	16.80	33.24	35.24	3.76	49.41	1.04	0.16	14.11	24.03
Asia	DPR Korea	NK-43		21.51	6.78	33.42	38.29	3.79	52.22	1.58	0.42	13.70	27.83
Asia	DPR Korea	NK-44	Kilju-Myongchon	16.15	7.59	38.79	37.47	4.35	56.72	1.05	0.46	13.68	28.72
Asia	DPR Korea	NK-45	Kilju-Myongchon	16.56	17.17	40.83	25.44	4.50	48.14	0.83	0.88	11.92	25.37
Asia	DPR Korea	NK-46	Kilju-Myongchon	21.22	27.56	28.07	23.15	2.67	33.08	0.55	1.62	13.30	17.15
Asia	DPR Korea	NK-47	Kilju-Myongchon	19.57	12.74	43.01	24.68	4.17	47.28	0.46	2.13	13.65	25.72
Asia	DPR Korea	NK-48	Kilju-Myongchon	22.45	16.16	32.99	28.40	3.29	42.50	0.99	0.42	14.19	22.61
Asia	DPR Korea	NK-49	Kocham	19.73	21.94	35.23	23.10	3.67	40.20	0.89	0.36	13.21	21.26
Asia	DPR Korea	NK-50	Kumya	31.19	8.40	33.83	26.58	3.29	41.56	0.64	0.28	14.64	25.12
Asia	India	GR-II-98		36.43	16.04	24.20	23.33	2.31	33.52	0.63	3.33	7.74	22.54
Asia	India	GR-IV-98		38.60	9.58	25.80	26.02	2.64	36.88	0.66	3.78	7.86	26.26
Asia	India	GR-VI-98A		26.83	42.60	18.30	12.27	1.76	18.32	0.34	3.39	6.76	10.60
Asia	India	GR-VI-98B		36.70	13.27	27.24	22.79	2.86	35.16	0.67	3.31	8.03	24.58
Asia	India	GR-GR-III-98		34.55	18.71	24.36	22.38	2.37	30.98	0.61	5.83	6.95	21.36
Asia	India	GR-VII-98-B		34.01	15.34	26.36	24.29	2.63	34.54	0.64	5.43	7.41	23.50
Asia	India	GR-GR-V-98		28.57	21.64	25.56	24.23	2.06	29.95	0.59	13.89	3.30	21.15
Asia	India	GR-GR-I-98		24.04	43.79	19.74	12.43	1.89	19.42	0.44	3.54	6.88	10.92
Asia	Mongolia	Hov-I-1-11/02	Atlay-Chandmani	29.84	6.11	34.09	29.96	2.07	39.71	1.51	0.66	20.10	21.28
Asia	Mongolia	Chdg-2-A-10/02	Choir-Niarga	36.47	6.24	26.43	30.86	2.94	41.72	0.53	0.54	11.56	27.27
Asia	Mongolia	Chdg-2-B-10/02	Choir-Niarga	38.08	8.04	25.44	28.44	2.70	39.23	0.49	0.75	10.71	26.21

TABLE 10.4 Chemical exergy of lignite coals. Coal compositions are as-received as listed in Ref. 3. (Continued)

Table 10.4 Chemical exergy of lignite coals. Coal compositions are as-received as listed in Ref. 3. (*Continued*)

Continent	Country	Primary field ID	Coal basin or field	Moisture	Ash	V.M.	F.C.	H	C	N	S	O	e^{ch} (MJ/kg)
								Wt.%					
Asia	Mongolia	Sivo-I-1-8/02	Choir-Niarga	43.55	4.28	26.11	26.06	2.37	38.72	0.42	0.35	10.31	27.85
Asia	Mongolia	Sivo-II-1-8/02	Choir-Niarga	44.22	8.04	22.42	25.32	1.98	35.45	0.34	2.42	7.55	26.30
Asia	Mongolia	Adnh-710-1-10/02	Choiybalsan	50.76	6.72	20.11	22.41	1.99	30.79	0.32	1.41	8.01	25.96
Asia	Mongolia	Adnh-720-1-10/02	Choiybalsan	35.73	5.26	28.31	30.70	2.46	43.01	0.37	0.79	12.38	26.85
Asia	Mongolia	Talb-I-1-2/03	Sukhbaatar	38.43	16.98	21.77	22.82	2.22	30.71	0.40	0.56	10.70	20.13
Asia	Mongolia	Talb-I-2-2/03	Sukhbaatar	38.49	17.31	21.70	22.50	2.20	30.48	0.45	0.56	10.51	19.99
Asia	Thailand	Ban Pu Coalfield	Li Basin, Ban Pu field	17.27	6.88	39.64	36.21	3.55	53.19	0.35	1.13	17.63	26.26
Asia	Thailand	Chiang Muan Coalfield, Lower Main Seam (LM)	Chiang Muan	15.66	5.62	43.43	35.29	4.35	54.36	1.24	0.23	18.54	27.04
Asia	Thailand	Chiang Muan Coalfield, Upper Seam (U2)	Chiang Muan	8.47	54.04	26.76	10.73	2.09	18.27	0.70	5.89	10.54	8.77
Asia	Thailand	Mae Lai Coalfield	Mae Lai	17.63	7.42	41.44	33.51	3.32	50.43	1.00	1.14	19.06	24.69
Asia	Thailand	Mae Lamao (Suje) Coalfield	Mae Lamao (Suje)	23.92	8.78	32.11	35.19	3.38	46.42	1.33	4.17	12.00	26.46
Asia	Thailand	Mae Moh Coalfield, J Zone	Mae Moh	34.80	15.98	30.13	19.09	2.32	33.96	0.84	3.72	8.38	22.28
Asia	Thailand	Mae Moh Coalfield, K Zone	Mae Moh	34.11	8.22	28.58	29.09	2.60	40.99	1.54	1.38	11.16	25.56
Asia	Thailand	Mae Moh Coalfield, Q Zone	Mae Moh	19.92	13.81	34.12	32.15	3.18	45.47	1.36	2.77	13.49	23.80
Asia	Thailand	Mae Than Coalfield	Mae Than	21.73	3.28	37.26	37.73	3.85	53.55	0.73	0.98	15.88	28.60
Asia	Thailand	Na Hong Coalfield	Na Hong	15.43	7.55	36.72	40.30	3.53	51.65	0.84	2.95	18.05	25.37
Asia	Thailand	Pa Kha Coalfield	Li Basin, Pa Kha field	13.34	1.97	37.38	47.31	4.62	65.16	1.15	0.48	13.28	31.84
Europe	Greece	GR-1		55.60	16.13	18.32	9.95	1.41	18.52	0.65	1.38	6.31	17.22
Europe	Greece	GR-2		55.28	17.80	18.42	8.50	1.29	17.09	0.59	1.53	6.42	15.61
Europe	Hungary	Visonta 1	Matra	48.17	15.17	22.55	14.11	2.03	24.36	0.39	1.35	8.53	19.81
Europe	Romania	Romania-2000		18.86	36.99	29.80	14.35	2.32	24.80	0.46	3.26	13.31	12.63
Europe	Serbia	Yugoslavia—1		44.78	13.26	30.16	11.80	1.73	25.86	0.67	2.90	10.80	19.25
Europe	Serbia	Yugoslavia—2		26.14	4.13	43.57	26.16	3.60	45.25	0.28	2.01	18.59	25.80
Europe	Serbia	Yugoslavia—3		16.32	16.26	46.02	21.40	3.20	44.35	0.74	3.46	15.67	22.20

Continent	Country	Primary field ID	Coal basin or field	Moisture	Ash	V.M.	F.C.	H	C	N	S	O	e^{ch} (MJ/kg)
								Wt.%					
Europe	Serbia	WOCQI SE-1	Kolubara	8.03	2.62	58.92	30.43	5.57	58.49	0.25	0.47	24.57	27.43
Europe	Serbia	WOCQI SE-2	Kolubara	34.49	6.76	36.28	22.47	3.63	40.29	0.41	0.36	14.06	26.41
Europe	Slovak Republic	SK-1	Novaky	31.60	7.25	43.04	18.11	3.19	40.66	0.67	2.56	14.07	25.40
Europe	Slovak Republic	11-Handlova	Handlova	12.27	34.93	29.24	23.56	2.80	34.12	0.48	2.80	12.60	16.15
Europe	Slovak Republic	12-Handlova	Handlova	12.38	23.67	34.38	29.57	3.53	44.01	0.51	0.81	15.10	20.75
Europe	Slovak Republic	13-Handlova	Handlova	18.96	6.04	37.68	37.32	3.99	52.65	0.60	1.15	16.61	27.30
Europe	Slovak Republic	14-Handlova	Handlova	15.32	10.02	37.56	37.10	4.23	54.15	0.68	0.95	14.65	27.13
Europe	Slovak Republic	15-Novaky	Novaky	26.51	11.21	33.71	28.57	3.07	42.53	0.85	2.18	13.65	24.22
Europe	Slovak Republic	16-Novaky	Novaky	22.76	7.48	36.42	33.34	3.38	48.34	0.88	2.77	14.39	26.38
Europe	Slovak Republic	17-Novaky	Novaky	34.80	18.07	26.84	20.29	2.33	31.32	0.60	2.03	10.85	19.92
Europe	Spain	ES-1	Cuenca Terciaria de As Pontes	5.98	42.09	21.05	30.88	2.15	39.36	0.77	0.91	8.74	16.12
North America	Guatemala	WOCQI GT-1		9.71	15.79	37.95	36.55	3.47	53.88	1.18	4.71	11.26	25.42

TABLE 10.4 Chemical exergy of lignite coals. Coal compositions are as-received as listed in Ref. 3. (*Continued*)

Continent	Country	Province/state	Coal basin or field	Wt.%						e^{ch} (MJ/kg)	Ref.
				Ash	H	C	N	S	O		
Asia	Japanese			12.1	5.71	68.2	0.99	0.19	12.26	29.71	[4]
Australia	Australia			8.7	5.31	76.3	1.54	0.46	7.31	32.75	
Australia	Australia			13.4	4.4	71.1	1.17	0.36	9.17	29.59	
Europe	German		Braunkohole	5.55	4.97	63.89	0.57	0.48	24.54	26.47	[5]
North America	Canada	Mississippi		29.76	3.91	49.35	1.21	0.79	14.98	20.41	
North America	Mexico		Ahmsa	13.3	4.9	76.1	1.11	0.90	3.7	32.53	
North America	Mexico		Cloete	29.3	4.0	61.6	0.97	0.95	3.2	26.00	
North America	Mexico		Micare 1	41.1	3.6	45.6	0.88	0.97	7.9	19.12	[6]
North America	Mexico		Micare 2	35.3	3.8	54.2	0.88	1.27	4.6	22.92	
North America	Mexico		Micare 3	35.0	3.8	53.2	0.83	0.99	6.2	22.37	
North America	Mexico		Palau	24.7	4.3	65.2	0.99	0.90	3.9	27.66	
North America	Mexico		Pocito Tec	19.4	4.5	71.8	1.06	1.04	2.2	30.56	
North America	Mexico		Tajo Tec 1	27.9	4.0	62.9	0.95	1.21	3.0	26.57	
North America	Mexico		Tajo Tec 2	26.8	4.1	64.2	1.02	0.86	3.1	27.10	
North America	United States	Pittsburgh		10.3	5.0	75.5	1.20	3.1	4.9	32.81	[5]
North America	United States		North Antelope Rochelle Mine	7.17	4.68	67.49	1.18	0.31	19.18	27.85	[7]
North America	United States		Black Thunder Mine	6.54	5.09	73.49	1.29	0.6	19.53	30.51	[8]
North America	United States		Antelope Coal Mine	3.0	6.1	72.5	0.8	0.3	17.2	31.48	[9]
North America	United States		Eagle Butte Mine	6.6	5.1	67.4	0.9	0.6	19.4	28.33	[10]
North America	United States		Cordero Rojo	5.56	3.60	50.6	1.35	0.52	15.3	20.95	
North America	United States	Illinois	Illinois Basin	9.56	4.64	67.0	2.22	3.10	8.11	29.08	[11]
North America	United States		Band Mill	7.22	4.95	77.8	2.32	0.75	5.56	33.14	
North America	United States		Blacksville	8.61	4.71	72.7	2.31	2.00	7.71	31.06	

TABLE 10.5 Chemical exergy of other coals with unknown rank. Coal compositions are given on a dry basis, except for Refs. 29 and 31 which are as-received.

Continent	Country	Province/state	Coal basin or field	Ash	H	C	N	S	O	e^{ch} (MJ/kg)	Ref.
North America	United States		Belle Ayr	4.83	6.29	51.4	0.80	0.29	36.4	22.36	[12]
North America	United States		Freedom mine	13.8	3.83	59.7	1.01	0.79	20.8	23.91	[13]
North America	United States		Polar River power station	20.4	3.61	53.2	0.85	0.70	21.2	21.12	
North America	United States		Buckskin	6.46	4.28	53.21	0.72	0.87	40.92	20.49	[14]
North America	United States	Pittsburgh #8	Bailey Mine	7.79	5.49	76.83	1.4	1.46	3.91	33.65	[15]
North America	United States	Ohio	Marshall County Mine	5.27	3.98	75.88	1.04	0.86	12.97	30.80	[16]
North America	United States	Wyoming	Rawhide Mine	7.5	5.2	72.3	0.95	0.51	21.0	30.04	[17]
North America	United States	Wyoming	Adaville	3.7	6.3	58.1	0.9	0.7	30.2	25.43	
North America	United States	Wyoming	Adaville	4.9	6.1	55.1	1.4	0.6	31.8	23.94	
North America	United States	Wyoming	Almy	7.2	5.4	60.0	1.2	0.2	26.0	25.32	
North America	United States	Wyoming	Lower Kemmerer	3.6	5.6	73.0	1.1	1.1	15.7	31.39	[18]
North America	United States	Wyoming	Lower Spring Valley	5.4	5.4	73.6	1.2	0.4	14.0	31.35	
North America	United States	Wyoming	Lower Willow Creek	8.7	5.4	71.7	1.3	1.1	11.8	30.92	
North America	United States	Wyoming	Upper Spring Valley	8.8	5.3	69.6	1.2	0.4	14.8	29.77	
North America	United States	Wyoming	Vail coal bed	4.6	5.8	69.9	1.5	0.4	17.8	30.15	
North America	United States	Wyoming	Willow Creek	8.0	5.3	72.8	1.2	0.9	11.8	31.19	

TABLE 10.5 Chemical exergy of other coals with unknown rank. Coal compositions are given on a dry basis, except for Refs. 29 and 31 which are as-received. (Continued)

Continent	Country	Province/state	Coal basin or field	Wt.%						e^{ch} (MJ/kg)	Ref.
				Ash	H	C	N	S	O		
North America	United States	Pennsylvania	Enlow Fork coal	7.36	5.4	76.16	1.35	1.58	8.14	32.98	[19]
North America	United States		Belle Ayr	4.87	6.18	57.5	0.7	0.36	30.4	24.97	[20]
North America	United States	Montana	Rosebud mine	10.1	4.5	67.2	0.78	0.52	17.7	27.65	[20]
North America	United States	Montana	Lignite	9.0	5.8	63.1	1.02	0.97	19.8	27.53	[21]
North America	United States	North Dakota	Falkirk mine	14.64	4.02	60.15	0.65	0.98	19.56	24.40	[21]
North America	United States	Wyoming	Wyodak coal	8.13	4.88	69.00	0.97	0.43	16.59	28.84	[22]
North America	United States	Utah	Sufco	8.4	4.61	77.1	1.29	0.36	8.29	32.17	[23]
North America	United States	Alabama	Blue Creek seam	9.52	4.41	81.68	1.5	1.44	0.2	34.50	[24]
North America	United States	New Mexico	El Segundo	5.4	5.4	77.1	1.3	1.0	15.2	32.64	[25]
North America	United States	North Dakota	Western mine	9.74	7.12	33.41	0.56	0.95	48.24	15.76	[26]
South America	Argentina	Rio Turbio	Rio Turbio 1	25.2	3.96	49.62	0.89	0.45	9.11	21.10	[26]
South America	Argentina	Rio Turbio	Rio Turbio 2	16.67	4.53	57.83	0.98	0.43	9.45	24.85	[26]
South America	Argentina	Pico Quemado	Rio Turbio 3	12.05	5.14	63.25	1.06	1.6	9.25	27.83	[26]
South America	Argentina	Pico Quemado	Pico Quemado 1	47.03	1.79	23.64	0.72	0.16	9.05	8.79	[26]
South America	Argentina	Pico Quemado	Pico Quemado 2	40.32	3.25	38.92	1.1	0.29	7.94	16.19	[26]
South America	Argentina	Pico Quemado	Pico Quemado 3	65.17	1.71	16.05	0.48	0.16	6.63	5.78	[26]
South America	Argentina	Pico Quemado	Pico Quemado 4	68.96	0.43	11.51	0.18	1.13	13.83	2.22	[26]

TABLE 10.5 Chemical exergy of other coals with unknown rank. Coal compositions are given on a dry basis, except for Refs. 29 and 31 which are as-received. (Continued)

Continent	Country	Province/state	Coal basin or field	Ash	H	C	N	S	O	e^ch (MJ/kg)	Ref.
						Wt.%					
South America	Brazil		3G Piano II run-of-mine	47.38	2.96	37.34	0.76	2.25	9.31	15.42	
South America	Brazil		3G Piano II clean coal	32.51	3.69	52.76	1.01	1.44	8.59	22.02	
South America	Brazil		Bonito Run-of-mine	69.76	1.64	21.5	0.29	1.78	5.03	8.04	
South America	Brazil		Bonito Clean coal	51.14	2.28	38.99	0.49	1.55	5.55	15.38	
South America	Brazil		UM II-Verdinho run-of-mine	63.67	1.95	26.68	0.49	2.11	5.1	10.44	
South America	Brazil		UM II-Verdinho clean coal	37.42	3.56	51.65	0.96	1.66	4.75	21.75	
South America	Brazil		Fontanela run-of-mine	54.85	2.06	36.45	0.47	2.34	3.83	14.44	
South America	Brazil		Fontanela clean coal	40.15	2.92	50.97	0.69	1.53	3.74	20.80	
South America	Brazil		Esperança Leste run-of-mine	52.44	2.84	36.51	0.78	2.02	5.41	15.17	[27]
South America	Brazil		Esperança Leste clean coal	36.59	3.53	51.80	1.05	1.37	5.66	21.65	
South America	Brazil		Barro Branco run-of-mine	53.91	2.32	36.41	0.71	2.72	3.93	14.80	
South America	Brazil		Barro Branco clean coal	42.97	3.26	46.21	0.91	2.70	3.95	19.61	
South America	Brazil		Mina 3 run-of-mine	44.92	3.09	45.43	0.90	1.75	3.91	18.91	
South America	Brazil		Mina 3 clean coal	29.44	3.95	60.84	1.17	1.12	3.48	25.68	
South America	Brazil		Morozino run-of-mine	48.69	2.69	41.72	0.90	2.65	3.35	17.27	
South America	Brazil		Morozino clean coal	34.38	3.40	56.55	1.18	1.40	3.09	23.50	
South America	Brazil		Cantao run-of-mine	55.86	2.44	35.1	0.77	2.44	3.39	14.41	
South America	Brazil		Cantao clean coal	42.76	2.81	49.02	0.99	1.37	3.05	19.95	
South America	Brazil		Morozin/Cantao clean coal	39.91	3.38	47.17	0.78	0.74	8.02	19.41	
South America	Brazil		Gabriela clean coal	46.05	3.29	44.32	0.89	0.90	4.55	18.48	
South America	Brazil		Car. Siderópolis clean coal	47.30	3.09	39.31	0.82	3.21	6.27	16.73	
South America	Chile	Arauco	Chile -3	6.56	5.54	71.70	1.09	3.57	7.86	31.95	
South America	Chile	Arauco	Manto Alto Lebu	5.74	5.48	73.57	1.76	3.84	7.15	32.69	
South America	Chile	Arauco	Manto Chico Lebu	2.43	5.54	77.08	1.95	1.30	8.82	33.46	
South America	Chile	Arauco	Manto Chiflon Lebu	33.88	4.01	49.87	1.14	1.89	6.57	21.56	[28]
South America	Chile	Arauco	Manto Huitrero Lebu	8.86	5.36	71.79	1.27	2.11	7.84	31.45	
South America	Chile	Arauco	Lebu #1	1.92	5.38	76.55	1.94	0.92	9.35	32.98	
South America	Chile	Arauco	Lebu #2	7.90	5.32	72.36	1.54	3.91	6.89	32.07	
South America	Chile	Valdivia-Los Lagos	Chile -1	5.80	4.02	55.57	1.06	0.28	13.15	23.35	
South America	Chile	Valdivia-Los Lagos	Mulpur 3004	44.12	2.63	29.17	0.66	0.43	9.73	11.77	

TABLE 10.5 Chemical exergy of other coals with unknown rank. Coal compositions are given on a dry basis, except for Refs. 29 and 31 which are as-received. (*Continued*)

Continent	Country	Province/state	Coal basin or field	Wt.%						e^{ch} (MJ/kg)	Ref.
				Ash	H	C	N	S	O		
South America	Chile	Valdivia-Los Lagos	Mulpur 3005	21.24	3.79	46.86	0.96	3.85	10.52	20.54	
South America	Chile	Valdivia-Los Lagos	Milahuillin	2.38	3.70	58.89	0.65	0.50	13.53	24.28	
South America	Chile	Valdivia-Los Lagos	Mulpur 3006	13.35	3.91	52.21	1.14	0.48	12.17	21.97	
South America	Chile	Magallanes	Chile -2	12.18	3.53	46.31	0.63	0.24	13.09	19.32	
South America	Chile	Magallanes	1	15.47	3.43	45.24	0.53	0.22	13.70	18.69	
South America	Chile	Magallanes	2A	19.48	3.15	41.48	0.58	0.33	12.82	17.04	
South America	Chile	Magallanes	2B	18.62	3.16	41.5	0.58	0.32	12.14	17.13	[28]
South America	Chile	Magallanes	3	22.9	3.27	40.56	0.54	2.62	9.85	17.47	
South America	Chile	Magallanes	4	21.12	3.55	42.72	0.60	0.44	11.93	17.99	
South America	Chile	Magallanes	5	11.13	3.73	48.40	0.62	0.25	13.85	20.25	
South America	Chile	Magallanes	6	13.02	3.72	47.38	0.62	0.20	13.79	19.83	
South America	Chile	Magallanes	Pecket #1	21.1	3.05	38.76	0.45	0.25	10.85	16.06	
South America	Chile	Magallanes	Pecket #2	20.3	3.02	39.42	0.46	0.27	11.82	16.20	
South America	Colombia	Cauca	Guachinte, Nivel 1	21.34	4.45	60.16	1.05	5.12	6.18	27.17	
South America	Colombia	Boyaca	Guaduas Formations Coal	8.53	4.85	78.04	1.57	0.93	5.05	33.51	
South America	Colombia	Boyaca	Guaduas	5.86	5.18	80.85	1.89	0.73	4.64	34.90	
South America	Colombia	Cesar	Los Cuervos	4.40	4.88	74.61	1.46	0.59	9.51	33.10	
South America	Colombia	Cordoba	Cerrito y	2.92	4.54	61.64	1.41	0.58	16.1	29.81	
South America	Colombia	Norte de Santander	Los Cuervos, Cuatro	5.50	5.28	78.46	1.57	0.62	6.90	34.21	
South America	Colombia	Norte de Santander	Los Cuervos, Superior	9.84	5.38	71.47	1.76	0.87	8.65	31.67	[29]
South America	Colombia	Norte de Santander	Los Cuervos, Veta Grande	4.95	5.19	81.04	1.52	0.60	5.54	35.00	
South America	Colombia	Guajira	Cerrejon, 70	1.39	5.18	78.72	1.52	0.39	9.71	34.48	
South America	Colombia	Guajira	Cerrejon, 75	1.01	5.33	77.70	1.73	0.47	10.10	34.47	
South America	Colombia	Guajira	Cerrejon, 115	0.53	5.20	75.91	1.70	0.63	12.14	33.59	
South America	Colombia	Norte de Santander	Cerrejon, 130	4.27	5.13	73.82	1.15	0.33	11.31	32.68	
South America	Colombia	Cundinamarca	Los Cuervos, Min. San Rafael	5.17	5.69	78.79	1.58	0.51	6.91	34.66	
South America	Colombia	Cordoba	Guaduas, Veta Grande	11.41	5.29	70.15	1.58	1.57	7.91	31.27	
South America	Colombia	Cauca	Cerrito y Cienaga de Oro, Inferior	3.37	4.27	60.07	1.43	0.47	16.87	28.94	
South America	Colombia	Rio Turbio	Mosquera, H 3	5.58	4.83	66.92	1.59	1.13	14.42	30.08	

Table 10.5 Chemical exergy of other coals with unknown rank. Coal compositions are given on a dry basis, except for Refs. 29 and 31 which are as-received. (Continued)

Continent	Country	Province/state	Coal basin or field	Ash	H	C	N	S	O	e^{ch} (MJ/kg)	Ref.
South America	Peru	Mina La Victoria	BP2400, Alto Chicama	5.86	1.08	86.0	0.56	0.27	1.16	33.88	
South America	Peru	Mina La Victoria	BP2500, Alto Chicama	2.17	1.03	88.18	0.54	0.28	2.03	34.91	
South America	Peru	Banos de Chimu	BP2900, Alto Chicama	8.73	0.79	82.90	0.32	0.44	3.60	31.49	
South America	Peru	Alto Chicama	BP3100, Mina Huarachal	3.57	0.99	88.13	0.65	0.53	1.72	34.40	
South America	Peru	Alto Chicama	BP3200, Mina Patro Sntiago	61.57	1.97	24.99	0.73	0.67	7.95	9.57	
South America	Peru	Mina La Galgada	BP0101	30.95	0.75	59.22	0.60	0.49	3.71	22.36	[30]
South America	Peru	Mina La Limena 2	BP0102	28.87	0.48	55.96	0.45	0.67	6.55	21.27	
South America	Peru	Mina La Limena 1	BP0103	22.28	0.41	66.37	0.47	0.38	4.14	25.25	
South America	Peru	Mina Cocobal	BP0104	10.07	0.83	81.45	0.71	0.34	2.23	31.44	
South America	Peru	Mina Villon	BP0105	11.39	2.04	80.70	0.79	0.92	1.56	32.11	
South America	Peru	Mina Caraz	BP0106	25.77	0.91	63.93	0.67	0.44	3.19	24.69	
South America	Peru	Mina San Roque	BP0107	16.22	1.43	72.67	0.73	0.61	2.75	29.06	
South America	Peru	Mina Sta	BP0108	26.72	0.31	46.84	0.62	1.50	8.23	19.48	
South America	Peru	Alto Chicama	BP0501	3.22	1.15	87.95	0.55	0.36	2.05	34.57	
South America	Venezuela	Mérida	Quebreda El Palmital	1.87	6.07	76.82	1.72	0.84	9.19	35.06	
South America	Venezuela	Mérida	Carretera Las Dantas La Vega	16.85	5.41	62.29	1.31	6.21	5.90	29.47	
South America	Venezuela	Mérida	Rio Escalante	3.46	5.63	69.61	1.39	0.91	12.25	32.66	
South America	Venezuela	Mérida	Qeubreda Palmichosa	4.86	4.61	62.95	1.43	3.42	11.48	30.94	
South America	Venezuela	Táchira	Rio Pajitas	15.42	5.01	65.94	1.22	5.97	4.98	30.25	
South America	Venezuela	Táchira	Finca Familia Arellano	7.67	6.60	69.98	1.24	0.43	9.61	33.19	
South America	Venezuela	Táchira	Caliche	4.55	6.09	79.33	1.47	2.61	5.02	35.75	[31]
South America	Venezuela	Táchira	Las Adjuntas	1.43	4.60	73.78	1.67	0.62	11.42	33.03	
South America	Venezuela	Táchira	Las Adjuntas	1.80	5.15	78.35	1.59	0.59	8.70	34.70	
South America	Venezuela	Táchira	Campamento KOPEX	0.91	4.35	62.40	1.30	0.27	14.11	31.44	
South America	Venezuela	Táchira	Hato de la Virgen	1.93	7.10	81.99	1.61	0.08	5.47	37.42	
South America	Venezuela	Táchira	Casadero	12.74	6.55	72.47	1.06	0.47	6.02	32.98	
South America	Venezuela	Táchira	La Pajarita	3.61	8.00	79.68	1.53	0.57	5.89	37.46	
South America	Venezuela	Táchira	Cerro Capote	2.83	5.91	74.88	1.13	1.15	9.49	34.57	
South America	Venezuela	Zulia	Paez Minas Norte	1.83	5.45	80.02	1.47	0.40	8.05	35.30	
South America	Venezuela	Zulia	Paez Paso Diablo	0.41	5.49	80.89	1.55	0.36	8.26	35.77	

TABLE 10.5 Chemical exergy of other coals with unknown rank. Coal compositions are given on a dry basis, except for Refs. 29 and 31 which are as-received. (*Continued*)

Categories	Biomass type	Wt.%						e^{ch} (MJ/kg)	Ref.
		C	H	O	N	S	Ash		
Algae	Low ash algae	50.4	6.9	29	7.1	0.5	6.1	23.33	[32]
Bark	Balsam bark	54	6.2	39.5	0.2	0.1	2.6	22.34	[33]
Bark	Beech bark	51.4	6	41.8	0.7	0.11	7.8	19.79	[33]
Bark	Birch bark	57	6.7	35.7	0.5	0.1	2.1	24.39	[33]
Bark	Elm bark	50.9	5.8	42.5	0.7	0.11	8.1	19.29	[33]
Bark	Eucalyptus bark	48.7	5.7	45.3	0.3	0.05	4.8	18.95	[33]
Bark	Hemlock bark	55	5.9	38.8	0.2	0.1	2.5	22.46	[33]
Bark	Maple bark	52	6.2	41.3	0.4	0.11	4	21.15	[33]
Bark	Pine bark	53.8	5.9	39.9	0.3	0.07	1.9	22.08	[33]
Bark	Poplar bark	53.6	6.7	39.3	0.3	0.1	2.2	22.86	[33]
Bark	Spruce bark	53.6	6.2	40	0.1	0.1	3.2	22.01	[33]
Bark	Tamarack bark	57	10.2	32	0.7	0.11	4.2	27.83	[34]
Bark	Oak bark	49.70	5.40	39.30	0.20	0.10	5.30	20.46	[33]
Branches	Grape marc	54	6.1	37.4	2.4	0.15	7.8	21.12	[33]
Branches	Pepper plant	42.2	5	49	3.2	0.57	14.4	13.98	[33]
Branches	Alfalfa stems	47.17	5.99	38.19	2.68	0.2	5.27	20.32	[35]
Branches	Cotton residue	47.03	5.96	38.42	1.79	0.19	6.61	20.18	[35]
Branches	Marabú	48.6	6.3	43.6			1.5	21.07	[35]
Branches	Oak wood (large branch)	48.57	6.81	42.23	2.39		2.07	21.41	[35]
Branches	Oak wood (medium branch)	48.62	6.52	42.28	2.58		3	21.08	[35]
Branches	Oak wood (small branch)	48.76	6.35	42.08	2.81		4.05	20.94	[35]
Branches	Pepper plant	36.11	4.26	41.86	2.72	0.49	14.44	13.96	[35]
Branches	Soplillo	48.8	6.5	43.2			1.5	21.07	[35]
Core	Olive pits	52.8	6.6	39.4	1.1	0.07	3.1	22.23	[33]
Core	Palm kernels	51	6.5	39.5	2.7	0.27	5.2	21.01	[33]
Core	Plum pits	49.9	6.7	42.4	0.9	0.08	1.4	21.47	[33]
Core	Olive kernel	54.6	6.8	36.1	0.8		1.7	24.11	[35]
Core	Olive kernel	52.44	6.17	37.85	1.32	0.09	2.13	22.49	[35]
Core	Olive pits	52.8	6.69	38.25	0.45	0.05	1.72	23.16	[35]
Core	Olive stone	49	6.1	42	0.8		2.2	20.79	[35]

TABLE 10.6 Chemical exergy of biomasses. Compositions are on a dry basis as listed in the cited references except for Ref. 3 in which case the elements are reported on a dry ash free basis with the ash content reported on a dry basis.

Categories	Biomass type	Wt.%						e^{ch} (MJ/kg)	Ref.
		C	H	O	N	S	Ash		
Core	Palm kernels	48.34	6.2	37.44	2.62	0.26	5.14	21.05	[35]
Food	Dried grains-solubles	50.24	6.89	33.42	4.79	0.77	3.89	22.98	[35]
Food	Oil palm fruit bunch	45.9	5.8	40.1	1.2		4.53	19.46	[35]
Food	Olive cake	53.7	6.7	36.2	0.6		2.8	23.64	[35]
Food	Shea meal	48.56	5.86	37.7	2.88	0.66	5	20.69	[35]
Food	Wet grains	52.53	6.6	32.28	5.35	0.66	2.58	23.59	[35]
Food Wastes	Chicken litter	60.5	6.8	25.3	6.2	1.2	37.8	16.28	[33]
Food Wastes	Meat-bone meal	57.3	8	20.8	12.2	1.69	24	20.87	[33]
Food Wastes	Olive residue	58.4	5.8	34.2	1.4	0.23	7.2	22.71	[33]
Food Wastes	Pepper residue	45.7	3.2	47.1	3.4	0.6	8.2	14.64	[33]
Food Wastes	Sugarcane bagasse	49.8	6	43.9	0.2	0.06	2.1	20.38	[33]
Food Wastes	Bagasse	44.80	5.35	39.55	0.38	0.01	11.27	18.47	[34]
Food Wastes	Millet grain waste	40.56	5.24	45.50	0.40		8.40	16.36	[34]
Food Wastes	Sugarcane bagasse	45.48	5.96	45.21	0.15		3.20	19.07	[34]
Food Wastes	Sugarcane bagasse	45.48	5.96	0.15	45.21		3.2	23.06	[35]
Food Wastes	Sugarcane bagasse	43.79	5.96	43.36	1.69		5.2	18.58	[35]
Food Wastes	Sugarcane bagasse	47.2	7	43.1			2.7	21.03	[35]
Grass	Arundo grass	48.7	6.1	44.5	0.6	0.13	3.4	19.77	[33]
Grass	Bana grass	50.1	6	42.9	0.9	0.13	9.8	18.80	[33]
Grass	Buffalo gourd grass	46.1	6.5	44.5	2.6	0.27	4.7	19.03	[33]
Grass	Kenaf grass	48.4	6	44.5	1	0.15	3.6	19.52	[33]
Grass	Miscanthus grass	49.2	6	44.2	0.4	0.15	3	19.96	[33]
Grass	Reed canary grass	49.4	6.3	42.7	1.5	0.15	8.9	19.10	[33]
Grass	Sorghastrum grass	49.4	6.3	44	0.3	0.05	4.2	20.07	[33]
Grass	Sweet sorghum grass	49.7	6.1	43.7	0.4	0.09	4.7	19.88	[33]
Grass	Switchgrass	49.7	6.1	43.4	0.7	0.11	5.1	19.82	[33]
Grass	Switch grass	46.68	5.82	37.38	0.77	0.19	8.97	19.94	[35]
Husk	Coffee husks	45.4	4.9	48.3	1.1	0.35	2.8	17.19	[33]
Husk	Cotton husks	50.4	8.4	39.8	1.4	0.01	3.4	23.18	[33]
Husk	Mustard husks	45.8	9.2	44.4	0.4	0.2	4.1	21.9	[33]

TABLE 10.6 Chemical exergy of biomasses. Compositions are on a dry basis as listed in the cited references except for Ref. 3 in which case the elements are reported on a dry ash free basis with the ash content reported on a dry basis. (Continued)

Categories	Biomass type	C	H	O	N	S	Ash	e^{ch} (MJ/kg)	Ref.
				Wt.%					
Husk	Olive husks	50	6.2	42.1	1.6	0.05	2.3	20.78	[33]
Husk	Palm fibers-husks	51.5	6.6	40.1	1.5	0.3	8.3	20.48	[33]
Husk	Rice husks	49.3	6.1	43.7	0.8	0.08	18	16.69	[33]
Husk	Soya husks	45.4	6.7	46.9	0.9	0.1	5.4	18.61	[33]
Husk	Sunflower husks	50.4	5.5	43	1.1	0.03	3.1	19.9	[33]
Husk	Rice husk	38.50	5.20	34.61	0.45		21.24	16.24	[34]
Husk	Rice husk bran	38.92	5.12	36.77	0.55		18.64	16.17	[34]
Husk	Rice husk char (750°C)	31.51	2.50	19.34	1.45		45.20	11.54	[34]
Husk	Rice husks Patni-23	38.92	5.10	37.89	2.17	0.12	15.80	16.14	[34]
Husk	Coffee husk	47.5	6.4	43.7			2.4	20.43	[35]
Husk	Rice husk	38.5	5.2	34.61	0.45		21.24	16.24	[35]
Husk	Rice husk	38.2	5.6	33.7			22.5	16.62	[35]
Leaves	Pine needles	46.4	6	41.7	1.4	0	4.5	19.72	[32]
Leaves	Casuarina equisetifolia leaf	46.12	6.9	42.64	1.18		3.93	20.54	[35]
Leaves	Ipil ipil	48.3	6.8	42.5			2.4	21.26	[35]
Leaves	Lantana camara leaf	45.01	6.68	43.79	2.02		7.26	19.73	[35]
Seaweed	Soquel poin giant brown kelp	27.80	3.73	23.69	1.63	1.05	42.10	11.45	[34]
Shell	Almond hulls	50.6	6.4	41.7	1.2	0.07	6.1	20.33	[33]
Shell	Almond shell	50.3	6.2	42.5	1	0.05	3.3	20.61	[33]
Shell	Coconut shell	51.1	5.6	43.1	0.1	0.1	3.2	20.23	[33]
Shell	Groundnut shell	50.9	7.5	40.4	1.2	0.02	3.4	22.35	[33]
Shell	Hazelnut shell	51.5	5.5	41.6	1.4	0.04	1.5	20.78	[33]
Shell	Pistachio shell	50.9	6.4	41.8	0.7	0.22	1.4	21.58	[33]
Shell	Walnut blows	54.9	6.7	36.9	1.4	0.11	2.4	23.47	[33]
Shell	Walnut hulls and blows	55.1	6.7	36.5	1.6	0.12	2.9	23.45	[33]
Shell	Walnut shell	49.9	6.2	42.4	1.4	0.09	2.8	20.61	[33]
Shell	Akhrot shell	49.81	5.64	42.94	0.41		1.20	20.52	[34]
Shell	Coconut coir pith	43.36	4.98	44.87	1.63		5.16	17.21	[34]
Shell	Coconut shell	50.22	5.70	43.37	0.00		0.71	20.71	[34]
Shell	Coconut shell char (750°C)	88.95	0.73	6.04	1.38		2.90	32.43	[34]

TABLE 10.6 Chemical exergy of biomasses. Compositions are on a dry basis as listed in the cited references except for Ref. 3 in which case the elements are reported on a dry ash free basis with the ash content reported on a dry basis. (Continued)

Categories	Biomass type	C	H	O	N	S	Ash	e^{ch} (MJ/kg)	Ref.
				Wt.%					
Shell	Groundnut shell	48.59	5.64	39.49	0.58		5.70	20.28	[34]
Shell	Almond hulls	47.53	5.97	39.16	1.13	0.06	6.13	20.29	[35]
Shell	Almond shell	48.8	5.9	43.7	0.5		1.1	20.38	[35]
Shell	Almond shell	49.3	5.97	40.63	0.76	0.04	3.29	20.86	[35]
Shell	Coconut shell	50.22	5.7	43.37			0.71	24.55	[35]
Shell	Hazelnut shell	50.9	5.9	42.8	0.4		1.1	21.21	[35]
Shell	Olive kernel shell	53.2	6.7	36.3	0.5		3.3	23.44	[35]
Shell	Peanut shell	47.4	6.1	44.4	2.1		1.7	20.02	[35]
Shell	Pistachio shell	50.2	6.32	41.15	0.69	0.22	1.41	21.60	[35]
Shell	Pistachio soft shell	45.53	5.56	47.17	1.74		14.21	18.25	[35]
Shell	Sunflower seed shell	51.7	6.2	41.1	1		3.6	21.93	[35]
Shell	Hazelnut shell	51.7	5.6	42.2	0.5		1.3	21.22	[36]
Stalk	Corn stover	45.48	5.52	41.52	0.69	0.04	6.73	18.83	[35]
Stalk	Cotton stalk	47.07	4.58	42.1	1.15		5.1	18.35	[35]
Stalk	Tobacco stalk	49.3	5.8	44.5	0.4		2.4	20.35	[36]
Straw	Alfalfa straw	49.9	6.3	40.8	2.8	0.21	5.3	20.28	[33]
Straw	Barley straw	49.4	6.2	43.6	0.7	0.13	5.3	19.76	[33]
Straw	Corn straw	48.7	6.4	44.1	0.7	0.08	7.7	19.13	[33]
Straw	Mint straw	50.6	6.2	40.1	2.8	0.28	10.8	19.18	[33]
Straw	Oat straw	48.8	6	44.6	0.5	0.08	5.9	19.12	[33]
Straw	Rape straw	48.5	6.4	44.5	0.5	0.1	4.7	19.72	[33]
Straw	Rice straw	50.1	5.7	43	1	0.16	20.1	16.15	[33]
Straw	Straw	48.8	5.6	44.5	1	0.13	10.8	17.64	[33]
Straw	Wheat straw	49.4	6.1	43.6	0.7	0.17	7.1	19.25	[33]
Straw	Cotton stalks	39.47	5.07	38.09	1.25	0.02	16.10	16.26	[34]
Straw	Jawar straw	42.10	5.60	43.38	0.04		8.88	17.48	[34]
Straw	Millet straw	43.71	5.85	45.16	0.01		5.27	18.26	[34]
Straw	Rice straw	35.68	4.62	39.14	0.28		20.28	14.21	[34]
Straw	Wheat straw	42.95	5.35	46.99	0.00		4.71	17.29	[34]
Straw	Corn straw	44.73	5.87	40.44	0.6	0.07	7.65	19.03	[35]

TABLE 10.6 Chemical exergy of biomasses. Compositions are on a dry basis as listed in the cited references except for Ref. 3 in which case the elements are reported on a dry ash free basis with the ash content reported on a dry basis. (Continued)

Categories	Biomass type	Wt.%						e^{ch} (MJ/kg)	Ref.
		C	H	O	N	S	Ash		
Straw	Rape straw	46.17	6.12	42.47	0.46	0.1	4.65	19.72	[35]
Straw	Rice straw	38.24	5.2	36.26	0.87	0.18	18.67	16.09	[35]
Straw	Sugar cane straw	43.5	6.1	41.1			9.2	18.73	[35]
Straw	Wheat straw	42.95	5.35	46.99			6.9	17.25	[35]
Wood	Plywood	53.6	5.6	37.7	0.4	0	2.7	22.27	[32]
Wood	Treated wood	48.6	5.7	38.9	1.1	0.1	5.6	20.42	[32]
Wood	Untreated wood	49.6	5.9	41.9	0.3	0.1	2.2	20.82	[32]
Wood	Alder-fir sawdust	53.2	6.1	40.2	0.5	0.04	4.2	21.49	[33]
Wood	Bamboo whole	52	5.1	42.5	0.4	0.04	0.9	20.58	[33]
Wood	Christmas trees	54.5	5.9	38.7	0.5	0.42	5.1	21.70	[33]
Wood	Demolition wood	51.7	6.4	40.7	1.1	0.09	6.9	20.60	[33]
Wood	Fir mill residue	51.4	6	42.5	0.1	0.03	0.5	21.44	[33]
Wood	Forest residue	52.7	5.4	41.1	0.7	0.1	3.2	20.75	[33]
Wood	Furniture waste	51.8	6.1	41.8	0.3	0.04	3.6	21.01	[33]
Wood	Land clearing wood	50.7	6	42.8	0.4	0.07	16.5	17.43	[33]
Wood	Mixed wastepaper	52.3	7.2	40.2	0.2	0.08	8.3	21.3	[33]
Wood	Oak sawdust	50.1	5.9	43.9	0.1	0.01	0.3	20.79	[33]
Wood	Oak wood	50.6	6.1	42.9	0.3	0.1	0.5	21.24	[33]
Wood	Olive wood	49	5.4	44.9	0.7	0.03	3.2	19.12	[33]
Wood	Pine pruning	51.9	6.3	41.3	0.5	0.01	2.7	21.52	[33]
Wood	Pine sawdust	51	6	42.9	0.1	0.01	0.1	21.35	[33]
Wood	Poplar	51.6	6.1	41.7	0.6	0.02	2.1	21.31	[33]
Wood	Sawdust	49.8	6	43.7	0.5	0.02	1.1	20.62	[33]
Wood	Spruce wood	52.3	6.1	41.2	0.3	0.1	0.5	22.00	[33]
Wood	Willow	49.8	6.1	43.4	0.6	0.06	1.6	20.62	[33]
Wood	Wood	49.6	6.1	44.1	0.1	0.06	0.2	20.84	[33]
Wood	Wood residue	51.4	6.1	41.9	0.5	0.08	5.4	20.45	[33]
Wood	Wood yard waste	52.2	6	40.4	1.1	0.3	20.4	17.15	[33]
Wood	Wood-agricultural residue	52.4	6	41.2	0.4	0.04	3.3	21.24	[33]

TABLE 10.6 Chemical exergy of biomasses. Compositions are on a dry basis as listed in the cited references except for Ref. 3 in which case the elements are reported on a dry ash free basis with the ash content reported on a dry basis. (*Continued*)

Categories	Biomass type	Wt.%						e^{ch} (MJ/kg)	Ref.
		C	H	O	N	S	Ash		
Wood	Wood-almond residue	50.9	5.9	42.5	0.6	0.08	6.9	19.67	[33]
Wood	Wood-straw residue	51.7	6.3	41.5	0.4	0.13	8.2	20.12	[33]
Wood	Bamboo wood	48.76	6.32	42.77	0.20		1.95	20.89	[34]
Wood	Block wood	46.90	6.07	43.99	0.95		2.09	19.83	[34]
Wood	Casuria wood char (950°C)	77.54	0.93	5.62	2.67		13.24	28.34	[34]
Wood	Casurina wood	48.50	6.24	43.12	0.31		1.83	20.68	[34]
Wood	Chaparra wood	46.90	5.08	40.17	0.54	0.03	7.28	18.97	[34]
Wood	Cottongin waste	42.66	6.05	49.50	0.18		1.61	17.81	[34]
Wood	Douglass firewood	50.64	6.18	43.00	0.06	0.02	0.10	21.44	[34]
Wood	Dry subabul wood	48.15	5.87	44.75	0.03		1.20	20.01	[34]
Wood	Eucalyptus globulus wood	48.18	5.92	44.18	0.39	0.01	1.32	20.13	[34]
Wood	Eucalyptus wood	46.04	5.82	44.49	0.30		3.35	19.17	[34]
Wood	Eucelyptus wood char (950°C)	76.10	1.33	11.10	1.02		10.45	27.84	[34]
Wood	Fresh subabul wood	46.24	5.80	46.59	0.25		1.12	19.09	[34]
Wood	Jujuba wood	47.63	6.12	43.78	0.15		2.32	20.16	[34]
Wood	Mango wood	46.24	6.08	44.42	0.28		2.98	19.55	[34]
Wood	Neem wood	48.26	6.27	43.46	0.08		1.93	20.59	[34]
Wood	PlyWood	48.13	5.87	42.46	1.45		2.09	20.19	[34]
Wood	Red wood char (800–1725°F)	78.80	3.50	13.20	0.20	0.20	4.10	31.20	[34]
Wood	Sawdust	47.13	5.86	40.35	0.65	0.16	5.85	19.95	[34]
Wood	Subabul wood char (950°C)	83.61	1.95	10.48	0.01		3.95	31.43	[34]
Wood	Western hemlock wood	50.40	5.80	41.40	0.10	0.10	2.20	21.04	[34]
Wood	Bamboo wood	48.76	6.32	42.77	0.2		1.95	20.89	[35]
Wood	B-wood	50.26	6.91	39.66	1.03		1.85	22.35	[35]
Wood	Esparto plant	46.94	6.44	43.56	0.86		2.2	20.29	[35]
Wood	Forest residue	53.16	6.25	40	0.3	0.09	0.2	22.70	[35]
Wood	Hybrid poplar	50.18	6.06	40.43	0.6	0.02	2.7	21.31	[35]
Wood	Willow wood	49.9	5.9	41.8	0.61	0.07	1.71	20.94	[35]
Wood	Beech wood	49.5	6.1	40.9	0.3		0.5	21.10	[36]

TABLE 10.6 Chemical exergy of biomasses. Compositions are on a dry basis as listed in the cited references except for Ref. 3 in which case the elements are reported on a dry ash free basis with the ash content reported on a dry basis. (*Continued*)

Categories	Biomass type	C wt.%	H wt.%	O wt.%	N wt.%	S wt.%	Ash wt.%	e^{ch} (MJ/kg)	Ref.
Various	Animal waste chars	35.1	5.3	38.7	2.5	0.4	18.0	14.91	[34]
Various	Greenhouse-plastic waste	70.9	11.2	16.4	1.5	0.01	31.8	24.29	[37]
Various	Missippi hyacinth digested slurry	31.7	3.82	23.2	1.98	0.00	39.3	12.86	[34]
Various	Municipal solid waste	45.6	6.0	26.5	0.8	0.3	20.8	20.46	[32]
Various	Municipal solid waste char	54.9	0.8	1.8	1.1	0.2	41.2	19.77	[34]
Various	Refuse-derived fuel	53.8	7.8	36.8	1.1	0.47	26.1	17.92	[37]
Various	Sewage sludge	14.2	2.1	10.5	1.1	0.7	71.4	5.18	[34]
Various	Sewage sludge	50.9	7.3	33.4	6.1	2.33	46.3	11.94	[37]
Various	Tire	73.8	6.8	9.0	0.3	1.3	8.8	33.5	[32]

TABLE 10.7 Chemical exergy of wastes from human activities. Compositions are dry basis from Ref 32 and 34 and are dry-ash free for Ref. 37

References

1. Kotas TJ. *The Exergy Method of Thermal Plant Analysis*. Paragon Publishing: London (2012).
2. Song G, Xian J, Zhao H, Shen L. A unified correlation for estimating specific chemical exergy of solid and liquid fuels. *Energy* 40(1):164–173 (2012).
3. United States Geological Survey. World Coal Quality Inventory Data, 2019; https://www.usgs.gov/centers/eersc/science/world-coal-quality-inventory-data?qt-science_center_objects=0#qt-science_center_objects (accessed November 2020).
4. Watanabe H, Otaka M. Numerical simulation of coal gasification in entrained flow coal gasifier. *Fuel* 85(12–13):1935–1943 (2006).
5. Channiwala SA, Parikh PP. A unified correlation for estimating HHV of solid, liquid and gaseous fuels. *Fuel* 81:1051–1063 (2002).
6. Lin CC, Parga JR, Drelich J, Miller JD. Characterization of washability of some Mexican coals. *Coal Preparation* 20(3–4):227–245 (1999).
7. Swanson M, Laudal D. *Subtask 7.4-Power River Basin Subbituminous Coal-Biomass Cogasification Testing in a Transport Reactor*. University of North Dakota (2009); https://www.osti.gov/biblio/986897 (accessed December 20, 2022).
8. Taghiei MM, Huggins FE, Mahajan V, Huffman GP. Evaluation of cation-exchange iron for catalytic liquefaction of a subbituminous coal. *Energy and Fuels* 8(1):31–37 (1994).
9. Guffey FD, Bland AE. Thermal pretreatment of low-ranked coal for control of mercury emissions. *Fuel Processing Technology* 85(6–7):521–531 (2004).
10. Boysen JE, Kang TW, Cha CY, Berggren MH, Jha MC. Development of an advanced process for drying fine coal in an inclined fluidized bed. *Western Research Institute Technical Report* (1989). doi:10.2172/5586826.
11. Laudal DL, Pavlish JH, Galbreath KC, Thompson JS, Weber GF, Sondreal E. Pilot-scale evaluation of the impact of selective catalytic reduction for NOx on mercury speciation. *University of North Dakota Technical Report* (2000). doi: 10.2172/824958.
12. Zhuang Y, Zygarlicke CJ, Galbreath KC, Thompson JS, Holmes MJ, Pavlish JH. Kinetic transformation of mercury in coal combustion flue gas in a bench-scale entrained-flow reactor. *Fuel Processing Technology* 85(6–7):463–472 (2004).
13. Pavlish JH, Holmes MJ, Benson SA, Crocker CR, Galbreath KC. Application of sorbents for mercury control for utilities burning lignite coal. *Fuel Processing Technology* 85(6–7):563–576 (2004).
14. Shamsi A, Shadle LJ, Seshadri KS. Study of low-temperature oxidation of buckskin subbituminous coal and derived chars produced in ENCOAL process. *Fuel Processing Technology* 86(3):275–292 (2004).
15. Slezak A, Kuhlman JM, Shadle LJ, Spenik J, Shi S. CFD simulation of entrained-flow coal gasification: coal particle density/size fraction effects. *Powder Technology* 203(1):98–108 (2010).
16. Bownocker JA, Dean ES. Analyses of the coals of Ohio. *Ohio Division of Geological Survey* 34 (1929).
17. Calkins WH. Determination of organic sulfur-containing structures in coal by flash pyrolysis experiments. *American Chemical Society, Division of Fuel Chemistry* 30(4):450–465 (1985).
18. Glass GB. Description of Wyoming coal fields and seam analyses. *Wyoming State Geological Survey* (1982).
19. Henderson AK, Swanson ML, Hurley JP, Watne TM. Characterization of hot-gas filter ash under PFBC operating conditions. No. DOE/FETC/C-98/7307; CONF-9705236-. North Dakota University, Energy and Environmental Research Center, Grand Forks, ND (United States) (1998).

20. Haussmann GJ, Kruger CH. Rapid pyrolysis and combustion of pulverized Montana rosebud subbituminous coal. *Symposium (International) on Combustion* 22(1):223–230 (1989). doi: 10.1016/S0082-0784(89)80028-1.

21. Kolker A, Mroczkowski SJ, Palmer CA, Dennen KO, Finkelman RB, Bullock JH Jr. Toxic substances from coal combustion—a comprehensive assessment, phase II: element modes of occurrence for the Ohio 5/6/7, Wyodak and North Dakota coal samples. *US Geological Survey (US)* (2002).

22. Shadle LJ, Monazam ER, Swanson ML. Coal gasification in a transport reactor. *Industrial and Engineering Chemistry Research* 40(13):2782–2792 (2001).

23. Karacan CÖ. Integration of vertical and in-seam horizontal well production analyses with stochastic geostatistical algorithms to estimate pre-mining methane drainage efficiency from coal seams: Blue Creek seam, Alabama. *International Journal of Coal Geology* 114:96–113 (2013).

24. Aresco SJ, Haller CP, Abernethy RF. Analyses of tipple and delivered samples of coal: (collected during fiscal year 1956). Vol. 5332. US Department of the Interior, Bureau of Mines (1957).

25. Zygarlicke CJ, Stomberg AL, Folkedahl BC, Strege JR. Alkali influences on sulfur capture for North Dakota lignite combustion. *Fuel Processing Technology* 87(10):855–861 (2006).

26. Brooks WE, Finkelman RB, Willett JC, Torres IE. World coal quality inventory: Argentina. US Geological Survey Open File Report 2006:1241 (2006).

27. Oliveira ML, Ward CR, Sampaio CH, Querol X, Cutruneo CM, Taffarel SR, Silva LF. Partitioning of mineralogical and inorganic geochemical components of coals from Santa Catarina, Brazil, by industrial beneficiation processes. *International Journal of Coal Geology* 116:75–92 (2013).

28. Hackley PC, Warwick PD, Guillermo AH, Rosenelsy MC. World coal quality inventory: Chile. World coal quality inventory: South America. USGS, Open File Report 1241:90–131 (2006).

29. Tewalt SJ, Finkelman RB, Torres IE, Simoni F. World coal quality inventory: Colombia. US Geological Survey Open File Report 2006:1241 (2006).

30. Brooks WE, Finkelman RB, Willett JC, Gurmendi AC, Yager TR, Carrascal R, Mucho R. World coal quality inventory: Peru. US Geological Survey Open File Report 2006:1241 (2006).

31. Hackley PC, Finkelman RB, Brooks WE, Gonzáles E. World coal quality inventory: Venezuela. US Geological Survey Open File Report 2006:1241 (2006).

32. Janajreh I, Raza SS, Valmundsson AS. Plasma gasification process: modeling, simulation and comparison with conventional air gasification. *Energy Conversion and Management* 65:801–809 (2013).

33. Vassilev SV, Baxter D, Andersen LK, Vassileva CG. An overview of the chemical composition of biomass. *Fuel* 89(5):913–933 (2010).

34. Channiwala SA, Parikh PP. A unified correlation for estimating HHV of solid, liquid and gaseous fuels. *Fuel* 81(8):1051–1063 (2002).

35. Yin C-Y. Prediction of higher heating values of biomass from proximate and ultimate analyses. *Fuel* 90(3): 1128–1132 (2011).

36. Demirbaş A. Conversion of biomass using glycerin to liquid fuel for blending gasoline as alternative engine fuel. *Energy Conversion and Management* 41(16):1741–1748 (2000).

37. Vassilev SV, Baxter D, Andersen LK, Vassileva CG. An overview of the chemical composition of biomass. *Fuel* 89(5):913–933 (2010).

38. Grummel ES, Davis IA. A new method of calculating the calorific value of fuel from its ultimate analysis. *Fuel* 12:199–203 (1933).

Index

Note: Page numbers followed by *f* denote figures; by *t*, tables; by *n*, footnotes.